# Polymer Nanofibers
## Building Blocks for Nanotechnology

## RSC Nanoscience & Nanotechnology

*Series Editors:*
Professor Paul O'Brien, *University of Manchester, UK*
Professor Sir Harry Kroto FRS, *University of Sussex, UK*
Professor Ralph Nuzzo, *University of Illinois at Urbana-Champaign, USA*

*Titles in the Series:*
1: Nanotubes and Nanowires
2: Fullerenes: Principles and Applications
3: Nanocharacterisation
4: Atom Resolved Surface Reactions: Nanocatalysis
5: Biomimetic Nanoceramics in Clinical Use: From Materials to Applications
6: Nanofluidics: Nanoscience and Nanotechnology
7: Bionanodesign: Following Nature's Touch
8: Nano-Society: Pushing the Boundaries of Technology
9: Polymer-based Nanostructures: Medical Applications
10: Metallic and Molecular Interactions in Nanometer Layers, Pores and Particles: New Findings at the Yoctolitre Level
11: Nanocasting: A Versatile Strategy for Creating Nanostructured Porous Materials
12: Titanate and Titania Nanotubes: Synthesis, Properties and Applications
13: Raman Spectroscopy, Fullerenes and Nanotechnology
14: Nanotechnologies in Food
15: Unravelling Single Cell Genomics: Micro and Nanotools
16: Polymer Nanocomposites by Emulsion and Suspension
17: Phage Nanobiotechnology
18: Nanotubes and Nanowires: 2$^{nd}$ Edition
19: Nanostructured Catalysts: Transition Metal Oxides
20: Fullerenes: Principles and Applications, 2$^{nd}$ Edition
21: Biological Interactions with Surface Charge Biomaterials
22: Nanoporous Gold: From an Ancient Technology to a High-Tech Material
23: Nanoparticles in Anti-Microbial Materials: Use and Characterisation
24: Manipulation of Nanoscale Materials: An Introduction to Nanoarchitectonics
25: Towards Efficient Designing of Safe Nanomaterials: Innovative Merge of Computational Approaches and Experimental Techniques
26: Polymer–Graphene Nanocomposites
27: Carbon Nanotube-Polymer Composites
28: Nanotechnology for the Conservation of Works of Art
29: Polymer Nanofibers: Building Blocks for Nanotechnology

*How to obtain future titles on publication:*
A standing order plan is available for this series. A standing order will bring delivery of each new volume immediately on publication.

*For further information please contact:*
Book Sales Department, Royal Society of Chemistry, Thomas Graham House, Science Park, Milton Road, Cambridge, CB4 0WF, UK
Telephone: +44 (0)1223 420066, Fax: +44 (0)1223 420247
Email: booksales@rsc.org
Visit our website at www.rsc.org/books

# *Polymer Nanofibers*

## *Building Blocks for Nanotechnology*

**Dario Pisignano**
*Dipartimento di Matematica e Fisica "Ennio De Giorgi", University of Salento and CNR-Istituto Nanoscienze, Lecce, Italy*
*Email: Dario.pisignano@unisalento.it*

RSC Publishing

RSC Nanoscience & Nanotechnology No. 29

ISBN: 978-1-84973-574-2
ISSN: 1757-7136

A catalogue record for this book is available from the British Library



Published by The Royal Society of Chemistry,
Thomas Graham House, Science Park, Milton Road,
Cambridge CB4 0WF, UK

Registered Charity Number 207890

For further information see our web site at www.rsc.org

Printed in the United Kingdom by CPI Group (UK) LTD, Croydon, CR0 4YY

# Preface

Nanoscience and nanotechnology hold great promises, and consequently they have been the subject of a tremendous research effort. The global investment by governments since the year 2000 is currently around USD 100 billion. The nanotechnology-related budget of the Seventh Framework Programme for Research and Technological Development (2007-2013) of the European Union, only considering transnational cooperative projects, has been of € 3.5 billion.[1] Strengthening science through frontier research and improving industrial competitiveness are widely believed to be boosted by nanosciences and nanotechnologies. Nowadays, several applications and devices enabled by nanotechnologies have become mature, and are gradually flooding the market in the industrial sectors of catalysis, reinforced materials and nanocomposites, optoelectronics, biotechnologies, and so on. However, the same nanotech community also admits that some issues still need to be addressed in order to fully exploit the undoubted, enormous potential of nanomaterials. Among critical challenges, one should mention the still relatively limited *throughput* of nano-production technologies, the urgent need to combine nanoparticle fabrication technologies with the highest possible *chemical and biological flexibility*, thus eventually developing unique and low-cost fabrication platforms that could impact on different industrial fields, and, last but not least, regulatory issues related to *nanoparticle toxicology*. *Polymer nanofibers* are a relatively recently developed class of nanomaterials, which can offer solutions to many of these critical issues, thus potentially giving actual perspectives of socio-economic impact to future nanotechnologies.

While one-dimensional or quasi-one dimensional nanostructures can be produced with a large variety of materials and methods, technologies for producing polymer nanofibers are almost unique in being effectively

RSC Nanoscience & Nanotechnology No. 29
Polymer Nanofibers: Building Blocks for Nanotechnology
By Dario Pisignano

Published by the Royal Society of Chemistry, www.rsc.org

up-scaled in perspective. The electrospinning method (Chapter 2) is certainly the best example of such versatility and throughput, allowing to process up to several liters of polymer solution under continuous runs at laboratory scale. Overall, techniques for the realization of polymer nanofibers are compatible with production processes for pre-industrial and, in perspective, industrial production. Furthermore, the variety of functional organic materials which can be shaped or synthesized as nanofibers is astonishing. Usable polymers span from thermoplastics to photocurable compounds, from conjugated polymers to biomacromolecules, and, in addition, an enormous assortment of dopants or surface modifiers can be embedded in organic nanofiber architectures or used to functionalize them, including inorganic nanoparticles, carbon nanotubes or graphenes, low molar-mass light-emitting molecules, cell-directing proteins, growth-factors, and so on, as described through this book. Finally, polymer nanofibers can be made of biocompatible or biologically inert materials. For these reasons, not only polymer nanofibers are very intriguing from a scientific point of view, but their production technologies also exhibit a clear, remarkable potential to bring laboratory processes to the market. Therefore, these nanomaterials are opening new horizons for applications and industrial production by organic-based nanotechnologies. Today, polymer nanofibers are among the few promising candidates available for achieving intelligent *multifunctional* nanomaterials (so-called *third-generation* nanomaterials),[2,3] assembling in a controlled way and allowing one to perform multiple (biological, structural, electronic, photonic, filtering, molecular delivery and recognition, etc.) functions simultaneously, thus impacting on cross-sectorial applications. Reliable molecular nanosystems (*fourth-generation* nanomaterials and nanotechnologies)[2] can also be based on polymer nanofibers in perspective, especially in the framework of genetic therapies and of hierarchically-assembling supramolecular electronics.

One should also mention that, in order to fully and safely exploit this potential, minimizing the environmental impact of nanofiber production technologies and taking into account the possible health risks associated to the use and handling of nanofiber materials will be certainly mandatory in the near future. In particular, carrying out nano-toxicology studies and collecting satisfactory related statistics necessarily require long times, and at the present state of research, polymer nanofibers have not yet been considered as subject of large-scale studies which could later inspire suitable regulatory actions. It is well-known that a few nasty artificial and natural materials, mainly inorganic such as silica and asbestos, can take a microfiber form. On the other hand, organic nanofibers are largely produced in nature under generally mild and physiological, environmentally-friendly conditions at cellular and extracellular level. While a lot of both synthetic and biological polymers, as mentioned above, show clear and well-established properties of biocompatibility, biodegradability, or biological inertia, as for all newly developed production processes the precautionary principle should prevail until complete data will be available about nanofiber-body interactions in living organisms. Evidences, coming from studies which successfully apply polymer nanofiber scaffolds in tissue engineering and for stem cell control (Chapter 7), can contribute to shade light in this respect.

This book aims to give a comprehensive presentation of polymer nanofibers, including both the relevant physical aspects at the base of nanofibers formation and of their structural, superficial, optical, electrical and biological properties, and the main technological and engineering issues of fabrication methods. The progress of research on polymer nanofibers, gradually becoming mature in last two decades, has become especially relevant and rapid in most recent years. Polymer nanofibers are today a particularly hot topic in the scientific and technical literature, transversal to many fields of nanotechnology, and therefore attracting a particularly broad audience of researchers, engineers, and, potentially, end users.

The highly interdisciplinary composition of such very broad audience, the natural differences existing among various readerships (academics, industrialists, postgraduates, students, etc.), together with the intrinsic cross-cutting nature of the presented technologies, put some constraints to the kind and level of technical language to use for making reading transversally accessible (a need which is indeed typical of nanotechnology and nanoscience lectures and essays in general). The constant efforts are therefore of carefully relating the materials physico-chemical properties, as also due to or modified by the specific processing and fabrication technologies, to the resulting functional behaviour of polymer nanofibers, and of providing a unique source of information using a fair but consistent level of technical language. From the point of view of scientists, nanotechnology needs a peculiar (but incredibly fruitful) effort in learning and using a multidisciplinary language and way of thinking, which often go well beyond those, generally quite monothematic and highly specialized, gained during undergraduate studies. Working as nano-scientists and nano-technologists means, and requires, handling multiple scientific languages and approaches, and characterization methods transversal to physics, chemistry and biology, which is the only way to fully exploit the potentialities opened by nanostructures and nanomaterials. We believe that this exploration is facilitated by having close at hand a few basic concepts from different disciplines, which are needed to understand the logical flow originating the design of innovative experiments, manufacturing approaches, and theoretical models in nanosciences. This is the way in which this book is organized, aiming to provide some introductory descriptions of polymer physics and phenomenology, device operation, and properties of molecular systems, that are required to better catch the concepts underlying the production, characterization and utilization of polymer nanofibers. Of course, reaching the level of detail that can be found in other dedicated reviews and books is beyond our scope.

The coverage of applications in the different fields as arising from peculiar polymer nanofiber properties (surface-related, mechanical, electric, optoelectronic, and biological) complete the book, whose single Chapters 4-7 can be exploited as reference for implementing a "from materials to applications" path, or for supporting the demands of multi-disciplinary courses mixing physics, materials science and engineering, chemical engineering or biotechnology. In fact, moving from the choice of a suitable material or class of materials aiming at a particular application, passing through processing to tailor physical properties, is the most frequent and relevant sequence of tasks performed by today materials and nano-scientists. In some cases, more than one material can address the desired application or device, and are processable with analogous effectiveness. Safety considerations should prevail and drive the choice of experimentalists towards less toxic molecules and solvents.

Overall, a double track is therefore designed, trying to linearly follow a logical flow from general, basic material or nanostructure properties to polymer nanofiber characteristics and, in parallel, to document latest research. The most of introductory material concerning polymer physics and engineering, with special attention paid to the phenomena which are transversally relevant to different nanofiber fabrication technologies such as glass-transition phenomenology, basic flow, and solution properties, is grouped in Chapter 1. This introduction is kept as sharp as possible aiming to recall general polymer properties without entering specialized details, thus allowing the reader to quickly consult the definitions of physical quantities used in experiments and models on polymer nanofiber fabrication. Chapters 2 and 3 describe fabrication technologies of polymer nanofibers, including electrospinning, self-assembly and polymerization methods, hard and soft templating, serial writing and phase separation. Finally Chapters 4–7, dedicated to specific applications, are opened by or include Sections about relevant material properties of organic compounds, to which nanofiber architectures impart a peculiar behaviour.

Attention has been also paid to make formalisms as easy as possible to readers from different scientific areas. I tried my best to make the book accurate. Of course, inaccuracies are always possible and I will be grateful to any reader and colleague pointing out possible problems. Whenever possible, the International System has been used as metric system (for units and for shaping equations), with a few exceptions however when different units of measurements are definitely established in a given literature field. Each physical quantity has been described by a dedicated symbol or character throughout the book. Symbols are listed in a Table at the end of the book, where the reader also finds the reference of the first Section where each symbol is introduced.

The hope is that readers with different scientific backgrounds can enjoy this volume, and that this can be an occasion for new inspiration and scientific cross-fertilization, for researchers and engineers working in different fields having the use of polymer nanofibers as common denominator.

Dario Pisignano

# References

1. http://ec.europa.eu/research/fp7/pdf/fp7-factsheets_en.pdf
2. M. C. Roco, *AIChE J.*, 2004, **50**, 5–9.
3. European Commission 7th Framework Programme, *Work Programme on Nanosciences, Nanotechnologies, Materials and new Production Technologies,* 2012.

# Contents



RSC Nanoscience & Nanotechnology No. 29
Polymer Nanofibers: Building Blocks for Nanotechnology
By Dario Pisignano

Published by the Royal Society of Chemistry, www.rsc.org

**Chapter 7 Biological Functional Properties and Their Applications**

# Acknowledgements

I wish to acknowledge many friends and colleagues for helpful discussions and for their suggestions, critical review and useful material provided during the preparation of this book. I would like to thank Prof. Laura De Lorenzis, Prof. Alfonso M. Gañán Calvo, Prof. Unyong Jeong, Dr. Onofrio Maragò, Prof. Charles R. Martin, Dr. Francesco Quochi, Dr. Hussam Razzaq, Dr. Stefano Roddaro, Dr. Cagri Tekmen and Prof. David A. Weitz. A special thanks goes to Dr. Andrea Camposeo, Dr. Israel Greenfeld, Dr. Luana Persano, Prof. Darrel H. Reneker, Prof. Gregory C. Rutledge and Prof. Eyal Zussman for their invaluable suggestions and support. Dr. Abdullah Aslamaci, Dr. Matthias Brenner, Dr. Lynn Davis, Mr. Francesco De Donato, Prof. Benjamin R. Horrocks, Prof. Liberato Manna, Dr. Sergio Marras, Prof. Chad A. Mirkin, Prof. Werner E. G. Müller, Prof. Jose M. Moran-Mirabal and Dr. Eva Pokorna are also gratefully acknowledged for some of the figures of the book. I am grateful to Miss Silvia Rizzo for her continuous help in editing my pages and collecting literature files. It was a great pleasure to work with Mrs. Rosalind Searle and Dr. Merlin Fox from RSC Publishing during the preparation of this book.

Finally, I also wish to acknowledge the agencies and sources of funds that have been supporting the research of my group on polymer nanofibers. In particular, I acknowledge the European Research Council for supporting, under the European Union's Seventh Framework Programme (FP7/2007-2013), the ERC Starting Grant "NANO-JETS" (grant agreement n. 306357).

RSC Nanoscience & Nanotechnology No. 29
Polymer Nanofibers: Building Blocks for Nanotechnology
By Dario Pisignano

Published by the Royal Society of Chemistry, www.rsc.org

CHAPTER 1

# *Soft Matter Nanotechnologies*

## 1.1 Nanotechnology and Nanomaterials

Nanotechnology means the capability of controlling, assembling, manipulating and characterizing matter at atomic and molecular length-scale. Materials that are realized and processed in this way, and that exhibit at least one dimension below 100 nm, are known as nanomaterials, and their production and use fully belong to the nanotechnology domain according to the definition from the National Nanotechnology Initiative*.[1] However, as a matter of fact, both in the literature and in industrial applications, the word "nanomaterial" is applied less rigorously (in practice for a large variety of objects and structures having a dimension below 1 μm). This is partially because of the appeal, at times overenthusiastic, gained by the prefix "nano" in the last two decades, but also, reasoning about a more fundamental viewpoint, because some material physico-chemical properties and characterization tools actually start to "feel" the effects of size reduction at microscopic scales which are roughly in the 100 nm–1 μm range. Polymer nanofibers are an instructive example of both these facts. Claiming for *nano*fibers in reports dealing with fibers of diameter between 100 nm and 1 μm, though certainly not fully accurate, is almost standard in

RSC Nanoscience & Nanotechnology No. 29
Polymer Nanofibers: Building Blocks for Nanotechnology
By Dario Pisignano

Published by the Royal Society of Chemistry, www.rsc.org

* The National Nanotechnology Initiative, founded in 2000, is a huge inter-agency organization in the United States, which today includes 26 Federal agencies, most of which have dedicated budgets for nanotechnology research and development. This organization coordinates shared goals and priorities and facilitates interfacing with industry, spanning a very wide range of research and regulatory issues. USD 1.8 billion are provided to the National Nanotechnology Initiative by the 2013 Federal Budget, with a total investment as high as USD 18 billion since 2001.[1] The current strategic plan of the organization can be downloaded at http://www.nano.gov/sites/default/files/pub_resource/2011_strategic_plan.pdf and makes mention, among many others, of structural fibers and of cellulosic nanofibers derived from renewable vegetable biomaterials.

specialized literature as well as in many technical applications and industries. On the other hand, as briefly recalled in the following, quantum mechanical effects are very difficult to observe in organic materials. Consequently, the association between nano-properties and charge-carrier confinement mechanisms, which is at the base of quantum size effects such as those found, for instance, in inorganic crystalline semiconductors, basically fails when analyzing polymer fibers. Here, size reduction leads instead to a group of properties, including ultra-high surface-to-volume ratio, often enhanced molecular orientation along the fiber longitudinal axis, possibly improved charge transport, optical anisotropy, controllable porosity at nanoscale *etc.*, all of which can partially arise even above 100 nm. Overall, for organic materials and structures, being more generous in using the "nano" prefix is still not fully rigorous, but likely a bit more acceptable than when dealing with inorganics because the attention is not focused on quantum size effects.

Regardless of quantitative definitions, nanosciences and nanotechnologies are among the most interdisciplinary scientific adventures undertaken by mankind to date, benefiting from and making an impact on such diverse fields and technological applications as electronics, photonics, biology and biotechnology, chemistry, basic and applied physics, materials sciences and so on. The ubiquitously cited source of inspiration for the implementation and development of nanotechnologies originates in the world-famous talk ("*There's plenty of room at the bottom*") given by physicist R. P. Feynman on December 29, 1959, at the annual meeting of the American Physical Society at the California Institute of Technology.[2] In that speech, sub-titled "*An invitation to enter a new field of physics*", the future recipient of the Nobel Prize is fascinated by the enormous and still unexplored possibilities opened once the writing, reading and processing of information at the nanoscale is made possible by newly developed miniaturization techniques, or within biological systems. Living organisms are indeed a particularly fertile source of inspiration, as most biochemical phenomena occurring at the cellular level can be somehow recognized as "nanotechnology at work". Feynman goes beyond this and he imagines future microscopic machines and factories able to manufacture objects, control the arrangement of atoms and molecules and allow new physical effects arising only at the small scale to be observed, thus greatly overwhelming (but certainly stemming from) the concepts of classical chemical synthesis, and mimicking the way in which nature works.

Nowadays, more than fifty years later, a continuously improved control on matter at small length-scales has been made possible by the advancement of lithographic, etching, assembly and surface functionalization technologies. The new properties observed in matter at the nanoscale include many quantum effects due to the confinement of electrons or of excitons within nanostructures, together with a great variety of phenomena related to the enormously increased surface-to-volume ratios of nanostructures compared to bulk materials. The control of scientists over atoms and molecules extends over a lot of fundamental aspects, including the size and chemical composition of synthesized and lithographically made nanoparticles and nanostructures, the crystallographic organization of many semiconductor materials and the supramolecular packing of molecular systems, and over several surface properties such as polarization, wettability from liquids, capability of selectively binding chemical species, and others. These issues will be addressed later in this book for the specific case of polymer nanofibers. Crucial milestones of this exceptional progress have been the introduction of nanoelectronic junctions,[3,4] the invention of scanning tunneling microscopy (STM)[5] and of atomic force microscopy (AFM),[6] the demonstration of photonic crystal structures allowing the behavior of light to be controlled with great accuracy,[7] the discovery of fullerenes,[8] the huge development of the research on carbon nanotubes,[9–13] on epitaxially grown quantum dots[14] and nanowires,[15–19] on colloidal nanocrystals[20–22] made of inorganic materials (whose research covers a large part of what is today called *nanochemistry*),[23–25] and, recently, the advent of graphene.[26–28]

In parallel with the development of nanotechnologies, chemistry and chemical physics have been revolutionized as well. The concept of a *macromolecule*, namely a molecular object with high molar mass, often complex, including several classes of both natural and synthetic compounds, was originally proposed by H. Staudinger in the 1920s, for which he was awarded the Nobel Prize in Chemistry in 1953.[29] Staudinger has also foreseen that macromolecules, for their peculiar structure and consequently specific rheology, are the entities at the base of existence and physical properties of fibers in nature.[30] Since the 1940s–1950s, polymer chemistry has become a strategic component of industrial production processes and an increasingly vigorous research field. The inventions of nylon by W. Carothers in 1935 and of polymerization catalysts by K. Ziegler and G. Natta (awarded the Nobel Prize in Chemistry in 1963) have laid the foundations of the plastics industry, and thermoplastic polymers constitute a major class of materials applied in diverse sectors, including biotechnology, optoelectronics, coatings, textiles, and so on. In the last three decades, important breakthroughs have concerned the introduction of new and optimized synthetic routes, the development of novel technologies relying on polymers that have unsaturated bonds, *i.e.* the discovery and the exploitation of conjugated polymers and of other organic semiconductors,[31] and the new ideas stemmed from supramolecular chemistry[32] and by the broad concept of *soft matter*†. The latter concepts group an extraordinarily varied set of organic systems in which physical phenomena are governed by relatively weak inter-component interactions, or in other words occur at energy scales comparable to thermal energies. Soft matter and supramolecular architectures are characterized by an extraordinarily various

† On many occasions, De Gennes has stigmatized the use of the term *complex fluids*, which North American scientists often call molecular systems, because many of these materials do not show an evident flow behavior, but mainly because the mental association with "complexity" can discourage students and young researchers from undertaking the study of this fascinating subject.

behavior, based on some key factors which include the steric hindrance of molecular systems, the variable strength and directionality of the intermolecular interactions, the degree of flexibility and possible adjustability of the supramolecular configuration and entanglements. At almost all levels, disorder plays a crucial role, and transport phenomena are largely based on diffusional processes, as will be described later in this book. Self-assembly and molecular recognition capabilities are other key features, which make these systems strategic for nanofabrication, catalytic applications and chemical and biochemical sensing. D. J. Cram, J.-M. Lehn and C. J. Pedersen were awarded the Nobel Prize in Chemistry in 1987 for the development and use of molecules with structure-specific interactions of high selectivity.[33] P.-G. De Gennes was awarded the Nobel Prize in Physics in 1991 for the introduction of the concept of soft matter and its applications to liquid crystals and polymers.[34] Today nanostructures and nanosystems are bringing new and exciting opportunities of scientific and technological development to these research fields. Many important ideas, such as the possibility of creating controlled supramolecular systems by means of peculiar self-assembly properties induced in organic materials through suitable molecular design and engineering, have been strengthened in this framework. Especially in the last decade, these intersecting topics have strongly attracted an increasing interest in the research, academic and industrial communities. The merging of the world of molecules with nanosciences has generated the field of *soft matter nanotechnologies* or simply *soft nanotechnologies*,[35] in which organic materials and assemblies are used to build new constructs, and as building blocks of photonic and electronic devices, novel tools for biotechnology and life sciences, new nanocomposite materials, *etc*. The *Soft Matter* journal[36] was launched in 2005, and a beautiful overview about the origin, challenges and future perspectives of soft nanotechnologies has been published in volume 143 of the *Faraday Discussions* from the Royal Society of Chemistry.[37,38]

Soft materials and devices based on organic compounds require an entirely dedicated set of specific nanofabrication methods, in order to effectively interface molecular systems to external agents (photons, electrical connections, other chemical compounds, living organisms, *etc.*) without damaging the functionality of organics. Furthermore, many organic materials exhibit additional criticalities compared to their inorganic counterparts. Playing with organic molecules and devices, scientists learn that the reproducibility of experiments *is* an issue, because the low energy scales of interactions involved in molecular phenomena make these very susceptible to possibly uncontrolled variations of external conditions.[37] Furthermore, charge-carrier transport and some optical properties are often largely sub-optimal. Refractive indexes are generally low ($\cong$ 1.4–2.5 in the visible spectral range), which can limit the application of some organics for realizing optical waveguides or more generally for gathering photons within photovoltaic and lasing devices. Thermal stability is an issue for almost all organic systems. Mechanical strength can be insufficient for some applications. At a more fundamental level, the degree of confinement achievable by excitons is strictly related to their nature and degree of intrinsic delocalization. For this reason, nanostructures able to confine excitons effectively are much easier to realize in conventional inorganic semiconductor materials (where Wannier excitons are highly delocalized) than in molecular solids (where Frenkel excitons represent the situation where an electron and a hole are much closer to each other).[39] A good idea of the difference in the two cases is given by the Bohr radius of the exciton, which is $R_B = 4\pi\varepsilon_0\varepsilon_r\hbar_P^2/m_0q^2$, where $\varepsilon_0$ and $\varepsilon_r$ are the vacuum and the material dielectric constant or permittivity ($\varepsilon_0 \cong 8.854 \times 10^{-12}$ $A^2$ $s^4$ $kg^{-1}$ $m^{-3}$ in International System units), respectively, $m_0$ is the reduced mass of the exciton, $q$ is the unit charge of electrons and holes and $\hbar_P = h_P/2\pi$ ($h_P$ is the Planck constant = $6.62617 \times 10^{-34}$ J s). The dielectric constant is therefore particularly important in determining the effect of screening of the Coulomb attraction, which keeps together the electron and the hole, and materials having lower values of $\varepsilon_r$, such as organics, consequently have a much more localized exciton wavefunction. The binding energies of Frenkel excitons will be correspondingly much larger (say ~0.1–1 eV) than those (say, $10^{-2}$ eV) of Wannier excitons. For comparison, one can think of the exciton Bohr radius as ranging roughly from 1 Å to a couple of nm,[40] and from 1 nm to many tens of nm,[16] for organic and inorganic semiconductors, respectively. As a consequence, as mentioned above, it is difficult to see quantum effects in electronically active organics, and especially demanding from the point of view of nanofabrication. We will rarely describe quantum phenomena throughout this book. Overall, there are many reasons why functional organics have not been considered as candidates to replace conventional inorganic materials, at least in several industrial sectors. On the other hand, molecules open unprecedented perspectives by virtue of their peculiar mechanical and rheological properties, ease of embedment in printable, stretchable or bendable device platforms, the almost unlimited flexibility allowed by chemical synthesis and the capability expressed by organic materials concerning molecular recognition, non-linear optical properties, controllable switching behavior, *etc.* All these features are shared by many classes of molecules, allowing researchers and engineers to have maximum freedom of choice when selecting the most suitable or promising compounds in view of different targeted applications.

Today, the need to tackle some challenging aspects of nanotechnology, including the potentially hazardous nature of synthesized nanoparticles for human health (at the base of the recent field of nanotoxicology), and more generally the environmental impact of nanotechnology manufacturing processes, is becoming more and more urgent. In this respect, many countries are launching dedicated initiatives and research programs aiming to define the risks associated with nanotechnology research and industry, and to prevent long-term toxicological damage by the introduction of specific regulatory aspects. Examples of guidelines come from the Recommendation of the European Commission of February 7, 2008, on a code of conduct for responsible nanosciences and nanotechnologies research.[41] Generally initiated with experimental campaigns focused on inorganic nanoparticles or carbon nanomaterials, these programs should soon also take into consideration soft matter nanostructures and in particular polymer nanofibers. In fact, nanofibers are not free from potential hazards due to their lightweight nature, which make them susceptible to being inhaled, and to the large variety of biochemical actions possibly driven once they are in the body, depending on the specific polymer used, on the fiber dimensions, hydro- and liposolubility, *etc.* Some studies have shown that elongated particles such as fibers with higher aspect ratios are more susceptible to deposition in the human respiratory system (and especially into so-called higher generations of the respiratory tree, namely in deeper regions of the lungs) compared to more isotropic particles, at least for fiber diameters above about 200 nm.[42,43] *In vitro* studies with cell cultures on cellulosic nanofibers seem encouraging in this respect, showing the absence of appreciable genotoxicity and influence on the resulting cell morphology,[44] and significantly lower cytotoxicity and inflammatory response compared to carbon nanotubes.[45] However, some nanofibers derived from cotton have been reported to be genotoxic in plant cells and, to a lesser extent, in animal cells.[46] Furthermore, polymer nanofibers production technologies, as many of those described in this book, often involve the use of organic solvents, which may cause additional concern about their impact on health and environment. This issue is also very important in view of the use of nanofibers in biotechnology and pharmaceutical systems, because of the possible solvent residues trapped within the polymer nanostructures. Until a sufficient amount of knowledge is available about the environmental and health impact of nanomaterials, the principles of sustainability and precaution should be applied, in an effort to anticipate potential hazards as much as possible. Fortunately, various strategies exist to increase substantially the safety of these processes, also in the rapidly growing domain of so-called green chemistry. Nanofibers production technologies relying on aqueous polymer solutions certainly deserve to be further developed. This is also encouraged by the existence of novel formulations of polymers, including conjugated materials, which are processable by aqueous solutions, thus possibly allowing the consumption of volatile organic solvents to be significantly reduced. Furthermore, the development of manufacturing processes using natural polymers finds in nanofibers research an ideal field of application, as inspired by the fibrous shape taken by many of these polymers in nature. Indeed, many methods for producing nanofibers are already largely tested with natural polymers, as described in detail in Chapters 2 and 3.

We close this introduction by pointing out that, especially for young scientists, nanosciences and nanotechnologies are an intriguing and vast field of research, offering unlimited possibilities to design and implement new experiments and models. How much these approaches will be able to keep their promises and to fulfill the enormous expectations that they have generated in the research community, and in society at large, has still to be assessed. Addressing these issues will likely require more and long-lasting investments in research, thus guaranteeing a wide and certainly fertile space for carrying out innovative experiments, implementing novel models and transferring new generated knowledge and developed technologies to industry. In particular, among all the different classes of nano-objects and nanosystems, soft matter nanostructures such as polymer nanofibers have unique appealing features of chemical and functional flexibility, relatively low cost of production, ease and assortment of processing, potentiality of interfacing with biological systems and of being applied to many branches of science and technology. For all these reasons, we believe they are particularly suitable for stimulating an enthusiastic response in the next generations of interdisciplinary researchers.

## 1.2 Nanotechnology Routes from Inorganic to Organic Materials

The electronics industry has followed a well-defined trend of miniaturization over the last decades. This has been formalized by means of a world-famous scaling law, formulated in 1965 by G. E. Moore. The law predicts that the number of transistors inside an integrated circuit would double roughly every two years (originally the prediction was of a doubling per year),[47] which corresponds to decreasing the lateral size of the features of an electronic component at a rate of approximately a factor of two every four years. Computing performances can improve at even higher rates, doubling about every 18 months. The count of transistors currently present in microprocessors has reached a value in the billion range and component feature size is of a few tens of nm. So far, this trend has been driven by top-down microfabrication techniques on inorganic semiconductor materials, and especially by optical lithography. This approach consists in patterning semiconductors by selectively exposing a semiconductor wafer to a radiation, and then removing undesired features by wet- or dry-etching. However, in the last decades the technical limitations and increasing costs of massive parallel production carried out by conventional lithographies have motivated the development of alternative nanofabrication approaches, inspired by chemistry and biology, and often working by bottom-up concepts. This means we need to start considering molecules as building blocks of patterns and devices, thus assigning a new, pivotal role to soft matter and to the capability of realizing controlled supramolecular assemblies and nanostructures. In this framework, polymer nanofibers are attracting a continuously increasing interest, because their physico-chemical properties make them intriguing functional elements for an astonishingly various range of applications. For their elongated shape and lateral dimension below 100 nm, they can be utilized as sacrificial polymeric features in lithographic processes and nanoelectronics. A special consideration is deserved by the surface-to-volume (or surface-to-weight) ratio of these strongly elongated nanoparticles. Polymer nanofibers with material density around 1.0–1.5 $\times$ $10^3$ kg m$^{-3}$ and fiber radius, $r$, of 100 nm would have a specific surface area as high as about (13–20) $\times$ $10^3$ m$^2$ kg$^{-1}$. These data are more often and conveniently presented in units of m$^2$ g$^{-1}$, namely as 13–20 m$^2$ g$^{-1}$. The specific surface area would increase to 130–200 m$^2$ g$^{-1}$ for fibers with a radius of 10 nm (*i.e.* scaling as $r^{-1}$, see also Figure 4.17)‡. For this ultra-high surface-to-volume ratio, they can be strategic for all the applications where a very large interface is needed for exchanging chemical signals with the surrounding environment, such as in sensing and catalysis. Their relevant flexibility in terms of chemical composition can be further enhanced by blending and encapsulation approaches, allowing nanofibers with mixed polymeric architectures, core-shell nanostructures and composite nanomaterials also involving inorganic nanoparticles to be obtained. Optical, electrical and mechanical properties of nanofibers may be significantly improved compared to the bulk material, because of the intrinsic anisotropy and the possibly higher degree of molecular orientation within the nanostructures. In this way, active polymer nanofibers may even be applied as active media in transistor devices, as described in detail in Chapter 6. The full exploitation of these properties, however, is achievable only by a careful design taking into account peculiarities and requirements of soft materials, which with no exception impact on fabrication methods developed to obtain polymer nanofibers, on possible causes of deterioration of organics during processing and operation and ultimately on performances when molecular systems are embedded in samples and devices. In other words, nanotechnology-based production processes have to be *specialized to the soft matter case*

‡ The specific surface of polymer nanofiber samples can be determined experimentally in various ways. A possibility is extracting directly average geometrical information by electron microscopy or AFM and then computing the external surface of fibers. However, these approaches can have some difficulty to probe the internal structure of complex nanofiber assemblies. An effective method is the so-called Brunauer–Emmett–Teller (BET) technique,[48] which is very widely used in research and industry. This technique exploits the adsorption of an inert gas, such as nitrogen, measured upon varying the relative pressure of the gas. The BET adsorption isotherm can be derived under the hypothesis of adsorbed molecules which are not mutually interacting and adsorbed layers which are not mutually interacting. In addition, a constant heat of adsorption is assumed between the gas molecules and the polymer surface. The resulting isotherm reads: $Y = [(c_{BET} - 1)/v_{m,ads}\, c_{BET}]\, X + (1/v_{m,ads}\, c_{BET})$, where $Y = V_0^{-1}[(p_S/p) - 1]^{-1}$, $X = p/p_S$, $c_{BET} = \exp[(E_1 - E_O)/k_B T]$ is the so-called BET constant, $v_{m,ads}$ is the volume of an amount of gas corresponding to an adsorbed monolayer, $p$ and $p_S$ are respectively the pressure and the saturation pressure of the gas, $V_0$ indicates the volume of the adsorbed gas, $E_1$ and $E_O$ are respectively the heat of adsorption of the first and of the other monolayers and $k_B$ is Boltzmann's constant (1.38066 $\times$ $10^{-23}$ J K$^{-1}$). The above $Y(X)$ linear behavior, observed for $X$ values between 0.05 and 0.35, can be used to determine the surface area, which is given by $v_{m,ads} A_{ads} N_{AV} / V_M$, where $A_{ads}$ is the cross-sectional area of the adsorbing species, $N_{AV} \cong 6.022 \times 10^{23}$ is Avogadro's number and $V_M$ is the molar volume of the adsorbate gas ($\cong$ 22.41 L for an ideal gas at normal conditions of temperature and pressure). $c_{BET}$ and $v_{m,ads}$ are obtained by the slope and the intercept of the $Y(X)$ isotherm with the $Y$ axis. Several studies have applied the BET method to polymer nanofibers.[49–51] The pore size distribution can be characterized as well for mesoporous materials.

in order actually to be effective when applied to organic and polymeric compounds. Micro- and nanolithographic manufacturing, which is at the base of the fabrication of miniaturized devices across all industrial applications and technologies, is an excellent example of how this specialization can be pursued and shares many concepts with polymer nanofiber production methods.

### 1.2.1 Lithographies

Optical or photo-lithography is still the dominant patterning technique utilized at an industrial scale in micro- and optoelectronics, by virtue of its large-area and parallel operation, high accomplished resolution and well-established and optimized procedural protocols.[52] As for all the other exposure-based lithographic approaches, the working principle of photo-lithography is the variation of a well-defined physical property (*i.e.* the *solubility* in a suitable development solvent) of a sensitive polymeric film (the *resist*), due to the interaction with an incident beam of radiation (here, UV-photons, in other methods electrons, X-rays,[53] *etc.*)[§]. Patterning is carried out because the exposure to impinging radiation is made spatially selective, by means of photo-masks presenting features that are either transparent or opaque at the illuminating light wavelength, $\lambda$, or, alternatively, by serially drawing on the resist surface by means of a narrow beam of radiation as in laser photon writing or in electron-beam lithography (EBL). The variation of the resist solubility is in turn due to the changes of molecular structure and nanoscale conformation of the resist material, which occur upon exposure. This latter point is indeed the crucial operation step of all exposure-based lithographies. The variation of configuration at the molecular level can be induced by two possible basic mechanisms, namely either (i) the breaking of molecular bonds because of the incident particles, and the consequent enhancement of solubility of the exposed material in a suitable development solution (positive resist), or (ii) the further polymerization and cross-linking of molecules by means of the energy delivered by the beam, which would determine a reduced solubility of the exposed area (negative resist). Incidentally, photo-induced cross-linking of polymeric materials is also used with analogous mechanisms in many methods to produce polymer nanofibers, such as in some variants of reactive electrospinning (Section 2.3), of nanofluidics (Section 3.2) and of block copolymers nanopatterning (Section 3.3.2). New perspectives for the realization of functional polymer nanoparticles by means of exposure-based methods have also been opened by continuous flow lithography (CFL), proposed by P. S. Doyle and coworkers.[54–59] In CFL, the particle shape is defined by projection photo-lithography on negative resist compounds flowing in

§ The sensitivity of the resist material (in International System units of J $m^{-2}$, C $m^{-2}$ *etc.* depending on the kind of radiation) is defined as the exposure dose at half thickness, obtained by measuring the height profile of the resist film after irradiation and development.

microfluidic channels. This allows ultimately self-standing micro-objects to be fabricated. Chemical anisotropy can be achieved in the particles by polymerizing across the liquid–liquid interface separating different solutions co-flowing in the microchannels, and multifunctional particles can be produced by hydrodynamic focusing lithography.[59]

Regardless of the specific exposure method, on the basis of their operation principles it is immediately evident that a careful design of the underlying chemistry (of both radiation-sensitive species and eventually of so-called chemical amplifiers) is needed for realizing suitable resist compounds, and that, consequently, photo-lithography and other exposure-based lithographies work only on a very restricted class of molecular systems**. In conclusion, exposure lithographies generally offer poor chemical flexibility in terms of the target materials, this approach being applicable only to a few photo- (or electron-) sensitive polymers, while being incompatible with most organic matter, which can support bottom-up nanofabrication methods or the production of one-dimensional or quasi-one-dimensional functional polymer nanostructures. Many reviews exist in the literature about fundamental issues, points of strength and limitations of different lithographic approaches, to which readers can refer for more details.[64,65]

The crowning achievements of research efforts in looking for alternative lithographic approaches that could be more suitable to pattern organic materials should be attributed to G. M. Whitesides and coworkers, who proposed in the 1990s a novel and rich set of patterning techniques, collectively named soft lithographies.[66] Soft lithography relies on the use of elastomeric elements to transfer the pattern from a master structure (in turn realized by conventional lithographies, generally on standard inorganic materials such as silicon, metals, *etc.*) to the target materials, and exhibits an unequalled chemical flexibility enabled by its own gentle approach to patterning. The process named replica molding (REM), namely the generation of elastomeric replicas of the previously fabricated master structures, is the enabling step of soft lithography.[67,68] For REM, a pre-polymer in the liquid state is cast onto

** Another intrinsic limitation of optical lithographies, making them hardly applicable to sub-100 nm patterning of most soft matter systems, functional polymers, *etc.*, consists in the spatial resolution of the features that can be produced. In fact, the minimum feature transversal size achievable by photolithography is limited by light diffraction as stated by the well-known Rayleigh's law: $\Delta x = k\lambda / N.A.$, where $k$ is a material- and process-dependent proportionality constant, $\lambda$ is the wavelength of the light used in the exposure step and $N.A.$ indicates the numerical aperture of the optical set-up used to produce the pattern on the resist surface. Therefore, due to the diffraction of light, the minimum lateral size of the exposed features is limited by the illuminating wavelength, hence in order to increase resolution light sources of progressively higher energies have been developed and utilized over the last decades,[60,61] together with complementary strategies such as phase-shifting masks,[62] chemically amplified resists[63] and immersion lithography. Today, the exploitation of light in the deep UV (with wavelength below 200 nm) allows integrated circuits with minimum feature size of the order of very few tens of nm to be produced. Serial methods such as EBL allow optical diffraction drawbacks to be tackled, but present a much lower throughput, higher costs and time expenditure. In all cases, realizing components with smaller size is more and more expensive and technologically demanding.

the master structure and polymerized *in situ*, generally by thermal curing. Polydimethylsiloxane (PDMS) has been largely applied as the most used elastomer for REM processes at resolutions down to the micrometer-scale, because of its optical transparency and surface conformability. After curing, the elastomeric replicas can then be peeled off from the master by slight reversible deformations allowed by their elastomeric character, providing the cured elements with some structural flexibility. The final result of REM is therefore a faithful, negative copy of the master pattern (the overall process is schematized in Figure 1.1a). The REM process is therefore a direct demonstration that soft matter nanotechnologies (besides producing devices for the generation, processing, transmission and detection of optical, electrical and chemical signals) provide effective tools to *replicate stored information*, here initially contained in the master features.[66] By virtue of their working principle, extreme experimental simplicity and cheapness, soft lithographies allow the fabrication costs and time spent for producing the master structures to be spread over tens or hundreds of patterned, soft replica samples, with almost no additional cost. For this reason, these methods are fully accessible to bio-organic research laboratories, and create practically no barrier to entry in terms of capital costs and equipment, which explains their extraordinary success. Daughter technologies stemmed from REM include micro-[69,70] and nanofluidic[71] lithographies based on elastomeric elements, and various soft molding approaches.[72] Using REM-fabricated elastomeric replicas of the initial master as second-generation patterns, these other techniques again produce a positive copy of the original features on the ultimate target material. In addition, many of these approaches can be entirely carried out at room temperature, thus being largely compatible with active polymers for optoelectronics and nanoelectronics. On these and many other materials, several soft lithography techniques can be performed with slight or no deterioration of molecular functionality. Of course, the soft nature of both elastomeric elements and the target materials are at the base of a few intrinsic issues of soft lithographies, mainly in terms of reproducibility and pattern distortion, which can occur especially when one uses patterns with high resolutions. In fact, silicone-based elastomers such as PDMS swell severely in most oil-soluble organic compounds[††],[73] and their surface energy and low Young's modulus ($E \cong$ 3–4 MPa, see the following Section 1.4.2) are not fully suitable for achieving sub-100 nm resolution when transferring the pattern onto final materials.[74–76] However, many of these drawbacks can be overcome by carrying out soft lithographies with other elastomers than PDMS. For instance, researchers at IBM have proposed a harder version of PDMS, obtained from trimethylsiloxy-terminated vinylmethylsiloxane-dimethylsiloxane and methylhydrosiloxane-dimethylsiloxane

†† Compounds soluble in aqueous solutions or in organic solvents are also called *water*-soluble (*i.e. hydro*-soluble) or *organo*-soluble, respectively.

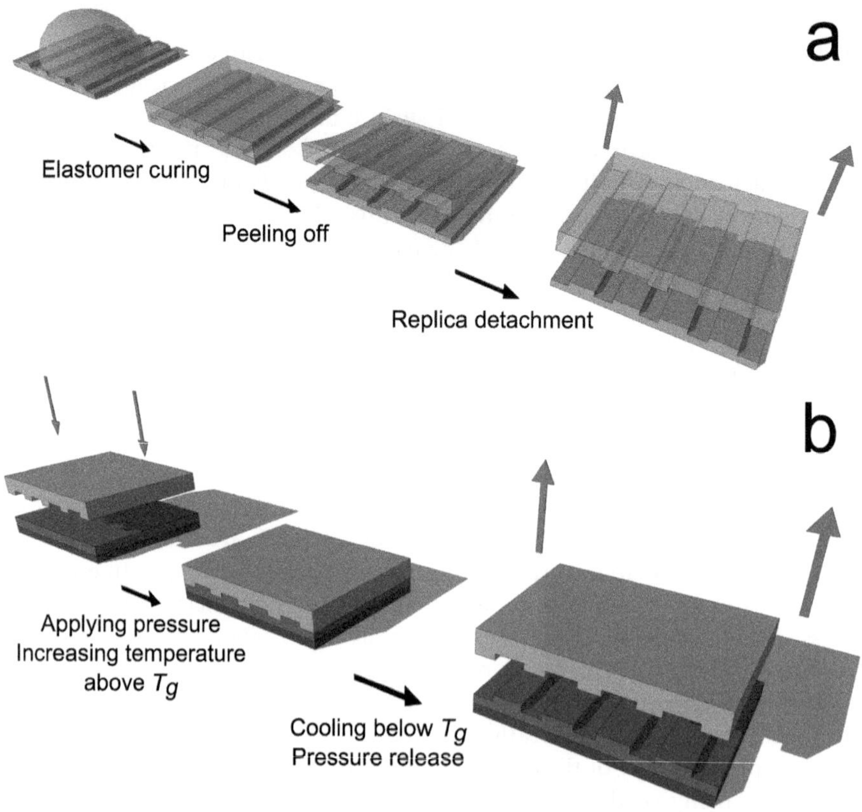


**Figure 1.1** Schemes of the REM (a) and the NIL (b) processes. Soft lithography and NIL are among the methods allowing polymer nanofibers deposited on substrates to be produced.

copolymers[‡‡,76] which exhibits a compression modulus up to 9 MPa. This elastomer can be used within composite stamps to reduce distortions during pattern transfer. More recently, J. M. DeSimone and coworkers have introduced the use of photo-curable perfluoropolyether (PFPE)-based elastomers, exhibiting high resistance to organic solvents and being suitable for high-resolution soft lithography,[74] having surface energies down to 14 mN $m^{-1}$. These materials, while largely retaining the PDMS capability of

‡‡ This is the first time that *copolymers* are mentioned in this book. Many examples of technologies for the fabrication of polymer nanostructures and nanofibers use copolymers and smartly exploit the peculiar phase-separation properties of these molecular compounds. More generally, these macromolecules are widely used in many fields of modern technologies and are often considered as a particularly intriguing tool in the framework of nanosciences. Differently from simple polymers (or *homopolymers*), which are based on the repetition of the same monomer units, copolymers are constituted by two or more species of repeating blocks, in turn composed of different monomers (A, B, *etc.*). In the case of diblock copolymers, one may indicate the molecules as poly(A)-*block*-poly(B), poly(A-*block*-B) or simply A*b*B by an abbreviated notation.

undergoing conformal contact on surfaces and allowing easy release from the masters, are compatible with organo-soluble molecules because of the intrinsically oleophobic nature of highly fluorinated materials. Consequently, these developments have opened the way to apply some soft lithography methods, such as nanofluidics based on elastomeric channels, to produce polymer nanofibers by means of organic solutions. The advantages of this approach include the extremely high accuracy in positioning the realized nanostructures on solid surfaces, thus facilitating their embedment within device architectures, high control of the density and spatial resolution of the nanofibers *via* the top-down produced master and very low material waste. More details on these technologies are presented in Section 3.2.

Furthermore, soft lithographies have been complemented by nanoimprint lithography (NIL) proposed by S. Y. Chou and collaborators,[77,78] which uses a hard mold to imprint cheaply micro- and nanostructures onto large areas of the target compounds, without exposure to radiation, development and etching processes (Figure 1.1b). NIL is likely the most effective nanotechnological example of exploitation of the glass transition (Section 1.4.1) and rheology (Section 1.4.2) behavior of polymer materials.[79] While being based on high-temperature processing and hence applied mostly to thermoplastic polymers in its early historical stage, nanoimprinting can be easily extended to work at low temperature (*i.e.* without leading polymers above their glass transition temperature), and to pattern compounds of relatively lower molar mass,[80–82] thus allowing one to utilize much more in depth the enormous variety of functional materials offered today by chemistry. Room-temperature NILs find an important application also in patterning light-emitting nanofibers,[83] as described in Section 5.6.

Many production approaches that allow organic quasi-one-dimensional nanostructures such as polymer nanofibers to be fabricated can be broadly classified as lithographic. As described in Chapter 2, electrospinning is nothing but a very rapid way to draw nanofibers (with spatial resolution down to the nm-scale and various degrees of positioning control) up to achieving a complete coverage of the target surface if needed, and it is basically an extension of the electrospray process, which in turn is one of the most widely used methods to deposit and pattern organic and biomolecular solutions. Direct writing methods deriving from conventional, even exposure-based, lithographic techniques have been also proposed to realize polymer nanofibers (Section 3.5). Other approaches, such as those described in Sections 3.1, 3.3, 3.4 and 3.6, are instead more based on chemistry, relying therefore on bottom-up, lithography-free fabrication concepts as ways to assemble macromolecules in spatially defined, elongated nano-objects.

## 1.3 One-dimensional Structures and Nanostructures

A first aspect to be clarified when dealing with one-dimensional or quasi-one-dimensional nanostructures is semantic. In the literature one reads about nanofibers, nanowires, nanotubes, nanocylinders, nanotubules, nanorods, nanoribbons and so on, and this may lead of course to some confusion. Here we briefly sketch some guidelines concerning the most often found nomenclature. Some of the most frequently encountered elongated nanostructures are schematized in Figure 1.2. First of all, one can assume that, among all these terms, "nanofibers" is the only one that describes properly *mechanically flexible nanostructures of extremely high length-to-radius ratio* (*i.e.* axial ratio). The "nano" prefix, together with its meaning gradations, has already been presented in Section 1.1, so one can expect that the diameter of "nano"-fibers should be roughly below 100 nm. However, how much the aspect ratio (and how flexible the structures) should be to call a microscopic object a nanofiber is of course something more questionable. A "nanofiber" that has a diameter of 50 nm can be from, say, a few hundreds of nm to tens of km in length.

The term "polymer nanowires" is often used almost equivalently to "polymer nanofibers" in many papers, although a meaning more shifted to describe crystalline or semicrystalline organic nanostructures is certainly to be preferred. This crystallinity would also imply that a "nanowire" is not as flexible, and hardly as long, as a "nanofiber" in most cases. A polymeric "nanotube" is also generally not too long (say, below 1 mm), because being hollow it hardly shows the structural stability needed for reaching extreme lengths. A "nanoribbon" is a strongly elongated nanoparticle in which the cross-sectional dimensions (*i.e.* those perpendicular to the longitudinal axis of the nanostructure) are quite different along the two mutually normal directions. In other words, a nanoribbon is a nanofiber or a nanowire with markedly rectangular cross-section. In this case, only the smaller dimension is

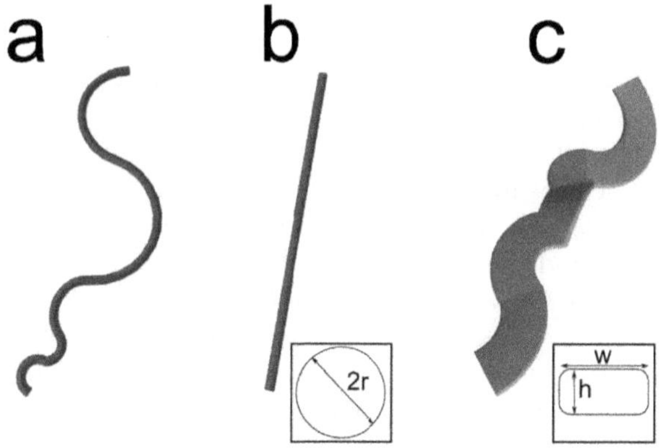


**Figure 1.2** Schemes of a few one-dimensional nanostructures that can be realized by polymeric materials. (a) Flexible nanofiber (typically made of amorphous polymers), (b) nanowire (frequently made of semicrystalline organics) and (c) nanoribbon. The insets in (b) and (c) show schematics of the corresponding cross-sections of the nanostructures. *r*: nanowire radius; *w*: nanoribbon width; *h*: nanoribbon thickness.

often below 100 nm, whereas the long side of the rectangle can be many μm or more.

Leaving aside the prefix "nano" for a moment, seeing to what extent fibers and fiber materials are widespread in modern technologies is stupefying. Entire industrial sectors such as textile and cloth fabrication, products for thermal and acoustic insulation, filters at all scales, nanocomposites increasingly used in buildings and aeronautics, electrical connections and even telecommunications fundamentally rely on fiber materials or fiber-based devices. Fiber optics (Figure 1.3a), using total internal reflection of photons in glass or polymer fibers having a high-refractive index core along their longitudinal axis, is the cornerstone of modern information and communication technologies, and their inventor C. K. Kao was awarded the Nobel Prize in Physics in 2009.[84] Carbon fibers with diameter in the range of 5–10 μm have become very important for automotives, aeronautics, buildings, *etc.* (Figure 1.3b).[85] In nature, living organisms are able to produce fibers by soft matter, such as the adhesive filaments in the byssus of mussels[86] and other bivalve molluscs, spider and cocoon silk, feathers, human hairs, *etc.*, and even by inorganic materials, such as in the spicules of many species of sponges,[87,88] in which mineralization is driven by polycondensation of silica favored by specific proteins such as silicatein (Figure 1.3c–f). It is natural to expect that inducing nanoproperties in fiber-shaped objects, exploiting at the same time the astonishing variety of functions enabled by chemistry, can lead to new and intriguing material systems, which in turn can open the way to both observing novel physical or biological phenomena and designing entirely new technological applications.

Elongated nanostructures and super-structures can be grown or synthesized, for instance, in the form of inorganic nanoparticles. Many different classes of inorganic nanowires, nanorods *etc.* are realized by nano-chemists, with specific shapes, chemical and crystalline compositions and ability of hierarchical self-assembly in super-lattices. Two examples are shown in Figure 1.4.[89,90] Several reviews are available that describe in a comprehensive way the fabrication technologies targeting inorganic one-dimensional nanowires and nanofibers.[91,92] Interested readers can find a summary of intriguing prospective applications, which also lists the key academic groups working in the field, in Ref. 93. In the rest of the book, we specialize instead in the case of polymer-based nanostructures and nanofibers.

## 1.4 Introduction to Polymer Physics and Engineering

This Section aims to provide some basic introductory concepts of polymer physics, which are needed to understand the working principles of polymer nanofibers manufacturing technologies. This background information includes the glass transition behavior and the fundamental definitions of rheological and flow properties. Most of the materials described in this volume are processed or used in a prevalently amorphous state, which is the glassy state of

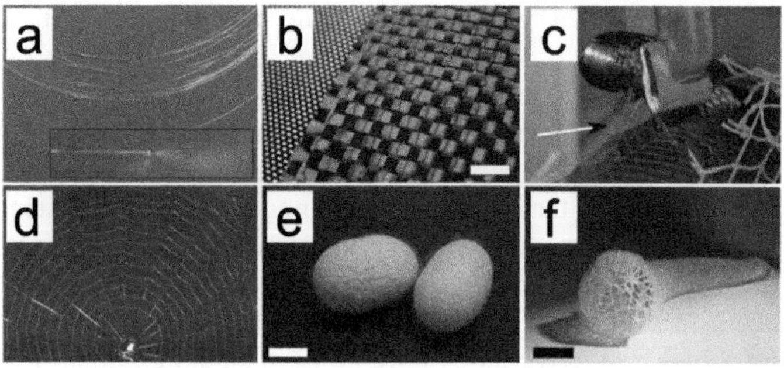


**Figure 1.3** (a) Bodies and termination (inset) of optical fibers. The external layer (coating) of these fibers is around 700 μm. (b) Fabrics made of carbon nanofibers. Scale bar = 1 cm. Courtesy of G. Angeloni S.r.l. (c) Byssus threads produced by mussels. Adapted with permission from Ref. 86, *Aquacult. Eng.*, 2010, **42**, 128–139. Doi: j.aquaeng.2010.02.001. Copyright © 2010, Elsevier B.V. (d) A spider web and (e) *B. mori* cocoons. Both are made of silk microfibers. Scale bar in (e): 1 cm. (f) Siliceous fibrous skeleton of the glass sponge, *Euplectella aspergillum*. Scale bar = 2 cm. Courtesy of Prof. W. E. G. Müller.

matter very frequently exhibited by polymeric systems. Rationalizing the basic concepts at the base of the liquid-to-glass transition is not only highly interesting from the fundamental viewpoint, but also particularly important to

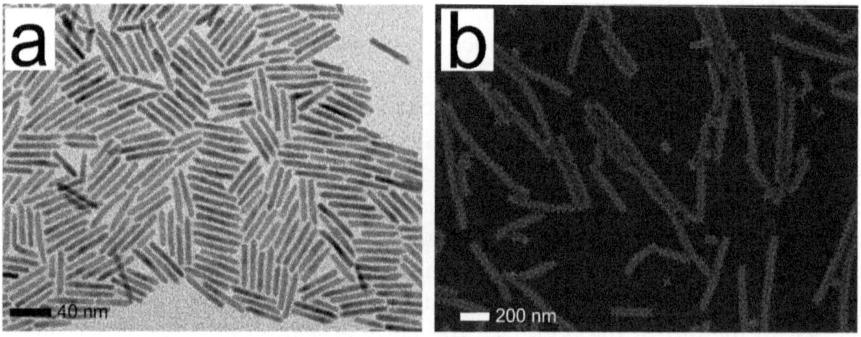


**Figure 1.4** One-dimensional inorganic nanoparticles and super-structures. (a) CdSe/CdS core/shell nanorods imaged by transmission electron microscopy (TEM) and (b) linear chain super-structures obtained by self-assembly of monodisperse colloidal octapod-shaped nanocrystals and imaged by scanning electron microscopy (SEM). Semiconductor nanorods are fabricated by a seeded-growth approach, starting from preformed CdSe nanocrystal seeds and using suitable CdS precursors in a hot solution. The resulting particle length is finely tunable up to 150 nm.[89] Linear chains of octapods form on timescales of a few hours when nanocrystals are dispersed in toluene, due to mutually attractive Van der Waals forces resulting from nanocrystal–nanocrystal interactions, which are slightly stronger than solvent–nanocrystal interactions, as is typical of self-assembling processes (see Section 3.1).[90] Courtesy of Prof. L. Manna.

properly design fabrication techniques and more generally experiments involving polymer molecules. A few of these concepts, such as those of steric hindrance and energy landscape, are indeed so general to be useful when dealing with both solutions and solid samples. A few general aspects of the glass transition phenomenology are recalled in Section 1.4.1. Incidentally, it should be mentioned that a few classes of nanostructures analyzed in the book present instead some features of long-range molecular ordering that clearly belong to the domain of crystalline materials. Examples include lasing oligophenyl nanowires grown by molecular epitaxy (Section 5.5), highly conductive polymer nanofibers with internal packing of molecules in ordered lamellae (Section 6.1) and piezoelectric nanofibers (Section 6.4.1). The crystalline fraction is of course higher, and may lead even to a complete monocrystal formation, for oligomer molecules that are able to arrange much more easily in an ordered state, whereas large macromolecules like polymers are generally too complex from the steric viewpoint to form completely crystalline materials, as is better explained in the following section. A frequent supramolecular configuration in polymers, which is of help in understanding the internal structure of polymer nanofibers as well, is mainly amorphous in its whole, but can include domains that comprise more ordered molecules (see, for instance, Figure 4.1). An important example is given by microcrystalline *lamellar* structures with $\pi$–$\pi$ stacking in poly(3-alkylthiophene)s (P3ATs), which will be described in detail in Section 6.1. Of course, if a crystalline fraction is present in the polymer, the ultimate degree of crystallization and the orientation of ordered molecular domains will strongly affect the mechanical, optical and electrical properties, a situation which is particularly relevant for polymer nanofibers as shown in many examples throughout the book. Differently from amorphous materials, crystals can be represented by the periodic repetition of a unit cell along different directions of space, and such highly ordered architecture brings about very well-characterized solid-state physical properties, including the electronic bandstructure for available and forbidden energies, and various kinds of anisotropy. Fundamentals of crystalline materials are presented in many excellent books with different degrees of complexity, such as those by Ashcroft and Mermin,[94] Bassani and Pastori Parravicini[95] and Pope and Swenberg,[96] the latter being focused on organic crystals.

### 1.4.1 Glass Transition Phenomenology

The glass transition phenomenology of polymers is ubiquitously present and exploited in modern technological processes involving these materials, including many methods by means of which polymer fibers and nanofibers are produced. In general, upon decreasing temperature (or, as an alternative, upon applying different methods that allow the molecular mobility and transport effectiveness to be reduced such as, for instance, by an isothermal increase of pressure),[97] a polymer is driven to approaching its glass transition.

Indeed, the transition onset is driven by a gradual decrease of the molecular mobility, correspondingly to the increase of many orders of magnitude of the compound viscosity and of its characteristic time of response to external stresses. Cooling at a sufficiently rapid rate, the largely disordered microscopic structure of the polymeric molecular systems can be frozen in what becomes a practically permanent amorphous configuration with external solid-like appearance. Early concepts have identified so-called *glass-forming* materials as those capable of generating such microscopically disordered assemblies. Nowadays, the kinetic nature of the glass transition is well established, since a pivotal role is played by the competition between the externally imposed cooling rate and the characteristic response time by which a molecular system tries to recover its own thermodynamic equilibrium. Incidentally, it should be recalled that several methods can be used to prime the glass transition, including polymerization and hydrolysis reactions, condensation from vapor phase, bombing by heavy particles on crystalline samples, *etc.*, although the most intuitive and widely applied approach is still starting from an ordinary liquid polymer and then decreasing temperature, taking care to avoid crystallization, which leads the system to a *super-cooled* state.[98–100] Experimental cooling rates that one can easily achieve in a laboratory vary in quite a broad range (from, say, $10^{-3}$ K $s^{-1}$ to many K $s^{-1}$), and much higher cooling rates are accessible by simulations. The super-cooled and ultimately glassy states correspond to keeping the polymeric material in a fluid state, and hence *processable* by various technologies, although the temperature is well below the melting point. At the same time, starting from the glassy state and gradually increasing the molecular mobility, as by means of an increase of temperature, leads to a continuous liquefaction process (Figure 1.5), instead of a sudden transition at a well-defined characteristic temperature (as in ordinary solid-liquid transition).

The peculiar phenomenology of glass transition is at the base of the astonishingly widespread use of amorphous solids, and especially of polymeric materials, in such diverse fields of technology ranging from buildings to plastic industry, from telecommunications to optoelectronics and microelectronics. The use of glass has been fascinating mankind for thousands of years, inspiring artists and offering important tools to scientists and technicians. A glycerol-like molecule acting as antifreeze has even been found in the body fluids of Arctic insects (*E. Scudderiana*), in which liquid bodies remain in a super-cooled state and allow them to survive at temperatures below −30 °C.[101] In particular, the technologies developed to produce polymer nanofibers largely rely on materials that are prepared, processed and shaped in their amorphous state.

To understand how these concepts can be quantified and related to technological applications, special attention is deserved by the *relaxation processes* of molecular systems in proximity of the glass transition.[102,103] A bit of mathematical formalism is generally used to keep this description compact. However, the reader more interested in how this applies in practice may skip the following few lines and go directly to the most intuitive description of the

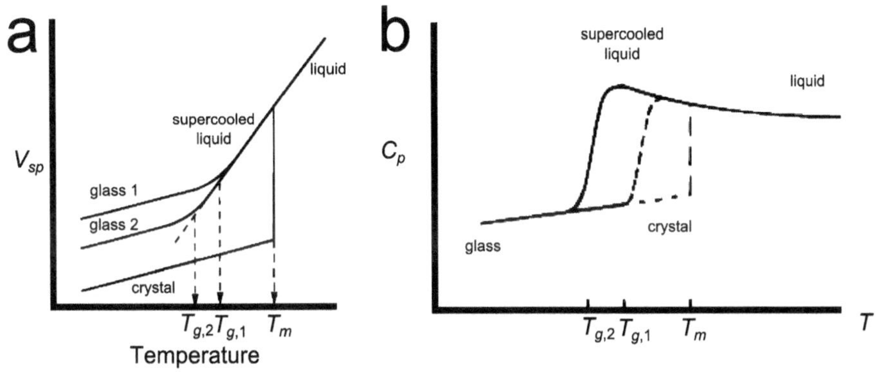


**Figure 1.5** (a) Scheme comparing the variation of the specific volume ($V_{sp}$) as a function of temperature, for a liquid that can both crystallize and form a glass. The experimental cooling rate can lead to distinct states. In the figure, the glass labeled number 2 is obtained by using a slower cooling rate compared to glass 1, leading correspondingly to a lower value of the glass transition temperature ($T_{g,2} < T_{g,1}$). The glass transition temperature, $T_g$, is here found by extrapolating the curve in the glassy state towards higher temperatures up to the super-cooled liquid line, and finding the corresponding intercept. $T_m$: melting point at which the liquid may crystallize. (b) Schematic diagrams of the corresponding temperature dependence of the specific heat, $C_p$, of the crystal (right dashed curve), of the liquid, of the liquid in the super-cooled state and of the two glasses 1 (middle dashed curve) and 2 (left, solid curve). Adapted with permission from Ref. 100, *J. Phys. Chem.*, 1996, **100**, 13200–13212. Doi: 10.1021/jp953538d. Copyright © 1996, American Chemical Society.

characteristic relaxation time, after Equation (1.2), without losing continuity. In general, relaxations can be described by a unified formalism, in which a generalized applied force, $F^*$ (related to a mechanical stress, an externally applied electric field, *etc.*), is introduced together with an associated generalized shift, $x$, that indicates the system response (corresponding deformation, resulting electrical polarization, *etc.*). $x$ takes non-zero values as the system is taken out of equilibrium by the applied force. Under the hypothesis of causality, time-independence and linearity of the physical system, $F^*$ and $x$ are related by:

$$x(t)=\chi_0\int_{-\infty}^{t}-\frac{d\Psi(t-t')}{dt}F*(t')dt', \qquad (1.1)$$

where $\chi_0$ is a generalized static susceptibility, and the relaxation function, $\Psi$, whose negative derivative ($-\mathrm{d}\Psi/\mathrm{d}t$) is the so-called Green's function of the system, represents the response of the system to an instantaneously removed force, $F^*(t) = \theta(-t)$ ($F^* = 0$ for $t > 0$ and $= 1$ for $t < 0$, *i.e.* $\theta$ is the Heaviside, or unit, step function). Therefore, $\Psi$ describes the molecular system coming back to the thermodynamic equilibrium after the external force is removed. The characteristic relaxation time is consequently defined as:

$$\tau = \int_0^{+\infty} \Psi(t)dt. \tag{1.2}$$

In polymers and other organic materials, the structural relaxation time is strictly related to the viscosity, $\eta$, which is affected by several parameters including the system temperature and pressure. Therefore, $\eta$ has a crucial role in directly affecting the molecular response to external solicitations, and in this way viscosity impacts significantly on many applications that exploit the glass-transition of molecular systems, including many soft lithographic processes (Section 1.2.1) and several routes used for fabricating polymer nanofibers (Sections 2.3 and 3.2). The structural or main relaxation is usually characterized by a broad spectral width, and the shape of the relaxation function exhibits a quite varied behavior, depending on each molecular system. The relaxation function remains almost unchanged over large temperature ranges for many compounds, whereas it shows sudden variations within narrow temperature ranges in other materials. The frequently observed independence of the shape of the relaxation function of viscoelastic materials on $T$ (allowing to obtain master curves where the different relaxation responses overlap, by simply shifting the curves collected at different temperatures) is named *time-temperature super-position principle*,[104] because a given response by the system can be accomplished both by waiting for a longer time and by increasing temperature. This is an important principle finding many applications in melt-based polymer processing. Overall, the structural dynamics of macromolecules is largely dominated by inter-molecular, *steric* interactions, hence their glass transition and flow behavior belong to the field of interest of *supramolecular physical-chemistry*.

The glass transition of polymers can be experimentally investigated by several, often complementary, techniques, including light-scattering, magnetic resonance, differential scanning calorimetry (DSC), viscosimetry and dielectric spectroscopy. Given the importance of kinetic aspects in determining the glass transition behavior (Figure 1.5), even the glass transition temperature, $T_g$, is better defined on kinetic bases, namely specifying a few, conventionally accepted experimental conditions in which the transition should be induced and observed. For instance, in calorimetric measurements $T_g$ is defined as the temperature at which a sudden change is observed in the heat capacity, $C_p$ (with a specific value of the cooling rate). In viscosimetry, one can consider the temperature at which the viscosity reaches a value of $10^{13}$ P (1 P = $10^{-1}$ Pa s). In dielectric spectroscopy, one defines $T_g$ as the temperature at which the characteristic time, $\tau$, of the structural relaxation becomes larger than a typical observation interval (conventionally $10^2$ s).

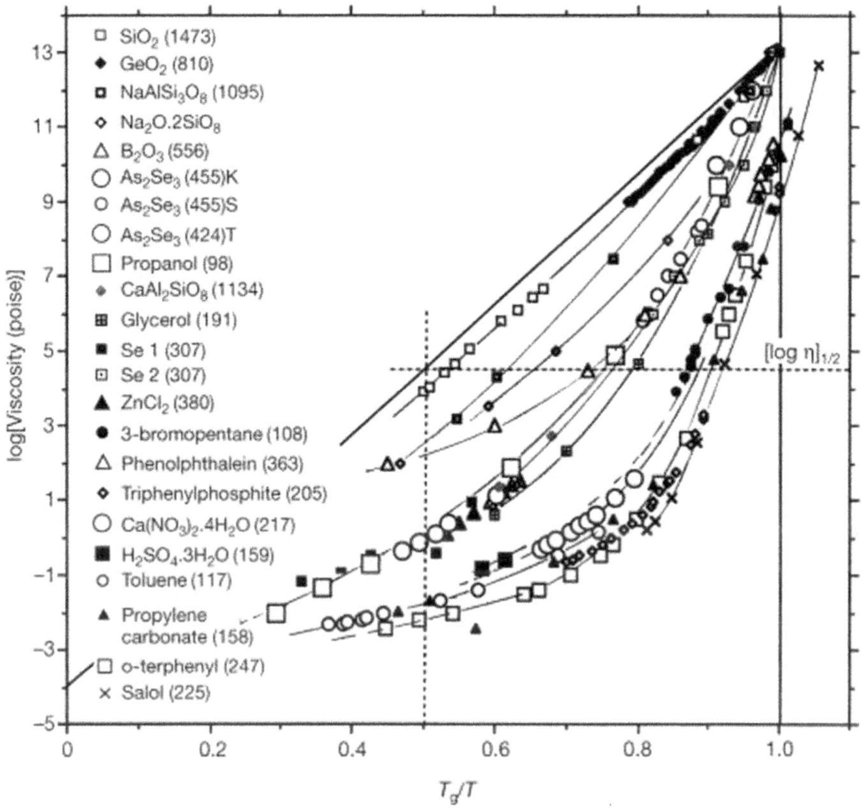


**Figure 1.6** Angell plot showing the dependence of viscosity on temperature for many molecular systems near the glass transition. The continuous and dashed lines are guides for the eye. Reproduced with permission from Ref. 110, *Nature*, 2001, **410**, 663–667. Doi: 10.1038/35070517. Copyright © 2001, Nature Publishing Group.

For its enormous impact on material sciences and fabrication technologies, the microscopic rationalization of the phenomena at the bases of the glass formation at microscopic level is nowadays one of the most active research fields of condensed matter physics. An effective classification of glasses has been proposed by C. A. Angell,[105–110] based on a so-called fragility index, $m_F$. This index is defined as the slope at $T_g$ of the curve describing the logarithm of the viscosity, $\eta$, *versus* the reciprocal temperature normalized by the transition temperature (Angell plot, Figure 1.6):

$$m_F = \left.\frac{dLog(\eta)}{d(T_g/T)}\right|_{T=T_g}. \tag{1.3}$$

In short, $m_F$ quantifies how much the structural relaxation process differs from a simply temperature-activated, Arrhenius behavior (namely from a process defined by an activation energy not depending on temperature, as better described in the following). This parameter is termed fragility index since it indicates how much the system is stable nearby the glass-transition temperature. Materials that soften rapidly, that are less stable, and that exhibit fast variations of their physical properties (viscosity, heat capacity, *etc.*) upon increasing temperature are described as fragile, whereas more stable materials, showing slower, less evident variations of their physical properties, are described as strong.

This concept is very useful for depicting how different polymer materials behave depending on their peculiar microscopic configurational and energetic landscape. The so-called strong materials (characterized by low values of $m_F$) would present a potential energy (M) profile, at microscopic length scale, with a low number of accessible configurations. On the contrary, exploring many configurations would mean for a polymer to exhibit large variations of physical properties at the glass transition, namely to be fragile. These different behaviors can be due to various factors, which may disfavor or block the exploration of adjacent configurations by the system. Angell explains how these factors can be thermodynamically or kinetically determined. A "thermodynamically strong" molecular system is characterized by spaced out energy minima, a "kinetically strong" molecular system has energy minima that, although possibly denser than in thermodynamically strong materials, are separated by high barriers, and are therefore difficult to reach, as schematized in Figure 1.7. For example, topological constrains, such as those due to molecular entanglements (scheme in Figure 1.8) and steric hindrance, can set kinetic limitations to the system to explore different configurations, unless entanglements are released easily. In terms of the chemical structure of the materials, a system forming non-directional bonds is generally able to explore more easily a wide range of molecular configurations. Systems with quite complex molecular structures and weak Van der Waals-type inter-molecular bonds are therefore generally fragile. Instead, molecular glasses whose molecular rearrangements would need to break directional bonds, such as for instance hydrogen-bonds, are definitely labelled as strong. Recent studies have also highlighted a striking resemblance between the trend of viscosity-volume ($\eta$ *vs.* $V_0$) variations for fluids of increasing fragility and the trend of relaxation time-concentration ($\tau$ *vs.* $c$) variations for colloidal suspensions of increasing particle hardness.[111,112] In other words, colloids made of softer particles seem to behave analogously to stronger liquids.

Fragile polymeric materials, since they exhibit larger and quicker changes, particularly of their viscosity, near $T_g$, are excellent prototype systems for the experimental investigation of the glass transition. From the point of view of technical applications, these compounds are very suitable to develop heating-based nanofabrication processes, including many methods used to produce nanofibers. In particular, as is well known the simplest model that one can

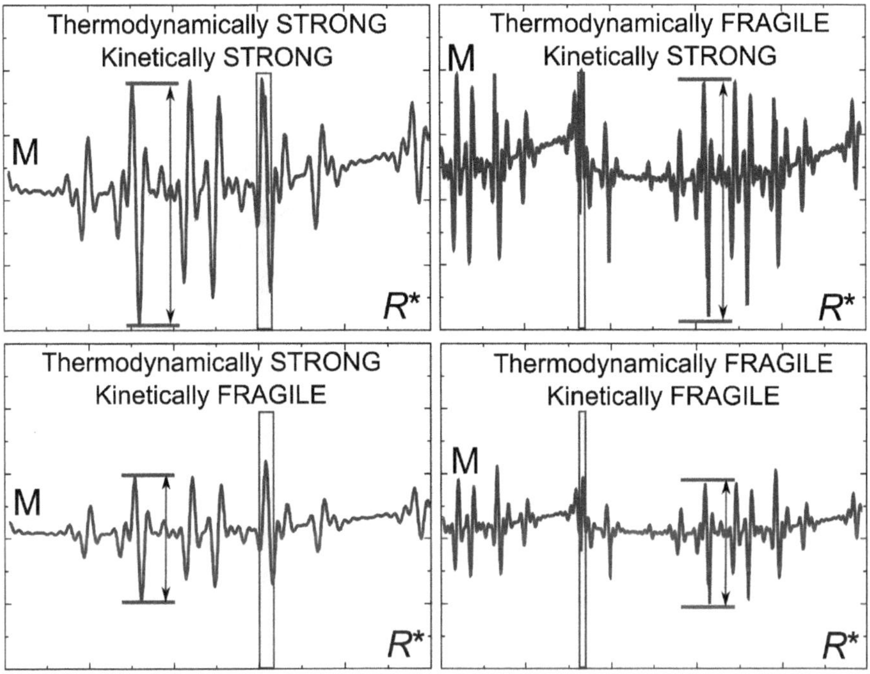


**Figure 1.7** Scheme of the behavior of the potential energy (M) *versus* a generalized coordinate, $R^*$, in the space of configurations for a fragile or a strong material, in a thermodynamic or kinetic sense. The thermodynamic fragility is determined by the density of M minima – higher in the two right plots – whereas the kinetic fragility decreases upon increasing the M barriers – which are higher in the two top plots. The shadowed boxes indicate the typical distance between adjacent minima in the energy landscape, whereas the vertical arrows highlight a typical energy barrier for the different molecular systems.

consider for the temperature dependence of the characteristic time of a relaxation process is a thermally activated (Arrhenius) behavior, with activation energy $\Delta E$:

$$\tau = \tau_0 \exp(\Delta E / k_B T). \tag{1.4}$$

However, as anticipated above, the glass transition phenomenology involving structural processes in polymer materials rarely follows simple Arrhenius laws, and especially near $T_g$ (where decreasing the temperature by 10 K can produce an increase of $\tau$ and $\eta$ of, say, three orders of magnitude), the dramatic increase of the viscosity and of the characteristic structural relaxation time is much better described by the Vogel–Fulcher–Tamman (VFT) equation:[113–115]

**Figure 1.8** Schematics of polymer chains in various coiled conformations. Entanglements are highlighted by dashed squares.

$$\eta = \eta_0 \exp\left(\frac{D_S T_0}{T - T_0}\right), \quad (1.5)$$

where $D_S$ is named the strength parameter and $T_0$ is the temperature at which the divergence of the relaxation time is predicted§§. The strength parameter is directly related to the fragility of the system (lower values of $D_S$ correspond to more fragile liquids). The VFT equation holds for many compounds and over large temperature ranges, and different models have been proposed to account for it. Indeed, Equation (1.5), sketching the temperature dependence of polymer viscosity, can be a valuable tool to predict the softening behavior of molecular systems while at the fluid state above the glass transition. Although such prediction is of course approximate, it can be useful in designing melt-polymer processing technologies. Therefore, for both fundamental and practical reasons, rationalizing Equation (1.5) by a proper microscopic description of polymeric systems has gathered a lot of interest.

Many excellent reviews exist about relaxational phenomena in amorphous systems and about the main models formulated throughout the years for describing the glass transitions.[98,100,102–104] In the following we recall only a couple of approaches that have had particular historical importance in the

§§ Another expression often found in the literature is the following (William–Landel–Ferry equation):[104,116] $\mathrm{Log}\left[\frac{\eta(T)}{\eta(T_g)}\right] = -\frac{C_1(T-T_g)}{C_2+(T-T_g)}$, where $C_1$ and $C_2$ are constants. This expression is equivalent to the VFT law by setting $\eta(T_0)=\eta(T_g)\exp(-C_1)$, $D_S = C_1 C_2 / T_0^2$ and $T_0 = T_g - C_2$.

framework of glass transition theories, and that are especially sound for the simplicity of the proposed physical pictures, which make them accessible also to a non-specialized audience of physicists and chemists. A particularly straightforward and intuitive approach is that based on the concept of *free volume*, originally proposed by M. H. Cohen and D. Turnbull in 1959,[117–119] and among the first able partially to rationalize experimental findings on amorphous compounds. The free volume, $v_f$, per molecule in a polymer system can be defined as the available volume within the cage formed by the surrounding molecules (*i.e.* the specific volume, $\overline{v}$, available per molecule), minus the critical volume ($v_C$) enabling the motion:

$$v_f = \overline{v} - v_C. \quad (1.6)$$

The free volume model describes the diffusive motion of a molecule in the system, occurring on distances of the order of the molecular size given that a free volume larger than a critical volume, $v_0$, is available. Instead, if not enough free volume is available in the system, the diffusive dynamics slows down to the glass transition***. One can easily figure out that the amount of free volume available can be significantly affected by the specific supramolecular packing in which the polymer system arranges, which in turn can depend on the used processing technologies, on the eventual solvents and of course on the specific sample geometry and dimensionality (bulk, film, fibers, nanofibers, *etc.*). Other theoretical models are based on thermodynamic considerations relying on the idea of *cooperative* motions.[120] Given the existence of mutually independent regions in the system, within each region in the polymer a molecule would not be able to rearrange its own configuration without involving all the other molecules. This approach assumes the existence of a critical size of the cooperative region, needed for an effective rearrangement, corresponding to a well-defined number of molecules and related to the molar configurational

*** The constraints imposed by the conservation of the total free volume and of the total number of molecules ($N$) easily lead, under the hypothesis of mutual statistical independence of the particles, to a Boltzmann distribution for the free volume. This is rationalized by assuming that the molecular movement is successful with a frequency proportional to the number of molecular objects that have enough available free volume. The frequency of success for the movement is moreover proportional to the diffusion coefficient, $D$, for which one gets the following Doolittle's equation: $D \propto \exp(-v_0/\overline{v})$. In addition, the ratio of the average free volume over the total average volume clearly has to increase upon increasing temperature above $T_g$. Under a few simplifying hypotheses, the VFT formula is straightforwardly obtained for the diffusion coefficient and, through the Stokes–Einstein ($D \propto \eta^{-1}$) and the Debye–Stokes ($\tau \propto \eta$) equations, the VFT law is immediately extended to the viscosity and to the structural relaxation time. Of course, such derivation only depicts a physically reasonable but greatly oversimplified behavior. Several variants and improvements to the model have been proposed in the last decades, such as the inclusion of a possible dependence of $v_0$ on the chemical composition of the system and on pressure.[104] In addition, a straightforward extension of the model is able to explain fractional Debye–Stokes–Einstein laws, by simply considering two different critical volumes for the translational and the rotational motion of molecules.[102]

entropy. In this way, this model allows also a derivation of the VFT formula, Equation (1.5)††† to be obtained.

To conclude this section, we emphasize that the glass transition phenomenology in polymer nanofiber systems is still basically unexplored. In some experiments, polymer nanofiber systems are found to exhibit a lower glass transition temperature than cast films or bulky samples.[124] This finding has been attributed to the very large surface-to-volume ratio of nanofibers, and to the consequently plasticizing role played by surrounding air molecules. More generally, the variation of $T_g$ for so-called nanoconfined systems has been greatly debated in the last years, and largely studied particularly in polymeric thin-films.[125–137] Nanoconfinement should be considered together with interfacial effects in order to elucidate properly the mechanisms at the base of observed $T_g$ variations in both thin-films and polymeric nanoparticles, and possible contributions originating from sample history, physical ageing and degradation, annealing, *etc.* should all be carefully taken into account. This results in a complex scenario in which polymer nanofibers, and particularly core-shell nanofibers allowing size and interfacial effects to be played with at the same time, can provide a very useful tool for the next experimental investigations. Meanwhile, pioneering studies on semicrystalline polymer fibers have started to show a decrease in the melting temperature, $T_m$, upon reducing the fiber radius in the μm[138,139] and the sub-μm[140] ranges.

## 1.4.2 Rheology and Flow Behavior

The glass transition phenomenology of polymers gives a clear idea of how amorphous systems behave in terms of their temporal response, namely of the relaxation towards equilibrium occurring after an external (thermal, mechanical, electrical, *etc.*) stress. This picture is made even more evident

††† The configurational entropy is defined as the difference of entropy in the amorphous material and the entropy in the corresponding crystal ($S_{conf} = S_{liquid} - S_{crystal}$), and it can be also expressed in terms of the heat capacity of the liquid in excess of that of the corresponding crystal, $\Delta C_p$: $S_{conf} = \int_{T_0}^{T} \frac{\Delta C_p}{T'} dT'$. According to the Adam–Gibbs model[120] the critical size of the cooperative region for rearrangement can be written as $z^* = N_{AV}\, s_{c,\min}/S_{conf}$, where $s_{c,\min}$ stands for the minimum configurational entropy of the cooperative region (assumed independent of temperature). For instance, in the simplest situation where only two possible configurations are considered, one has $s_{c,\min} = k_B \ln 2$ according to well-known basics of statistical mechanics. Furthermore, the average probability, W, *for a rearrangement, namely the probability for a region to undergo a transition between its two states, clearly has to depend on temperature. In particular, this rearrangement may follow a thermally activated behaviour,* $W = A\exp(-z * \Delta M/k_B T)$, where $\Delta M$ is the energy barrier that hinders a cooperative rearrangement, per monomer unit, and $A$ is a constant. The characteristic relaxation time (which is proportional to $W^{-1}$) is finally found to follow the VFT law. Overall, this approach therefore relies on strongly intuitive, basic physical concepts for describing the polymeric system. A number of other models have been formulated through the years, based on the idea of cooperative molecular motions nearby the glass transition.[121–123]

by taking into account the *rheology* of materials, which allows one to clearly draw a distinction between the limit cases of Newtonian (viscous) liquids and Hookean (elastic) solids, and to characterize polymeric systems as intermediate materials that exhibit so-called *viscoelastic* behavior. Here, we will introduce the rheological concepts of immediate usefulness for understanding the molecular and macroscopic properties at the base of fabrication processes of nanofibers. The aim is to present the material physical properties and their mutual relationships, which are needed to understand the basic rheology at the base of the formation of nanofibers, and of a few related modeling approaches such as those reviewed in Section 2.5 about electrospinning. For these reasons and for the sake of simplicity, we will mainly limit ourselves to the case of linear viscoelasticity, which means staying within the range of sufficiently small stresses and material displacements. A more complete discussion, also including non-linear viscoelasticity, can be found in excellent reference texts by Ferry,[104] Graessley[141] and Bird, Armstrong and Hassanger.[142]

### 1.4.2.1 Linear Viscoelastic Response and Material Functions

*Rheology* refers to the study of the relationship existing between a *stress* in a material and the corresponding flow or deformation. The stress, in turn, may depend on some kind of force that is externally applied to the system. More specifically, the stress is dimensionally a force per unit area, and it is generally transmitted from a layer of material undergoing a deformation to adjacent layers because of their mutual contact, by means of many classes of intermolecular interactions (electrostatic, Van der Waals, steric effects related to molecular entanglements, *etc.*), which act at the nanoscale. An important classification of possible material deformations distinguish between *simple shear* and *elongation* flow.[141] A simple shear deformation corresponds to a relative motion of adjacent layers, sliding one over another without stretching phenomena, as in the case of a liquid flowing in a pipe. On the contrary, an elongational (also named extensional) flow is characterized by a stretching motion along the direction of deformation. In this case, adjacent layers may exhibit a variety of behaviors depending on the specific material or sample conditions, ranging from absence of relative motion with respect to one another,[141] to so-called steady shearfree flow,[142] where the mutual distance of neighboring elements can increase exponentially over time. Though the case of simple shear flow is generally simpler and adopted as an introductory example in textbooks, elongational aspects are particularly important for processes based on stretched polymer melts or solutions, such as in electrospinning and other drawing methods to produce nanofibers. In the following, we introduce first the general definitions of strain and stress by leaning on the case of shearing motion, and then briefly present some peculiarities of elongational flows.

*Simple shear flow.* The classical, introductory example for visualizing simple shear and starting to quantify the relevant physical quantities of shear stress and strain is given by a homogeneous parallelepiped body of material, which is held in place at its bottom side and undergoes a deformation ($\Delta L$) at its upper side as a consequence of a force ($F_0$) applied parallel to the material top surface, *i.e.* along the $x$-direction as schematized in Figure 1.9. Here, the *shear stress* ($\sigma$) is simply given by $F_0/A$, where $A$ is the area (= parallelepiped length times the thickness along the $z$-direction) of the surface where $F_0$ is tangentially applied and, as previously mentioned, is transmitted from one layer of material to the layer immediately underneath. In a more general and mathematically rigorous definition taking into account all the three spatial directions (labeled as $x_1 = x$, $x_2 = y$ and $x_3 = z$), the stress is a symmetric tensor ($\boldsymbol{\sigma}$), which is defined in any point of the material and whose $ij$ component is given by the $j$-th component of the force, considered per unit area and acting on an infinitesimal surface element ($dA_i$) whose normal is identified by $dx_i$.[142] According to this general definition, in our example the shear stress, originating from a force vector parallel to the material surface, is actually the off-diagonal, $yx$-component of the overall stress tensor.

The corresponding *strain* ($\gamma$) is the ratio, $\Delta L/H$, where $H$ is the transversal thickness of the parallelepiped and does not change during the motion in our example. The strain quantifies how large is the relative deformation along the $x$-direction. Once again, $\gamma$ here is actually the $yx$-component of a strain tensor ($\boldsymbol{\gamma}$), whose $ij$-element is defined by the components of the infinitesimal deformation or material displacement, $\delta L$:[142]

$$\gamma_{ij} = \frac{\partial}{\partial x_i}\delta L_j + \frac{\partial}{\partial x_j}\delta L_i. \tag{1.7}$$

Indeed, in our deformed body, the deformation takes place only along $x$, with amplitude increasing linearly with the $y$ coordinate, namely $\delta L_x = (\Delta L/H)\, y$, $\delta L_y = \delta L_z = 0$.

During the flow, the deformation changes over time, therefore one also defines a deformation rate or *shear rate* ($\dot{\gamma}$) as $d\gamma/dt = d\,(\Delta L/H)/dt = (d\,\Delta L/dt)/H = v_x/H$, where $v_x$ is the maximum velocity along the $x$-direction in Figure 1.9. If the velocity, like $\delta L_x$, drops linearly from the top side ($y = H$) to the bottom side ($y = 0$) of the deforming body, $v_x/H$ clearly defines the velocity gradient moving through the material along the $y$-direction. This $\dot{\gamma}$ is the $yx$ component of the rate-of-strain (or rate-of-deformation) tensor ($\dot{\boldsymbol{\gamma}}$), defined by the components of the velocity as follows:

$$\dot{\gamma}_{ij} = \frac{\partial}{\partial x_i}v_j + \frac{\partial}{\partial x_j}v_i. \tag{1.8}$$

The basic relationships between stress and strain are defined for the two idealized, limit cases of a Newtonian liquid and of a Hookean solid during shearing motion as:

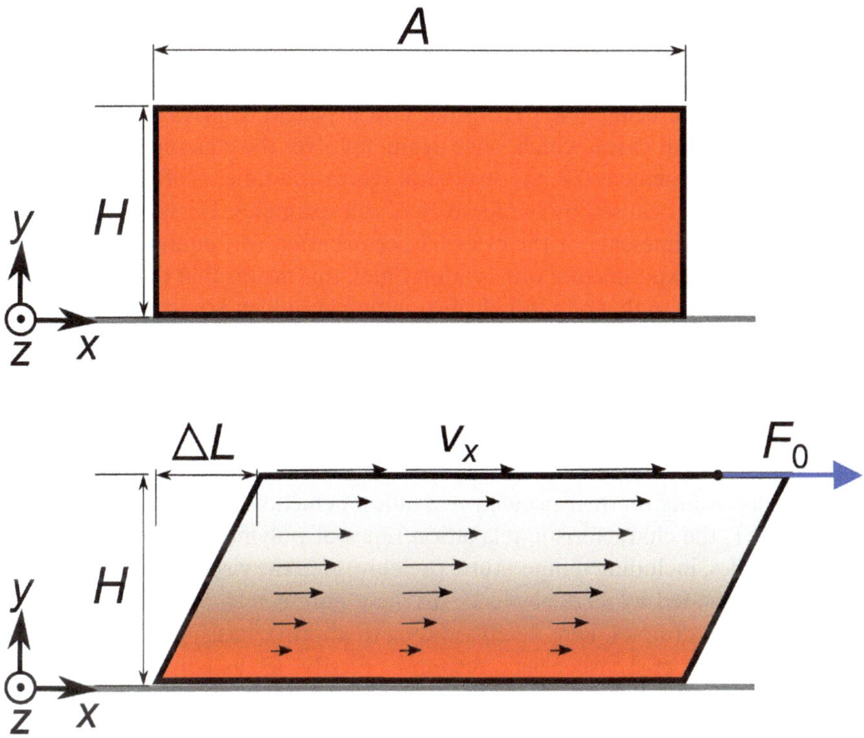


**Figure 1.9** A parallelepiped body in its pristine (top panel) and shear-deformed (bottom panel) form. The force, $F_0$, is applied parallel to the top surface of area $A$ (= length along $x$ times thickness along $z$). In the bottom scheme the linear color gradient exemplifies the velocity gradient (red: lower velocity, white: higher velocity), $v_x/H$, which is also indicated by arrows of length increasing along $y$.

$$\sigma(t) = \eta\dot{\gamma}(t) \text{ or, more generally, } \sigma_{yx}(t) = -\eta\dot{\gamma}_{yx}(t) \text{ Newtonian liquid} \quad (1.9a)$$

$$\sigma(t) = G\,\gamma\,(t) \text{ or, more generally, } \sigma_{yx}(t) = -G_S\,\gamma_{yx}\,(t) \text{ Hookean solid} \quad (1.9b)$$

In the previous expressions, $\eta$ and $G_S$ are the viscosity and the *shear modulus*, respectively. Equation (1.9a) immediately leads to Newton's law for viscous drag in a fluid ($F_0 \propto \eta\ v$), which conveys the message of why we call "Newtonian" liquids whose viscosity does not depend on $v$[‡‡‡], whereas

‡‡‡ Many of the liquid systems analyzed in this book, and particularly most of the polymer solutions and melts used in techniques for fabricating polymer nanofibers, are instead non-Newtonian, namely they exhibit a viscosity that depends on velocity or, equivalently, on the shear rate. Some fluids show a so-called shear-thinning behavior, in which viscosity decreases upon increasing $\dot{\gamma}$. Others are shear-thickening, and viscosity increases upon increasing $\dot{\gamma}$. Many models are available for these different behaviors as reviewed in detail in Ref. 142.

Equation (1.9b) reformulates Hooke's law for an elastic body deformed by an external force ($F_0 \propto \Delta L$)[§§§].

The important step in order to understand the viscoelastic response of real materials stands in being able to consider how a system behaves in comparison with these two limit cases, which, once again, involves the characteristic time of molecular rearrangement (*i.e.* the structural relaxation time, $\tau$) after an external stress (such as that caused by the force $F_0$ of our example). If $\tau$ is much shorter than the typical timescale of the external perturbation (or, equivalently, much shorter than our experimental observation time), this means that the behavior of our system is close to that of a rapidly responding liquid, and the response will be mainly viscous (*i.e.* well-described in terms of the material viscosity). On the contrary, a molecular rearrangement characteristic time much longer than the perturbation timescale or than the observation window will lead the elastic, solid-like response to prevail. Real materials all have some degree of *viscoelasticity*, *i.e.* locate between the two limit cases and exhibit a complex response depending on their peculiar relaxation behavior. Since, as described in Section 1.4.1, the characteristic relaxation times of polymers are influenced by many variables, including temperature, pressure, *etc.*, the viscoelastic response is further complicated by its dependence on these parameters.

A couple of examples, here described on a qualitative basis, generally help in getting the point of a viscoelastic response. Firstly, let us consider the stress relaxation occurring in a material (for instance the usual parallelepiped body which has height, $H$, of Figure 1.9) after a sudden shear deformation, $\Delta L(t) = \Delta L_0\,\theta(t)$, where $\theta(t)$ is a step function ($= 0$ for $t < 0$ and $= 1$ for $t > 0$). The temporal behavior of the deformation and the possible corresponding material responses are sketched in Figure 1.10a and b, respectively. We already know that the strain, defined on the base of the deformation, is $\gamma(t) = \gamma_0\,\theta(t)$ with $\gamma_0 = \Delta L_0/H$. For a Newtonian liquid, in which the stress is proportional to the temporal derivative of the deformation, one would expect a spike response at $t = 0$ and then a vanishing stress for $t > 0$ (red line in Figure 1.10b), since the

§§§ In a more complete, tensor form, one would write $\boldsymbol{\sigma} = \mathbf{G}\,\boldsymbol{\gamma}$ or $\boldsymbol{\gamma} = \mathbf{J}\boldsymbol{\sigma}$, where $\mathbf{G}$ and $\mathbf{J}$ are $(6 \times 6)$ matrices ($\mathbf{J}$ is the so-called compliance matrix) to generalize Hooke's law (Equation (1.9b)). Indeed, reducing $\boldsymbol{\gamma}$ and $\boldsymbol{\sigma}$ to columnar $(6 \times 1)$ vectors, which is a simplification made possible by the symmetric nature of these tensors and generally called Voigt notation in mathematics, one has: $\begin{pmatrix}\gamma_1\\ \gamma_2\\ \gamma_3\\ \gamma_4\\ \gamma_5\\ \gamma_6\end{pmatrix} = \begin{pmatrix} J_{11} & J_{12} & J_{13} & J_{14} & J_{15} & J_{16}\\ J_{21} & J_{22} & J_{23} & J_{24} & J_{25} & J_{26}\\ J_{31} & J_{32} & J_{33} & J_{34} & J_{35} & J_{36}\\ J_{41} & J_{42} & J_{43} & J_{44} & J_{45} & J_{46}\\ J_{51} & J_{52} & J_{53} & J_{54} & J_{55} & J_{56}\\ J_{61} & J_{62} & J_{63} & J_{64} & J_{65} & J_{66}\end{pmatrix} \begin{pmatrix}\sigma_1\\ \sigma_2\\ \sigma_3\\ \sigma_4\\ \sigma_5\\ \sigma_6\end{pmatrix}$, where $\begin{pmatrix}\gamma_1\\ \gamma_2\\ \gamma_3\\ \gamma_4\\ \gamma_5\\ \gamma_6\end{pmatrix} \stackrel{Def}{=} \begin{pmatrix}\gamma_{xx}\\ \gamma_{yy}\\ \gamma_{zz}\\ \gamma_{yz}\\ \gamma_{xz}\\ \gamma_{xy}\end{pmatrix}$ and $\begin{pmatrix}\sigma_1\\ \sigma_2\\ \sigma_3\\ \sigma_4\\ \sigma_5\\ \sigma_6\end{pmatrix} \stackrel{Def}{=} \begin{pmatrix}\sigma_{xx}\\ \sigma_{yy}\\ \sigma_{zz}\\ \sigma_{yz}\\ \sigma_{xz}\\ \sigma_{xy}\end{pmatrix}$.

This notation will be useful when introducing piezoelectricity of polymer nanofibers in Section 6.4.1.

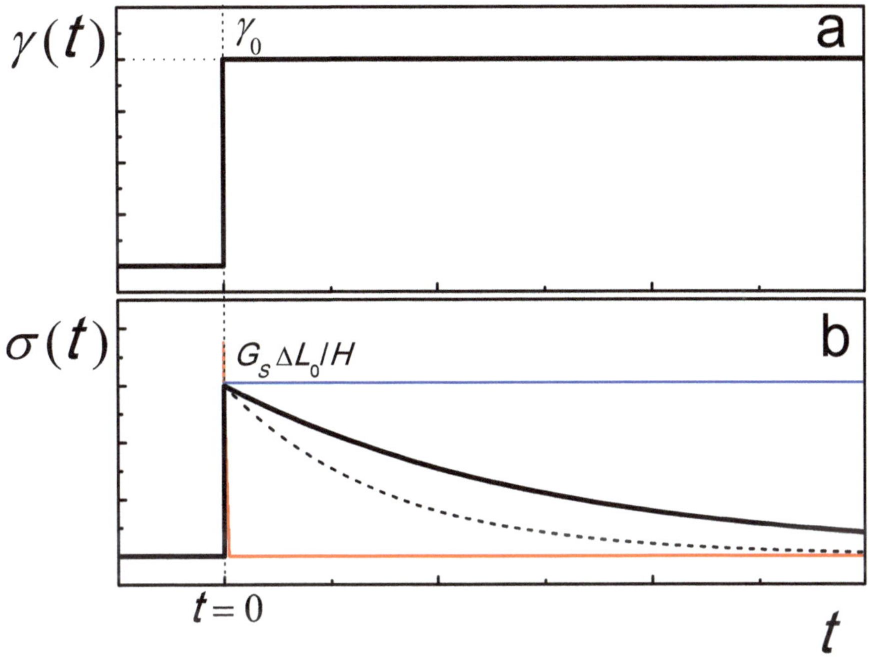


**Figure 1.10** Temporal behavior of an idealized, step-like shear deformation, $\gamma(t) = \gamma_0$ $\theta(t)$ (a), and the corresponding response of material stress (b) for a Newtonian liquid (red line), a Hookean solid (blue line) and a viscoelastic material (continuous black curve) at temperature, $T = T_1$. The viscoelastic response at temperature, $T_2 > T_1$, is also shown for comparison (dashed black curve).

strain is constant for positive times. Instead, for a Hookean solid, in which the stress is proportional to $\gamma$, one would expect a step-like temporal behavior, with $\sigma(t) = G_S\, \Delta L_0/H$ for $t > 0$ (blue line in Figure 1.10b). A viscoelastic material (continuous black curve in Figure 1.10b) will respond, at short times after the onset of the external force (*i.e.* for experimental timescales much shorter than the characteristic times of molecular rearrangements), like a solid, *i.e.* the stress will exhibit a step across $t = 0$, rising to a value around $G_S \Delta L_0/H$. At longer times, $\sigma(t)$ will no longer be constant as in a perfectly elastic solid, but will decrease down to zero, thus approaching the corresponding value of the Newtonian liquid for very long time values ($t >> 0$). The viscoelastic behavior will be more and more liquid-like as time increases, becoming comparable to and finally much longer than the characteristic rearrangement timescale ($\tau$) of the molecular system. Upon increasing temperature, $\tau$ will decrease, typically depending on the particular glass-transition behavior of the material (for instance by the VFT law, Equation (1.5)), and this means that the dropping dynamics of $\sigma(t)$ will be correspondingly faster (dashed black curve in Figure 1.10b). For each temperature, one may define a time-dependent shear modulus that describes this relaxational behavior, by generalizing Equation (1.9b):

$$G_S(t) = \sigma(t)/\gamma_0 = \sigma(t)/(\Delta L_0/H). \quad (1.10)$$

As a second exemplary case, let us reverse the input and the output of our problem, namely let us consider as input a constant stress ($\sigma_0$, for instance due to an external load) applied to our material object between two instants of time, $t = t_0$ and $t = t_1$ (Figure 1.11a), and as output the correspondingly caused deformation. In this case, according to Equation (1.9a) a Newtonian liquid will flow with constant shear rate ($\dot{\gamma} = \sigma_0/\eta$) for $t_0 \leq t \leq t_1$, and then will stabilize at a constant strain, $\gamma_L = \sigma_0 (t_1 - t_0)/\eta$, thus suffering an irreversible long-term deformation (red line in Figure 1.11b). *Vice versa*, according to Equation (1.9b) a Hookean solid will undergo only a reversible deformation, $\gamma_S = \sigma_0/G_S$ for $t_0 \leq t \leq t_1$, being able to store and giving back elastically the energy it receives (blue line in Figure 1.11b). A viscoelastic material (continuous black curve in Figure 1.11b) will respond with a characteristic creep and recovery dynamics.[143] The creep stage exhibits a sudden increase resembling the solid-like step response at $t = t_0$, and also includes a retarded response component, which may be referred to as *anelastic*, and which greatly contributes to the resulting strain dynamics (green line in Figure 1.11b). The recovery upon unloading leads the system to the liquid-like response at long times ($t > t_1$), as physically expected. This final stage also defines the amount of deformation irreversibly set in the material. Upon increasing temperature and correspondingly reducing the characteristic relaxation times, the slopes of $\gamma(t)$ both in the ascendent (creep) and in the descendent (recovery) stages will increase in modulus because of the faster molecular rearrangements. In this case, the most suitable way to generalize Equation (1.9b) is by introducing a time-dependent *shear creep compliance*, $J(t)$, as follows:

$$J(t) = \gamma(t)/\sigma_0. \quad (1.11)$$

As one can easily see, this creep/recovery dynamics is very frequently observed and of enormous importance for a large variety of nanotechnology and prototyping approaches involving polymer processing. Among these, one should especially mention template-assisted methods, polymer molding and NIL. Aiming to get an irreversibly transferred deformation in a polymer system, one has to reach a regime of so-called *terminal flow*, in which $J(t)$ takes a linearly growing dependence on time, as would be in the case of a viscous liquid (for which $J \cong t/\eta$). The system temperature, molecular weight ($MW$)**** and

**** For a statistical distribution of macromolecules having different molecular weights, namely a different number of repeat units as resulting from a given synthetic route, one defines an average molecular weight, $MW$, to describe the polymer sample. Different quantities are found in the literature, including the number-average $MW$, $\overline{MW}_n = \frac{\sum_i n_i m_i}{\sum_i n_i}$, and the weight-average $MW$, $\overline{MW}_w = \frac{\sum_i w_i m_i}{\sum_i w_i}$. In these expressions, $n_i$ are the different populations of molecules having a mass $m_i$, and $w_i$ are the corresponding weight fractions of each population (*i.e.* $w_i = n_i m_i$).

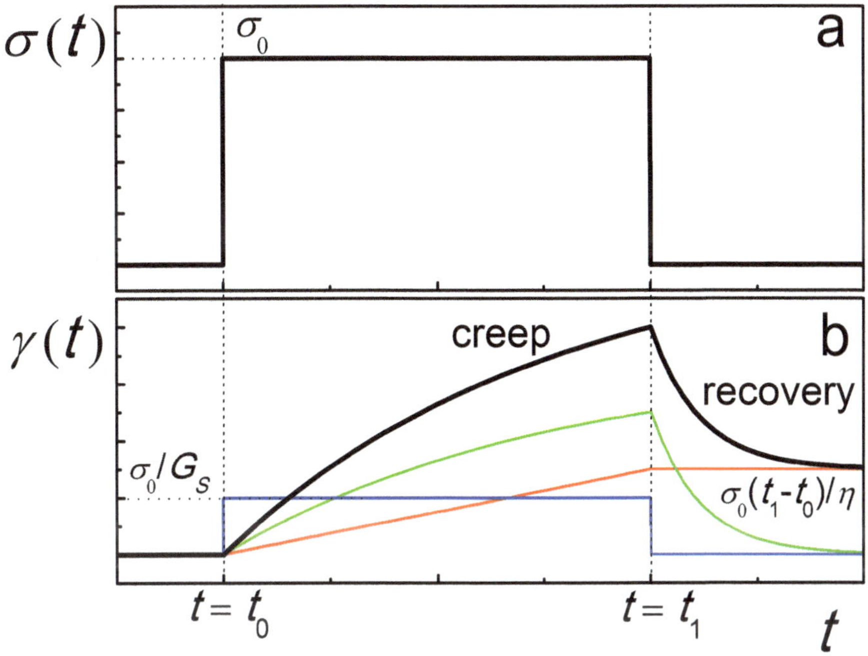


**Figure 1.11** Temporal behavior of a step-like stress, $\sigma_0$, applied between $t_0$ and $t_1$ (a), and of the corresponding $\gamma(t)$ (b) for a Newtonian liquid (red line), a Hookean solid (blue line) and a viscoelastic material (continuous black curve) showing a creep-recovery dynamics. The anelastic component to the viscoelastic response is also shown (green line).

degree of steric hindrance affecting molecular entanglements and the eventually applied pressure all critically influence the viscoelastic response and the effectiveness of the polymer in reaching the terminal flow.

*Elongational flow and related elastic properties.* Elongational deformations are those occurring in fibers and nanofibers when these are deformed by means of a tensile stress. Furthermore, the concept of tensile stress applied to a fluid body of polymer is especially important for nanofibers production processes, and particularly for electrospinning and other drawing methods where it is related to viscoelastic forces exerted in polymers during fabrication, as described in Section 2.5.2. In this case, deformation and strain rates are very large and non-linear effects can arise, with the elongational viscosity often increasing with $\dot{\gamma}$.[104] Here we mainly limit our discussion to the regime of linearity, *i.e.* of small deformations, and consider only isotropic materials for the sake of simplicity. Let us consider a unit cube (*i.e.* with initial unit side length, $L_0$) of material which is deformed by a *tensile stress* pulling along two of its faces as in Figure 1.12. According to our previous definition, also the tensile stress is dimensionally a force ($F_0$) per unit area. It represents a diagonal element of the stress tensor and it is calculated by the force per unit *cross-sectional* area (instead of the

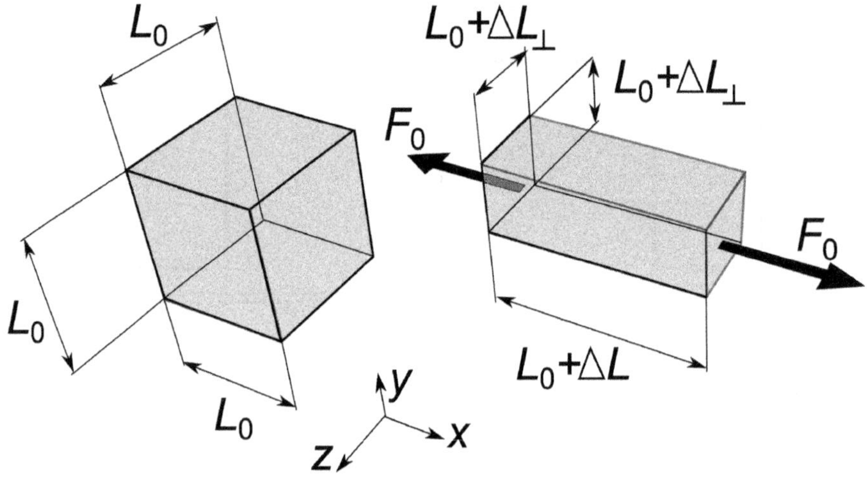


**Figure 1.12** Deformation of a cubic body under an applied tensile stress. The force, $F_0$, is applied along the $x$-direction. $\Delta L > 0$ and $\Delta L_\perp < 0$ in our figure (stretching due the tensile stress). The case in which $\Delta L_\perp$ is the same along the two directions perpendicular to the direction of elongation, as in this figure, is named *simple extension*.[104]

area of the surface where $F_0$ is tangentially applied as in the case of the shear stress shown in Figure 1.12). For instance, the area to be considered in the example of Figure 1.12 is the cube cross-section lying in the $yz$-plane, *i.e.* perpendicular to the applied force. The resulting stretching deformation in the material involves the direction parallel to the applied stress, and also the perpendicular directions along which a contraction is observed. Considering the coordinates of the sides of the cube under deformation which move along the $x$-, $y$- and $z$-directions, respectively, a simple way to describe the uniaxial elongational flow in the so-called shear-free case is by the equations: $v_x = \mathrm{d}(\Delta L)/\mathrm{d}t = \dot{\varepsilon}x$, $v_y = \mathrm{d}(\Delta L_\perp)/\mathrm{d}t = -(\dot{\varepsilon}/2)\,y$ and $v_z = \mathrm{d}(\Delta L_\perp)/\mathrm{d}t = -(\dot{\varepsilon}/2)\,z$. Here, $\dot{\varepsilon}$ is the *elongational rate* ($\mathrm{s}^{-1}$), and the velocity component along the $x$- (or $y$-, or $z$-) direction is uniform for all the points on the side parallel to the $yz$-plane (or $xz$-, or $xy$-plane, respectively). In general, a distinctive feature of elongational flows is that the separation distance between two initially close elements of fluid increases exponentially over time, which is at the base of fiber spinning methods. In these techniques, one often talks about the so-called *draw ratio*. If the total volume of the fluid body is conserved during elongation, the draw ratio can be calculated by the ratio of the initial and the final cross-sectional area $[L_0^2/(L_0+\Delta L_\perp)^2$ in Figure 1.12]. On the contrary, for solvent-based techniques such as electrospinning (Chapter 2), the total volume is not conserved during the flow due to the concomitant solvent evaporation. This has to be taken into account in determining the draw ratio, which is given by the cross-sectional area reduction factor *times* the polymer concentration of the used solution.[144] As an alternative, one can consider the ratio of the velocities of the fluid body in its initial and in its final part, which can be suitably corrected to take into account solvent evaporation as well.[145] The achievable draw ratios vary enormously (from the order of unity to, say, $10^5$) depending on elongational conditions, spinning methods and used polymer solutions or melts.

In solids, if a deformation is elastic a certain amount of mechanical energy is reversibly stored in the material. For solid-like polymeric materials, the analogy of Equation (1.9b) for a tensile deformation is effectively expressed as:[146] $\sigma = E\,\gamma_L$, where $\sigma$ indicates the tensile stress, $E$ is the *Young's modulus* (which is therefore a tensile modulus) and $\gamma_L$ is the *longitudinal strain*,[143] the latter being the ratio between the longitudinal length increase ($\Delta L$) and the initial length ($L_0$) along the same direction ($x$-direction in Figure 1.12). Of course, in the case of a more complex (and temporally delayed) response to an instantaneous tensile deformation, $\Delta L(t) = \Delta L_0\,\theta(t)$, also the Young's modulus can be generalized to a time-dependent expression analogous to that in Equation (1.10), *i.e.* one can define $E(t) = \sigma(t)/(\Delta L_0/L_0)$.[104] The Young's modulus also offers a simple way to quantify the amount of mechanical energy, $U_M$, stored in materials as a consequence of an elastic deformation. For example, for a fiber of radius $r$, stretched by a deformation $\Delta L$, without crack formation, the stored mechanical energy is given by:[147]

$$U_M = \frac{1}{2}\pi r^2 E\frac{(\Delta L)^2}{L_0}. \tag{1.12}$$

Another often useful material parameter is *Poisson's ratio*, $\nu_P$, which is defined as the ratio between the perpendicular (transverse) deformation ($\Delta L_\perp$ in Figure 1.12) and the longitudinal deformation ($\Delta L$): $\nu_P = -\Delta L_\perp/\Delta L$. In the general, viscoelastic case, similar considerations as in the case of shear flow hold for the temporal response of polymeric fluids at molecular scale, and the resulting macroscopic response.

### *1.4.2.2 Modeling Linear Viscoelasticity. The Maxwell Model*

Once the viscoelastic response of polymer materials is figured out qualitatively, one needs appropriate tools to develop a quantitative model. Firstly, one should bear in mind that the response described by the time-dependent modulus, as in any kind of response function for a casual, time-independent and linear system, can of course be reformulated in the frequency domain. Having, as a problem input, a sinusoidal deformation at a given angular frequency ($\omega$), $\gamma(t) = \gamma_0 \sin\omega t$, the corresponding stress, $\sigma(t)$, in the material will have a component in phase and a component in quadrature with respect to the input:

$$\sigma(t)/\gamma_0 = G'(\omega)\sin\omega t + G''(\omega)\cos\omega t. \quad (1.13)$$

In this expression, $G'$ and $G''$ are the *storage modulus* and the *loss modulus* of the substance, and indicate the weight of the elastic and of the dissipative component of the response at a particular frequency value.[148] The relative weight of the two components in the material response is immediately given by their ratio, $G''/G' = \tan\delta$, where $\delta$ indicates the phase angle between the stress and the strain. It is very easy to check that $G' = 0$ ($\delta = \pi/2$) for a Newtonian liquid (Equation (1.9a)) and $G'' = 0$ ($\delta = 0$) for a Hookean solid (Equation (1.9b)). The dependence of $G'$ and $G''$ on $\omega$ is further evidence of the distinguishing feature of viscoelastic materials, that is exhibiting a solid-like behavior at short times (high $\omega$) and a liquid-like behavior at long times (low $\omega$) compared to the characteristic temporal scale of molecular rearrangement. So, for $\omega$ below a characteristic value $\omega_C$ (cross-over frequency), the external stimulus is much slower than the material relaxation time, the system is mainly dissipative and $G'' > G'$, whereas the opposite situation holds for the region $\omega > \omega_C$, where the elastic response prevails (Figure 1.13). As usual, the response is critically influenced by temperature. For instance, upon decreasing temperature one expects that the relaxational behavior at the molecular scale becomes slower, leading the system towards a more solid-like response and shifting the cross-over frequency towards lower values in the moduli-$\omega$ plot (dashed curves in Figure 1.13).

For the general case of a viscoelastic compound and of an arbitrary (but not large enough to lead the system out of linearity) deformation input, $\gamma(t)$, by using a mathematical formalism analogous to that in Equation (1.1), one has:

$$\sigma(t) = \int_{-\infty}^{t} \frac{dG(t-t')}{dt} \gamma(t,t')dt'. \quad (1.14)$$

Here, one explicitly underlines the dependence of the deformation by *two* time instants, $t$ and $t'$, which highlights that the material may exhibit a memory of its previous history. Hence, $\gamma\ (t, t')$ indicates the deformation at a time $t'$, *relative* to a previous instant ($t$) taken as reference, so that $\gamma\ (t, t) = 0$. $G(t)$ once again describes the general stress response to an instantaneous deformation occurring at $t = 0$, *i.e.* to $\gamma\ (t) = \gamma_0\ \theta(t)$.

Of course, in order to describe properly the behavior of a polymer, one would need to have the expression of $G(t)$ and its dependence on the material characteristics, which means modeling the system and drawing a physically plausible correlation between its nano- or microscale properties and its macroscopic response. A brilliant example is the well-known model of viscoelasticity due to J. C. Maxwell,[149] which has been used for longer than a century by several generations of physicists and engineers. Nowadays, Maxwell's picture is still very popular and it is often deployed to describe the dynamics of viscoelastic materials during nanofabrication processes. For

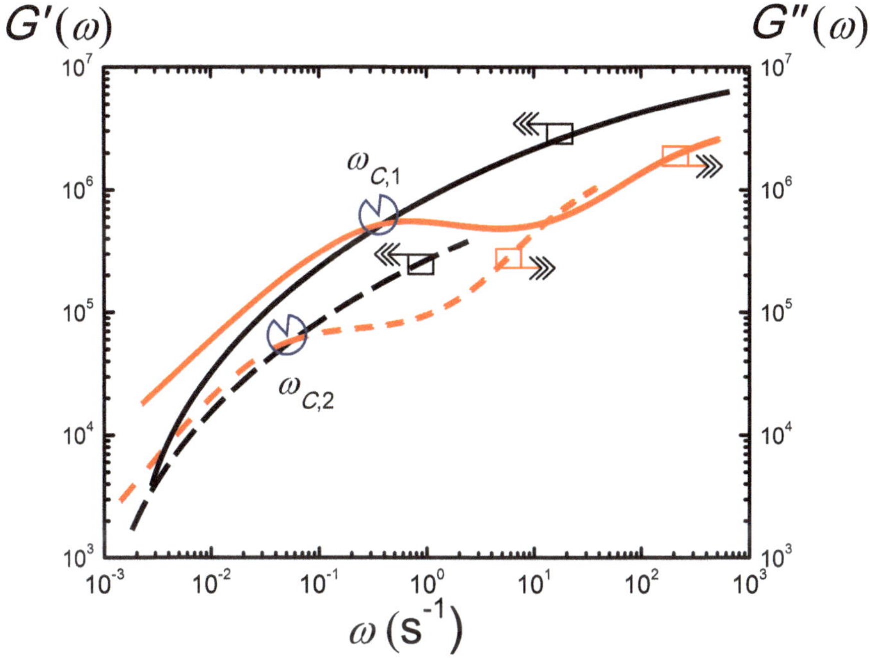


**Figure 1.13** Typical behavior of the storage (black lines, left vertical axis) and loss moduli (red lines, right vertical axis) for a viscoelastic material. Curves correspond to temperature, $T = T_1$ (solid curves, cross-over frequency $\omega_{C,1}$) and $T = T_2 < T_1$ (dashed curves, cross-over frequency $\omega_{C,2}$). Curves are vertically shifted for better clarity, and cross-overs are highlighted by circles. Arrows on curves indicate the respective axes. Further intersections of the $G'$ and $G''$ curves are possible at frequency values higher than the displayed range.

example, a nice application to the formation of polymer nanofibers may be found in models of electrospinning in Section 2.5.2.

The Maxwell model describes the stress relaxation. It draws a one-dimensional element of viscoelastic material as a *dashpot* made by a piston dipped in a viscous fluid, and linked to a perfectly elastic body, as a *spring* (Figure 1.14). The dashpot and the spring are positioned along a straight segment, thus forming a hybrid, linear dumbbell. This means that the two (dissipative and elastic) forces at the base of the idealized behaviors of Newtonian liquids and Hookean solids act simultaneously and their superposition determines viscoelasticity. In the Maxwell model, the dashpot and the spring are connected in series to each other. Other combinations are of course possible, such as the Kelvin or Voigt model,[104,146] which describes the viscoelastic material as a parallel of a dashpot and a spring. According to the Maxwell model, the dynamics roughly develops as follows: once excited by a sudden longitudinal strain, $\gamma_S$, the spring consequently undergoes a tensile stress, $\sigma_S$, instantaneously activated by that deformation according to Hooke's

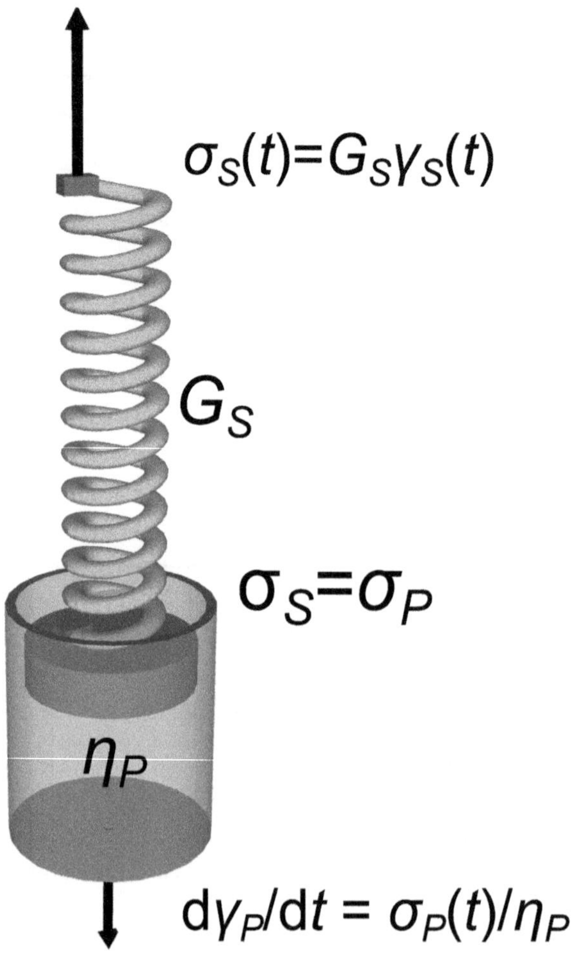


**Figure 1.14** Scheme of the Maxwell model. Spring and dashpot in series and main constitutive parameters. The two stresses, $\sigma_S$ and $\sigma_P$, are equal, whereas the total deformation is given by the sum of the contributions due to the spring and the piston ($\gamma_P + \gamma_S = \gamma$).

law, *i.e.* $\sigma_S(t) = G_S\gamma_S(t)$, where $G_S$ indicates the elastic modulus. The force is transmitted to the piston, which starts moving and dissipating energy because of the viscosity, $\eta_P$, of the fluid in the dashpot. Such a drag of course determines a temporal delay in the response of the dashpot, whose shear rate, $\dot{\gamma}_P$, varies over time proportionally to the stress, $\sigma_P$, on the piston, by Equation (1.9a): $\dot{\gamma}_P(t) = \sigma_P(t)/\eta_P$. Now, one can put $\sigma_P = \sigma_S = \sigma$ and $\gamma_P + \gamma_S = \gamma$ because the spring and the piston are bound together in series and contribute to the overall displacement. Therefore, the resulting viscoelastic dynamics is described by a simple linear differential equation:

$$\dot{\gamma}=\dot{\gamma}_P+\dot{\gamma}_S=\frac{\sigma_P(t)}{\eta_P}+\frac{1}{G_S}\frac{d\sigma_S(t)}{dt}=\frac{\sigma(t)}{\eta_P}+\frac{1}{G_S}\frac{d\sigma(t)}{dt}. \tag{1.15}$$

which summarizes the Maxwell model. For instance, after an instantaneous deformation, $\gamma(t) = \gamma_0\,\theta(t)$ (as in Figure 1.10a, for which $\dot{\gamma} = 0$ for $t > 0$), this equation simply becomes: $d\sigma(t)/dt = -G_S\,\sigma(t)/\eta_P$, whose solution is a stress relaxation described by an exponential decay: $\sigma(t) = \sigma_0\exp(-t/\tau_C)$. Here, $\tau_C = \eta_P/G_S$, and placing the initial stress at $t = 0$, $\sigma_0$, equal to $G_S\gamma_0$, allows one properly to fit the solid-like behavior at short times. This exponential decay can be a quite reasonable analytic expression for the continuous black curve in Figure 1.10b. It can be demonstrated[104,142] that the Maxwell model leads to the following expression, which is in the form of Equation (1.14):

$$\sigma(t)=\int_{-\infty}^{t} G_S\frac{d}{dt}\left[\exp\left(-\frac{t-t'}{\tau_C}\right)\right]\gamma(t,t')dt'. \tag{1.16}$$

In particular, comparing Equations (1.14) and (1.16) the relaxation function becomes: $G(t) = G_S\exp(-t/\tau_C)$, which completes our introductory description of viscoelasticity. The complete version of Equation (1.15) in tensor form reads:[142]

$$\dot{\boldsymbol{\gamma}}=\frac{\boldsymbol{\sigma}(t)}{\eta_P}+\frac{1}{G_S}\frac{d\boldsymbol{\sigma}(t)}{dt}. \tag{1.17}$$

Finally, one should bear in mind that the Maxwell model *does not* describe creep phenomena, such as those displayed in Figure 1.11b. Putting a constant stress, $\sigma_0$, in Equation (1.15) simply leads to a constant shear rate, $\dot{\gamma} = \sigma_0/\eta_P$, as in a Newtonian liquid, without reproducing any real viscoelastic behavior. For describing creep, one has to use other approaches, such as the Kelvin or Voigt model.

### 1.4.3 Polymer Solutions

In the previous sections, organic solutions are already mentioned a few times, and many of the techniques that allow polymer nanofibers to be produced, as described in Chapters 2 and 3, rely on systems composed of macromolecules in solution. Investigating the properties of polymer solutions is the subject of an entire sub-field of chemical physics, and we give here only introductory background information. A *solution* involves two or more chemical species that are intimately mixed, *i.e.* in which the mutual interaction between the components is at the molecular level. In other words, a solution is homogenously composed of only one phase. Polymer solutions comprise at least one polymeric material and the solvent (water, organic solvents, *etc.*) in which the polymer is *dissolved*. Of course, further molecular species can be optionally added to this basic mixture, such as, for example, a second polymer (the two involved polymers then forming a *blend*), a salt to increase the overall conductivity, other molecular dopants *etc*. The exact composition of the solution and the relative concentrations of the different chemicals will depend indeed on the physico-chemical properties that one wants to achieve for each specific targeted application.

A *dispersion* is instead a mixture in which the less abundant compound is dispersed, but not molecularly dissolved, in the other component. Examples are a dispersion of a solid phase (powder, nanoparticles, nanocrystals, nanotubes, *etc*.) in a solvent, which is called a *suspension*, or a dispersion of an immiscible liquid phase in a second liquid, which is called an *emulsion*. Milk and mayonnaise are familiar examples of emulsions. *Aerosols* are dispersions of tiny liquid droplets or solid particles in a continuous gaseous phase. The science of aerosols is particularly relevant in order to design filter elements able to remove droplets and particles from air, which can be performed with very high efficiency by polymer nanofibers (Section 4.3.1). Finally, another way to indicate homogeneously mixed dispersions, emulsions or aerosols of nano- or microparticles is *colloids*. Colloidal dispersions of inorganic nanocrystals or organic nanofibers are familiar examples for nanotechnologists.

In fact, most of the methods for producing polymer nanofibers rely on solution systems, although dispersions are also used for some specific purposes, particularly for loading particles or liquid reservoirs within nanofibers as described in detail in Sections 2.3, 2.3.3 and 7.4. When dealing with suspensions or emulsions, achieving a homogeneous distribution of the dispersed phase in the continuous liquid phase and avoiding aggregation effects of the dispersed particles or coalescence of the carried droplets are often critical issues to be carefully addressed. Similar considerations may apply effectively to dissolve polymers in solutions. Different degrees of solubility and different conformations taken by macromolecules in solutions originate from the particular interactions existing at the nanoscale in each polymer-solvent system. An extensive literature exists in the field. Given the importance of solution systems in techniques for the realization of polymer nanofibers, a few important aspects are summarized in the following. The reader may notice that a few concepts already introduced, particularly those of steric hindrance and entanglements, are also useful in the present framework.

Polymers of technological interest are generally dissolved in water, which is a polar liquid, or in organic solvents, which can be apolar or polar, with the different degrees of polarity depending on the magnitude of the permanent electric dipoles eventually existing on molecules. What solvent is effective to dissolve a molecular compound will depend of course primarily on polarity, but also on other classes of inter-molecular interactions. In all cases, there will be a variation of the Gibbs free energy, $\Delta U_G$, as a consequence of mixing polymer and solvent molecules:

$$\Delta U_G = \Delta H_0 - T\Delta S_0, \tag{1.18}$$

where $\Delta H_0$ and $\Delta S_0$ indicate the variations of enthalpy and entropy, respectively. A favorable free energy balance ($\Delta U_G < 0$) will indicate that dissolving the polymer in the solvent is a thermodynamically favored process. In the following, we refer to the Flory–Huggins interaction theory for solutions, which is likely the most known and intuitive model describing these systems.[150–152]

To evaluate the variation in enthalpy, $\Delta H_0$, it is convenient to suppose that each repeat unit in the flexible polymer chain is surrounded by a number, $Z_0$, of interacting species ($Z_0$ is practically the coordination number related to each monomer composing the polymer macromolecules). In particular, each polymer segment can interact with other segments or with the molecules of solvent. How many segments and how many solvent molecules are involved in interacting with each monomer of the polymer chain, given that their total number (segments + solvent molecules) has to be $Z_0$? To distinguish among the two species, one can simply refer to fractions coming from the solution specimen considered in its total. Let $N_s$ and $N_p$ indicate the total number of molecules of the solvent and of the polymer, respectively, and $\phi_s$ and $\phi_p$ the corresponding volume fractions occupied by the two species. The order of magnitude of $\phi_s$ is of $N_s/(N_s + N_p\, N_r)$, where $N_r$ is the number of repeat units in the polymer chain, and each monomer is supposed to have a size comparable to that of a single molecule of solvent. Analogously, $\phi_p$ is given by $N_p\, N_r/(N_s + N_p\, N_r)$. Once the overall volume fractions and the above approximations have been introduced, one can assume that the polymer segments interacting with each particular monomer are $Z_0\phi_p$, and that the solvent molecules interacting with the same monomer will be $Z_0\phi_s$. Of course, the sum of these two numbers ($Z_0\phi_p + Z_0\phi_s$) gives the total coordination number, $Z_0$, of species surrounding the monomer, because $\phi_p + \phi_s = 1$ as is evident from the definitions above. Then, the dimensionless Flory–Huggins interaction parameter, $\chi_{FH}$, can be defined as follows:

$$\chi_{FH} = \frac{Z_0 \Delta H_{s-p}}{k_B T}, \tag{1.19}$$

where $\Delta H_{s-p}$ indicates the variation of energy resulting from displacing a segment of polymer or a molecule of solvent from a particular point in the fluid (or, in other words, from a site of a lattice in the fluid), and re-filling the corresponding site with a different species, as would be the result of intimately mixing the two compounds at the molecular scale. The overall enthalpy variation upon mixing will be then given by:

$$\Delta H_0 = \chi_{FH} k_B T \times N_s \phi_p = \chi_{FH} k_B T \times N_p N_r \phi_s. \tag{1.20}$$

In the previous expressions, $N_s \phi_p$, or equivalently $N_p\ N_r\ \phi_s$, provides the total number of contacts or pairs (polymer segment-solvent molecule) interacting in the solution, and their product by the single-pair displacement energetic variation yields $\Delta H_0$. As is clear from Equation (1.20), the role of the interaction parameter is very important for assessing how much the polymer and the solvent like each other. The smaller $\chi_{FH}$, the lower will be the enthalpy increase upon mixing, the more favorable will be the related energy budget and the better will be the solubility. A complete table of values of interaction parameters for many polymer–solvent systems can be found in Ref. 153.

Concerning the second part in the right term of Equation (1.18), as one can easily figure out from a physical viewpoint, dissolving polymer molecules will lead to an increase of entropy ($\Delta S_0 > 0$), namely to a more disordered system in which a higher number of microscopic configurations are accessible. This also allows to be seen immediately, from Equation (1.18), why increasing temperature can favor the preparation of solutions, decreasing the resulting $\Delta U_G$. In particular, by simple approximations, $\Delta S_0$ upon mixing can be estimated as:

$$\Delta S_0 = k_B \left( N_s \ln \phi_s + N_p \ln \phi_p \right). \tag{1.21}$$

The balance between the two terms, $\Delta H_0$ and $T\Delta S_0$, will therefore determine the final sign of $\Delta U_G$. Depending on this issue, a solvent can be *good* or *poor* for a particular polymer. A good solvent is a system that polymeric molecules like to interact with ($\Delta U_G < 0$), and this leads the macromolecular chains to take extended, elongated configurations. In other words, solvent–polymer interactions at the molecular scale are stronger than polymer–polymer, cohesive inter-molecular forces. Instead, in a poor solvent, polymer molecules preferentially adopt coiled, compact configurations ($\Delta U_G > 0$ as an effect of mixing). The ultimate, thermodynamically unfavorable mixing situation will be that of a polymer that is completely insoluble in a given solvent. The polymer molecules will then prefer to stay close to other polymer molecules, which will likely lead to a suspension of polymeric aggregates or self-assembled particles as detailed in Section 3.1.1. Inter-molecular interactions are therefore pivotal in determining solubility. In general, as mentioned above and in previous sections, forces between molecules can comprise Van der Waals interactions such as those related to induced dipoles, polar interactions such as those related to permanent dipoles and hydrogen bonds. Each of these contributions can be quantified by suitable solubility parameters, which can help considerably in choosing the most suitable solvent or mixture of solvents to dissolve a given polymer.[154] The particular condition in which $\Delta H_0 = T\Delta S_0$, *i.e.* when $\Delta U_G = 0$, defines what is called a $\theta$-solvent for a given polymer at a given temperature (or, as an alternative of a $\theta$-temperature for a given polymer–solvent system), which means that solvent–polymer and polymer–polymer equal each other. We will see in the next chapters that inter-molecular interactions, and the conformations taken by macromolecules, affect many physical properties of the solution, which are often inherited by the resulting polymer nanofibers, including for instance mechanical, optical and electrical properties.

As described in Section 1.4.2, entanglements between molecular chains play an important role in determining viscoelastic properties, both in solutions and in melts, and these are in turn crucial to influence the reliability and throughput of many methods for producing polymer nanofibers and especially of electrospinning. Entanglements are of course favored at higher concentrations of the polymer in solution, and for polymeric species of higher average molar mass (Section 1.4.1), namely statistically having longer macromolecular chains (*i.e.*, a higher number of repeat units).

## References


1. www.nano.gov.
2. R. P. Feynman, *There's plenty of room at the bottom*, presented at the *Annual Meeting of the American Physical Society*, California Institute of Technology, 1959 (available at: http://calteches.library.caltech.edu/1976/1/1960Bottom.pdf).
3. A. Aviram and M. R. Ratner, *Chem. Phys. Lett.*, 1974, **29**, 1–49.
4. Y. Cui and C. M. Lieber, *Science*, 2001, **291**, 851–853.
5. G. Binnig, H. Rohrer, Ch. Gerber and E. Weibel, *Phys. Rev. Lett.*, 1982, **49**, 57–61.
6. G. Binnig, C. F. Quate and Ch. Gerber, *Phys. Rev. Lett.*, 1986, **56**, 930–933.
7. J. D. Joannopoulos, R. D. Meade and J. N. Winn, *Photonic Crystals – Molding the Flow of Light*, Princeton University Press, Princeton, 1995.
8. H. W. Kroto, J. R. Heath, S. C. O'Brien, R. F. Curl and R. E. Smalley, *Nature*, 1985, **318**, 162–163.
9. K. Kratschmer, L. D. Lamb, K. Fostiropoulos and R. D. Huffman, *Nature*, 1990, **347**, 354.
10. S. Iijma, *Nature*, 1991, **354**, 56–58.
11. R. H. Baughman, A. A. Zakhidov and W. A. de Heer, *Science*, 2002, **297**, 787–792.
12. M. Zhang, S. Fang, A. A. Zakhidov, S. B. Lee, A. E. Aliev, C. Williams, K. Atkinson and R. H. Baughman, *Science*, 2005, **309**, 1215–1219.
13. D. Tasis, N. Tagmatarchis, A. Bianco and M. Prato, *Chem. Rev.*, 2006, **106**, 1105–1136.
14. P. Bhattacharya, S. Ghosh and A. D. Stiff-Roberts, *Annu. Rev. Mater. Res.*, 2004, **34**, 1–40.
15. M. H. Huang, S. Mao, H. Feick, H. Q. Yan, Y. Y. Wu, H. Kind, E. Weber, R. Russo and P. D. Yang, *Science*, 2001, **292**, 1897–1899.
16. M. Law, J. Goldberger and P. Yang, *Annu. Rev. Mater. Res.*, 2004, **34**, 83–122.

17. J. Wang, M. S. Gudiksen, X. Duan, Y. Cui and C. M. Lieber, *Science*, 2001, **293**, 1455–1457.
18. H. Kind, H. Yan, B. Messer, M. Law and P. Yang, *Adv. Mater.*, 2002, **14**, 158–160.
19. X. Duan, Y. Huang, R. Agarwal and C. M. Lieber, *Nature*, 2003, **421**, 241–245.
20. P. Alivisatos, *Science*, 1996, **271**, 933–937.
21. Y. Yin and A. P. Alivisatos, *Nature*, 2005, **437**, 664–670.
22. R. Krahne, G. Morello, A. Figuerola, C. George, S. Deka and L. Manna, *Phys. Rep.*, 2011, **501**, 75–221.
23. G. A. Ozin and L. Cademartiri, *Small*, 2009, **5**, 1240–1244.
24. G. A. Ozin and A. C. Arsenault, *Nanochemistry: A Chemical Approach to Nanomaterials*, Royal Society of Chemistry, Cambridge, UK, 2005.
25. G. A. Ozin, *Adv. Mater.*, 1992, **4**, 612–649.
26. K. S. Novoselov, A. K. Geim, S. V. Morozov, D. Jiang, Y. Zhang, S. V. Dubonos, I. V. Grigorieva and A. A. Firsov, *Science*, 2004, **306**, 666–669.
27. K. S. Novoselov, A. K. Geim, S. V. Morozov, D. Jiang, M. I. Katsnelson, I. V. Grigorieva, S. V. Dubonos and A. A. Firsov, *Nature*, 2005, **438**, 197–200.
28. A. K. Geim and K. S. Novoselov, *Nat. Mater.*, 2007, **6**, 183–191.
29. H. Staudinger, *Macromolecular Chemistry, Nobel Lecture*, Dec. 11, 1953. Available online at: http://www.nobelprize.org/nobel_prizes/chemistry/laureates/1953/staudinger-lecture.pdf
30. H. Staudinger, W. Heuer, E. Husemann and I. J. Rabinovitch, *Trans. Faraday Soc.*, 1936, **32**, 323–332.
31. A. J. Heeger, *J. Phys. Chem. B*, 2001, **105**, 8475–8491.
32. J.-M. Lehn, *Angew. Chem. Int. Ed.*, 1990, **29**, 1304–1319.
33. J.-M. Lehn, *Angew. Chem. Int. Ed.*, 1988, **27**, 89–112.
34. P.-G. De Gennes, *Soft Matter, Nobel Lecture, Dec. 9, 1991.* Available online at: http://www.nobelprize.org/nobel_prizes/physics/laureates/1991/gennes-lecture.pdf
35. I. W. Hamley, *Angew. Chem. Int. Ed.*, 2003, **42**, 1692–1712.
36. http://pubs.rsc.org/en/journals/journalissues/sm.
37. R. A. L. Jones, *Faraday Discuss.*, 2009, **143**, 9–14.
38. G. M. Whitesides and D. J. Lipomi, *Faraday Discuss.*, 2009, **143**, 373–384.
39. R. Farchioni and G. Grosso (ed.), *Organic Electronic Materials*, Springer-Verlag, Berlin, Heidelberg, 2001.
40. F. Sunjoo Kim, G. Ren and S. A. Jenekhe, *Chem. Mater.*, 2011, **23**, 682–732.
41. http://ec.europa.eu/nanotechnology/pdf/nanocode-rec_pe0894c_en.pdf.
42. W. C. Su and Y. S. Cheng, *J. Aerosol Sci.*, 2006, **37**, 1429–1441.
43. R. Sturm and W. Hofmann, *J. Hazard. Mater.*, 2009, **170**, 210–218.

44. S. Moreira, N. B. Silva, J. Almeida-Lima, H. A. O. Rocha, S. R. B. Medeiros, C. Alver, Jr. and F. M. Gama, *Toxicol. Lett.*, 2009, **189**, 235–241.
45. M. J. D. Clift, E. J. Foster, D. Vanhecke, D. Studer, P. Wick, P. Gehr, B. Rothen-Rutishauser and C. Weder, *Biomacromolecules*, 2011, **12**, 3666–3673.
46. R. de Lima, L. Oliveira Feitosa, C. Rodrigues Maruyama, M. Abreu Barga, P. Cristina Yamawaki, I. J. Vieira, E. M. Teixeira, A. C. Corrêa, L. H. Caparelli Mattoso and L. Fernandes Fraceto, *Int. J. Nanomedicine*, 2012, **7**, 3555–3565.
47. G. E. Moore, *Electronics*, 1965, **38**, 114–117.
48. S. Brunauer, P. H. Emmett and E. Teller, *J. Am. Chem. Soc.*, 1938, **60**, 309–319.
49. J. M. Deitzel, J. Kleinmeyer, D. Harris and N. C. Beck Tan, *Polymer*, 2001, **42**, 261–272.
50. X. Xu, L. Yang, X. Xu, X. Wang, X. Chen, Q. Liang, J. Zeng and X. Jing, *J. Control. Release*, 2005, **108**, 33–42.
51. D. Hussain, F. Loyal, A. Greiner and J. H. Wendorff, *Polymer*, 2010, **51**, 3989–3997.
52. Z. Cui, *Micro-Nanofabrication*, Higher Education Press, Beijing, PR China, 2005.
53. F. Cerrina and C. Marrian, *MRS Bull.*, 1996, **21**, 56–62.
54. D. Dendukuri, D. C. Pregibon, J. Collins, T. A. Hatton and P. S. Doyle, *Nature Mater.*, 2006, **5**, 365–369.
55. D. C. Pregibon, M. Toner and P. S. Doyle, *Science*, 2007, **315**, 1393–1396.
56. D. Dendukuri, S. S. Gu, D. C. Pregibon, T. A. Hatton and P. S. Doyle, *Lab Chip*, 2007, **7**, 818–828.
57. J.-H. Jang, D. Dendukuri, T. A. Hatton, E. L. Thomas and P. S. Doyle, *Angew. Chem. Int. Ed.*, 2007, **46**, 9027–9031.
58. K. W. Bong, D. C. Pregibon and P. S. Doyle, *Lab Chip*, 2009, **9**, 863–866.
59. K. W. Bong, K. T. Bong, D. C. Pregibon and P. S. Doyle, *Angew. Chem. Int. Ed.*, 2010, **49**, 87–90.
60. R. F. W. Pease, *J. Vac. Sci. Technol. B*, 1992, **10**, 278–285.
61. B. La Fontaine, *SPIE Professional*, 2010, Doi: 10.1117/2.4201010.09.
62. M. Levenson, *Phys. Today*, 1993, **46**, 28–36.
63. S. A. MacDonald, C. G. Willson and J. M. J. Fréchet, *Acc. Chem. Res.*, 1994, **27**, 151–158.
64. Y. Xia, J. A. Rogers, K. E. Paul and G. M. Whitesides, *Chem. Rev.*, 1999, **99**, 1823–1848.
65. B. D. Gates, Q. Xu, M. Stewart, D. Ryan, C. G. Willson and G. M. Whitesides, *Chem. Rev.*, 2005, **105**, 1171–1196.
66. Y. Xia and G. M. Whitesides, *Angew. Chem. Int. Ed.*, 1998, **37**, 550–575.
67. Y. Xia, E. Kim, X.-M. Zhao, J. A. Rogers, M. Prentiss and G. M. Whitesides, *Science*, 1996, **273**, 347–349.

68. Y. Xia, J. J. McClelland, R. Gupta, D. Qin, X.-M. Zhao, L. L. Sohn, R. J. Celotta and G. M. Whitesides, *Adv. Mater.*, 1997, **9**, 147–149.
69. E. Kim, Y. Xia and G. M. Whitesides, *Nature*, 1995, **376**, 581–584.
70. E. Kim, Y. Xia and G. M. Whitesides, *J. Am. Chem. Soc.*, 1996, **118**, 5722–5731.
71. C. De Marco, E. Mele, A. Camposeo, R. Stabile, R. Cingolani and D. Pisignano, *Adv. Mater.*, 2008, **20**, 4158–4162.
72. D. Pisignano, G. Maruccio, E. Mele, L. Persano, F. Di Benedetto and R. Cingolani, *Appl. Phys. Lett.*, 2005, **87**, 123109.
73. J. N. Lee, C. Park and G. M. Whitesides, *Anal. Chem.*, 2003, **75**, 6544–6554.
74. J. P. Rolland, E. C. Hagberg, G. M. Denison, K. M. Carter and J. M. DeSimone, *Angew. Chem. Int. Ed.*, 2004, **43**, 5796–5799.
75. E. Delamarche, M. Schmid, B. Michel and H. Biebuyck, *Adv. Mater*, 1997, **9**, 741–746.
76. H. Schmid and B. Michel, *Macromolecules*, 2000, **33**, 3042–3049.
77. S. Y. Chou, P. R. Krauss and P. J. Renstrom, *Science*, 1996, **272**, 85–87.
78. S. Y. Chou, P. R. Krauss, W. Zhang, L. Guo and L. Zhuang, *J. Vac. Sci. Technol. B*, 1997, **15**, 2897–2904.
79. H.-C. Scheer, H. Schulz, T. Hoffmann and C. M. Sotomayor Torres, in *Handbook of Thin Film Materials*, ed. H. S. Nalwa, Academic Press, New York, 2002, vol. 5: Nanomaterials and Magnetic Thin Films.
80. D.-Y. Khang, H. Yoon and H. H. Lee, *Adv. Mater.*, 2001, **13**, 749–752.
81. P. S. Hong and H. H. Lee, *Appl. Phys. Lett.*, 2003, **83**, 2441–2443.
82. E. Mele, F. Di Benedetto, L. Persano, R. Cingolani and D. Pisignano, *Nano Lett.*, 2005, **5**, 1915–1919.
83. F. Di Benedetto, A. Camposeo, S. Pagliara, E. Mele, L. Persano, R. Stabile, R. Cingolani and D. Pisignano, *Nat. Nanotechnol.*, 2008, **3**, 614–619.
84. C. K. Kao, *Rev. Mod. Phys.*, 2010, **82**, 2299–2303.
85. J.-B. Donnet, T. K. Wang, J. C. M. Peng and S. Rebouillat (ed.), *Carbon Fibers*, Marcel Dekker, New York, 3rd edn, 1998.
86. M. Brenner and B. H. Buck, *Aquacult. Eng.*, 2010, **42**, 128–139.
87. J. N. Cha, K. Shimizu, Y. Zhou, S. C. Christianssen, B. F. Chmelka, G. D. Stucky and D. E. Morse, *Proc. Natl Acad. Sci. USA*, 1999, **96**, 361–365.
88. X. Wang, M. Wiens, H. C. Schröder, S. Hu, E. Mugnaioli, U. Kolb, W. Tremel, D. Pisignano and W. E. G. Müller, *Adv. Eng. Mater.*, 2010, **12**, B422–B437.
89. L. Carbone, C. Nobile, M. De Giorgi, F. Della Sala, G. Morello, P. Pompa, M. Hytch, E. Snoeck, A. Fiore, I. R. Franchini, M. Nadasan, A. F. Silvestre, L. Chiodo, S. Kudera, R. Cingolani, R. Krahne and L. Manna, *Nano Lett.*, 2007, **7**, 2942–2950.

90. K. Miszta, J. de Graaf, G. Bertoni, D. Dorfs, R. Brescia, S. Marras, L. Ceseracciu, R. Cingolani, R. van Roij, M. Dijkstra and L. Manna, *Nat. Mater.*, 2011, **10**, 872–876.
91. Y. Xia, P. Yang, Y. Sun, Y. Wu, B. Mayers, B. Gates, Y. Yin, F. Kim and H. Yan, *Adv. Mater.*, 2003, **15**, 353–389.
92. Y.-Z. Long, M. Yu, B. Sun, C.-Z. Gud and Z. Fanc, *Chem. Soc. Rev.*, 2012, **41**, 4560–4580.
93. P. Yang, R. Yan and M. Fardy, *Nano Lett.*, 2010, **10**, 1529–1536.
94. N. W. Ashcroft and N. D. Mermin, *Solid State Physics*, Thomson Brooks/Cole, New York, 1976.
95. G. F. Bassani and G. Pastori Parravicini, *Electronic States and Optical Transitions in Solids*, Pergamon Press, Oxford, 1975.
96. M. Pope and C. E. Swenberg, *Electronic Processes in Organic Crystals and Polymers*, Oxford University Press, New York, 2nd edn, 1999.
97. G. Li, H. E. King Jr., W. F. Oliver, C. A. Herbst and H. Z. Cummins, *Phys. Rev. Lett.*, 1995, **74**, 2280–2283.
98. J. Jäckle, *Rep. Prog. Phys.*, 1986, **49**, 171–231.
99. M. Giordano, D. Leporini, and M. Tosi (ed.), *Non Equilibrium Phenomena in Supercooled Fluids, Glasses and Amorphous Materials*, World Scientific, Singapore, 1996.
100. M. D. Ediger, C. A. Angell and S. R. Nagel, *J. Phys. Chem.*, 1996, **100**, 13200–13212.
101. K. B. Storey and J. M. Storey, *Scientific American*, 1990, **62**, 92–97.
102. R. Richert and A. Blumen (ed.), *Disorder Effects on Relaxational Processes*, Springer-Verlag, Berlin, 1994.
103. R. Richert, *J. Phys. Condens. Matter*, 2002, **14**, R703–R738.
104. J. D. Ferry, *Viscoelastic Properties of Polymers*, Wiley, New York, 3rd edn, 1980.
105. R. Böhmer and C. A. Angell, in *Disorder Effects on Relaxational Processes*, ed. R. Richert and A. Blumen, Springer-Verlag, Berlin, 1994.
106. C. A. Angell, *J. Non-Cryst. Solids*, 1991, **131–133**, 13–31.
107. R. Bohmer and C. A. Angell, *Phys. Rev. B*, 1992, **45**, 10091–10094.
108. C. A. Angell, *Proc. Natl Acad. Sci. USA*, 1995, **92**, 6675–6682.
109. C. A. Angell, *Science*, 1995, **267**, 1924–1935.
110. L.-M. Martinez and C. A. Angell, *Nature*, 2001, **410**, 663–667.
111. C. A. Angell and K. Ueno, *Nature*, 2009, **462**, 45–46.
112. J. Mattsson, H. M. Wyss, A. Fernandez-Nieves, K. Miyazaki, Z. Hu, D. R. Reichman and D. A. Weitz, *Nature*, 2009, **462**, 83–86.
113. H. Vogel, *Phys. Z*, 1921, **22**, 645.
114. G. S. Fulcher, *J. Am. Ceram. Soc.*, 1925, **8**, 339–355.
115. G. Tamman and W. Hesse, *Z. Anorg. Allg. Chem.*, 1926, **156**, 245–257.
116. M. L. Williams, R. F. Landel and J. D. Ferry, *J. Amer. Chem. Soc.*, 1955, **77**, 3701–3707.
117. M. H. Cohen and D. Turnbull, *J. Chem. Phys.*, 1959, **31**, 1164–1169.
118. D. Turnbull and M. H. Cohen, *J. Chem. Phys.*, 1961, **34**, 120–125.

119. D. Turnbull and M. H. Cohen, *J. Chem. Phys.*, 1970, **52**, 3038–3041.
120. G. Adams and J. H. Gibbs, *J. Chem. Phys.*, 1965, **43**, 139–146.
121. F. Garwe, A. Schönals, H. Lockwenz, M. Beiner, K. Schröter and E. Donth, *Macromolecules*, 1996, **29**, 247–253.
122. K. L. Ngai, *J. Chem. Phys.*, 1998, **109**, 6982–6994.
123. C. León, K. L. Ngai and C. M. Roland, *J. Chem. Phys.*, 1999, **110**, 11585–11591.
124. X. Zong, K. Kim, D. Fang, S. Ran, B. S. Hsiao and B. Chu, *Polymer*, 2002, **43**, 4403–4412.
125. J. L. Keddie, R. A. L. Jones and R. A. Cory, *Faraday Discuss.*, 1994, **98**, 219–230.
126. J. A. Forrest, K. Dalnoki-Veress, J. R. Stevens and J. R. Dutcher, *Phys. Rev. Lett.*, 1996, **77**, 2002–2005.
127. G. B. DeMaggio, W. E. Frieze, D. W. Gidley, M. Zhu, H. A. Hristov and A. F. Yee, *Phys. Rev. Lett.*, 1997, **78**, 1524–1527.
128. M. Arndt, R. Stannarius, H. Groothues, E. Hempel and F. Kremer, *Phys. Rev. Lett.*, 1997, **79**, 2077–2080.
129. A. Huwe, F. Kremer, P. Behrens and W. Schwieger, *Phys. Rev. Lett.*, 1999, **82**, 2338–2341.
130. R. D. Priestley, C. J. Ellison, L. J. Broadbelt and J. M. Torkelson, *Science*, 2005, **309**, 456–459.
131. C. J. Ellison and J. M. Torkelson, *Nature Mater.*, 2003, **2**, 695–700.
132. R. D. Priestley, L. J. Broadbelt, J. M. Torkelson and K. Fukao, *Phys. Rev. E*, 2007, **75**, 061806.
133. P. Rittigstein, R. D. Priestley, L. J. Broadbelt and J. M. Torkelson, *Nature Mater.*, 2007, **6**, 278–282.
134. G. B. McKenna, *Eur. Phys. J. Spec. Top.*, 2007, **141**, 291–301.
135. Z. Yang, Y. Fujii, F. K. Lee, C.-H. Lam and O. K. C. Tsui, *Science*, 2010, **328**, 1676–1679.
136. C. Zhang, Y. Guo and R. D. Priestley, *Macromolecules*, 2011, **44**, 4001–4006.
137. H. K. Nguyen, M. Labardi, S. Capaccioli, M. Lucchesi, P. A. Rolla and D. Prevosto, *Macromolecules*, 2012, **45**, 2138–2144.
138. Y. Liu, C. H. Li, S. Chen, E. Wachtel, T. Koga, J. C. Sokolov and M. H. Rafailovich, *J. Polym. Sci. Part B: Polym. Phys.*, 2009, **47**, 2501–2508.
139. A. Arinstein, Y. Liu, M. H. Rafailovich and E. Zussman, *EPL*, 2011, **93**, 46001.
140. W. Wang and A. H. Barber, *Nanotechnology*, 2010, **21**, 225701.
141. W. W. Graessley, in *Physical Properties of Polymers*, Cambridge University Press, Cambridge, 3rd edn, 2004.
142. R. B. Bird, R. C. Armstrong and O. Hassager, Dynamics of Polymeric Liquids, Wiley, New York, 2nd edn, 1987, vol. **1**.
143. G. R. Strobl, *The Physics of Polymers*, Springer-Verlag, Berlin, 1997.

144. D. H. Reneker, A. L. Yarin, H. Fong and S. Koombhongse, *J. Appl. Phys.*, 2000, **87**, 4531–4547.
145. S.-C. Wong, A. Baji and S. Leng, *Polymer*, 2008, **49**, 4713–4722.
146. D. I. Bower, *An Introduction to Polymer Physics*, Cambridge University Press, New York, 2002.
147. R. S. Yang, Y. Qin, L. Dai and Z. L. Wang, *Nat. Nanotechnol.*, 2009, **4**, 34–39.
148. G. M. Kavanagh and S. B. Ross-Murphy, *Prog. Polym. Sci.*, 1998, **23**, 533–561.
149. J. C. Maxwell, *Phil. Trans. Roy. Soc.*, 1867, **A157**, 49–88.
150. M. L. Huggins, *J. Chem. Phys.*, 1941, **9**, 440.
151. P. J. Flory, *J. Chem. Phys.*, 1941, **9**, 660–661.
152. P. J. Flory, *J. Chem. Phys.*, 1942, **10**, 51–61.
153. R. A. Orwoll and P. A. Arnold, in *Physical Properties of Polymers Handbook*, ed. J. E. Mark, Springer Science+Business Media, LLC, New York, 2nd edn, 2007.
154. J. Burke, in *AIC Book and Paper Group Annual*, ed. C. Jensen, American Institute for Conservation of Historic and Artistic Works, Washington, DC, 1984.

CHAPTER 2
# *Electrospinning*

## 2.1 Electrostatic Spinning

The story of modern electrostatic spinning is the classic case of a scientific rediscovery. Indeed, coupling electrical phenomena and fluids has been fascinating physicists for centuries, as reported in many studies. Some of these works are very famous, others are known only by a very restricted audience of specialists. However, it is in these early contributions that one has to recognize the foundations of electrohydrodynamic science and technology. Involved scientists include, among others, W. Gilbert (1544–1603), G. M. Bose (1710–1761), J. A. F. Plateau (1801–1883), Lord Rayleigh (1842–1919), Sir C. V. Boys (1855–1944) and Sir G. I. Taylor (1886–1975).[1] For instance, Gilbert has noticed that a sessile water drop on a dry surface is deformed, exhibiting an upward conical protrusion, upon approaching a piece of rubbed amber. With hindsight, this finding is really fascinating, showing clearly a deformed liquid–air interface under the influence of an external electric field as in later electrospinning methods.[2] A book of the famous "*Course of Theoretical Physics*" by L. D. Landau and E. M. Lifshitz is also dedicated to electrodynamic phenomena in continuous media.[3]

In particular, the first electrospinning-related patents were issued to J. F. Cooley (entitled "*Apparatus for electrically dispersing fluids*")[4] and to W. J. Morton ("*Method of dispersing fluids*"),[5] both in 1902 in the United States. The method involving an "electrical discharge from liquid points" was first studied in 1914 by J. Zeleny,[6] and a few developments were reported in 1915–1917.[7,8] The first, famous patent ("*Process and apparatus for preparing artificial threads*"), describing the realization of fibers by electrified jets of cellulose acetate solutions and issued to A. Formhals, dates back to 1934.[9] This

RSC Nanoscience & Nanotechnology No. 29
Polymer Nanofibers: Building Blocks for Nanotechnology
By Dario Pisignano

Published by the Royal Society of Chemistry, www.rsc.org

has been followed by several other related patents from the same inventor, issued from 1939 to 1944.[10–13] Other contributions that deserve to be mentioned are described in patents by C. L. Norton ("*Method of and apparatus for producing fibrous or filamentary material*"), issued in 1936,[14] and by H. L. Simons ("*Process and apparatus for producing patterned nonwoven fabrics*"), issued in 1966.[15] Acrylic fibers with diameter below 1 μm have been electrospun and reported in a pioneering study by DuPont researchers in 1971, using dimethylformamide (DMF) solutions.[16] However, electrospinning technologies have received poor or no interest until the 1990s, when the possibility of easily producing strongly elongated polymer or ceramic structures with sub-μm diameter by means of applied electric fields has drawn new attention concomitantly to the vigorous development of nanoscience and nanotechnology. A set of pioneering papers from the group of D. H. Reneker at the University of Akron, published in the years 1995–1996, is at the base of such renewed interest.[17,18] The technique immediately met with success. Less than five years later, electrospinning as an inexpensive method for producing electronic polymer nanofibers has been mentioned many times in the Nobel Lecture of A. G. MacDiarmid.[19] Recently, the appeal of electrospinning has increased enormously, spreading to a very large number of research laboratories and companies worldwide, mainly because of the low cost and experimental simplicity of this technique, which set almost no barrier to entry in terms of needed capital investment or sophisticated know-how. Dedicated books[20–23] and outstanding review articles[18,24–28] have been written, and special monothematic issues have been published by several international scientific journals. Dedicated congresses include, among numerous other events and workshops around the world, the "International Conference of Electrospinning", whose first and second editions have been in Melbourne, Australia (2010), and in Jeju Island, South Korea (2012), respectively.

The electrostatic spinning, or electrospinning, process includes different, sequential stages, which are described in the following.

### 2.1.1 Droplet Formation

Firstly, a polymer solution is placed into a pipette or a syringe, and then driven to the tip of a needle or a spinneret (also called an orifice or a *die*)* at a given (usually constant) flow. In most of the reported experiments, the internal diameter of this needle or capillary is of the order of hundreds of μm, however larger orifices or even free liquid surfaces can be used, because it is the formation of tiny local protrusions on the external liquid–air interface that plays a crucial role in the process, much more than the diameter of the fluidic conduit delivering the polymer solution. In addition, as described later (Section 2.5), gravity plays an almost irrelevant role in electrospinning, therefore the

* Some authors prefer to use "tip" or "capillary" instead of spinneret, because spinneret is already used to describe conventional spinning of textile fibers.[29]

orientation of the spinneret-deposition plane axis can be chosen without regard for gravity (horizontal or vertical). A line through the orifice and perpendicular to the collecting surface will be the central axis of the trajectory of the electrified jet as shown in Figure 2.1, which has become an electrospinning classic.[20] The concentration of the polymer solution is generally quite high (5–30% in weight), significantly higher than in deposition

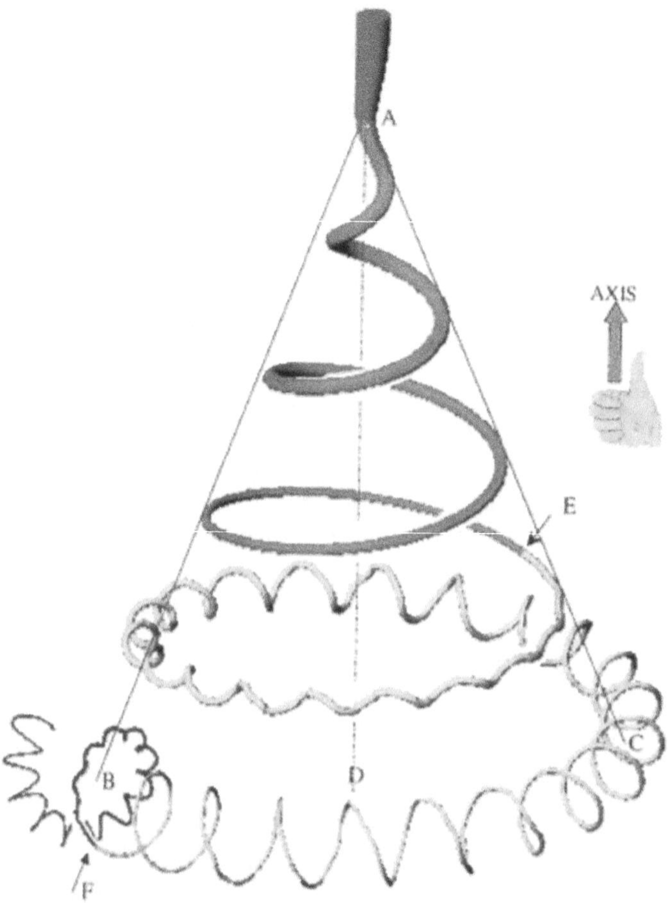


**Figure 2.1** (a) Schematics of the typical trajectory of the electrospun jet, evidencing the successive onset of three bending instability regimes characterized by spiraling loops with increasing frequencies. A: Transition from the straight part of the jet to the first bending instability region. D: Projection of the A point along the vertical direction. The segment AD defines the central axis of the jet path. The segments AB and AC define the envelope of the trajectory. E: onset of the second bending instability. F: onset of the third bending instability. Reproduced with permission from Ref. 20, *Polymeric Nanofibers*, American Chemical Society, Washington, DC, 2006. Doi: 10.1021/bk-2006-0918.ap001. Copyright © 2006, American Chemical Society.

methods such as spin-coating. Indeed, in electrospinning the solution viscoelastic properties are very important. Molecules must be entangled. Too dilute solutions often lead to electrospraying without nanofiber formation.

The first step of the process generally leads to a pendant droplet at the termination of the fluidic conduit, namely at the tip. An electric (positive or negative) bias ($V$, 2–100 kV) between the solution and the (often grounded) collecting surface is then applied, by connecting electrodes to a high-voltage generator. The voltage applied between the spinneret and collector may be either positive or negative.[30] Positive $V$ values at the spinneret work in most cases, and negative voltages lead to only slight differences. The first effect of increasing the voltage is the accumulation of an excess charge on the surface of the solution droplet.

The physical origin and behavior of such an excess charge deserves a bit of explanation. In the polymer solution, the excess charge is given by the motion of ions, which migrate towards the surface of the fluid in response to the applied electric field ($F$). The mobility of these ions is on the order of, say, $10^{-6}$–$10^{-7}$ $m^2$ V $s^{-1}$.[18,30–32] Hence, a polarization may be induced whenever an ionic solution is placed in a volume of space where an external electric field is present. The applied field tends to move positive and negative ions in opposite directions, and consequently to cause a non-zero electric charge in certain regions of the fluid surface. In solutions employed for electrospinning, conductivity ($\chi$) is usually due to the migration of ions present as dissolved salts or as residues of synthesis or processing procedures. Ionic impurities are often present in organic solvents as well as in water. For many polymer solutions, conductivity values range between 0.05 and 100 mS $m^{-1}$, and often increase upon addition of polymer molecules to the pristine solvents.[33] The ionic conductivity may be measured by methods such as conductimetry and dielectric spectroscopy.[34] The relatively low mobility causes the drift motion of ions to be quite slow. The order of magnitude of the resulting velocity, $v_d$, due to the electric field can be estimated by the well-known Stokes law relating the friction coefficient ($\xi$) in the fluid to the particle hydrodynamic radius, $R$ ($\sim 10^{-10}$ m):

$$v_d = \frac{qF}{\xi} = \frac{qF}{6\pi R\eta}, \tag{2.1}$$

where $q$ is the particle charge and $\eta$ is the (dynamic) viscosity (Sections 1.4.1 and 1.4.2.1), which can take values in the range of, say, $10^{-1}$–$10^3$ Pa s for electrospinning solutions, depending on the solvent and polymer properties and on the solution concentration. According to the previous equation, the mobility of ions is given by the ratio, $q/(6\pi R\eta)$. In this way, a maximum velocity of the order of $10^{-1}$ m $s^{-1}$ can be estimated for ions under typically applied electric fields of the order of $10^4$–$10^5$ V $m^{-1}$.[35] Once formed and accumulated on the surface of the pendant droplet, the excess charge becomes the origin of electric fields both complicated and important in determining the subsequent flow dynamics.

### 2.1.2 Jet Formation

The surface tension at the droplet surface competes with the increasing electrostatic repulsion of the accumulating surface charges. Upon increasing the applied voltage, electric forces ultimately prevail and an electrically charged solution jet is extruded. During the onset of this process, the droplet at the spinneret progressively becomes more and more elongated. The corresponding shape of the apex is named Taylor cone by a series of pioneering works published in the 1960s by Taylor about drops and shapes taken by liquid interfaces under applied electric fields.[36–39] It can be demonstrated that the Taylor cone, having a half angle of 49.3°, corresponds to a particular, so-called self-similar solution of the problem given by an axisymmetric, infinite ionic fluid body kept at an electrostatic potential, $\Phi_0$, and with the apex of the liquid surface at some distance from an equipotential plane kept at a different value of potential (Figure 2.2).[40] Other, non-self-similar solutions exist, which do not tend to a cone half angle of 49.3° (but instead to a half angle around 30°), and which better describe experimental observations coming from electrospinning experiments.[40] These solutions are more fully described in Section 2.5.

Giving an approximate estimate of the electric field and surface charge density needed to de-stabilize a droplet when the electrostatic repulsion overcomes surface tension is something feasible for students in their second year of introductory physics courses. In terms of accumulated surface charges, the electric field on the charged spherical surface, which is at a distance, $R_T$, from the sphere center, is:

$$F = \frac{1}{4\pi\varepsilon_0}\frac{4\pi R_T^2 \sigma_C}{R_T^2} = \frac{\sigma_C}{\varepsilon_0}, \tag{2.2a}$$

where $\sigma_C$ is the surface charge density and $\varepsilon_0$ is the dielectric constant. $4\pi R_T^2\sigma_C$ is, of course, the total charge accumulated at the surface. Furthermore, as stated above the force due to the electric field and exerting an electrostatic pressure ($\varepsilon_0 F^2/2$) at the surface has ultimately to overcome the surface tension ($\sigma_s$), which is the basis of Laplace pressure:

$$p = 2\sigma_s / R_T. \tag{2.2b}$$

In electrospinning experiments, $R_T$ indicates the apex of the Taylor cone better than the radius of the entire droplet. Equating the electrostatic and Laplace pressures, the threshold electric field on the solution may be estimated as:[17]

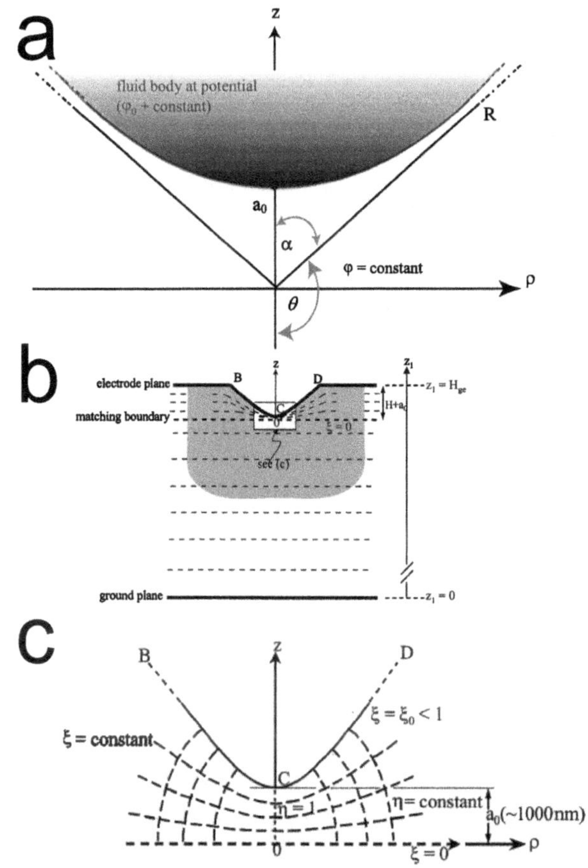


**Figure 2.2** (a) Scheme of an axisymmetric infinite fluid body representing the apex of a droplet deforming under an applied electric field. The droplet is kept at potential, $\Phi_0 = \varphi_0 + \text{constant}$. $a_0$ is the distance between the apex of the liquid surface and an equipotential plane kept at a different, constant value of potential ($\varphi$). ($R$, $\theta$) and ($\rho$, $z$) in this figure indicate the possible systems of coordinates (spherical or cylindrical, respectively) used to describe the spatial distribution of the potential in the region around the fluid body. Here $\alpha = \pi - \theta$. The self-similar, $a_0$-independent solution for the potential leads to obtain $\alpha = 49.3°$. (b, c) Modeling the droplet-air interface as a hyperboloid of revolution (BCD), which can be described by a prolate spheroidal coordinate system, ($\xi$, $\eta$) in (c). In these coordinates, curves with $\xi$ = constant represent hyperboloids, whereas curves with $\eta$ = constant represent ellipsoids. The fluid–air interface is described by the equation $\xi = \xi_0$, where $\xi_0$ is a constant $<1$. Again, $a_0$ is the distance between the apex of the hyperboloid and an equipotential plane at $z = 0$. The axis, $z_1$, in (b) represents the vertical coordinate from a grounded plane (at $z_1 = 0$). $z_1 = H_{ge}$ is the coordinate of the top electrode, and H is the height of the drop. Dashed lines in (b) and dashed hyperboloids ($\xi$ = constant) in (c) indicate the equipotential surfaces. Adapted with permission from Ref. 40, *J. Appl. Phys.*, 2001, **90**, 4836–4846. Doi: 10.1063/1.1408260. Copyright © 2001, American Institute of Physics.

$$F = \left(\frac{4\sigma_s}{\varepsilon_0 R_T}\right)^{1/2}. \tag{2.2c}$$

Another equivalent way to express the previous equation is through an electric Bond number, *i.e.* through a dimensionless parameter, $(\varepsilon_0 R_T F^2)/\sigma_s$, providing the ratio of electrostatic and Laplace pressures and indicating larger droplet deformations when this quantity approaches unity.[41] The threshold surface density and total accumulated charge can then be calculated by considering together Equations (2.2a) and (2.2c).[42] At threshold, one gets for the total charge: $4\pi R_T^2 \sigma_C \geq 8\pi\left(\sigma_s \varepsilon_0 R_T^3\right)^{1/2}$, which is called the Rayleigh condition. After extrusion, the jet moves from the flow-modified Taylor cone towards the collector. The volume and, to a certain extent, the exact shape of the droplet terminating at the Taylor cone, which may sometimes oscillate with time,[33] are therefore determined by the competition between the delivery of polymer solution by the flow from the syringe, and the concomitant withdrawal of fluid by the jet.

The velocity of the jet ($v_j$) can be measured directly by several methods, and in particular by direct bright-field imaging and time-lapse photography,[43] laser Doppler velocimetry[44] and particle imaging velocimetry (PIV).[45] $v_j$ is in the range from roughly $10^{-1}$ m s$^{-1}$ to a few m s$^{-1}$, and in general it is well above that of ions in the liquid ($v_j >> v_d$). Consequently, charges can be considered as almost fixed in the fluid, as assumed in many relevant models (see Section 2.5)†. The longitudinal acceleration during the jet motion is extremely high. The maximum velocities of up to a few m s$^{-1}$ are reached within centiseconds, therefore accelerations can be easily estimated to be in the order of hundreds of m s$^{-2}$. According to our definitions in Section 1.4.2.1, the order of magnitude of the deformation rate, $\dot{\gamma}$, of the electrified jet body can be roughly estimated as $v_j/r$, which can be as high as $10^5$–$10^7$ s$^{-1}$. By this mechanism, the effect of the electrical force is transferred by the ionic charges to the polymer mass

† It is important to point out that the comparison of the velocities of ionic charges in the solution and of the deformation of the electrospun fluid body changes significantly when different parts of the jet are considered. In particular, in the region of the Taylor cone the situation can be practically opposite to that in the flying jet. In fact, in the pendant droplet $v_d$ can still be in the range of $10^{-1}$ m/s, whereas the velocity of the motion of the liquid can be comparable or slower. In other words, instead of velocities one can compare the hydrodynamic relaxation time in the cone region with the charge relaxation time. The hydrodynamic relaxation time is introduced later in Equation (2.5). Having ionic charges moving faster than the deformation of the solution body means that the electrified droplet at the spinneret can be considered as a conductor where charges are only at the external surface. This is the reason why the apex, where the Taylor cone provides the onset of the extruded jet, can be effectively modeled as an equipotential surface[40] separating air and an internal region of liquid where no electric field is present. These two situations (of ionic charges either with practically infinite mobility or fixed with respect to the jet) describing the two successive regions of the electrified jet clearly show the richness of the phenomenology of electrospinning. The most successful theoretical models are those that take advantage of simplifying conditions valid in specific regions of the jet.

throughout the jet, which is consequently coiled and stretched. This very rapid, electrically driven motion is counteracted by the solution viscoelastic forces, which act to stabilize the jet as described later. By the velocity $v_j$, the collector, which is typically at a distance, $h$, from a few cm to a few tens of cm from the spinneret, is reached by the jet in a time of the order of 0.1 s.

Overall, this stage of the process can be seen as generating a continuous current ($I$), which is carried by the jet, thus flowing from the needle to the collector. The current is, $I = Q\rho$, where $Q$ indicates the jet flow rate and $\rho$ is the effective volumetric charge density‡. $I$ may vary in a quite broad range, from the order of nA to hundreds of μA, depending on the polymer solution and experimental parameters.[18,30,48,49] Especially in the early, rectilinear part of the jet trajectory, the flow conditions inside the jet can be considered with excellent approximation to be laminar, *i.e.* with straight flow lines. This can be easily appreciated by estimating the Reynolds number in the liquid conduit, which compares the magnitude of inertial and viscous forces in a fluid volume:

$$\mathrm{Re} = 2\delta r v_j / \eta, \tag{2.3}$$

where $\delta$ is the density. A rough calculation with physically reasonable values of the involved quantities ($\delta \cong 10^3$ kg m$^{-3}$, $v_j = 0.1$–$1$ m s$^{-1}$, $\eta \cong 10^{-1}$–$10^3$ Pa s, and diameter of the fluid body decreasing from 10–100 μm at the spinneret where the velocity is lower, down to 10–100 nm in correspondence of the highest $v_j$ values) would set a maximum order of magnitude of Re around $10^{-2}$. As is well known, such low values of Re indicate substantially irrelevant inertial effects in the flow, and absence of internal turbulence.

### 2.1.3 Jet Instabilities

While the jet follows a straight or slightly bent segment in the first part of its trajectory, it then generally undergoes several different instabilities, which sequentially set in during the flight, as shown in the diagram of Figure 2.1 and in

‡ In principle, the electric current, $I$, associated to a jet can include several different contributions, such as (i) a direct conduction component of charges, possibly moving with respect to the frame of reference of the jet because of ohmic transport driven by the external electric field, (ii) a volumetric advection component, given by charges distributed in the fluid volume and considered at rest with respect to the jet as explained above, and (iii) a surface advection component, given by the surface charge density carried by the jet. The relative importance of the different components to current changes during the various phases of the jet flight. As the jet become thinner and thinner, it is physically clear that the resistance linked to its cross-sectional area progressively increases, hence advection components become dominant.[46] In addition, the internal charges, if present, move quickly to the surface of the jet. These considerations are used in important models of electrospinning (Sections 2.5.2 and 2.5.3). An effective volumetric charge density can be used to summarize both the bulk and the surface advection components, (ii) and (iii). Finally, we recall that corona discharges have been observed and photographed at the tip of the Taylor cone, before the jet is formed, and in cylindrical regions around the jets during electrospinning.[47]

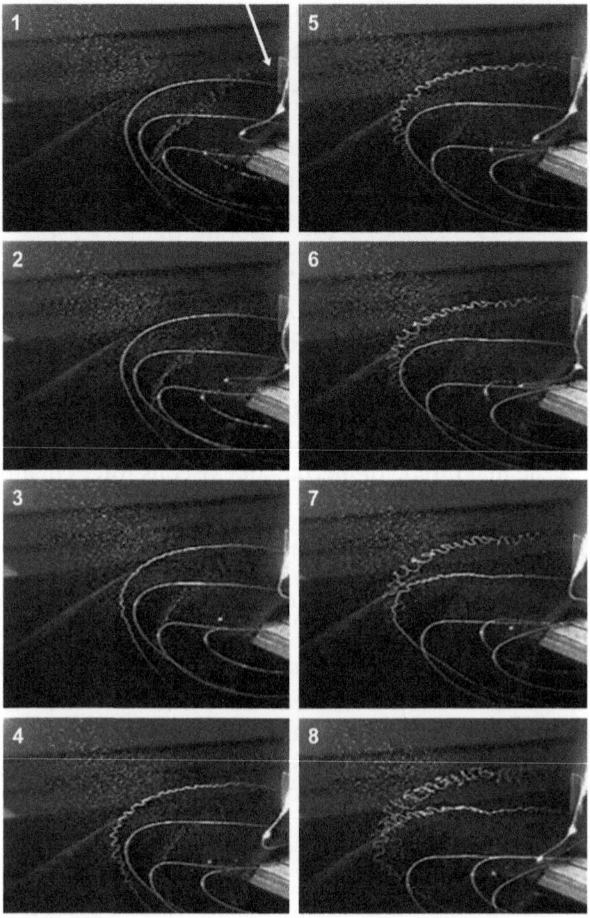


**Figure 2.3** Photographs showing an oblique view of a jet of polyisobutylene in a viscous mixture of acetone and paraffin oil. The different bending instability stages are clearly visible. In this sequence, frames are captured by a video camera at a rate of 30 per second, and the diameter of the largest coils is about 20 cm. Reproduced with permission from Ref. 50, *Polymer*, 2008, **49**, 2387–2425. Doi: 10.1016/j.polymer.2008.02.002. Copyright © 2008, Elsevier Ltd.

the photographic frames in Figure 2.3.[20,50,51] It is largely in the instability phase that the reduction of the jet diameter by several orders of magnitude occurs, ultimately leading to the formation of fibers with sub-μm transversal size[§].

§ It has been pointed out,[52] however, that electrospinning experiments in which this instability is limited by exploiting the linear part of the jet or removed by using proper secondary electrodes[53,54] do not result in nanofibers whose diameter is significantly larger than in conventional electrospinning. These apparently contradictory evidences have further suggested how much electrospinning and its dynamics are complex and how much they need to be investigated in depth to be fully rationalized.

Indeed, by some polymer solutions, electrospinning can produce fibers that are astonishingly thin. Nylon is the classic example of successfully electrospun polymer,[55] allowing fibers or filament segments with diameter down to the scale of 1 nm to be realized, which corresponds to six or seven macromolecules arranged across the filament cross-section.[56]

Since the jet is electrically charged, it is not surprising that instability can easily arise as a consequence of the fact that a collection of charges (here basically at rest in the jet reference system as discussed above) is not able to sit in a stable equilibrium position (Earnshaw's theorem).[35] Several models agree indeed in attributing the main cause of bending instabilities to electric charges (Section 2.5). In fact, the phenomenology of electrospun jets is very rich and its instability regimes are indicated by an equally rich nomenclature. As experimentally observed by means of high-speed (*i.e.* 2000 fps) imaging,[35,57] the jet generally exhibits a clear whipping behavior (*whipping instability*) and then a series of spiraling loops (which is the onset of the *bending instability*) with increasing diameter. The envelope of these spirals constitutes a conical surface. As highlighted by a few models,[48,58] whipping instabilities indeed cause bending and stretching and are characterized by deformation of the centerline of the jet. In other words, they are not symmetric upon rotations around the jet longitudinal axis (*i.e.* they are *not axisymmetric*). Overall, one concludes that whipping and bending of the jet trajectory are somehow related effects, and indeed whipping and bending instability are often used as almost equivalent terms in most of the literature. In particular, bending, though not always observed,[33] is very frequent, and especially favored for polymers with relatively low *MW* and relatively poor viscoelastic behavior, since viscoelasticity indeed acts as a stabilizing force for the extruded jet, as evidenced by theoretical models (Section 2.5). Indeed, it is generally assumed that a prerequisite for electrospinning is that the polymer solution exhibits sufficient molecular entanglements to assure good viscoelastic behavior.

Upon continuing to follow the jet trajectory, one may find that a *second bending instability* grows while the jet oscillates at a shorter length-scale, *i.e.* with smaller loops spiraling roughly around the extrapolation of the main trajectory path. A third stage of instabilities is also possible before reaching the collector, and consists in even smaller loops, again spiraling with smaller wavelengths (higher frequency) around extrapolated, previous trajectory segments. During all these phases, the concomitant solvent evaporation and jet elongation favor the continuous reduction of the diameter of the fluid filament, which finally evolves into solid micro- or nanofibers.

Though the flow conditions *within* the jet during the whipping stage are practically never taken into consideration and are very difficult to investigate experimentally, one should bear in mind that in principle bending instabilities may also generate variations of the internal flow lines. Slight inertial features can arise in liquid flowing along curved trajectories**, as a consequence of radially acting forces (scheme in Figure 2.4).[59] Another clear consequence of such forces is that, at some points during the flight of the jet, *splaying* phenomena may also take place, determining the separation of secondary paths of polymer solution from the main trajectory. The occurrence of splaying implies that centrifugal/electrical forces acting along the radial direction with respect to the longitudinal jet axis overcome cohesive forces among neighbor elements of the fluid body. Splaying events are frequently observed in experimental systematics performed to calibrate electrospinning, and reported in detail for polyethylene oxide (PEO)[35] and other polymers.[18] Especially in early studies, splaying has led to attribution of the extraordinary reduction of the jet diameter found in experiments to the rapid and repeated, seeming division of the main extruded flow in many sub-jets.[18] This effect has been initially thought to determine an extended cone of dividing sub-filaments, with its apex originating in the main jet. Each such sub-filament, acquiring its own radial velocity component, has been considered as ultimately resulting in a very thin collected fiber. However, with the rapid advent of more sophisticated, high-frame-rate imaging tools allowing the jet to be visualized during its motion, multiple sub-filaments have been interpreted as an artifact due to the fast bending and whipping instabilities, whose envelope corresponds indeed to the cone that one images by using high enough exposure times.[35,57] Today, it is generally recognized that splaying

** In fact, a circular tube of liquid bending with a radius of curvature, $r_B$, much larger than its cross-sectional dimension, undergoes a flow change with respect to a straight channel, *i.e.* it shows the onset of secondary (so-called "Dean") flows induced by centrifugal forces locally acting on the liquid (Figure 2.4).[59] A fluid element of mass $m$ within the bending jet would experience internal centrifugal forces depending on its specific position within the polymer solution filament traveling from the spinneret to the collector. For a jet bending with radius of curvature of the order of centimeters, these forces ($\sim m v_j^2/r_B$) can be in the attoNewton range for a mass contained in a tiny volume of, say, $10^{-24}$ m$^3$ (*i.e.* in a cube of side of 10 nm), and the generated secondary flows would have velocity of the order of Re $V_j\, r/r_B$ (*i.e.* up to the order of 1–10 μm/s).[59] This phenomenon may be described with help of a dimensionless parameter, the so-called Dean number, $N_D$, defined by:[60] $N_D = (2\delta r v_j/\eta) \times (r/r_B)^{0.5}$, *i.e.* by the product of the Reynolds number and the square root of the ratio of the jet radius to its bending radius of curvature. In other words, $N_D$ roughly compares the magnitude of centrifugal ($\sim \delta r^3 v_j^2/r_B$) and viscous ($\sim \eta\, v_j\, r$) forces acting on the fluid volume and it can be estimated to be up of $10^{-4}$ for an electrospun jet in the whipping region. This value appears to be very small, but on the other hand one has to consider that such secondary flows would occur within a jet of very small diameter, and that the situation is further complicated by the simultaneous evaporation of solvent. While too low for affecting significantly the process of nanofiber formation, these secondary velocities may ultimately impact on the resulting orientational configuration of polymer molecules within the produced nanofibers. In a time interval as short as a few centiseconds, a secondary flow velocity of 1 $\mu$m/s would produce a potential movement of a fluid element within the jet of tens of nm, which could potentially contribute in slightly re-configuring the molecular microenvironment, especially in the more external layers of the fluid where also solvent kinetics plays a more relevant role.

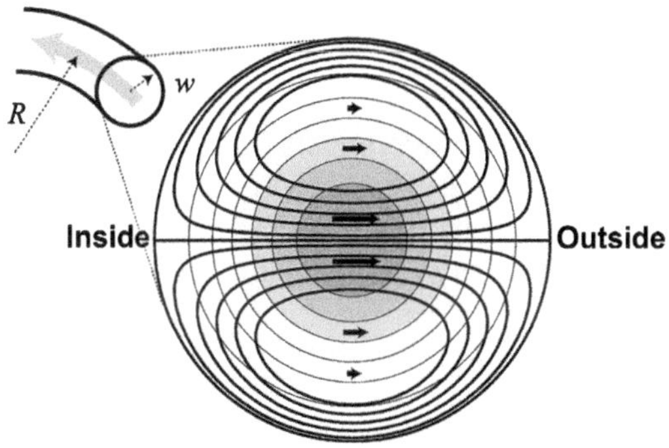


**Figure 2.4** Schematics of secondary Dean flow resulting from centrifugal forces on fluid flowing through a slightly curved pipe. In the cross-section shown in this scheme, centrifugal forces act on fluid elements in the plane of the page (here with greatest magnitude in the center of the channel, where the primary flow is supposed to be fastest). These inhomogeneous forces drive the secondary flows (here in the plane of the page). Reproduced with permission from Ref. 59, *Rev. Mod. Phys.*, 2005, **77**, 977–1026. Doi: 10.1103/RevModPhys.77.977. Copyright © 2005, The American Physical Society.

events can be made relatively rare if a proper process optimization is performed. Finally, it is worthwhile to mention that one can also take advantage of splaying in order to intentionally obtain branched fiber nanostructures.

Another class of instabilities possibly underwent by electrified jets is the so-called *varicose instability*, which, differently from whipping/bending, is symmetric with respect to the jet longitudinal axis (*axisymmetric*). This mode can be drawn by keeping fixed the jet centerline while varying the jet cross-sectional area normal to the longitudinal axis. Since this can be reflected in the shape of the resulting nanofibers, axisymmetric instabilities, including among others the *Rayleigh* or Plateau–Rayleigh instability, can lead to generally undesired variations of diameter along the fiber length being obtained, and eventually fibers exhibiting repeated or even periodic beads. The Rayleigh instability is better introduced in the following Section 2.5.3.

Finally, when the jet path is stopped by the counterelectrode, *buckling* effects can take place, which result in distinctive patterns rich in periodic coils often formed by deposited nanofibers.[35,50,61] Mechanical buckling is a familiar phenomenon that one observes, for instance, when a jet of honey falls on a surface under the effect of gravity. Buckling instabilities are related to the sharp longitudinal compression of the impinging jet reaching a solid surface, and the produced pattern will depend on the jet velocity, geometry and viscoelastic characteristics. For electrospun jets, buckling frequencies are in the range of MHz (*i.e.* much higher than those characteristic of bending instabilities), and the radius of the resulting coils measures generally from a few μm to a few tens of μm.[50] Therefore, buckling patterns are typically found in collected nanofibers as high-frequency, small-curvature radius loops superimposed to slowly varying, ample curves due to jet bending and whipping. In some cases, the buckling coils have roughly constant diameter (Figure 2.5a), whereas in other experiments more complex, zigzag features are observed in the deposited nanofibers (Figure 2.5b). All these features have been investigated in depth in experiments using inclined and moving collectors, and by varying the distance between the spinneret and the deposition surface (Figure 2.5c–h).[61]

### 2.1.4 Solvent Evaporation

Issues related to solvent evaporation are very relevant in the electrospinning processes and have repeatedly attracted the attention of researchers throughout the years.[62–65] Solvent evaporation has been described by means of a non-linear mass diffusion-transfer model, which catches the features of the drying process in dilute polymer solution jets, evidencing that significant inhomogeneities of solvent concentration can generate within the electrified flow.[66] An especially interesting effect is the formation of an external shell in the jet, resembling a polymer skin due to the more rapid solvent evaporation from the outer flowing layers. Further complicated by additional effects such as the possible collapse of such skin as a consequence of the subsequent, slower solvent evaporation and volume reduction of the internal, wrapped body of the jet, this phenomenon can generate polymer nanofiber structures with elliptical cross-section or even with different shapes such as ribbons and belts (Figure 2.6).[62] Recently, also radial buckling features observed in tubular polymer fibers realized by coaxial electrospinning (Section 2.3.1) have been rationalized in terms of local skin collapses due to solvent evaporation from the inner part of the deposited nanostructures.[64] While difficult to access experimentally, given their importance in affecting the ultimate structure of fibers solvent evaporation effects have been included in theories describing electrospun jets (Section 2.5).[65] In particular, mathematical models have been formulated that demonstrate that, in order to form the external skin, solvent evaporation rates have to be large, and higher than a critical value.[63] In addition, it can be reasonably figured out that compositional, orientational or morphological gradients (depending on the solution used) of the resulting polymer nanofibers can be associated to different rates of solvent evaporation occurring at different radial coordinates of the jet.

In particular, among other properties, nanofiber porosity is especially related to solvent evaporation. Some general design guidelines can help in predicting the resulting porosity and surface morphology of fibers, for solvent-assisted fabrication methods where phase-separation between immiscible components is likely to occur, such as electrospinning. These principles have been outlined in detail in the literature and summarized by G. C. Rutledge and coworkers, and are based on estimating three characteristic times involved in the process.[43,67,68] A *drying time*, $t_D$, can be defined as:

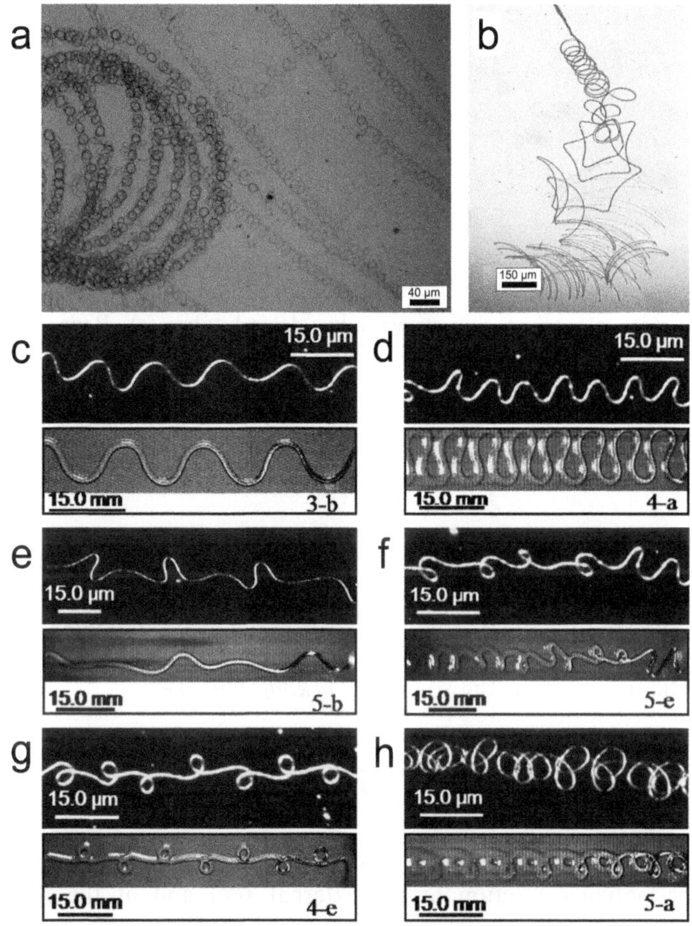


**Figure 2.5** (a) Buckling coils on electrospun nanofibers of Nylon 6. The diameter of the coils is about 15 μm. (b) Complex nanofiber features obtained by varying the distance from the spinneret to the collector. The distance is changed by a constant rate from the top to the bottom of the micrograph, resulting firstly in a straight segment, then in buckling coils, and ultimately in zigzags. The final deposition visualized at the bottom of the micrograph clearly suggests a pendulum-like motion of the electrified jet. (a, b): Adapted with permission from Ref. 50, *Polymer*, 2008, **49**, 2387–2425. Doi: 10.1016/j.polymer.2008.02.002. Copyright © 2008, Elsevier Ltd. (c)–(h) Various buckling phenomenology as shown by electrified jets of PEO, spun into water and collected on glass slides (top panels), and corresponding, strongly similar patterns formed by gravity-driven syrup jets, which are about 1000 times larger than the electrospun patterns (bottom panels, from Ref. 51). Adapted with permission from Ref. 61, *Polymer*, 2007, **48**, 6064–6076. Doi: 10.1016/j.polymer.2007.08.002. Copyright © 2007, Elsevier Ltd. Bottom panels adapted with permission from Ref. 51, *J. Fluid Mech.*, 2006, **569**, 89–111. Doi: 10.1017/S0022112006002503. Copyright © 2006, Cambridge University Press.

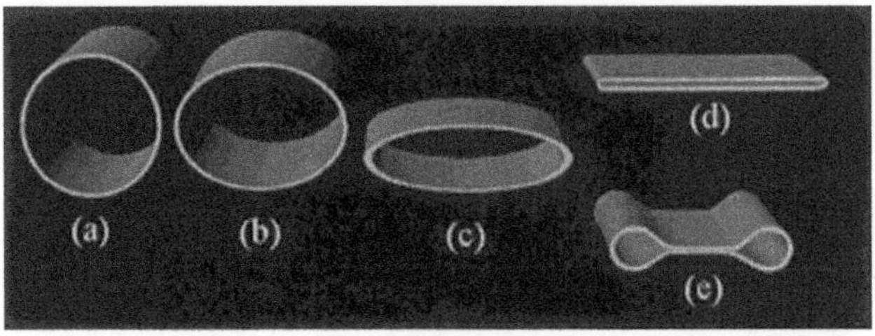


**Figure 2.6** Sequence schematizing the collapse of the skin formed at the surface of an electrospun jet (a–c), which can lead to the formation of polymer ribbons (d) or even of ribbons with two lateral tubes (e). Reproduced with permission from Ref. 62, *J. Polym. Sci., Part B: Polym. Phys.*, 2001, **39**, 2598–2606. Doi: 10.1002/polb.10015. Copyright © 2001, John Wiley & Sons, Inc.

$$t_D = \frac{V_0}{A W_{EV}}, \tag{2.4a}$$

where $V_0$ is the initial liquid volume, $A$ indicates the initial vapor-drop surface area and $W_{EV}$ (here expressed in m s$^{-1}$) is the evaporation rate. For instance, $W_{EV}$ is reported to be about 2–3 μm s$^{-1}$ for DMF at room temperature, and to decrease for higher values of environmental relative humidity.[68] The second characteristic parameter is the so-called *buckling time*, $t_B$:

$$t_B = \frac{D(\phi_p - \phi_{p0})^2}{W_{EV}^2}, \tag{2.4b}$$

where $D$ indicates the polymer-solvent mutual diffusion coefficient and $\phi_p$ and $\phi_{p0}$ are the polymer fraction at the external skin and at the fluid core, respectively. Furthermore, if phase separation between two immiscible components takes place within the electrified jet, one should also take into account a characteristic *phase separation time*, $t_{PS}$. $t_B$, $t_D$ and $t_{PS}$ can be of the same order of magnitude (say, 0.1–1 s), and the resulting fiber morphology can be reasonably predicted depending on which particular condition is fulfilled by the electrospun solution.[43,68] In particular, solid or porous fibers are obtained depending on whether the ratio, $t_{PS}/t_D$, is greater than or less than 1, respectively. Indeed, when phase separation phenomena are absent or very slow, the fiber interior is solid. Then, for $t_{PS}/t_D > 1$ (solid fibers), one can have two possible situations depending on the $t_D/t_B$ ratio:

- if $t_D/t_B > 1$, the fibers should be buckled, with the presence of surface corrugation;
- if $t_D/t_B < 1$, the fibers should be smooth.

When phase separation phenomena originating internal porosity are instead present and fast enough ($t_{PS}/t_D < 1$) one obtains porous fibers and considers the following cases:

- if $t_{PS}/t_B > 1$, the fibers should be buckled;
- if $t_{PS}/t_B < 1$, the fibers should be smooth.

A phase diagram illustrating these regimes is reported in Ref. 43.

### 2.1.5 Molecular Orientation

Understanding to what extent electrospun nanofibers present internal porosity or specific morphologies, and eventually to what extent they are made by partially or significantly uniaxially oriented molecules is crucial for a wide variety of applications since these features directly affect the mechanical (Section 4.1), optical (Sections 5.2.1 and 5.2.2) and electronic (Sections 6.1.1 and 6.2) characteristics. In general, higher orientational anisotropies at the molecular scale are indeed associated to a portfolio of enhanced physical properties, which is one of the main motivations for using polymer nanofiber systems in technology. A first consideration that helps in rationalizing the orientation of macromolecules in electrospun nanofibers is based on longitudinal strain rates or velocity gradients, $\partial v_j/\partial z$, underwent by the polymer during the process. This strain rate gives a measure of the variation of the jet velocity along the axial direction. Indeed, strain rates comparable to or higher than the reciprocal relaxation time, $\tau^{-1}$, of the molecules favors the transition of random coils into stretched and relatively aligned molecular assemblies.[69,70] This is certainly the case of electrospun jets. In fact, given the enormous and fast elongation of the filament during the flight of the jet, the strain rate within electrospun fluids is estimated to take values up to the order of $10^3$–$10^5$ s$^{-1}$. In fluid jets, a convenient way to estimate relaxation timescales is through the following equation:

$$\tau = \left(\frac{\delta r^3}{\sigma_S}\right)^{1/2}, \tag{2.5}$$

where $\delta$, $r$ and $\sigma_s$ indicate the solution density, the jet radius and the surface tension, respectively. With physically reasonable values for the various quantities entering the right member of Equation (2.5) ($\delta \cong 10^3$ kg m$^{-3}$, $\sigma_s \cong$ 50 mN m$^{-1}$), one estimates the relaxational timescale for a fluid jet at the beginning of its trajectory to be in the millisecond range (*i.e.* $\tau^{-1} \cong 10^3$ s$^{-1}$, to be compared with deformation rates also of $10^3$ s$^{-1}$ or slightly higher). More generally, in polymer solutions the reciprocal relaxation times are $\cong 10^2$ s$^{-1}$, therefore in practically all cases one can conclude that strain rates in electrospun jets are significantly higher than $\tau^{-1}$.[35,71,72]

In other but substantially physically equivalent words, how macromolecules can be stretched and aligned in the electrospinning flow can be understood by considering the values of the Deborah number (De) typical of the jet.[45] The Deborah number is an adimensional parameter defined as:[59,73]

$$\mathrm{De} = \frac{\tau}{\tau_{flow}}, \tag{2.6}$$

where $\tau_{flow}$ is the reciprocal strain rate, and indicates the characteristic time at which the jet radius contracts. Here, $\tau_{flow}$ ($10^{-5}$–$10^{-3}$ s) takes the role of the "experimental timescale" or "observation time window" that we have introduced in Section 1.4.2. Indeed, De values <1 or >1, respectively, indicate that the polymer has or does not have enough time to rearrange its molecular conformation after flow-induced stretching. For instance, Deborah numbers in the range of 900–8500 are estimated for electrospun DNA molecules, which indicate relevant, persistent stretching due to the jet extrusion.[45]

From an experimental viewpoint, evidences of molecular alignment and stretching within nanofibers come from birefringence,[74,75] electron paramagnetic resonance (EPR)[76] polarized infrared absorption,[77–79] wide angle X-ray diffraction (XRD),[75,80,81] Raman spectroscopy[82–86] and other kinds of optical measurements. Chain packing can be also probed by AFM.[87,88] The usually adopted system of coordinates used to represent the orientation of polymer molecules with respect to a main, preferential axis of the sample is sketched in Figure 2.7.[89] The alignment of polymer chains with respect to the fiber longitudinal axis (whose directions define the angle Φ in Figure 2.7) can be quantified by considering the following orientational order parameter or Herman's orientation function, $S$:

$$S = \frac{1}{2}\left(3 < \cos^2 \Phi > -1\right), \tag{2.7a}$$

where <...> brackets indicate the average over the entire ensemble of polymer chains:

$$< \cos^2 \Phi > = \frac{\int_0^\pi f_P(\Phi) \cos^2 \Phi \sin \Phi d\Phi}{\int_0^\pi f_P(\Phi) \sin \Phi d\Phi}. \tag{2.7b}$$

In the previous expression, $f_P$ indicates the population of polymer chains at each orientation. For backbones perfectly aligned along the preferential orientation axis, $Z$, one has $\Phi = 0$ for all the molecules, and $S = 1$. For a fully random ensemble, the ensemble average of $\cos^2\Phi$ is 1/3, and $S = 0$. Figure 2.7 also shows the position of a microscopic probe exploited to analyze the system at nanoscale. For example, in EPR the microscopic probe is given by electron magnetic moments. In these experiments, blend nanofibers of emeraldine-base polyaniline (PAN) doped with camphorsulfonic acid in PEO show a moderate increase of the Pauli susceptibility, a decrease of the EPR linewidth and a slight increase of the Lorentzian character of the measured lineshape, which suggests an increased structural order compared to control cast films.[76] In optical measurements, systems are mainly investigated by exploiting optical transition dipole moments, $\vec{\mu}$. In fact, it is physically reasonable that the alignment of molecules within the nanofiber (or within any kind of anisotropic solid-state sample) may somehow be related to the orientation of absorbing or emissive chromophores. These are an easily accessible (by means of absorbed or emitted

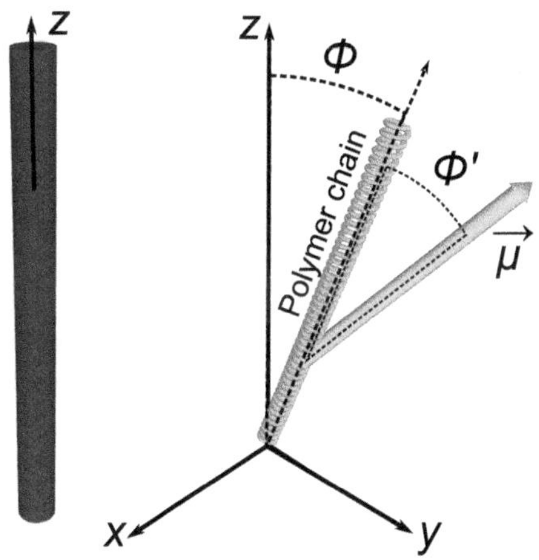


**Figure 2.7** Scheme of the orientation of a polymer backbone, and of a transition dipole ($\vec{\mu}$) with respect to the polymer chain, within a nanofiber whose longitudinal axis is along the $Z$ direction. Here, $X$, $Y$ and $Z$ indicate the axes taken as reference in the macroscopic sample. $\varphi$ defines the angle between the direction of the polymer chain and the $Z$ direction, and depends on the assembly or nanofabrication process used to produce the nanofiber, and on the characteristics of the molecular system used (steric hindrance, supramolecular interactions, *etc.*) and flow (De number). The dipole transition moment is oriented at an angle, $\varphi'$, with respect to the main chain axis (for instance, for polyfluorenes $\varphi'$ is of about 20°).[89] Other azimuthal angles, not shown here, complete the description of the polymer chain and dipole transition moment directions.

photons) part of the polymer molecules and as such constitute valuable probes to evaluate the overall degree of order. The alignment of transition dipoles with respect to the polymer chain axis (whose directions define the angle $\Phi'$ in Figure 2.7) can be quantified by the orientational order parameter, $S'$:

$$S' = \frac{1}{2}\left(3 < \cos^2 \Phi' > -1\right), \tag{2.8}$$

where the ensemble average is performed over the distribution of emitters. The overall expression relating the optical emission or absorption to the degree of alignment of molecules along $Z$ is therefore:[89]

$$SS' = \frac{1}{4}\left(3 < \cos^2 \Phi > -1\right)\left(3 < \cos^2 \Phi' > -1\right). \tag{2.9}$$

Together with the polarization properties of optical signals in absorption, emission and light scattering experiments, the previous expressions are often used to study the properties of internal molecular alignment within both electrospun nanofibers and polymer nanostructures realized by other methods (Chapter 3).

Both molecular and transition dipole moment orientation can be probed within the solid state samples, especially upon combining structural information with polarized absorption measurements.[89,90] Small-area electron diffraction experiments are also effective since they allow the orientation of crystalline regions within polymer nanofibers to be probed.[91] Most of these techniques are already well established from the study of polymer alignment in thin-films. Improved analysis methods, which make use of combined polarized Raman spectroscopy and UV-visible absorption data, are especially worthwhile to be mentioned as valuable tools to investigate the orientational properties of light-emitting polymer systems.[90] We will come back in depth to optical properties of polymer nanofibers in Chapter 5.

## 2.2 Process Parameters

Electrospinning is a process putting together (i) an inherently complex, multicomponent material such as the polymer solution, (ii) a fluidic and electrical equipment most often including syringes, needles, motorized elements for carefully controlling flow-rates and generators supplying high voltages and (iii) the environment external to the process, which may vary from just air atmosphere to locked chambers operating with controlled gases or in vacuum conditions, stabilized temperature, *etc.* Consequently, the overall process (though operationally simple and also quite robust upon slight variations of the total set of involved parameters) is very complex in itself in terms of the physical behavior of the electrospun solution, and it depends on so many experimental variables that fully reliable approaches allowing one to predict the specific geometry of nanofibers coming out from a single experiment are still basically missing. It has to be taken into account that a few parameters are directly chosen by the operator (we will call them *primary*), whereas other parameters (which we will name *secondary*) in turn depend on other properties of the polymer solution, or of the processing environment. Secondary parameters can critically affect the process results though are not always easily controlled. For example, the polymer concentration in the prepared solution is an independent, primary parameter whose value is generally directly controlled, whereas the solution viscosity is secondary. Indeed, the value of the solution viscosity, which may heavily influence the final geometry of electrospun samples, is most often determined indirectly, since it is in its turn affected by the chosen concentration and by the processing temperature, $T$. Analogous considerations hold for many other involved variables. Here, the process parameters are listed according to the usually employed classification,[17] *i.e.* grouping them on the basis of their general dependence on:

- *material properties* of the used polymer solution, such as the initial polymer concentration ($c_{p0}$, wt.% or moles $L^{-1}$), and the molecular weight ($MW$, a.m.u. or g moles$^{-1}$) of each spun polymer species. These are primary parameters. Other, secondary material properties are the solution density [$\delta$ ($c_{p0}$, $T$), kg m$^{-3}$], viscosity [$\eta(c_{p0}$, $T$), Pa s], relaxation time [$\tau$ ($c_{p0}$, $T$), s], dielectric constant or relative permittivity ($\varepsilon_r = 1 + \chi_r$, where $\chi_r$ is the electric susceptibility) and conductivity [$\chi(c_{p0}, T)$, S m$^{-1}$]. These parameters can also be influenced by the environment pressure, though with generally insignificant changes in the narrow range of pressure variations typical of electrospinning experiments. Finally, additional (secondary) properties are of course mainly related to the used chemicals (considered as separated molecular species), such as the degree of plasticity and molecular steric hindrance of the polymer chains, the solvent boiling point (K), vapor pressure (atm, bar or Pa)[††], diffusion coefficients (m$^2$ s$^{-1}$) through the polymer and through external atmosphere, the solution pH and the surface tension ($\sigma_s$, N m$^{-1}$);

- *set-up settings*, such as the applied voltage ($V$, V), the solution flow rate ($Q$, L s$^{-1}$ or m$^3$ s$^{-1}$), the nozzle-to-collector distance ($h$, m), the pipette orifice or the needle internal diameters ($\varPhi$, m) for equipment using pipettes or needles to extrude the solution. These are all primary parameters, and their proper choice is often a major part of electrospinning experiments. The solution flow rate always has a significant impact on the morphological characteristics of the electrospun products, including the resulting fiber diameter. The influence of $V$ is generally more subtle and may vary depending on the used range of values and polymer solutions, so both the situations of nanofiber diameters increasing or decreasing upon increasing the applied voltage bias are possible. The outer electric field in the process is of course secondary, being determined by the applied voltage and set-up geometry [$F \cong V/h$ (V m$^{-1}$)], and strongly enhanced near tips and sharp corners generally present at the spinneret. The applicable electric field is limited by the dielectric breakdown in the processing atmosphere. It has also been pointed out that the extruded jet could be more directly affected by the diameter of the terminating tip of the Taylor cone (which in turn depends on the solution properties and on the applied electric field) rather than by $\varPhi$.[71] Another, peculiar secondary property depending on both material and set-up parameters is the volumetric charge density [$\rho$ ($c_{p0}$, $V$, $Q$, $h$), C L$^{-1}$ or C m$^{-3}$] in the jet. This variable is fruitfully used in some models[65] and demonstrated to have a significant effect on the resulting fiber diameter;[71]
- other (primary) issues due to *ambient conditions*, such as the chemical composition and the eventual presence of gas flow in the atmosphere in which experiments are carried out, the relative (%) humidity, temperature ($T$, °C or K), pressure ($p$, atm, bar or Pa), *etc.* Ambient humidity is especially relevant, because while it has to be not too high to allow electrospinning to be carried out successfully (reasonable values are of 30%–40%), on the other hand the presence of aqueous vapor can be utilized to

†† The boiling point and solvent vapor pressure are particularly important since the rapid evaporation from solvent-rich regions (which can be detrimental when aiming for smooth and uniform collected fibers) can be fruitfully exploited to obtain nanofibers with regular porosity[92] and surface texturing as explained in Sections 2.1.4 and 4.2.1. Due to the general importance of the porosity properties of nanofibers, and to the potential use of these nanostructures for filtration and nanophotonics, these effects are recalled later in the book when dealing with specific applications (Sections 4.3.1 and 5.2.4).

produce controllable porous features on the surface of polymer nanofibers (Section 4.2.1).[93,94] For this reason, electrospinning has evolved rapidly towards the use of process chambers in which atmosphere or at least humidity and temperature can be controlled, and from which gas can be safely evacuated. This set-up is also a valuable alternative to fume hoods to guarantee an effective protection of operators against evaporating organic solvents.

Various studies have been aimed at rationalizing how the electrospinning process depends on its many variable parameters. For instance, a systematic investigation of the influence of the applied voltage, solution flow rate ($Q \cong$ 50–500 μL $\text{min}^{-1}$), polymer concentration and molecular weight and nozzle-to-collector distance in the range 10–30 cm is performed in Ref. 33 and analyzed in terms of a series of power-law dependencies of the jet current ($I$) and volumetric charge density ($\rho$) on the different parameters ($I \propto x^m$ with $m >$ 1 for $x = V$ and $m < 1$ or even slightly negative for $x = Q$). These dependencies are combined with an exponentially decreasing variation of $I$ and $\rho$ upon increasing $h$ ($I \propto e^{-h/h*}$, where $h*$ is a solution-dependent constant). In fact, jet currents for a variety of polymers and solvents are found to exhibit a monotonous increase upon increasing the external electric field,[49] whereas the dependence on the flow rate only shows a weaker increase or decrease, thus generally leading to a decreasing dependence of $\rho$ on $Q$ according to the rough relation, $\rho = I/Q$, already introduced in Section 2.1.2. The exponential decrease of the current upon increasing $h$ may be straightforwardly rationalized by the concomitant reduction of the electric field ($V/h$). Finally, for PEO compounds, values of $m$ between 0.5 and 0.9 are found for dependencies of $I$ and $\rho$ on $c_{p0}$ and $MW$.

According to the model of electrospinning by Reneker and Yarin (Section 2.5.2),[35,65] $\rho$ and $h$, together with the relaxation time, initial jet radius, and elongational viscosity are the parameters that more heavily influence the resulting jet diameter.[71] Some effects are also registered for $c_{p0}$, $V$ and for the solution density and the solvent vapor pressure, whereas only a minor influence is found for the relative humidity, solution surface tension and vapor diffusivity in air.[71] The fiber diameter tends to increase upon increasing the polymer solution concentration, with a behavior that has been described by a power law.[49] Furthermore, a clear effect of $V$ on the formation of beads along deposited nanofibers has been observed.[49] Finally, one should mention that additives such as pyridine or palladium(II)diacetate, which increase the solution conductivity without heavily affecting viscosity and surface tension, can help nanofibers without beads or with ultra-thin diameters (< 10 nm) to be obtained.[95] Importantly, the effects of additives on optical, electrical and other properties of the resulting nanofibers has to be carefully taken into account when electrospinning is used to realize functional nanostructures.

The variety of polymers and solvents reported to date for electrospinning is huge, including a very wide range of both natural and synthetic materials. The large number of process variables together with their frequent mutual dependence makes listing a unique set of parameters for electrospinning a specific polymer largely illusory and maybe even useless. As a matter of fact, the process for each polymer species has to be optimized in one's own laboratory, which always has specific environmental conditions in terms of temperature, humidity, *etc.* The polymer solution history and age also play a role since physical properties can vary over time. Experimentalists interested in processing a specific polymer should carefully check the related literature for their particular material or refer to available models for drawing a suitable starting point in choosing their process parameters, bearing in mind that not only the values of the electrospinning variables, but also the technological approach (Section 2.3) and the almost ubiquitous reproducibility issues (Section 2.4.1) can make the difference.

## 2.3 Process Engineering

The electrospinning process has been the subject of very intense and continuously growing research efforts. As a consequence, many aspects of this technology are undergoing more and more varied and optimized engineering stages, aiming to achieve specific nanofiber shapes or architectures, multifunctionality of electrospun samples, ordered assemblies, higher productivity, *etc.* In many cases, exploring different collector geometries, such as pre-arranged metal patterns or rotating mandrels, is found to be a successful strategy to obtain particular nanofiber structures and assemblies that have special importance from the point of view of applications, such as aligned arrays or tubular capillaries (Section 2.3.2). By these and other methods, electrospinning has generated many daughter techniques, some of which are listed here.

*Two-* (or *multi-*)*stream* (or *jet*) electrospinning[96–103] simply consists in using two or more extrusion flows (pipettes, needles, *etc.*) at the same time, thus producing more than one jet. Combining this idea with rotating collectors[98–102] or extra-cylindrical electrodes[97] may help in minimizing the mutual interaction between jets, or at least in stabilizing the jets. Translating collectors can also be utilized to implement depositions.[98] While this method has been initially mainly aimed at increasing the process throughput by exploiting parallel jets of the same polymer solution, a more intriguing advantage has to be recognized in the possibility of depositing different populations of nanofibers simultaneously onto the same collector, thus fabricating multicomponent samples. Indeed, this method is especially interesting to fabricate composite sheets, membranes or matrices of fibers, which could display two or more functional properties that are difficult to obtain by a single species of polymer nanofibers. Typical examples are the lack of satisfactory structural and viscoelastic properties of compounds that on the other hand exhibit electrical conduction or light emission as in the case of conjugated polymers, or well-controlled hydrophilicity, or effective drug-release of embedded molecules.

Also related to the use of secondary electrodes, *focused* electrospinning can be implemented even of single jets by means of electrostatic lenses put along the path of the jet from the spinneret to the grounded collector, with the aim of controlling and limiting the area where nanofibers are deposited.[104]

*Layer-by-layer* electrospinning[98] is a complementary way to generate multicomponent samples, which consists in sequentially depositing different populations of nanofibers onto the same collector. The result is of course a multilayered sample, which can show interesting features of multifunctionality or optical, electrical, biochemical or mechanical anisotropy.

*Electro-blowing*[105] is a version of electrospinning whose set-up also incorporates a gas distributer. In a typical implementation, this can consist of conduits surrounding the spinneret and generating a controlled flow of gas all around the jet. The temperature of the gas flow can be changed as well by appropriate heaters before blowing. Such additional stream can assist the jet by additional shearing forces, thus further pulling and tending to stabilize the extrusion process. Electro-blowing can turn out to be strategic for producing nanofibers by some high-viscosity materials such as hyaluronic acid.[105]

*Reactive* electrospinning[106–110] involves a chemical, often cross-linking, reaction occurring at the Taylor cone or along the jet. Such a reaction can favor the solidification process during the solvent evaporation, or produce a new desired compound (say, *C*) starting from two (*A* and *B*) or more chemically different solutions injected into the spinneret. In this case, the set-up would of course need separated fluidic ducts to control (possibly independently) the injection flow-rates of the different reacting species. The method generally benefits a lot from letting the reagents mix as close as possible to the die, thus avoiding premature reactions, formation of agglomerates and possibly spinneret clogging. Reactive electrospinning also includes techniques in which one enhances the cross-linking while the jet travels towards the collector, by means of an external polymerization agent such as a UV lamp or laser light driving photo-polymerization reactions (*reactive photo*-electrospinning).[107,110] During the polymerization process, a rapid increase of viscosity occurs because of cross-linking. A few relevant equations describing polymerization kinetics are given in Section 3.2. Overall, these methods are usually deployed to obtain nanofibers by materials (*C* according to the above nomenclature) whose solutions would be too difficult to electrospin directly. Demonstrated applications include the fabrication of fibers from highly viscous solutions of hyaluronic acid (Section 7.2),[106] from chitosan or *N*-carboxyethylchitosan and Ag nanoparticles[109] and the reactive photo-electrospinning of polyethyleneimine[110] cinnamate-functionalized poly(-methyl methacrylate-*co*-2-hydroxyethyl acrylate)[108] and poly(2-hydroxyethyl methacrylate).[107] In other cases, in-flight polymerization is instead obtainable even without external cross-linking agents.[111,112]

*Melt* electrospinning produces fibers by driving a polymer through a region when heaters are present to lead the material above $T_g$, thus reducing viscosity at a value low enough to allow spinning.[113] Melt electrospinning exploits therefore in a smart way the phenomenology of glass transition exhibited by polymers as described in Section 1.4.1, and it is a valuable approach especially to spin those thermoplastic compounds that are difficult to dissolve in organic solvents at room temperature. After pioneering reports by L. Larrondo and R. St. J. Manley demonstrating polyethylene and polypropylene microfibers in 1981,[74,114,115] and more recent experiments by the Reneker group,[18] melt electrospinning has been extensively investigated by the group of F. Ko at Drexel University.[113] The influence of the electric field, of *MW* and of polymer tacticity on the resulting fiber morphology has been studied in depth for polypropylene, for which smaller diameters are obtained by using high values of the field ($V/h \cong 1.5 \times 10^6$ V m$^{-1}$), low molecular weights and isotactic compounds.[113] High precision deposition can also be implemented, analogously to conventional electrospinning, by using translation stages.[116] By melt electrospinning, achieving fibers with low ($\leq$ 1 μm) diameters is difficult because of the much larger number of molecular entanglements present per unit volume compared to solutions, and the process can be complicated by the electric field values, which are generally higher than in standard electrospinning, and by the generally needed vacuum conditions. On the other hand, it has to be underlined that reducing or avoiding the use of solvents as in this method can greatly contribute to make electrospinning methods more and more environmentally and biologically benign (although in melt spinning methods some environmental impact can be associated to the higher energy consumption due to heating). Avoiding using solvents can be especially advantageous for realizing biological scaffolds[116] and more generally for tissue engineering applications, as demonstrated in experiments where melt electrospinning is performed directly onto *in vitro* cultured cells, whose growth plate is used as collection target.[117]

*Centrifugal* electrospinning, also termed in some papers double-spinning, includes various kinds of combination with additional centrifugal forces resulting by conventional spinning units.[118–122] Advantages include good alignment of the produced nanofibers, and extra degrees of freedom available in the process concerning the stretching of the jet, which is controllable to some extent by means of the rotational speed of the spinner. Furthermore, this technique allows the nanofibers to be conveniently deposited and aligned on the concave surface of curved samples. *Forcespinning*™[123,124] and rotary jet-spinning (Figure 2.8)[125] have been proposed as well, meaning processes in which only centrifugal forces, without the need of electric fields, are exerted on a solution or on a melt to extrude nanofibers.

*Dual-opposite-spinnerets* electrospinning consists in placing two counter-facing spinnerets connected to opposite electric potentials, provided by high-voltage power supplies.[126,127] The two produced jets are observed to undergo the usual, complex helical whipping while attracting each other, and ultimately to collide and form a cluster of nanofibers. This method is suggested as a simple way to reduce the whipping and bending instability of electrified jets by

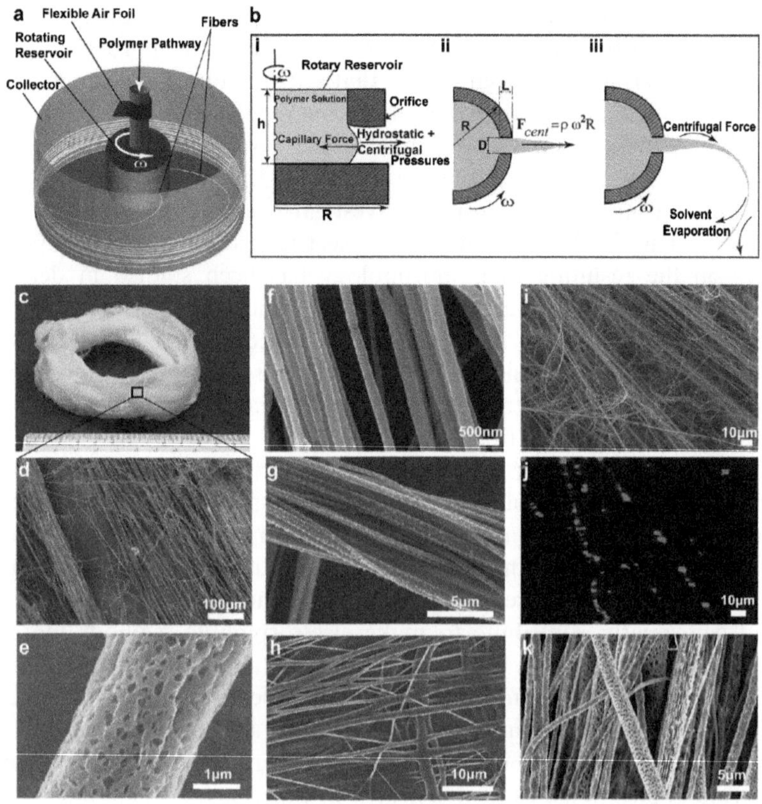


**Figure 2.8** Rotary jet-spinning. (a) A 700 μL reservoir with two side wall orifices having a diameter of 340 μm is continuously fed by a polymer solution, and rotates with angular speed, ω, about its vertical axis. The spun fibers are deposited on the external collector, which is static. (b) Schematics of the fiber formation process, consisting of (i) jet-initiation, (ii) jet-extension and (iii) solvent evaporation. Here, R is the radius of the rotating component, h is the height of the reservoir, and D and L are the diameter and the length of the orifices, respectively. The centrifugal force per solution volume priming the jet is $F_{cent} = \rho\omega^2 R$, where ρ indicates the solution density ($\cong$ 1.5 g $cm^{-3}$). (c) Resulting three-dimensional structure made of poly(lactid acid) nanofibers, spun from 8 wt.% $CHCl_3$ solutions. (d) Corresponding SEM micrograph. (e) Poly(lactid acid) fibers spun from 10 wt.% $CHCl_3$ solutions and with expedited solvent evaporation and relative humidity >55%, resulting in porous nanostructures. (f) PEO fibers spun from 5 wt.% aqueous solutions. (g) Polyacrylic acid (PAA) fibers spun from 8 wt.% aqueous solutions neutralized with sodium hydroxide. (h) PAA fibers spun from 8 wt.% aqueous solutions at 100% neutralization degree. (i) Gelatin fibers spun from 14 wt.% acetic acid solutions at 20% (v/v). (j) Confocal micrograph of fiber-encapsulated, light-emitting, 200 nm PS beads. (k) Fibers spun from an emulsion of gelatin in poly(lactid acid). Reproduced with permission from Ref. 125, *Nano Lett.*, 2010, **10**, 2257–2261. DOI: 10.1021/nl101355x. Copyright © 2010, American Chemical Society.

using highly concentrated polymer solutions, thus achieving a good control on thc final positioning and alignment of the fabricated nanofibers.[127]

Also the so-called *biased-AC electrospinning*,[128] superposing an alternate voltage bias[129] to the usual continuous voltage applied between the spinneret and the collector, is proposed as a way to favor the jet stability. When a purely alternate potential with sufficiently high frequency (say, tens of Hz) is applied, the jets can carry alternating regions with opposite charge, with both attractive and repulsive Coulombic interactions present at the same time throughout the extruded flow and an overall reduction of whipping instabilities.[129] Combining the alternate bias (at typical frequencies, $\nu$, between a few hundreds of Hz and $10^3$ Hz) with a continuous voltage bias baseline can lead to stable electrified jets as well.[128] In reported experiments with PEO solutions, possibly loaded with various types of nanoparticles, and with other polymers, the peak to peak amplitude of the alternate bias (9.3–12 kV) is kept at two to three times the continuous voltage (3.8–5.5 kV with an inter-electrode distance of 7 cm).[128] Dedicated models have been presented to describe AC electrospinning processes.[130]

*Two-phase* electrospinning generally consists in extruding concomitantly two solutions from a single spinneret, namely a polymer solution in an organic solvent and a relatively immiscible aqueous solution, which in turn can contain a water-soluble doping agent such as a protein or a hydrophilic drug.[131] *Emulsion* electrospinning is another way to refer to this concept.[132] A typical reported process includes the preparation of a dichloromethane solution of poly(ethylene-*co*-vinyl acetate) and of a phosphate-buffered saline (PBS) solution of a protein.[131] The two solutions are then mixed together and electrospun, which produces microfibers encapsulating liquid aqueous domains that serve as reservoirs of the loaded protein or drug molecules (Figure 2.9a, b). How much this method can be pushed to produce nanostructures would need to be further investigated. However, it immediately becomes clear that even microstructures realized in this way can be very useful for controlled delivery of drugs (Section 7.4), by reservoirs bursting, by diffusion of chemicals through the polymer walls or by biodegradation of the encapsulating material.[131,133] Electrospun emulsions have also been used in approaches to tailor the fiber morphology.[134] Emulsion-like separations between two components can of course also be observed in phase-separating polymeric blends. This class of co-electrospinning methods, using a single nozzle, is particularly simple to be implemented from the operation viewpoint, and has gathered an increasing interest in the scientific community.[135–142] "Islands at sea" geometries, with mutually disconnected filaments immersed in a continuous matrix along each single fiber, are also obtainable in this way.[137] Fibers with a nematic liquid crystal core and a polymer shell can be electrospun from a solution of 4-pentyl-4′-cyanobiphenyl and polylactic acid in chloroform/acetone,[143] and so on. A dedicated modeling effort has been directed to understand the working mechanisms of compound droplets and core-shell structure formation.[144] This has led it to be concluded that the embedment of the core material within the extruded jet is driven only by viscous traction (electric charges instead rapidly flow to the outer jet/gas interface) and that the presence of a core nozzle suitably positioned to deliver the internal solution (*i.e.* extending outside the shell conduit by about half the outlet radius) greatly facilitates the generation of core-shell nanofibers (Figure 2.9c).[138,144,145] These developments make this technique more and more similar to coaxial electrospinning, which uses concentric needles and is detailed in Section 2.3.1.

The *near-field* electrospinning stems from a 2006 paper by researchers at the Universities of California at Berkeley and of Xiamen.[146] The method was originally developed by the same group[147] and then it encountered increasing success in the years 2009–2011.[148–156] The first motivation for near-field techniques is based on the attempt of increasing the degree of control in positioning single fibers during the deposition stage, which is of course an important prerequisite for production lines integrating nanofibers within devices, especially for opto- and nanoelectronics. Since most of these devices present pre-defined excitation lines in the form of metal electrodes realized by lithography on a substrate, and/or other patterned areas, positioning nanofibers with high spatial accuracy and aligning them with respect to these patterns is crucial to complete the device architectures. In near-field electrospinning, drastically decreasing the distance between the spinneret and the collector (down to the range of 0.5–5 mm) allows the first stage of the electrified jet to be exploited, when the solution flow still follows a rectilinear

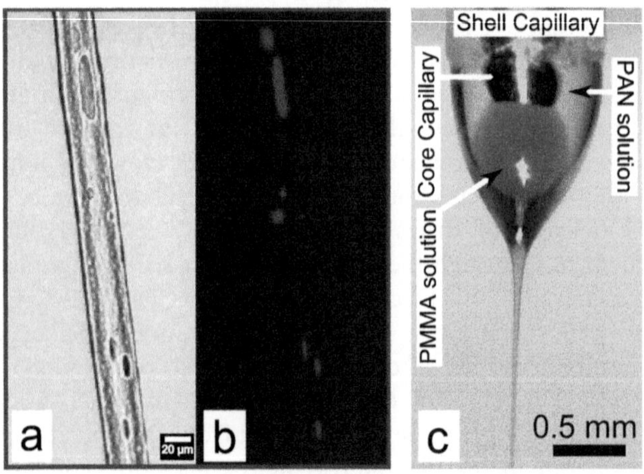


**Figure 2.9** Bright-field (a) and fluorescence (b) micrograph of fluorescently labeled protein encapsulated in reservoirs within a microfiber realized by two-phase electrospinning. Adapted with permission from Ref. 131, *Macromolecules*, 2003, **36**, 3803–3805. Doi: 10.1021/ma0217711. Copyright © 2003, American Chemical Society. (c) Compound droplet as typically formed at the spinneret of a two-phase electrospinning set-up. Adapted with permission from Ref. 145, *Adv. Mater.*, 2006, **18**, 348–353. Doi: 10.1002/adma.200501153. Copyright © 2006, Wiley-VCH Verlag GmbH & Co. KGaA.

path after leaving the Taylor cone. The absence of jet instability would enable a much more accurate positioning of the fibers than in conventional electrospinning,[149] and the deposition precision can be further enhanced by easily integrated, automatically controlled scanning tips,[146,157] or motorized translating collectors, allowing nanofibers to be drawn directly on surfaces as in serial writing. In fact, controlling the relative speed between the spinneret (tip, needle, *etc.*) and the deposition substrate is crucial to determine the resulting pattern of the realized polymer nanofibers. This can vary from linear drawing to a typical buckling pattern upon reducing the spinneret-collector relative velocity down to values comparable to the velocity of the spun jet, in a way substantially analogous to experiments with conventional electrospinning and moving collectors described in Section 2.1.3. Proper combinations of spinneret-collector relative velocities and distances (in the 0.5–10 mm range) also allow tiny non-woven mats to be deposited in a controlled way in determined regions.[155]

In its original formulation,[146] the near-field electrospinning operates without the use of assisting flows to inject the solution to the spinneret, but instead exploits single droplets pending from a tip as in dip pen methods (Figure 2.10). This method would make even easier implementing near-field electrospinning with scanning tips coupled to piezoelectric stages. Subsequent developments by continuous feeding flows involve the application of a sub-critical bias, *i.e.* below the threshold value needed spontaneously to initiate jetting but large enough to trigger a deformation of the terminating apex of the pending droplet, together with an initial mechanical pulling by microtips dipped into the polymer solution near the Taylor cone.[147]

Of course, in near-field electrospinning special attention must be paid in order to achieve solid fibers with sub-100 nm diameter notwithstanding the reduced needle-collector distance. A few tricks allow one to use flow conditions to permit the complete solvent evaporation and the remarkable cross-sectional shrinkage of the electrified jet, which are typical of electrospinning methods, thus somehow compensating the limited length available for the jet path. For example, one can use spinning tips of very small terminal diameter (a few tens of μm), which support minute volumes (down to fL) of liquid in the form of pendant drops. Besides helping by reducing as much as possible the starting diameter of the extruded jet, using sharp spinning tips also enhances significantly the local electric field. More generally, the *enhanced electric fields* (say, $10^7$ V m$^{-1}$) are characteristic of the near-field method because of the short inter-electrodes distance, and influence the process in several ways. On one hand, this can be exploited to reduce correspondingly the applied spinneret-collector bias (say, down to 0.5–1 kV). On the other hand, the orientation of molecular dipoles, which have an angle $\varphi'$ with respect to polymer backbones in turn aligning along the stretched electrified flow (Sec. 2.1.5) and ultimately in the deposited nanofibers, can be effectively improved by the high electric field values, and this effect can be useful for several applications.

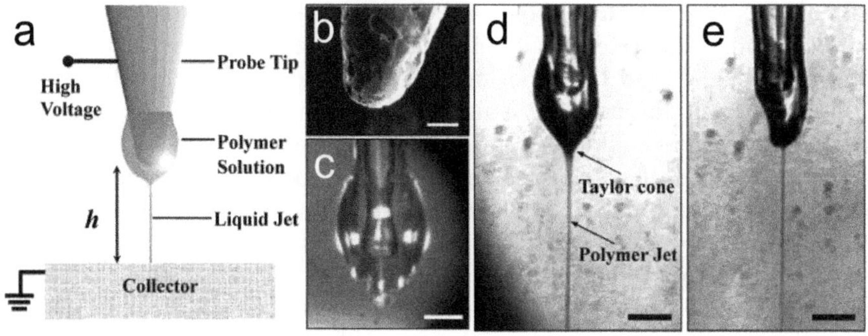


**Figure 2.10** (a) Scheme of near-field electrospinning. (b) SEM micrograph of the used tip. Scale bar = 10 μm. (c) Image of a 50 μm diameter polymer solution droplet pending at the tip. Scale bar = 20 μm. (d) Polymer jet and Taylor cone during near-field electrospinning. Scale bar = 25 μm. (e) Reduction of the volume of the pending droplet occurring as long as the jet carries out polymer solution. Scale bar = 25 μm. Adapted with permission from Ref. 146, *Nano Lett.*, 2006, **6**, 839–842. Doi: 10.1021/nl0602701. Copyright © 2006, American Chemical Society.

Near-field methods can be applied to a variety of systems, such as PEO[146,152] and its blends with titanium tetrabutoxide [Ti(OBu)$_4$], which are used to produce $TiO_2$ fibers upon post-electrospinning annealing,[148] PAN,[154] poly-caprolactone (PCL)[153] and poly(vinylidene fluoride) (PVDF).[151] Of course, this technique also exhibits some drawbacks that make it complementary and not really alternative to conventional electrospinning. In particular, near-field methods are undoubtedly less effective in terms of process throughput and they are not practicable to cover large (> cm$^2$) areas by nanofibers layers. However, advanced automation of the spinneret-deposition plane relative motion promises significant improvements in this respect. In the nanoelectronics framework, interesting results on highly conductive organic fibers written on large areas, by means of electrohydrodynamic printing based on short nozzle-substrate distances and motorized stages, have already been presented at the 2nd "*International Conference of Electrospinning*" in South Korea in 2012.[158]

To conclude this section, we would like to mention that, in addition to methods above, the variety of possible technological variants to implement electrospinning is still very wide. For instance, other possible approaches include various combinations with different nanofabrication and microfluidic techniques,[54] the use of elastomeric micro-nanofluidic architectures for generating multiple jets originating from parallel capillaries[159] or dual-syringe single-jet methods for producing hybrid nanofibers.[160]

## 2.3.1 Coaxial Architectures

Realizing core-sheath (or core-shell) polymer nanofibers can be important for an enormous variety of applications. In nanoelectronics, one can envisage fibers with an electrically conductive core made of a conjugated polymer, or even of a ceramic semiconductor or a liquid metal, and with an insulating sheath. In nanophotonics, core-shell nanofibers can be useful to investigate energy transfer phenomena (Sections 5.1.2 and 5.2.3) occurring at the interface between the internal and the outer material, to tune the resulting emission or absorption spectral features, or to constitute waveguides using two materials with different refractive indexes, or light-emitting or photovoltaic devices within the nanofiber geometry. Furthermore, the concomitant presence of two materials can provide the fibers with particular mechanical properties. Outer shells can be exploited for conferring specific functions, protecting the internal material from mechanical damage or from diffusion of external vapors or other contaminants, controlling the solubility of nanofibers in specific solvents, tailoring its release of drugs, making it more suitable for biological applications or more environmentally friendly, and so on. In tissue engineering, wrapping synthetic polymers within outer shells made of extracellular matrix (ECM) components can drastically improve interactions of nanofiber scaffolds with colonizing cells (Section 7.3). Finally, hollow nanofibers, which have a material of sufficient structural rigidity and strength to support an internal cavity as core, can be employed as nanofluidic channels, for catalysis or as a template of second-generation nanostructures. As a consequence of this extraordinary variety of possible applications, *coaxial electrospinning*, which allows core-sheath polymer nanofibers to be produced in a single run, has attracted enormous interest in the last years. A way to achieve core-sheath nanofibers consists in two-phase or emulsion electrospinning as described in Section 2.3. While this technique is based on emulsions of immiscible solutions and therefore can use a single spinning nozzle, coaxial or multi-axial electrospinning methods use more than one fluidic conduit to deliver polymer solutions at the die, which means that the different solutions do not need to be immiscible.

Coaxial electrospinning has been initially reported independently by several groups in the years 2003–2004.[161–165] A joint work from the teams of E. Zussman and A. L. Yarin at Technion and of J. H. Wendorff and A. Greiner at Philipps-Universität Marburg describes a polymer storage bath with two air inlets that serve to push the solution through two concentric openings and then into a bicomponent Taylor cone.[161] By this set-up, PEO-PEO, poly(dodecylthiophene)-PEO and Pd precursor-poly(L-lactide) (PLA) core-shell fibers are demonstrated.[161] Other reports by I. G. Loscertales *et al.* have also proposed a coaxial spinneret.[162,166] In its early stages of experimental development the coaxial approach has frequently led to jet break-ups, making the method useful to realize core-shell microcapsules and more similar to electrospray rather than to electrospinning for nanofiber production.[166] Later on, a rapid evolution to the form of a true electrospinning technology has been enabled by the use of sol-gel methods, and soon exploited to demonstrate hollow nanofibers by silica and other materials.[162,163] The team of Y. Xia at the University of Washington, now at Georgia Institute of Technology, has investigated in depth the co-electrospinning of immiscible solutions, which helps in reducing the jet instability.[24,163,167–169] Typically, a sol-gel precursor or a polymer such as poly(vinyl pyrrolidone) (PVP) is spun as shell material, jointly with a possibly doped, heavy mineral oil as core, upon a careful optimization of solution concentrations and viscosities. The extraction of the oily cores leads to hollow nanofibers (Figure 2.11a), which can also present several different functionalities, or even particles decorating their internal walls if these dopants are initially loaded in the oil phase.[168] Of course, the external surface of nanofibers can be functionalized as well, similarly to conventional electrospinning methods, by blending dopants in the shell material. Other pioneering experiments have been focused on the realization of core-shell nanofibers of poly(thiophene) and PVP, of poly[2-methoxy-5-(2′-ethylhexyloxy)-1,4 phenylenevinylene] (MEH-PPV) and PVP,[167] of polyacrylonitrile (PAC) and poly(acrylonitrile-*co*-styrene), of poly(aniline sulfonic acid) and polyvinyl alcohol (PVA), of *Bombyx mori* silk and PEO by the Rutledge group at M.I.T.[164] and of PCL and gelatin, PCL and collagen and PCL and bovine serum albumin by the group of S. Ramakrishna at the National University of Singapore.[165,170,171] Self-healing materials such as dicyclopentadiene and isophorone diisocyanate can also be encapsulated,[141] as well as rubbery degradable polymers such as poly(glycerol sebacate).[172]

It is especially noteworthy that extruding shell materials with good electrospinning behavior is also beneficial to assist the internal flow of the core compound, and this can help a lot to use materials (such as many conjugated polymers,[161,167] or metallo-organic precursors) that cannot be spun on their own because of their low molecular weight and insufficient amount of molecular entanglements. The coaxial method can also be useful to electrospin conjugated polymers by simply using an organic solvent, such as chloroform, as outer, rapidly evaporating extruded part, thus supporting the internal polymer jet and assisting the formation of conductive nanofibers.[173,174]

Nowadays (as deduced since early reports),[166] coaxial electrospinning is especially being applied to implement biochemical and biological encapsulation strategies. Core-shell nanofibers are realized with an internal part made of aqueous solutions or of biocompatible polymers, suitable for keeping cells alive or other biomaterials functional, and an external protective sheath layer rendering the overall micro- or nanostructure insoluble in water or better compatible with an organic solvent environment, or tailoring the resulting swelling and diffusional properties for controlled drug delivery. A few examples concerning these applications are described in Sections 7.4 and 7.7.5. The coaxial electrospinning of a PCL shell and an inner core of the oligomer liquid, bisphenol A ethoxylate dimethacrylate together with a photoinitiator has recently been performed, allowing to photo-cure the internal part and to realize nanostructures with superior mechanical properties compared to the corresponding bulk material (Section 4.1).[175]

Finally, coaxial electrospinning is not limited to the use of two fluids. Multifluidic approaches are also possible, allowing electrospun nanofibers

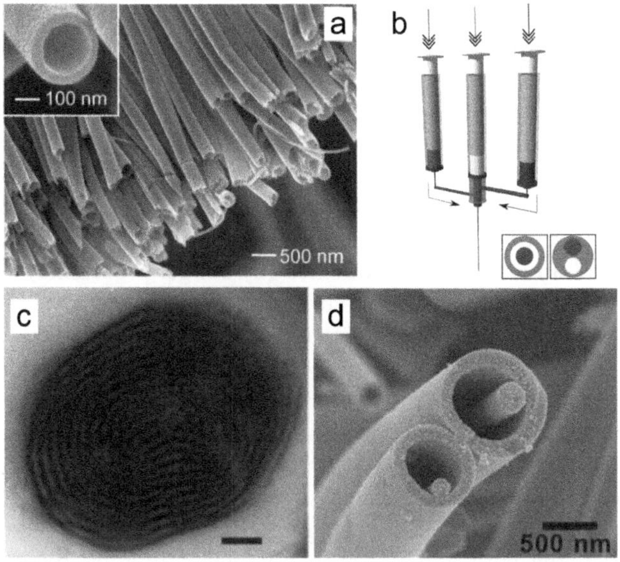


**Figure 2.11** (a) Hollow titania fibers, produced by co-electrospinning a mineral oil as core material (at an injection rate of 0.1 mL $h^{-1}$) and a metallo-organic precursor/PVP blend as sheath. The oil is then removed by octane, and then samples undergo calcination in air at 500 °C to obtain titania. Adapted with permission from Ref. 163, *Nano Lett.*, 2004, **4**, 933–938. Doi: 10.1021/nl049590f. Copyright © 2004, American Chemical Society. (b) Schematics of a spinneret unit for multiflow electrospinning. Coaxial and multi-axial spinnerets lead to nanofibers with different cross-sectional compositions as schematized in the two bottom insets. (c) TEM visualization of the cross-section of a fiber realized by coaxial electrospinning poly(styrene-*block*-isoprene-*block*-styrene) and a poly(-methyl methacrylate-*co*-methacrylic acid) random copolymer with higher $T_g$, and subsequent annealing. Adapted with permission from Ref. 180, *Nano Lett.*, 2006, **6**, 2969–2972. Doi: 10.1021/nl062311z. Copyright © 2006, American Chemical Society. (d) A nanowire-in-microtube structure realized by coaxial electrospinning of three fluids and subsequent removal of the middle spun layer. Used materials are PAN or $Ti(OBu)_4$ sol (inner layer), paraffin oil (middle layer serving as spacer during electrospinning and then removed) and PS or $Ti(OBu)_4$ sol (outer layer), respectively. $Ti(OBu)_4$ sol is calcinated at 450 °C after electrospinning, to yield titania. Adapted with permission from Ref. 182, *Langmuir*, 2010, **26**, 11291–11296. Doi: 10.1021/la100611f. Copyright © 2010, American Chemical Society.

with three or more concentric layers to be realized by means of a spinneret presenting three or more capillaries.[176] The fluidic inlets do not need to be concentric, but instead they can be assembled in a parallel, multi-axial fashion to realize structures with multiple cores, possibly made by different materials and protected by a unique external shell matrix (Figure 2.11b).[177,178]

Both coaxial and triaxial electrospinning give very interesting results when applied to block copolymers. These systems are introduced in Section 1.2.1 and the related patterning and template concepts are detailed in Section 3.3.2. Coaxial and triaxial electrospinning turns out to be a valuable platform for studying the peculiar assembly behavior and lamellar organization of phase-separated layers of these molecules within the cylindrical, confined volume edged by the nanofiber cross-section.[179,180] In particular, the realization of electrospun nanofibers with an internal part made of block copolymers and an external shell of silica (by sol-gel processing) allows these nanostructures to be provided with sufficient thermal stability to go through the annealing needed to induce ordered domain morphologies.[181] Another approach is to encapsulate block copolymers within another polymer having a higher glass transition temperature; in this case the subsequent annealing at temperature values above the $T_g$ of the block copolymer and below the $T_g$ of the shell polymer results in phase separation, up to the formation of periodic internal sub-structures in the fiber cross-section (Figure 2.11c).[180] Nanowire-in-microtubes can be obtained by using an additional middle fluid serving as spacer between the core and the shell polymer solutions (Figure 2.11d).[182] These methods put practically no limit on the kind of polymer nanostructures that one can design, in which two (optically, electrically, biologically, *etc.*) functional layers are both encapsulated by an external sheath and at the same time share an interface, across which the discontinuity of their physico-chemical properties (density, refractive index, optical spectra, electrical conductivity, porosity, chemical doping and so on) can be exploited for fine tailoring of the resulting behavior of nanofibers, thus targeting specific application fields.

## 2.3.2 Collector Geometries and Alignment of Nanofibers

Most of the time, the used electrospinning collector is just a flat, continuous metal surface allowing one an easy contact with one of the poles of the high-voltage generator. Such a deposition plane can also be covered by an electrically insulating material or even be the surface of a liquid. In these cases, nanofibers are generally collected as highly disordered non-woven mats (or simply *non-wovens*) namely as in the form of samples composed of many layers in which fibers are deposited sequentially, without forming mutual entanglements or kinks. A non-woven is therefore defined by a random distribution of fiber orientations in the plane of deposition, superimposed to a layer-by-layer stacking in the direction normal to the collector surface. In principle, if the electrified jet is stably produced and it follows its own trajectory from the spinneret to the collector without interruptions, an electro-spun non-woven can be constituted by a single, extremely long nanofiber, which is deposited with many bends up to form many overlapping layers of nanostructures. These mats can certainly be useful for a lot of bulky applications, for realizing membranes, filtration elements, engineered surfaces for catalysis, biocompatible substrates for tissue engineering, and so on. For some materials, non-wovens can be confined within desired regions of the coated surface, for instance by post-electrospinning photo-lithographic patterning assisted by photo-initiators loaded in nanofibers,[183] or by suitably choosing the relative velocities and distances between electrodes in near-field electrospinning.[155] Importantly, non-wovens can also be conformed to three-dimensional metal collectors in order to produce complex nanofiber-based surfaces and elements, such as fluidic connections, vessels of various types, valves *etc.*,[184,185] which are particularly important for bioengineering allowing some natural constructs of the human body to be mimicked effectively. Another intriguing collection geometry uses grounded tips and allows self-bundled yarns of electrospun nanofibers to be obtained (Figure 2.12).[186] Yarns with fiber twisting can also be produced by rotating the collector surface around an axis perpendicular to the deposition plane,[187] by rotating a collection needle in a multiple jet geometry,[188] by rotating a collecting point electrode with a single electrospun jet,[189] by rotating liquid vortices,[190] by means of a rotary funnel collector [191] or of secondary electrodes allowing the electric field distribution along the path of the jet to be tailored and varied over time,[192] and so on. Grounded tip collectors combined with near-field electrospinning have also been used to assemble hollow, coiled "pottery" structures by polymer fibers (Figure 2.13).[150]

However, for many other applications, some of which are mainly based on single-fiber properties, disordered mats or non-wovens are not really useful samples. For this reason, electrospinning has been engineered in order to achieve some degree of control on the position and on the mutual alignment of the deposited nanofibers.[193] In this regard, the flexibility of electrospinning in terms of geometry of the collection surface and of eventual secondary biased surfaces to be positioned nearby the jet path is an important advantage to

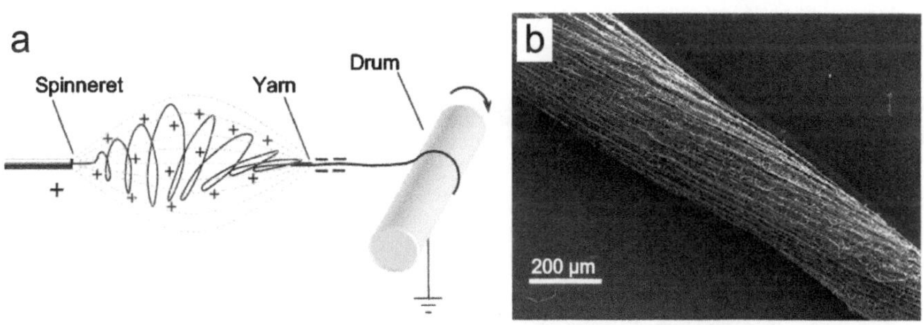


**Figure 2.12** (a) Schematics of the mechanism of the self-bundling electrospinning using a grounded metallic tip. The conductivity of the polymer solution is found to be crucial for this process and preferably ranging from 10 µS cm$^{-1}$ to 400 µS cm$^{-1}$. (b) Exemplary fiber yarn made of poly(3-hydroxybutyrate-*co*-3-hydroxyvalerate) (PHBV). Adapted with permission from Ref. 186, *Polymer*, 2008, **49**, 2755–2761. Doi: 10.1016/j.polymer.2008.04.015. Copyright © 2008, Elsevier Ltd.

obtain desired fiber assemblies.[193] This issue is especially critical when the electrospun nanostructures have to be embedded within the geometry of a photonic, microelectronic, sensing or biodiagnostic device, which very often present pre-defined, lithographically made architectures of metal electrodes, resist patterns, substrate layers, reservoirs, nanomicrochannels, and so on, with respect to which deposited nanofibers have to take a precise position and orientation. In other applications, one needs ordered assemblies of fibers that have to be orientated as uniformly as possible, thus forming uniaxially aligned arrays, or arrays of nanojunctions between mutually perpendicular fibers. Examples are when one wants to use nanofibers for realizing mechanically reinforcing nanocomposites, when macroscopic areas of light-emitting

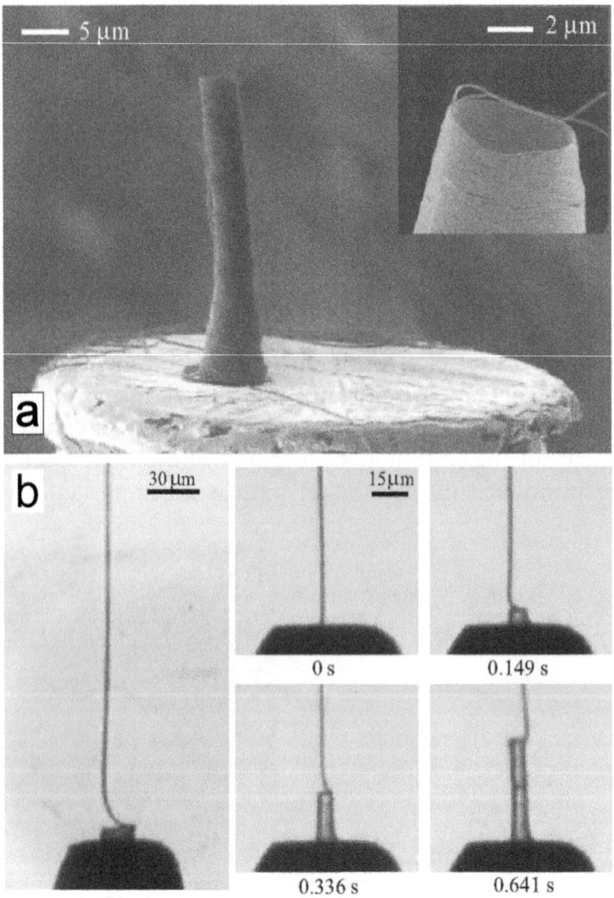


**Figure 2.13** (a) A hollow, coiled structure formed on the apex of a stainless steel conical tip used as collector in near-field electrospinning. (b) High-speed sequential photographs collected during the coiling process. Adapted with permission from Ref. 150, *Nano Lett.*, 2010, **10**, 2138–2140. Doi: 10.1021/nl100824d. Copyright © 2010, American Chemical Society.

nanofibers are required in order to emit uniformly polarized light[194] with sufficiently intense luminescence signal or when oriented arrays are employed to induce a similarly oriented morphology in living cells as in tissue engineering of muscles, neurons, *etc*. (Section 7.7). Aligned arrays of electrospun fibers can improve the proton conductivity and reduce the gas permeability of embedding polymer composites, which can be very useful in order to build proton exchange membranes for fuel cells as demonstrated by sulfonate polyimide.[195] Arrays of nanojunctions made by mutually intersecting and perpendicular polymer nanofibers loaded with reactants have been used as reactors with attoLitre volumes (corresponding to the zeptomolar, $10^{-21}$ mol, scale)[196] and as complex wearable sensors.[197] These extremely miniaturized reaction volumes would correspond to about $10^3$ molecules, and their arrangement in ordered arrays could open new perspectives for rapid screening of chemical reactions in a large variety of industrial applications. One should notice that these issues are relevant not only to the specific case of electrospun nanofibers, but also to elongated organic nanostructures realized by means of any other technology, such as those described in Chapter 3. In all these cases, producing ordered arrays becomes important, as well as evaluating quantitatively the mutual orientation of nanofibers in the arrays to assess fully the properties of morphological order of realized samples.

As a general concept, in order to realize uniaxially aligned arrays or to achieve positioning control, the crucial issue is how one *collects* the nanofibers. At the bare minimum, rapidly crossing the trajectory of the electrospun jet by a solid substrate can allow one to collect a single fiber. If conductive islands separated by a gap are pre-defined on the sample, moving the substrate in a plane roughly normal to the jet axis and parallel to the gap width often allows a fiber to be deposited perpendicularly to the edges of the electrodes on the substrate. This is due to the favorable local electric field distribution near the sample and to the quite large velocity that one has to impart to the crossing substrate ($\cong$ 0.1–1 m $s^{-1}$, which is comparable to that of the jet). However, this approach should be discouraged because of its poor reproducibility and, more importantly, because it cannot be considered good laboratory practice, unless the solid substrate is moved in the region of the electrified jet by means of fully automated and properly isolated motorized stages. In addition, manual approaches only work to position single electrospun nanofibers, without allowing any control on the mutual alignment of many fibers on a surface.

A much better solution has been invented in the Xia group and consists in carefully optimizing the landing conditions of electrospun nanofibers on *patterned* substrates which are kept immobile on the collecting plane.[198,199] The idea that textured collecting electrodes would allow electrospun non-wovens to be patterned and fibers to be oriented dates back to the 1966 patent by Simons.[15] In its modern formulation, the mechanism leading to the alignment of nanofibers in almost uniaxially oriented arrays between electrodes made by parallel conductive stripes has been highlighted in detail in Ref. 198. The orientation direction of fibers is the axis of the gap between the conductive stripes, *i.e.* it is perpendicular to the edges of the electrodes. This can be rationalized by considering the different forces exerted on the almost solidified jet in the ultimate part of its trajectory, immediately before fiber landing. Here, electrospun fibers undergo two classes of electrostatic forces, which are indicated as $F_1$ and $F_2$, respectively, in Figure 2.14. Briefly, $F_1$ is directly related to the electrospinning electric field and to how it distributes in intensity and direction near the collecting electrodes, whereas $F_2$ is given by the attractive interaction between the electric charges on the spun fiber and the corresponding image charges in the collecting electrodes. The net result is that the fiber is stretched across the gap between the two conductive stripes. In particular, numerical simulations show that the alignment of nanofibers improves significantly upon increasing the gap width in the range from 3 to 8 mm.[200]

Gaps between collecting electrodes can be of various types. Air gaps separating parallel metal elements are an obvious choice and result in collecting fibers suspended between the electrodes. However, for some polymers such as PVP, only fibers with diameter above about 150 nm can be effectively deposited on metal masks with a void gap, because the nanostructures need to be strong enough to support electrostatic, gravity and tension forces acting during and after landing.[199] More generally, the value of such a critical diameter for in-gap alignment would of course be closely dependent on mechanical properties at

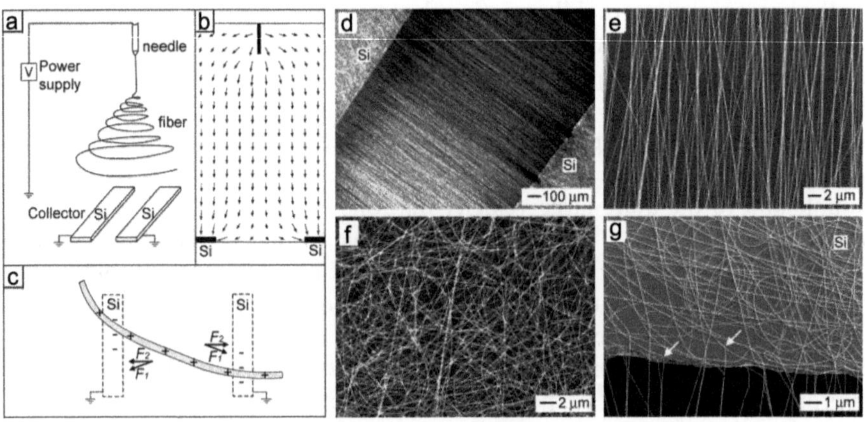


**Figure 2.14** (a) Schematics of electrospinning with parallel conductive stripe (here Si) collectors, used to produce uniaxially aligned arrays of nanofibers. (b) Calculated electric field strength vectors in the region between the spinneret and the stripe collector. The arrows indicate the directions of the electric field. (c) Electrostatic forces acting on nanofibers when landing across the gap. (d, e) Aligned PVP nanofibers electrospun across the gap. (f) The same fibers on the top of the Si stripe, where they take the usual random structure of electrospun mats. (g) SEM micrograph of the edge gap region. Adapted with permission from Ref. 198, *Nano Lett.*, 2003, **3**, 1167–1171. Doi: 10.1021/nl0344256. Copyright © 2003, American Chemical Society.

single-fiber level, stiffer electrospun materials allowing in principle free-standing conditions to be better supported between electrodes. However, lithographically patterned electrodes on solid insulating substrates easily allow this drawback to be overcome. In addition, using electrodes with slightly richer geometries and controlling the sequence of biasing of the different conductive elements allows multilayered architectures of electrospun nanofibers to be realized with increasing degree of complexity (Figure 2.15a–d).[199,201] Indeed, what is likely the most important advantage in orienting nanofibers by electrostatic methods based on collector patterns is that their working principle is really general and not related to a specific geometry. Framed collectors of various shapes can be used as well.[91] Radially aligned nanofibers can be electrospun by means of collectors composed of a central point electrode and a peripheral ring electrode (Figure 2.15e,f),[201] and so on.

A second family of methods for effectively aligning electrospun nanofibers relies instead on a mechanical approach. Specifically, a disk or a cylindrical element is rotated at high angular speed, $\omega$, by means of a mandrel, and used as a collecting surface for fibers.[202–204] Furthermore, a linear motor moving the cylinder parallel to its own longitudinal axis allows the aligned nanofibers to be distributed uniformly on the collection surface. Rotating surfaces of various extension, thickness and radius, $R_{col}$, can be used, ranging from wheel-shaped to cylinders. Disks have also been employed, on which the collecting surface rotates in a plane perpendicular to the extrusion axis, a method that has been combined with parallel trenches, etched in silicon and oriented radially in the rotating surface, to collect nanofibers roughly perpendicular to these stripes.[54,205] In general, rotating wheels tend to assure a higher orientational order in the deposited array of nanofibers, however the overall area of the produced samples is of course relatively small. Metal wires can also be employed, spaced evenly in a circular drum.[206]

In addition to the mutual alignment of fibers, the use of rotating collectors may have other consequences from which the realized samples can benefit. In particular, due to its high speed the collector surface strongly pulls the nanofibers during their landing, thus further contributing to the internal stretching of polymer chains. Larger diameters of the collector, resulting in higher linear speeds ($\omega R_{col}$) of the surface where nanofibers land, may help in this respect determining additional fiber stretching and alignment. Enhancing the anisotropy of the resulting nanostructures, this can be important to achieve improved polarization or piezoelectric performances, or higher tensile strength. Recently, the reduction of the whipping motion during electrospinning, obtained by lowering and making more uniform the electric field near the spinneret, has been proposed as a valid route to improve further the alignment of fibers, in combination with fast-rotating collectors.[207] These collectors are also especially interesting in order to fabricate tubular structures to be used for nerve guidance or in vascular tissue engineering.[98] In this framework, patterned collectors, both rotating and static, can be fabricated or assembled by simple soft lithography approaches such as REM (Section 1.2.1). Patterned

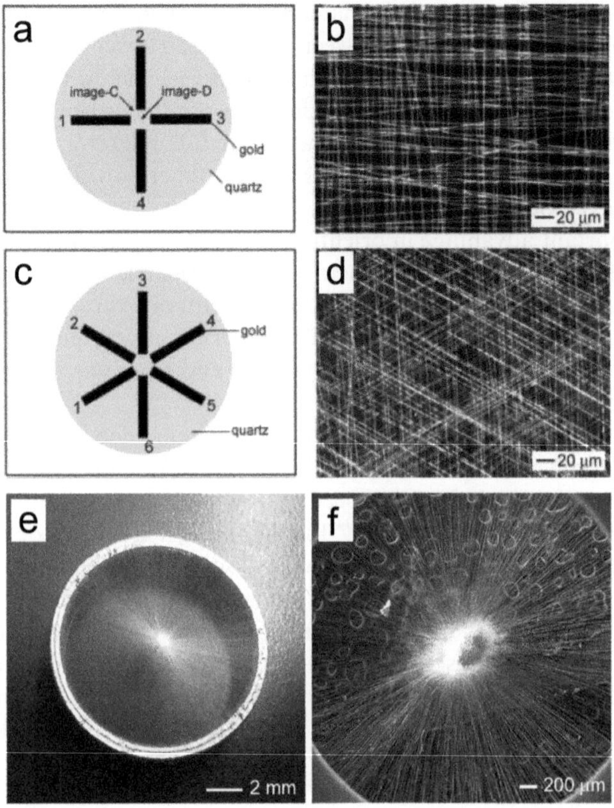


**Figure 2.15** (a–d) Schematics of patterned electrodes with four (a) or six (c) conductive elements, and corresponding multilayered structures of electrospun fibers of PVP (b and d, respectively), imaged by optical microscopy. These structures are fabricated by sequentially grounding each couple of counterfacing electrodes for 5 s. Adapted with permission from Ref. 199, *Adv. Mater.*, 2004, **16**, 361–366. Doi: 10.1002/adma.200306226. Copyright © 2004, Wiley-VCH Verlag GmbH & Co. KGaA. (e) Photograph of radially aligned nanofibers directly electrospun onto a ring collector, and (f) corresponding SEM micrograph. Visible circular features behind the nanofibers are not due to the actual sample morphology but are only dents on the conductive carbon tape. Adapted with permission from Ref. 201, *ACS Nano*, 2010, **4**, 5027–5036. Doi: 10.1021/nn101554u. Copyright © 2010, American Chemical Society.

collectors indeed allow their features to be transferred to the nanofiber aligned array or non-woven mat electrospun on top of them.[184] In this way, patterns with spatial resolution on the scale of 1–100 µm can be easily superimposed to the nanotopography of electrospun nanofibers, which could allow interplay and prevailing effects between micro- and nanomorphological features affecting various physical and biochemical phenomena to be investigated, such as surface wettability, optical properties, cell adhesion and growth *etc.*

At least two approaches exist that allow the landing position of electrospun fibers to be controlled by means of magnetic fields.[208,209] A possibility consists in preparing solutions incorporating magnetic nanoparticles, and hence in directing the so-doped electrified jet by means of externally applied magnetic fields.[208] For example, $F_3O_4$ particles tend to form lines along the magnetic field, and the resulting stress transmitted to the viscoelastic jet during electrospinning force doped nanofibers to align with high accuracy along the same direction.[208] A second method is based on two parallel strips of non-conductive ferrite magnets used as collector,[209] or as alternative on two bar magnets mutually separated by a gap and also electrically insulated from the usual metal-surface collector.[210] This approach allows aligned nanofibers to be produced even without magnetic nanoparticle dopants. The fibers align perpendicularly to the edge of the gap (Figure 2.16), similarly to what one can obtain by parallel conductive electrodes, but the underlying physics is different. If oriented perpendicular to the deposition plane, a magnetic field can also be used to decrease the radius of bending instability.[23] Here, the orientation mechanism can be rationalized by depicting the jet carrying its continuous current, $I$, as an electrically conductive wire that is forced to bend during its complex motion in the region where the magnetic field is present. It has been pointed out that a peculiar advantage of orienting methods based on magnetic fields, over electrostatic methods using for instance parallel conductive wires or pads as described above, is the possibility of retaining with high accuracy the direction of preferential alignment also for thick deposited mats of nanofibers.[210] In fact, by electrostatic alignment methods, long electrospinning runs produce samples with a gradient of orientational order in planes parallel to the collector surface, varying from uniaxially aligned arrays to practically random orientations while the assisting electrostatic field is gradually screened by deposited nanofibers. Some arrangements of electrodes exist that allow this shielding effect to be reduced as is also described in Section 2.4.[193,206,211]

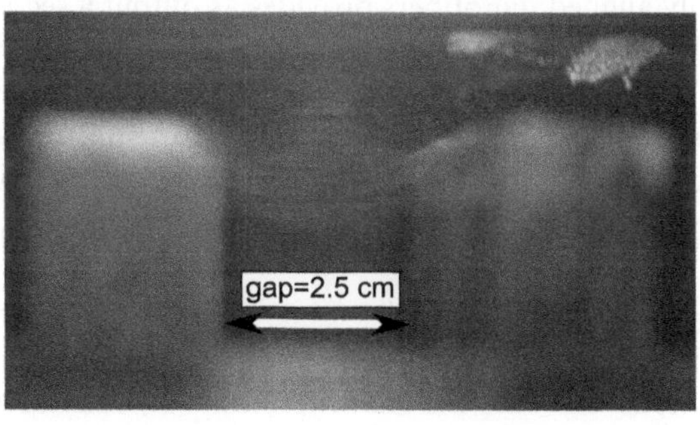


**Figure 2.16** Photograph of electrospun fibers aligning across the gap between two non-conductive ferrite magnets. Adapted with permission from Ref. 209, *J. Appl. Polym. Sci.*, 2008, **110**, 3368–3372. Doi: 10.1002/app.28896. Copyright © 2008, Wiley Periodicals, Inc.

Magnetic methods allow instead membranes up to many tens of μm thick to be easily obtained without loss of orientational order,[210] which can be an important advantage especially for producing reinforcement mats, photonic devices or scaffolds for tissue engineering benefiting from high anisotropy.

Concerning the characterization of the obtained arrays and of the degree of order achieved in them, a simple approach is investigating the sample morphology by SEM, and then extracting the usual order parameter, $S$, by image analysis:

$$S = \frac{1}{2}\left(3 < \cos^2 \varphi > -1\right). \tag{2.10}$$

In the previous expression, $\varphi$ indicates the angle between each single fiber of the array and a main axis identified as principal orientation direction of the sample (on the basis of the details of the fabrication process), and the <...> brackets indicate the average over the ensemble of fibers analyzed. For example, when parallel conductive stripes are used to collect an array of nanofibers, the principal orientation axis, with respect to which $\varphi$ is calculated, corresponds approximately to the direction perpendicular to the edge of the electrodes and coplanar with the collecting substrate (*i.e.* to the inter-electrode gap axis). The reader may note of course that the previous formula is identical to that used for quantifying the degree of alignment of polymer chains *within* a single nanofiber (Equation (2.7a)). Here, one has that for an ideally, perfectly oriented array, $S = 1$ and for a fully disordered mat, $S = 0$. Values of the order parameter, $S$, closely approaching unity (*i.e.* $> 0.9$) can be obtained for arrays of electrospun polymer nanofiber by means of a variety of methods as explained above. Methods such as fast Fourier transform (FFT) analysis [212,213] also greatly help to determine quantitatively the degree of mutual alignment of nanofibers in the array with highly automated procedures. Fourier analysis converts functions defined in the space domain into transformed functions in the frequency domain, and FFT algorithms are available in free software for image processing. Two-dimensional FFT performed on a SEM micrograph displaying roughly aligned nanofibers provides as output a second image made of gray-scale (or colored according to an intensity scale) pixels. The angular dependence of the brightness intensity of the FFT-generated pictures quantitatively indicates the percentage of fiber orientation at a particular angle (in the original micrograph). An example is given in Figure 2.17.

### 2.3.3 Use of Composite Solutions

Composite solutions or suspensions exploiting (i) mixtures of different solvents, (ii) blends of different polymer materials, (iii) low-molar-mass molecules as dopants or (iv) dispersed inorganic particles such as semiconductor or oxide nanocrystals are largely used in electrospinning. In particular, adding nanoparticles in the solution to be electrospun, together with the processed polymer, is a direct approach to obtain nanocomposite fibers.[214] These materials have an enormous variety of applications, ranging from fiber positioning control by magnetic dopants as described in the previous section,[208] to emission color tuning,[215–218] and to tissue engineering, especially concerning bone and cartilage reconstruction (Section 7.7.3). Various kinds of quantum dots and other nanoparticles (ZnSe, ZnO, CdSe/ZnS, Ag, $Fe_3O_4$, β-tricalcium phosphate),[208,215,219–223] nanorods (Au)[224,225] and nanowires (CdS),[226] and also latex particles[227] dispersed in solutions have been embedded within electrospun polymer nanofibers. However, the nanofabrication process can be disfavored by a number of possible difficulties related to particle aggregation in the spun solution, clogging of the spinneret, disuniform distribution of the dopants within the final nanofiber samples, *etc.* For these reasons, a careful optimization of the employed dispersion and

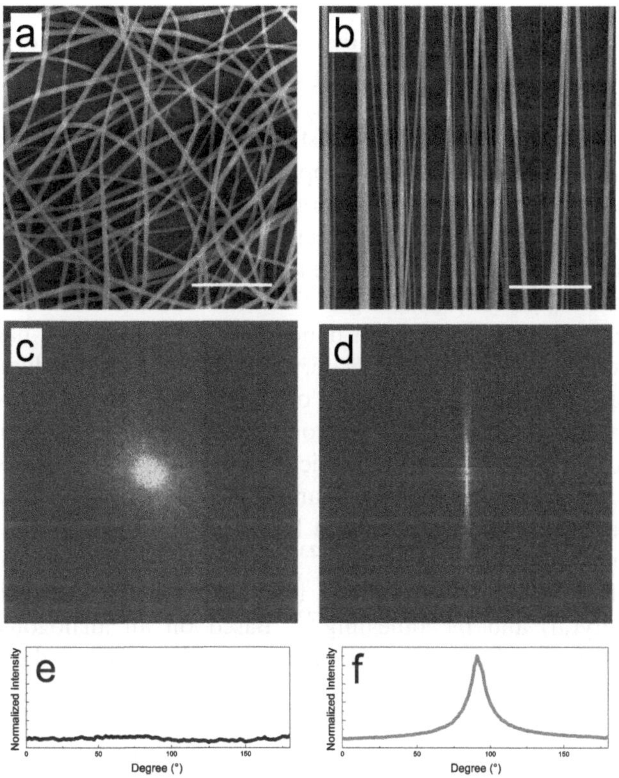


**Figure 2.17** Examples of SEM micrographs of random (a) and of aligned (b) nanofibers, and corresponding FFT output (c and d, respectively). The bottom panels (e and f) show the corresponding, radially integrated pixel intensity for each angle. Random and aligned ensembles of nanofibers respectively produce a normalized FFT intensity, which is (e) angle-independent and (f) strongly peaked (here at about 90°). Scale bars in (a) and (b): 10 μm.

process parameters is needed. Furthermore, a few pieces of experimental apparatus, some of which are specifically designed, have been proposed to circumvent these problems. For instance, electrospinning in its several versions can be combined with a single-screw[113] or with a twin-screw extruder with fully inter-meshing and co-rotating screws,[221] working in a mixing unit that is terminated by a biased spinneret die. The system could include injection and feed ports to introduce the liquid and solid components to be extruded, and it is configurable to enable the simultaneous conveying, melting, mixing of the components, shaping, *etc.* (Figure 2.18).[221] This approach, smartly combining complementary technologies, allows the dispersion of nanoparticles in the polymer matrix to be improved and has been applied to realize graded composite nanofiber scaffolds for tissue engineering (Section 7.7.3). By using coaxial or emulsion electrospinning methods (Sections 2.3 and 2.3.1) nanoparticles can be confined within one or more layers of the polymer nanofiber cross-section, and this degree of control on the particle placement within nanofibers can be enhanced by using copolymer systems organizing in lamellar layers, as demonstrated for magnetite nanoparticles in poly(styrene-*b*-isoprene) electrospun fibers.[179]

A way to circumvent problems of disfavored rheology and possible needle clogging related to the presence of nanoparticles in solutions is through *in-situ* synthesis, namely the generation of inorganic nanocrystals inside the polymer nanofibers only after the latter have been obtained. To this aim, one can use suitable molecular precursors, which are added to the electrospun solution and then decomposed by post-processing of fibers. This allows, for instance, light-emitting, PMMA-based electrospun nanofibers embedding CdS nanocrystals to be achieved, using cadmium thiolate precursors and then inducing crystal nucleation and growth thermally[228] or by means of electron-beams.[229] Alkoxide precursors to nickel and iron oxides have been electrospun in solutions with PVP to fabricate magnetic nanofibers by subsequent hydrolysis and calcination at 550 °C.[230] PVA nanofibers containing antibacterial Ag nanoparticles have been obtained from PVA/$AgNO_3$ nanofibers by a simple heat treatment or by photo-reduction.[231,232] Magnetically anisotropic nanofibers have been realized by spinning iron nitrate and cobalt nitrate with poly(vinyl butyral) and by annealing.[233] Based on an analogous concept, *in-situ* approaches to obtain nanocomposite nanofibers are based on exposing electrospun samples to a suitable gas or solution inducing the formation of nanoparticles. For instance, nanofibers electrospun from PVP/lead acetate ethanol/water solution and then exposed to $H_2S$ result in nanocomposites embedding PbS nanoparticles.[234] A similar process can be applied on PEO/cadmium acetate to obtain nanofibers doped by CdS nanocrystals.[235] Other versions of these methods are described in Section 3.3.

Finally, among other classes of nanoparticles, carbon nanotubes and graphene nanosheets deserve special attention for a number of reasons. They can be used to improve the conductivity of nanofibers, to enhance their mechanical strength and to serve as nanoscale templates inducing a higher

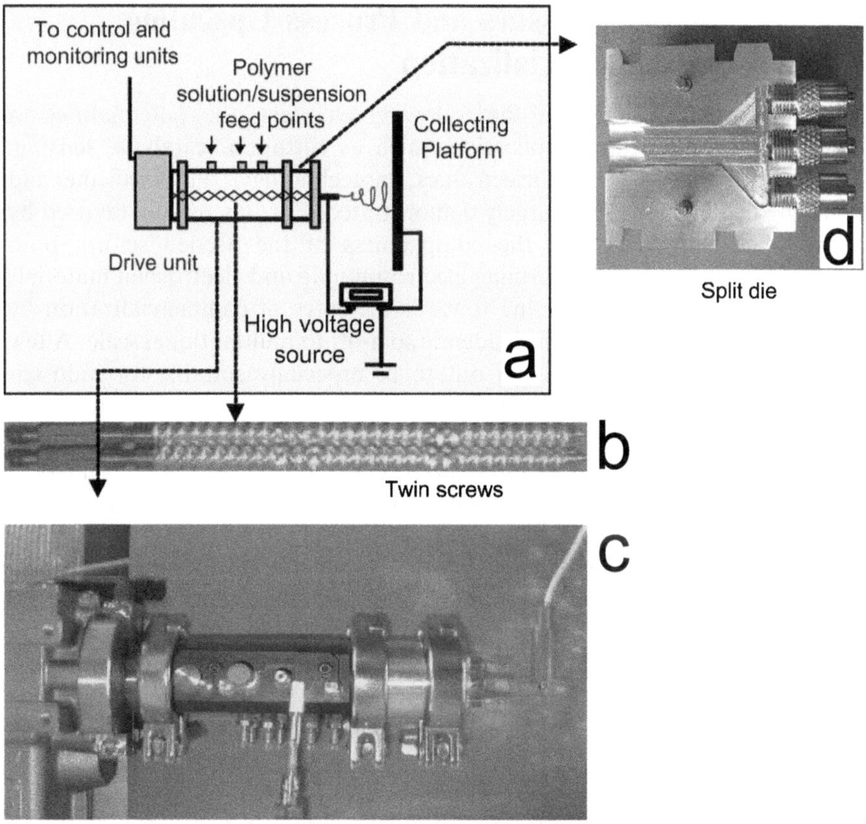


**Figure 2.18** Schematics (a) of the apparatus for twin screw extrusion combined to electrospinning for the production of composite nanofibers (features are not to scale). (b–d) Photographs of the twin-screw system (b), of the extrusion unit coupled to the spinneret (c) and of the die (d). Adapted with permission from Ref. 221, *Nanotechnology*, 2008, **19**, 165302. Doi: 10.1088/0957-4484/19/16/165302. Copyright © 2008, IOP Publishing Ltd.

orientation of polymer backbones during electrospinning. Achieving uniform dispersions is of course crucial to this aim, and the interfacial interaction with the polymer matrix can be improved by proper surface functionalization of the carbon nanofillers. Many procedures have been reported in the literature for both single-walled carbon nanotubes (SWCNTs)[236] and multi-walled carbon nanotubes (MWCNTs),[189,237,238] a few of which allow high concentrations (up to more than 30% in weight)[237] of well-aligned carbon nanotubes within electrospun polymer nanofibers to be reached. Graphene has been used as filler as well, for applications in ultra-fast photonics.[239]

## 2.4 Technological Issues and Process Upscaling Towards Industrialization

As described in this and in the following chapters, the potentialities of nanofibers for important applications such as filtration, catalysis, sensing, textiles, nanophotonics, nanoelectronics, biotechnology, nanomedicine and tissue engineering are today largely demonstrated. For this reason, favored by the technical simplicity and the compactness of the needed set-up, both dedicated equipment for performing electrospinning and electrospun materials and products are rapidly moving towards the stage of commercialization by many companies, ranging from academic spin-off to multinational scale. A few electrospinning technologies turn out to be especially suitable for industrialization, allowing the reproducibility and precision of the fabricated nanostructures to be improved or increasing the production throughput considerably, thus offering solid perspective for up-scaling. Near-field electrospinning (Section 2.3) is of course the leading strategy to achieve high precision, however other methods mentioned in the previous sections can also certainly be useful to achieve some degree of control in positioning nanofibers onto pre-defined, possibly patterned substrates to realize devices.

Of course, precise fabrication and deposition is just one issue. Other aspects to be considered carefully should include process safety, environmental impact and energy consumption as related to the use of volatile, often flammable compounds and of high voltage biases, effective and function-preserving packaging of the realized products, and so on. Quality control during nanomanufacturing is another important issue.[240] Electrospinning is largely (though not completely) free from many such complications.

Following the excitation raised by this technology at the academic level, the number of patents on electrospinning has been continuously increasing through the years (Figure 2.19a). More than 1,000 patent items have been filed in the years from 2000 to 2012.[241] Of these, almost 75% are international applications (Patent Cooperation Treaty, PCT), whereas the remainder are mainly European (more than 16%) and South Korean (more than 7%) (Figure 2.19b). Top patent applicants include multinationals such as DuPont (www.dupont.com) and Donaldson (www.donaldson.com), and leading universities and research bodies. Most of such a huge patent portfolio targets the development of nanofiber-based filtration media, which are described in detail in Section 4.3.1, or of improved production hardware and techniques. We recall that the first industrially produced electrospun fibers were realized well before electrospinning gathered the interest of Western academic researchers and scientists. Large-scale plants for the production of electrospun materials for gas filtration were installed in URSS in the 1950s (Section 4.3.1). Commercial products based on fibers with sub-μm diameter have been introduced since the early 1980s in the United States by Donaldson,[42] and so-called microdenier fibers (Section 4.3.3) have been spun by DuPont since 1989.[29]

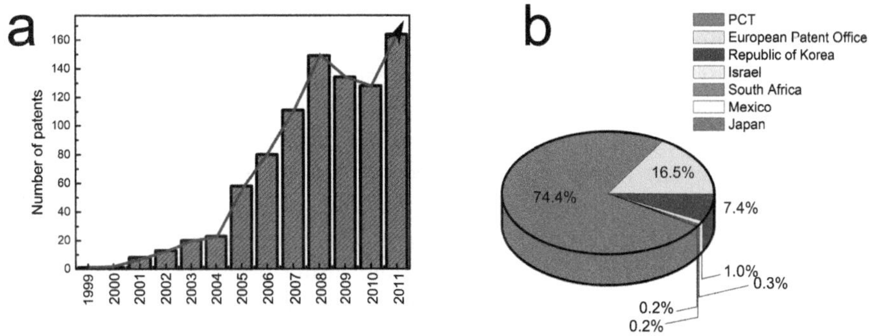


**Figure 2.19** (a) Statistics of patents about electrospun polymer nanofibers in the years 1999–2011. (b) Items per country (years 1999–2012). PCT: Patent Cooperation Treaty. Source: WIPO PATENTSCOPE (keywords: "Polymer" and "Nanofibers" and "Electrospinning").[241]

## 2.4.1 Production Yield and Reproducibility

A peculiar feature of electrospinning is its matchless high throughput compared to other nanofabrication technologies. This is indeed critically important in order to have actual possibilities of industrial exploitation of the realized nanostructures, because most applications of course require high production volumes, which in turn help in reducing the specific cost associated to fabrication. For a lot of nanotechnological and synthetic routes of production of nanoparticles, up-scaling is challenging, and yields can decrease upon increasing processed volumes. These factors strongly limit the transfer of such technologies to industry. On the other hand, current methods to realize conventional synthetic fibers have reached a tremendous production capability, up to hundreds of tonnes daily, which are not, and will likely never be, accessible to nanomanufacturing.

The production yield of fiber spinning technologies can be quantified according to several possible figures of merit. One consists in the volume of polymer solution that can be processed under continuous runs. A typical academic experiment for optimizing an electrospinning process or for producing a limited number of samples lead to areas of, say, 1–10 cm$^2$ to be coated by nanofibers, limited by the used collecting surfaces and by the deposition region accessible to jet bending and whipping. These experiments are generally performed with solution volumes in the range of milliliters delivered by syringes. However, even at laboratory scale, by properly configuring the delivering fluidic circuits electrospinning can allow up to several liters of solution to be processed without interruptions. Being in continuous operation over time intervals as long as possible is of course mandatory for up-scaling the process at pre-industrial and ultimately industrial level. For electrospinning on static or rotating collectors with deposition areas of the order of 10–10$^4$ cm$^2$, the process duration is usually limited by the thickness of the deposited samples, which by shielding the electric field can compromise the jet stability. Incidentally, special care has to be paid when one fabricates uniaxially aligned arrays of nanofibers, because many alignment methods (Section 2.3.2) tend to become ineffective upon increasing the sample thickness. These drawbacks could, however, be easily overcome by using moving collectors on conveyor belts,[242,243] which allows surfaces of many $m^2$ to be coated by keeping low enough the thickness of electrospun non-woven coatings. Other potentially scalable methods proposed for the realization of aligned electrospun nanofibers are based on wire drum collectors[206] and on three-pole configurations using a blade-cage collector set.[211] At larger scale, the limitation to continuous process operation is therefore mainly set by the volumes of the reservoirs containing the solution to be electrospun, and by the processing capacity of the motorized fluidic units delivering the solution to the spinneret.

Another possible way to quantify the throughput of electrospinning uses the amount of polymer material deposited in the form of nanofibers within a given time. Conversions between grams of fibers deposited and area ($m^2$) of substrate coated in the same time are quite easy to make, reasonable values of surface mass of electrospun nanofibers coatings being of 0.5–1 g $m^{-2}$. Hundreds of g $m^{-2}$ can be achieved by specifically optimized processes for fabricating non-wovens of high thickness. Concerning more specifically the process throughput, in principle a single electrified jet fed by a flow rate, $Q$, of 50 μL $min^{-1}$ (corresponding to about 60 mg of solution per minute) deposits polymer nanofibers at rates of the order of $10^{-3}$ g $min^{-1}$ (*i.e.* tens of g per day). This can be increased, at single jet scale, in free surface electrospinning or by using higher rates of the flow delivered to the spinneret by the pumping system. However, above some solution-dependent threshold, high $Q$ values can cause unstable jetting and needle clogging due to polymer accumulation at the die, which is the reason why flow rates have to be carefully optimized for each particular electrospun polymer and set of other processing parameters. Other estimates, reported for a four-jet apparatus, are of the order of 1–2 mg of solution extruded per minute, corresponding to an amount of deposited polymer nanofiber of the order of 0.1–0.2 mg (*i.e.* 0.1–0.2 g per day).[101] Twenty holes with 500 μm diameter drilled along two rows into the wall of a 130 mm long plastic tube, used as electrospinning spinneret, allow a throughput of 7–12 g per day to be reached.[244] A report in the early 2000s by researchers at Donaldson mentions 0.2–5 kg per day as throughput of electrospun nanofibers for non-woven mats to be embedded within cartridge dust collectors for air filtration, which corresponds to web surfaces larger than $10^4$ $m^2$ produced daily.[245] Nowadays, several types of apparatus for electrospinning at pre-industrial and industrial scale are commercially available, which can achieve comparable performances. For instance, Inovenso Ltd. (www.inovenso.com) supplies an apparatus that has a throughput of approximately five kg per day, on coated areas up to $5 \times 10^3$ $m^2$ per day.[246] A production of electrospun material up to the equivalent of 6.5 kg per hour (*i.e.* many tens of million $m^2$ per year) is reported for industrial plants realizing gas-filtering elements during the 1950s–1960s in the former USSR.[247,248] Other examples of production facilities, some with production capability of this same order, have been reported.[248,249]

Of course, as mentioned above multiple jets (Section 2.3)[96–102,249,250] can be produced by using many needles or pipettes in parallel, and directed onto the same collector to increase the deposition rate of non-woven coatings. However this approach can sometimes be operationally inconvenient, because of the related design complications, of interferences that can occur between jets (minimizable by inserting suitable secondary counterelectrodes near the paths of the jet) and of the higher probability of needle clogging by highly concentrated polymer solutions.[251] For these reasons, alternative methods to increase further the process throughput are certainly desirable. For instance, researchers at Terronics Development Co. (www.terronics.com) overcome the difficulty of combining many single-jet electrohydrodynamic atomizers by using a single liquid surface showing many Taylor cones and from which multiple jets are simultaneously achieved.[252] Along this strand of research, A. L. Yarin and E. Zussman report on multiple upwards jets in a needleless electrospinning set-up based on a bilayer of mutually immiscible liquids in a Pyrex dish.[251] In general, upward geometries have considerable advantages because they avoid droplets of polymer solution or polymer aggregates at the spinneret falling down on the sample during production, as would be in the case of downward jets. This is relevant because it has as a consequence an overall increase of defect-free samples realized in a given time, namely a higher throughput. A ferromagnetic suspension underneath the polymer solution is used to produce a stationary pattern of spikes by means of an external magnetic field, thus favoring the extrusion of multiple jets towards a suspended counterelectrode, which are produced when an electric field is also switched on. Applied to usual PEO solutions in ethanol/water mixtures,[251] this method allows tens of simultaneous jets per cm$^2$ of liquid surface (which is generally much denser than in other multiple-jet approaches mentioned in Section 2.3) to be obtained, thus leading to a significant improvement of throughput (by a factor of 12) compared to previous estimates for a set-up relying on separated nozzles.[35,65] In general, protrusion accidentally or intentionally produced (not only by means of ferromagnetic suspensions and magnetic fields, but also by metal spikes, bubbles, *etc.*) at the liquid–air interface are of course natural onset points for electrified jets. This has become the basis of the so-called needleless or free-surface electrospinning,[253] which has nowadays become a research field in itself and widely applied in equipment available on the market. The Nanospider™, developed and commercialized by the Czech company Elmarco (www.elmarco.com), produces non-wovens composed of fibers of various different polymers, with mean fiber diameters of 200 nm (standard deviation ~30%) with a throughput of up to 50 million m$^2$ per year.[254] This class of technologies, based on needleless electrospinning, has used a rotating (smooth or spiked, Figure 2.20a) cylinder wet by the polymer solution in a fluid reservoir, which serves to prime the formation of multiple, upward electrified jets.[253,255] The technology has been further developed in the last years (Figure 2.20b). It would be particularly interesting to test the applicability of this approach to many different materials, including compounds with relatively unfavorable viscoelastic properties such as conductive and light-emitting polymers, for which needle clogging is an especially relevant technological issue. Limitations of free-surface methods include the possible presence of defects deposited on samples and the eventually significant dispersion in size of the fabricated fibers (both issues, however, largely present even in conventional electrospinning and to be minimized by proper process optimization), together with an often reduced capacity of production of multicomponent polymer nanofiber structures, such as core-shell architectures.

**Figure 2.20** (a) Various spinning electrodes for free-surface electrospinning. Adapted with permission from Ref. 255, *MRS Proceedings*, 2009, **1240**, 1240-WW03-07. Doi: 10.1557/PROC-1240-WW03-07. Copyright © 2011, Cambridge University Press. (b) Modern development of the Nanospider$^{TM}$ technology. (a, b) Courtesy of Elmarco s.r.o.

Among other commercially available electrospinning systems, a centrifuge unit with three spin heads has been developed by Dienes Apparatebau (www.dienes.net),[256] exhibiting throughput much higher than conventional nozzle-based electrospinning. Finally, as described above the importance of using climate-controllable process chambers for optimizing electrospinning process parameters and improving reproducibility deserves to be recalled in this framework. Among many other home-made and commercially available examples, the company IME Technologies develops a controlled cabinet housing air conditioning system, electronics, *etc.* The processing chamber is designed to provide a $\pm 0.5$ °C accuracy in the control of temperature (between 20 and 45 °C) and $\pm 1\%$ in the control of humidity (between 10 and 90%).[257] These are reasonable accuracies and ranges in order to look for good sets of parameters to carry out electrospinning with desired polymer or blend systems at laboratory research level. More generally, the competition among different suppliers of spinning units, accessories and technologies is increasing rapidly following the promising market perspectives offered by research and production. In addition to the aforementioned enterprises, involved companies include Electrospinz (www.electrospinz.co.nz), Electrospunra (www.electrospunra.com), E-Spin Nanotech (www.espinnanotech.com), FibeRio (fiberiotech.com), Holmarc Opto-Mechatronics (www.holmarc.com), Kato Tech (www.keskato.co.jp), Linari Engineering (www.linaribiomedical.com), Mecc Co., Ltd. (www.mecc.co.jp), NaBond (www.electro-spinning.com), Nanostatics, Toptec (www.toptec.co.kr), Xanofi (www.xanofi.com) and Yflow (www.yflow.com). Upwards jets, rotating drums and winding-unwinding systems are all widely used elements of industrial electrospinning equipment. Two representative examples of commercial apparatus from Inovenso Ltd are shown in Figure 2.21. Companies involved in electrospun products are widely spread geographically and include, in addition to those previously mentioned,

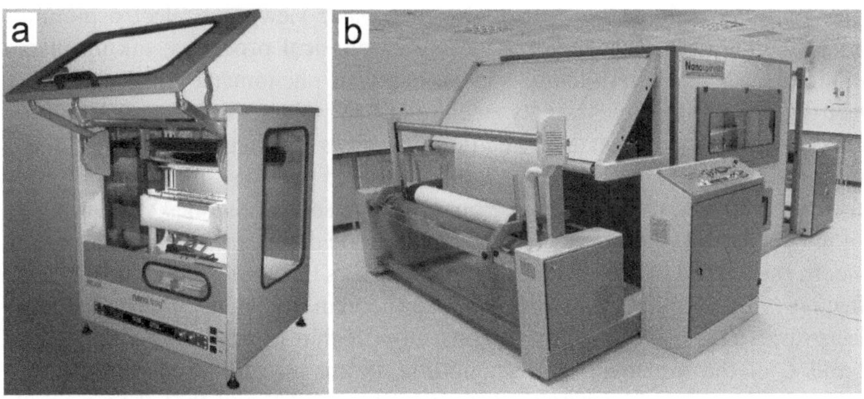


**Figure 2.21** Examples of laboratory-scale (a, NE300) and large-scale (b, Nanospinner416)[246] electrospinning production equipment, produced by Inovenso Ltd. Courtesy of Inovenso Ltd.

Ahlstrom Corporation (www.ahlstrom.com), Espin Technologies (espintechnologies.com), Esfil Tehno AS (www.esfiltehno.ee), Finetex Technology (www.finetextech.com), Hemcon Medical Technologies, Inc. (www.hemcon.com), Japan Vilene (www.vilene.co.jp), Johns Manville (www.jm.com), Kertak Nanotechnology (www.kertaknanotechnology.com), Nano109 (www.nano109.com), Nanofiber Solutions (www.nanofibersolutions.com), NanoSpun (www.nanospuntech.com), PolyNanoTec (www.polynanotec.com), SNS Nano Fiber Technology (www.snsnano.com) and Teijin Fibers, Ltd (www.teijinfiber.com). Other companies are listed in Section 4.3.1.

## 2.5 Process Theory and Modeling

The electrospinning process being so complex, predicting the most suitable parameters in order to obtain a desired nanofibers architecture is generally quite hard, and most experiments are still carried out on an empirical basis. It has often been observed that fully solid predictions for, say, spun nanofiber diameters are not available so far.[71] No theory exists to date that takes into account all the variables affecting electrospinning. For these reasons, notwithstanding many elegant and valuable modeling approaches that have been formulated through the years, continuing to investigate electrospinning from a theoretical point of view could still be strategic in perspective, possibly leading to further improved tools to achieve reliable predictions of the geometry, internal structure and composition of nanofibers resulting from each specific choice of the values of process parameters. At the bare minimum, modeling should aim to provide experimentalists with a likely starting point for calibrating processes, allowing one significantly to save time before subsequent optimization stages (most probably needed in any case). However, modeling is also fascinating and, in some respects, even more important to rationalize the fundamental physics of polymer solutions under intense electric fields and elongational conditions. From a basic viewpoint, electrospinning can be a unique test bench for a variety of physical processes, taking place simultaneously and interplaying, such as capillary phenomena at free surfaces, electrokinetics and viscoelasticity phenomena in polymer solutions. All these processes possibly exhibit relevant non-linearities in micro- and nanofluidic environments, which have been very rarely studied in electrospun systems.[59] In all cases, as introduced in the previous sections, several regimes of instability are underwent by electrified jets, in which electric charges play a fundamental role. In this regard, Reneker, Yarin and coworkers have proposed a simple and effective sketch to illustrate Earnshaw's theorem for point charges within electrospun jets (Figure 2.22a).[35] Let us suppose that three electric charges, $A$, $B$ and $C$ of the same sign and magnitude, $q$, are initially aligned along the longitudinal axis of the jet (with distances, $AB = BC$), and that then a small perturbation induces a lateral shift, $\delta d$, of the central charge, $B$. The intensity of the resultant of the two repulsive Coulombic interactions acting on $B$ by $A$ and $C$ is then $2\cos\Theta \times (q^2/4\pi\varepsilon_0\varepsilon_r d^2)$, where $\Theta$ indicates the angle between the perturbed *AB* and *CB* directions and the perpendicular to the jet axis, and *d* is the distance between *A* and *B* and between *C* and *B* ($d\cos\Theta = \delta d$), as shown in Figure 2.22a. The expression for the resultant force intensity also reads $2\delta d \times (q^2/4\pi\varepsilon_0\varepsilon_r d^3)$, hence the motion equation for *B* can be written as $m\mathrm{d}^2(\delta d)/\mathrm{d}t^2 = 2\delta d \times (q^2/4\pi\varepsilon_0\varepsilon_r d^3)$, where *m* is the mass of the particle. A solution of the

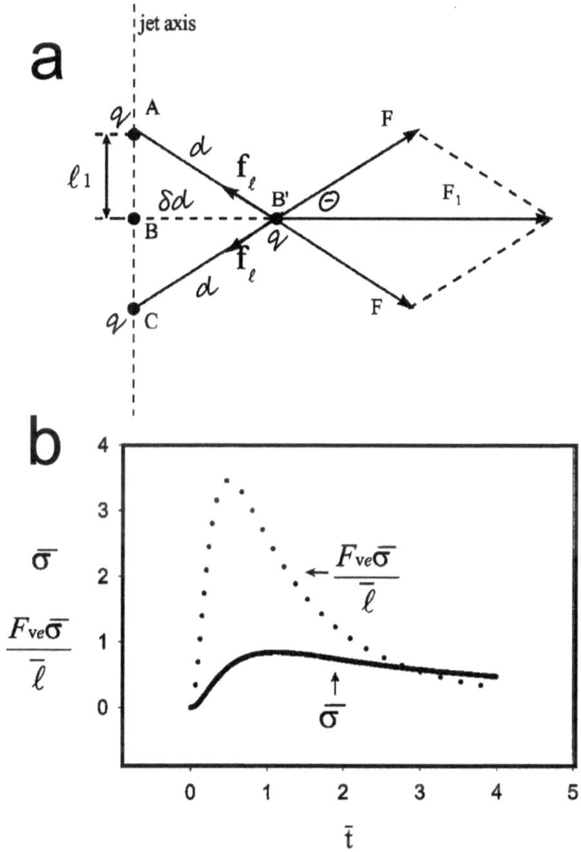


**Figure 2.22** (a) Scheme of the Earnshaw instability driven by three point charges (*A*, *B* and *C*) of the same sign in an electrospun jet. Here $F_1$ indicates the resultant of the two repulsive Coulombic forces (F) on the charge *B* in the point *B'*, from *A* and *C*. $f_l$ indicates the viscoelastic resistance counteracting the bending perturbation. (b) Longitudinal stress (here, $\bar{\sigma}$, which indicates the stress divided by the elastic modulus of the polymer solution; continuous line) in the rectilinear part of the electrospun jet, and longitudinal force ($F_{ve}\,\bar{\sigma}/\bar{l}$, where $\bar{l}$ is the distance between two point charges connected by a linear Maxwell dumbbell such as that in Figure 1.14, divided by the initial filament length; dotted line). In the horizontal axis, $\bar{t}$ is the time divided by the relaxation time ($\tau_C = \eta_P/G_S$, Section 1.4.2.2). Adapted with permission from Ref. 35, *J. Appl. Phys.*, 2000, **87**, 4531–4547. DOI: 10.1063/1.373532. Copyright © 2000, American Institute of Physics.

previous equation is clearly provided by an exponential law, $\delta d(t) = \delta d(t = 0) \exp[(q^2/2\pi m\varepsilon_0\varepsilon_r d^3)^{1/2}t]$, namely the initial perturbation rapidly grows and pushes charged particles away from each other,[35] with an exerted force that is transferred to the jet in which the charges are carried. Incidentally, from an experimental point of view this may result not only in instabilities but also in splaying events leading to the formation of branched polymer nanofibers as anticipated in Section 2.1.3.

In the following, we review the main models proposed to describe electrospinning, presenting them in chronological order in an effort to give the idea of how the sequence of the different theoretical approaches has been developed over time. Given the complex behavior of electrified jets, aiming to understand the main features of process models one should firstly familiarize oneself with the different stages of electrospinning (Sections 2.1.1–2.1.5), and with the corresponding geometry and main axial direction as sketched in Figure 2.1.

### 2.5.1 Early Models (1970s–1999)

A model of electrically driven liquids that is often cited in electrospinning literature is the so-called *leaky dielectrics*,[258] which takes note that these fluids are unlikely to behave as either perfect dielectrics or perfect conductors. In fact, on one hand electric charges are almost inevitably present and they are quite effectively carried in the liquids, thus preferably positioning on the external surfaces, whereas on the other hand the conductivity in the fluids is not so high to prevent the existence of a non-vanishing electric field tangential to these surfaces. In its turn, this component of field would cause a tangential stress leading to a deformation, possibly counterbalanced by viscous effects, which, for instance, may determine circulating patterns inside the fluid body. These considerations are supported by earlier results by Taylor.[36,39,259] On these bases, cylinders of fluids immersed in electric fields have been investigated in 1970–1971 by Saville, who has formulated a well-known linear stability analysis of uncharged jets[260,261] and of jets with constant surface charge density.[262] Though quantitatively inaccurate from a strictly predictive point of view, these pioneering studies capture some important characteristics, including the occurrence of both whipping instabilities involving the deformation of the cylinder centerline and of instabilities that are instead axisymmetrical along the jet, all features that are confirmed by much more recent models (Section 2.5.3).[58]

In general, the description of electrospun jets passes through quasi-one-dimensional models based on the slenderness of the system. For instance, modern models of electrospinning are also influenced by studies performed by A. M. Gañán-Calvo about steady cone-jet electrospraying systems of Newtonian liquids in the years 1997–1999.[263–266] These have led to pictures by means of electro-hydrodynamic equations that have identical features as in later approaches (Sections 2.5.3 and 2.5.4), and to the finding of a generally observed behavior for the jet electric current and characteristic diameter, the latter corresponding to the size of atomized droplets. Both these quantities are found to be proportional to the square root of the flow rate.[266] Recent developments, which may be of great interest for engineering core-shell nanofibers and optimizing electro-blowing processes, consider concomitant flow-focusing effects in the jet dynamics.[267–270]

The Spivak–Dzenis approach, proposed in the years 1998–2000, has the great merit of being the first modern model specialized to the electrospinning case.[271–273] It analyzes the regime of steady, straight stretching, which corresponds experimentally to the linear stage of the electrospun jet, focusing on the asymptotic behavior at distances sufficiently large from the spinneret. The fluid is schematized as a continuous slender body. Such an approximation means that the variations of the jet radius are slow along the axial direction, *i.e.* $|dr/dz| << 1$. Another simplifying assumption consists in taking the electric field substantially uniform along the jet longitudinal direction, *i.e.* neglecting the effect of the excess surface charges of the jet that could affect the axial component of the field. In addition, a power-law dependence of the viscosity on the shear rate, $\dot{\gamma}$ (the so-called Oswald–deWaele law)[274] is used to describe the non-Newtonian behavior of the polymer solution. Though still not describing bending instabilities, this work has been one of the first attempts to predict the dependence of the jet radius on the distance from the spinneret, which is obtained by numerically integrating the motion equation along the $z$-direction, thus paving the way for more sophisticated approaches.

## 2.5.2 Reneker and Yarin Models

A more general and detailed model of the electrospinning process has been formulated in the works by the Reneker and Yarin groups since 2000. This approach is also referred to as Reneker model in the literature.[35,65,71,275] First of all, taking into account the low mobility of ions usually present in electrospun polymer solutions (as mentioned in Section 2.1.1), the excess charge is schematized to be static in the frame of reference of the extruded fluid jet. The jet serves therefore as fluid medium to push away the mutually repulsing electric charges from the droplet pending at the tip of the capillary, once the solution surface tension is overcome by electrical forces at the spinneret. The main forces affecting the flight of the jet, which are accounted for in the model, include viscoelasticity, surface tension and electrostatic interactions with the external field and among excess charges in the liquid, whereas gravity, compressions due to the air drag and eventual aerodynamic effects, which may lead to instabilities, are considered as minor and are generally neglected.

The basic building block on which the authors formulate their model is a linear dumbbell connecting together two electrically charged masses, $A$ and $B$, by means of viscoelastic forces. The dumbbell is made by a serial dashpot-spring combination as in the Maxwell model (Section 1.4.2.2 and Figure 1.14).

The authors firstly analyze a single linear segment, which could reasonably schematize the electrospun fluid at the beginning of the electrospinning path, thus smartly modeling the jet in terms of the viscoelastic dumbbell. The overall behavior of the electrified jet will follow naturally as a combination of many of these dumbbells, positioned in a series and mutually coupled. Each dumbbell will then schematize a single rectilinear segment of the jet. A nice association between these elastic dumbbells and intermediate, stretchable structural units along the polymer chains (the so-called Rouse segments which are indeed sub-molecular portions represented by couples of beads inter-connected by linear springs)[276] is proposed and explained in Ref. 277.

For a single dumbbell, under the hypothesis of Maxwellian behavior (Section 1.4.2.2),[274,277,278] the uniaxial elongation, or, in other words, the motion of the mass (*B*) that is at the end of the filament and has velocity, $v_j$, is then described by the following equation:[35]

$$m\frac{dv_j}{dt} = -\frac{1}{4\pi\varepsilon_0}\frac{q^2}{l^2} - \frac{qV}{h} + \pi r^2\sigma, \tag{2.11}$$

where $m$ and $q$ are the mass and the electric charge of the two involved masses, which are at mutual distance $l$ (= $-dv_j/dt$), $\varepsilon_0$ is the dielectric constant, $V$ is the external bias applied over a distance of $h$ and $r$ indicates the jet radius. Both the Coulomb repulsion and the external electric field tend to increase $l$. The longitudinal stress, $\sigma$, tends instead to pull $A$ and $B$ close together, and the corresponding attractive viscoelastic force (= $\pi r^2\sigma$) in Equation (2.11) is therefore given by $\sigma$, here considered as tensile stress, times the jet cross-section as introduced in Section 1.4.2.1. According to the projection of Equation (1.17) along the jet longitudinal direction, $\sigma$ is related to the dumbbell deformation and material functions by:

$$\frac{d\sigma}{dt} = -G\left(\frac{v_j}{l} + \frac{\sigma}{\eta}\right), \tag{2.12}$$

where $G$ and $\eta$ indicate the elastic modulus and the viscosity of the polymer solution, respectively. Shearing forces are neglected compared to longitudinal forces, which is acceptable for small $r$ values, and at this stage surface tension effects are not yet included (these will be incorporated very soon as described in the following). The previous system of equations is solved numerically in Ref. 35 with the initial conditions of absence of motion and stress. However, an intrinsic difficulty in fully predicting the experimental behavior of electrified jets consists in the complex, and not completely known, rheology of concentrated polymer solutions. It has been pointed out that the values of viscosities relevant to the electrospinning process have to be taken to be quite a lot larger than zero-shear values, because of the non-Newtonian character of many polymer solutions.[35,277] Further complications are related to the continuous solvent evaporation occurring during the flight of the jet.

As a second step, in their model the authors describe in depth the jet dynamics, passing from the simplified case of a single dumbbell to a system of $N$ beads, each of mass $m$ and charge $q$, with each couple of neighbor beads being connected by viscoelastic segments. During a simulation run, $N$ is not constant, but increases as new beads are injected from the electrospinning needle into the jet flow. For each of the $N$ elements, one may write the corresponding versions of Equations (2.11) and (2.12), including the overall electrostatic interactions among beads and with the external field. At this stage, the authors also introduce in the motion equations a term accounting for the surface tension acting in correspondence of each bead and related to the local curvature radius of the jet. Indeed, the surface tension term tends to recover a straight segment in the presence of a bending along the jet. A complete picture of the $3N$ resulting coupled equations for the coordinates of each elemental bead along the jet is presented in Ref. 35. Also this system can be solved numerically, for instance introducing a small lateral oscillation, periodic along the $z$-direction at its initial state, and then following the jet evolution.[35] Once implemented with physically reasonable parameters[35] describing the polymer solution and with suitable initial perturbations, the model is able to reproduce very well the spiraling behavior observed during electrospinning experiments.

From a physical point of view, the onset of the bending instability is here identified with the condition of (i) longitudinal stress decreasing after reaching its maximum while (ii) the jet cross-section is sufficiently thin to make bending stiffness irrelevant and (iii) the longitudinal force at the same time takes small values (Figure 2.22b). Under these conditions, the Coulomb repulsion between electric charges overcomes viscoelastic forces ($= \pi r^2 \sigma$) that would tend to resist bending of the liquid surface. The Coulomb repulsion is therefore demonstrated to be the actual driving force for bending instabilities in electrically charged jets.[35] As a double-check, the authors find that zeroing the electric charge on the beads results in a straight jet without bending events. Finally, one has to take into account the effect of surface tension. This force tends to counteract bending, as the latter would necessarily lead to an increase of the fluid surface area, and limits the highest frequency of spiraling loops.

In Ref. 65 the authors also take into account effects due to the evaporation of solvent vapors and to the resulting solidification process occurring in nanofibers during the jet flight, obtaining an equation that describes the temporal variation of the jet volume:

$$\frac{\partial (f \lambda_E)}{\partial t} = -0.495 D \mathrm{Re}^{1/3} \mathrm{Sc}^{1/2} [c(T) - c_\infty] \pi \lambda. \quad (2.13)$$

In the previous expression, $f$ and $\lambda_E$ indicate the jet cross-sectional area ($\pi r^2$) and longitudinal length, respectively, $f\lambda_E$ corresponding therefore to the volume; $D$ is the vapor diffusion coefficient in air; Re is the Reynolds number relevant for a cylindrical object moving in air along the direction of its longitudinal axis ($2rv_j/\zeta_a$, with $\zeta_a$ air kinematic viscosity, given by the air absolute or dynamic viscosity divided by density); Sc is the so-called Schmidt number (Sc = $\zeta/D$); $c$ and $c_\infty$ are, respectively, the saturation vapor concentration (depending on temperature, $T$) of the solvent and the vapor concentration in the atmosphere (far from jet, for instance equal to the relative humidity in the atmosphere for water). Hence, the chemical composition of the jet changes over time with a monotonous relative increase of the polymer mass ratio, $c_p$, because of the concomitant evaporation of solvent reducing continuously the solvent mass fraction, $c_s$, namely:

$$c_p(t) = c_{p0}[(f\lambda_E)(t=0)/(f\lambda_E)(t)], \tag{2.14a}$$

$$c_s(t) = 1[(f\lambda_E)(t)/(f\lambda_E)(t=0)]c_{p0}, \tag{2.14b}$$

where $c_{p0}$ indicates the initial polymer mass fraction, and $(f\lambda_E)(t = 0)$ and $(f\lambda_E)(t)$ are the jet volume at the initial time and at time, $t$, respectively. The evaporation continues until $c_s$ becomes small enough (*i.e.* $\leq$10%)[65] to make the further diffusion of residual solvent molecules to the external jet surface through the polymer very inefficient. Finally, the solidification due to this process can be straightforwardly explained by the relation found between viscosity and polymer concentration,[279] $\eta \propto 10^{Bc_p^m}$, with $m$ assuming values between 0.1 and 1 and $B$ estimated to be of the order of 10. Depending on $c_p$, the viscosity of the jet, and also its elastic modulus, exhibit therefore a temporal dependence that can be included within the Maxwell rheological constitutive Equation (2.12), thus taking into account also evaporation/ solidification effects in the model.

Experimental results by PEO are widely and successfully employed to check the validity of the Reneker model.[35,65] Furthermore, this approach can be further developed to describe multiple jets,[96] and implemented by more complex versions of the Maxwell model,[274,277,278] also taking into account non-linear viscoelasticity.[96] As previously mentioned (Section 2.2) the method has been applied to discriminate the most important parameters affecting the jet diameter, showing a satisfactory, though generally qualitative, agreement with experimental results on several different polymers including PAN, polystyrene (PS) and Nylon-6.[71] The most relevant process parameters are recognized to be the volumetric charge density, $h$, the relaxation time, the initial jet radius and the solution elongational viscosity, and, more generally, a predominance of viscoelastic forces over surface tension is found for sufficiently concentrated and high-$MW$ polymer solutions, frequently used in electrospinning.[71]

As anticipated in Section 2.1.2, another work analyzes in depth the Taylor cone and the onset of jetting.[40] Here the surface of the polymer solution droplet is schematized as equipotential and described as a hyperboloid of revolution symmetric with respect to the central, vertical axis of Figure 2.1. Without entering mathematical details, which are fully presented in Ref. 40, we briefly recall that the procedure consists in determining the electrostatic potential, $\Phi_0$, at the surface of the hyperboloid, then finding the normal stress at the surface due to $\Phi_0$, and equating such a stress to that related to surface tension forces in order finally to achieve an equilibrium condition. In this way, as introduced in Section 2.1.2, one obtains values of the apex half angles ($\cong 30°$) in good agreement with experimental observations.[40]

## 2.5.3 Electrohydrodynamic Models

Another interesting approach has been proposed by the Rutledge group at M.I.T. and their collaborators,[46,48,57,58,280] who, going beyond the early studies by Saville, have reanalyzed in depth the instability regimes of a cylinder of fluids in an external electric field in the case of Newtonian liquids. The experimental set-up is modeled by a thin, metallic nozzle protruding in the space between two parallel metallic plates, the latter generating a largely uniform electric field as in a capacitor (Figure 2.23a). A few issues, such as the solvent evaporation, the solidification of fibers during the jet flight and, more importantly, the solution viscoelasticity, are, however, neglected in this simplified approach.

The system is schematized as a continuous slender body of circular cross-section, and a Taylor series expansion is proposed to express some relevant quantities, such as the velocity components, as functions of the radial coordinate in the region of space between the two parallel plates (say, $r'$).[58,281,282] The dynamics is described by the following four equations, which account for the conservation of mass, the conservation of charge, the momentum balance of a fluid element along the jet (*i.e.* the Navier–Stokes equation)[274] and the approximated electrostatic potential, $\Phi(z, r')$, among the capacitor plates, respectively (*electro-hydrodynamics*):

$$Q = \pi r^2 v_j \tag{2.15a}$$

$$I = 2\pi r v_j \sigma_C + \pi r^2 \chi F_z \tag{2.15b}$$

$$\frac{\partial v_j}{\partial t} + \frac{\partial}{\partial z}\left(\frac{v_j^2}{2}\right) = -\frac{1}{\delta}\frac{\partial p}{\partial z} + g + \frac{2\sigma_C F_z}{\delta r} + \frac{3\zeta_j}{r^2}\frac{\partial}{\partial z}\left(r^2\frac{\partial v_j}{\partial z}\right) \tag{2.15c}$$

$$\Phi(z,r) = \Phi_\infty - \left[\frac{1}{\varepsilon_0}\sigma_C r - \frac{(\varepsilon_r - 1)}{2}\frac{d(F_z r^2)}{dz}\right]\ln\frac{r}{h}. \tag{2.15d}$$

Most of the symbols used in the previous equations have been introduced previously (Sections 2.1.1, 2.1.2, 2.2 and 2.5.2 and List of Symbols). $F_z$ here indicates the axial component of the electric field, $p$ is the internal pressure in the fluid, $g$ is the gravitational acceleration, $\zeta_j$ is the kinematic viscosity in the

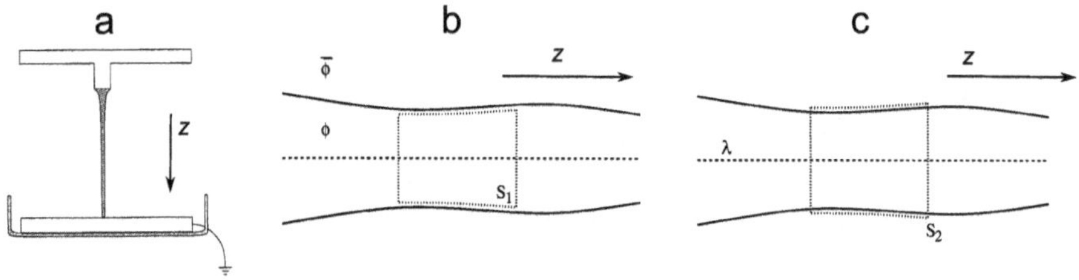


**Figure 2.23** (a) Set-up considered by the model: a metallic nozzle protrudes in the volume between two parallel metallic plates. The bottom plate is grounded. The $z$-axis indicates the vertical direction and is the initial axial direction of the jet. (b, c) Gaussian surfaces (dotted curves) leaning immediately below ($S_1$, b) and above ($S_2$, c) the surface of the electrospun jet (continuous curve). Here, $\phi$, $\bar{\phi}$ and $\lambda$ indicate the electrostatic potential inside and outside the jet, and the effective linear charge density in the jet, respectively. Adapted with permission from Ref. 58, *Phys. Fluids*, 2001, **13**, 2201–2220. Doi: 10.1063/1.1383791. Copyright © 2001, American Institute of Physics.

jet and $\Phi_\infty$ is the electrostatic potential at very large radial distance from the jet (*i.e.* due to the external applied field only). The dielectric constant in the free space outside the jet is approximated as $\varepsilon_0$.

It is worth considering the different terms in the equations. The jet radius ($r$) and local axial velocity ($v_j$) depend of course on the $z$ vertical coordinate among the plates (Figure 2.23a), and in principle on time, whereas $v_j$ is taken as constant across the cross-section of the jet at the leading order. In Equation (2.15b), the electric current ($I$) carried by the jet is considered as due to a bulk, ohmic conduction component in the liquid (rightmost term in the second member of the equation), plus a surface, advection component related to the jet carrying a surface charge of density, $\sigma_C$. By Equation (2.15a), the advection component may also be expressed as 2 $Q$ $\sigma_C/r$, which highlights how this contribution to the overall current directly relates to the jet flow rate. The fourth, rightmost term in the second member of Equation (2.15c) incorporates the fluid viscosity, and its expression results from a frequently used, Newtonian constitutive equation for the differences of the viscous stresses acting on the jet, $\sigma_{zz} - \sigma_{rr} = 3\eta(\partial v_j/\partial z)$. The third term in the second member of the same equation accounts instead for the tangential stress on the jet surface charges. Finally, the pressure, $p$, includes the effects of (i) the surface tension, $\sigma_s$, and of (ii) the electrostatic pressures on the surface of the jet, due to the differences of the electrostatic energy densities (both due to the external field, approximated as $F_z$, and to surface charges) outside and inside the liquid:

$$p \cong \frac{2\sigma_s}{r} - \frac{\sigma_C^2}{2\varepsilon_0} - \frac{\varepsilon_0(\varepsilon_r - 1)}{2} F_z^2 \tag{2.16}$$

A detailed derivation of the steady-state version of Equation (2.15c) is reported in Ref. 283. Incidentally, one may also notice that Equations (2.15a) and (2.15b) can be equivalently expressed in differential form, which is originally used in Ref. 58:

$$\frac{\partial}{\partial t}\left(\pi r^2\right)+\frac{\partial}{\partial z}\left(\pi r^2 v_j\right)=0; \tag{2.17a}$$

$$\frac{\partial}{\partial t}(2\pi r\sigma_C)+\frac{\partial}{\partial z}\left(2\pi r\sigma_C v_j+\pi r^2\chi F_z\right)=0. \tag{2.17b}$$

The last of the four equations describing the jet, Equation (2.15d), is obtained by considering an effective linear charge density along the jet. The expression of such charge density is found straightforwardly applying Gauss' law to two different Gaussian surfaces leaning, respectively, immediately below (Figure 2.23b) and above (Figure 2.23c) the jet surface, which is still axisymmetric at this stage. In this equation, one should evidence the presence of the jet aspect ratio ($r/h$, which is $<<1$) as argument of the logarithmic factor contributing to the electrostatic potential. The equation for the electric field at the jet is of course immediately derived once the potential is known, as $F_z=F_{z,\infty}-\left[\frac{1}{\varepsilon_0}\frac{d(\sigma_C r)}{dz}-\frac{(\varepsilon_r-1)}{2}\frac{d^2\left(F_z r^2\right)}{dz^2}\right]\ln\frac{r}{h}$, where $F_{z,\infty}$ is the spatially uniform external field.

The authors then perform a linear stability analysis solving the dynamics resulting from small perturbations of the jet radius or charge density, or of the electric field. These take expressions of the form, $r=r_0+\Delta r e^{\tilde{\omega}t+ikz}$, where $\tilde{\omega}$ indicates the growth rate of the instability, $k$ is the disturbance wavenumber along the $z$ direction, and the perturbation amplitude, $\Delta r$, is much smaller than $r_0$. In this way, three instability modes are identified (for an introduction to the main axisymmetric and non-axisymmetric instabilities observed in electrospinning, see Section 2.1.3):

- (i) an axisymmetric, *Rayleigh*-type[284] *instability*. This mode extends to electrified jets the classical Rayleigh instability, which is instead related to surface tension‡‡, and it is remarkably suppressed for sufficiently intense electric fields;

‡‡ The classical Rayleigh or Plateau–Rayleigh instability is induced by the liquid surface tension, which tends to minimize the area of the fluid–air interface. This is at the origin of the effect of jet breaking into droplets, which everyone can see in a stream of water falling from a tap. When perturbations with a given wavelength, $\lambda_R$, occur along the liquid jet, the overall jet surface area slightly decreases, therefore the instability would be favored by the action of surface tension and the stream could eventually break up into drops. After experimental observation by Savart and Plateau in 1833–1849,[285,286] Rayleigh found that an optimal condition for the propagation of the instability and hence for the generation of separated droplets from a vertically falling liquid stream is $\lambda_R \sim 9$r.[281,287] The characteristic timescale for perturbation growth and droplet formation can be estimated by Equation (2.5).[281] Research on Plateau–Rayleigh instabilities, especially when coupled to fiber fabrication technologies, still continues to yield exciting results. Recently, monodisperse particles with an impressive range of achievable sizes (2 mm–2 μm) have been realized by inducing Plateau–Rayleigh instabilities in heated drawn fibers.[288]

- (ii) a second axisymmetric mode named *conductive instability*, which is only found when the fluid conductivity is finite. Both (i) and (ii) are varicose modes according to the definition in Section 2.1.3;
- (iii) the typical, non-axisymmetric, whipping instability characteristic of electrospinning. This mode is in some sense "conductive", *i.e.* only possible in the case of finite conductivity of the solution. In particular, whipping can dominate dynamics under conditions of sufficiently high surface charge density. The non-axisymmetric behavior is included in the model by taking into account a dipolar component that can affect the effective linear charge density along the jet and hence the resulting electrostatic potential, and suitable force and torque balance equations.

An especially intriguing result consists in obtaining *phase diagrams* for each particular electrospinning experiment, in which the ratio of $\tilde{\omega}$ of the most unstable whipping mode to the most unstable varicose mode is plotted against $r$ and $\sigma_C$, for a particular polymer and electric bias (Figure 2.24).[48,58] As an alternative, electric field can be plotted *vs.* flow rate for a given polymer solution system, and the different regimes corresponding to stable or unstable jetting can be localized on the plot.[289] In Figure 2.24, whipping behaviors prevail in the top-right, bright region of the diagram, whereas varicose modes dominate in the bottom-left corner (smaller radius, lower $\sigma_C$). In addition, as previously mentioned one notices from Equation (2.15b) that $r$ and $\sigma_C$ are not fully independent parameters. For instance, for negligible conduction current, one has $I = 2Q\,\sigma_C/r$, which means that the operation point of a particular experiment would actually lean on a characteristic line with unit slope in the bi-logarithmic phase diagram (Figure 2.24). In the second paper of their series,[48] M. M. Hohman *et al.* refine their analysis of the jet behavior under steady conditions, taking into account how the electric field is complicated by fringes due to the extrusion needle that protrudes well into the region between the two parallel plates. In this way, the agreement with experimental profiles of electrospun jets is remarkably improved (Figure 2.25).

In Ref. 46, the authors extend their model to the regime of non-linear instability (still in the case of Newtonian liquids) to relate the main process parameters to the diameter of the resulting electrospun nanofibers. The surface advection component of current is considered to be dominant on conduction current as the jet cross-section becomes smaller and smaller, as explained in footnote ‡ in Section 2.1.2. This leads to an analytical expression of the terminal jet radius ($r_t$) as a function of set-up and process parameters, which has become quite popular:

$$r_t = \left\{ \sigma_s \varepsilon_0 \frac{Q^2}{I^2} \frac{2}{\pi[2\ln\kappa - 3]} \right\}^{1/3}. \tag{2.18}$$

The dimensionless parameter, $\kappa$, is given by the ratio, $r_B/r$, where $r_B$ is the radius of curvature of the whipping instability trajectory of the jet. The appealing suggestion here is of substantially identifying $r_t$ with the radius of the

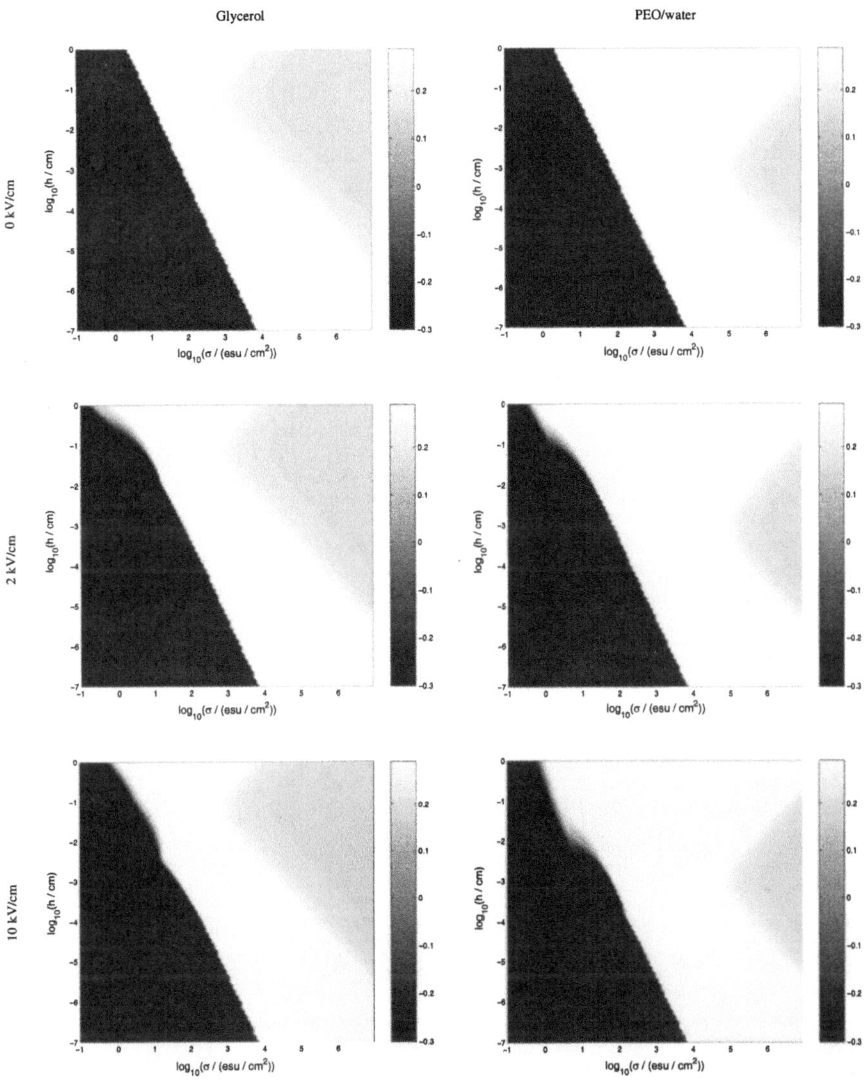


**Figure 2.24** Contour plots of $\log_{10}(h)$, where $h$ is expressed in cm and is the ratio of the growth rate of the most unstable whipping mode to the most unstable varicose mode, *vs.* $\log_{10}(\sigma)$, where $\sigma$ is expressed in electrostatic units $cm^{-2}$ (1 esu $cm^{-2} \cong 3.335 \times 10^{-6}$ C $m^{-2}$) and is the surface charge density. The plots are for various values of the electrospinning voltage bias, for glycerol and for PEO aqueous solutions, which differ in conductivity and viscosity. From these graphs, one can clearly appreciate that axisymmetric modes prevail at low charge density (namely in the bottom-left, dark regions of the diagrams), whereas whipping modes prevail at high charge density (namely in the top-right, bright regions). Reproduced with permission from Ref. 48, *Phys. Fluids*, 2001, **13**, 2221–2236. Doi: 10.1063/1.1384013. Copyright © 2001, American Institute of Physics.

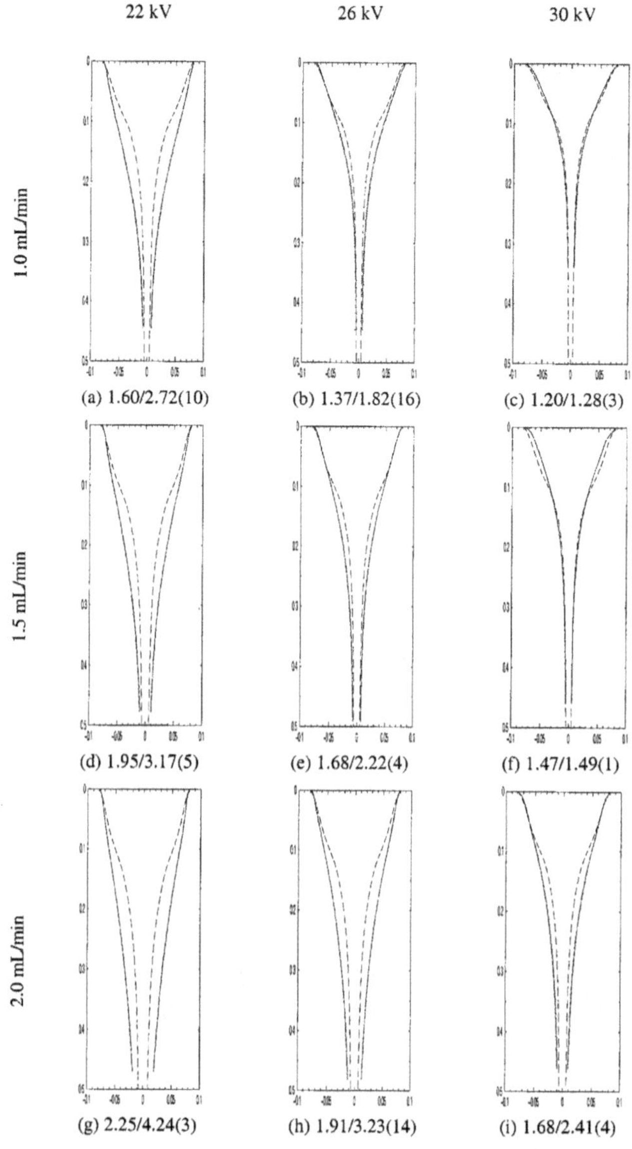


**Figure 2.25** Comparison of theoretical (dashed) and experimental (continuous) profiles for an electrospun glycerol jet, for various voltages and flow rates. Here the two plate electrodes (Figure 2.23a) are 6 cm apart, and the die protrudes by 7.2 mm from the top electrode. Beneath each plot there are the theoretical healing length and the experimental healing length, normalized to the outer diameter of the nozzle (here about 800 μm). The healing length is the axial interval over which the initial jet radius decreases by a factor of $1/e$. Reproduced with permission from Ref. 48, *Phys. Fluids*, 2001, **13**, 2221–2236. Doi: 10.1063/1.1384013. Copyright © 2001, American Institute of Physics.

resulting solid nanofibers, at least concerning the functional dependences on the ratio $Q/I$ [$r_t \cong (Q/I)^{2/3}$] highlighted by Equation (2.18). From Section 2.1.2, we recall that in other words the ratio, $Q/I$, is the jet effective volumetric charge density, $\rho$, which is indeed generally recognized among the most significant variables affecting the diameter of electrospun nanofibers.[71]

Of course, since solvent evaporation is not included in the model, comparing these findings to experiments would imply to assume that the evaporation does not significantly affect the jet dynamics before the terminal radius is reached, so that the final radius of the solid fibers can be directly obtained from $r_t$. A simple correction can be performed for the polymer concentration to further improve this point of the description,[46] since assuming that the solvent evaporation occurs only in the last part of the trajectory (*i.e. after* the terminal radius is reached) would determine a relative increase of $c_p$ at this final stage. Interestingly, the 2/3 scaling is indeed found for the diameter of dry fibers of PCL spun from differently concentrated solutions, and for samples of PEO and PAC.[46] The agreement of these experimental results with the model would suggest a weaker role of solvent evaporation, and that the Newtonian approximation can actually be exploited for a variety of electrospun solutions larger than expected.

A difficulty highlighted in Ref. 48 and subsequently pointed out by J. J. Feng[283] (next section) stands in the somehow phenomenological choice of the proper boundary conditions at the nozzle. In addition, Reneker and coworkers also observe[71] that neglecting the viscoelasticity of the polymer solution inevitably leads to overestimating the importance of the surface tension, $\sigma_s$, in affecting the terminal radius of the jet.

### 2.5.4 More Recent Approaches (2002–2011)

Feng has investigated the first stage of the electrospun jet, where the flow passes through a straight trajectory,[283,290] by using a set of equations substantially identical to Equations (2.15a–d) but making results more robust against the particular choice of the conditions at the nozzle. Furthermore, stemming from the early suggestions from the Spivak–Dzenis model (Section 2.5.1)[271–273] the analysis by Feng goes beyond the case of simple liquids by introducing a non-Newtonian rheology, a case that is certainly interesting to be analyzed and relevant when dealing with real electrospun solutions. C. P. Carroll and Y. L. Joo have extended the analysis of electrospinning to viscoelastic solutions as well, focusing on the initial, straight section of the jet and on axisymmetric instabilities.[291–293] Numerical simulations based on Equations (2.15a–d) can also be implemented by means of partial differential equation modules as in COMSOL® Multiphysics,[294] software that has also been used to describe multijet electrospinning.[295]

In general, to date models are not intended as fully predictive tools allowing a comprehensive description of the resulting nanostructures to be given, but instead as valuable methods to restrict the domain of process parameters to be experimentally tested. A crucial problem for more careful modeling of the process still relies in correctly matching the flow conditions in the electrified jets, which can be reasonably approximated by quasi-one-dimensional conditions, with those in the pending droplet at the spinneret, where the injection generates instead a more complex fluid transport. Furthermore, low memory effects in the nozzle-to-jet transition may significantly affect the dynamics. Recent and interesting developments in this respect come from researchers from the Yarin (now at University of Illinois at Chicago) and Zussman groups at Technion-Israel Institute of Technology, who have studied the different conditions of droplets attached to a conducting surface in the presence of an applied electric field, and matched the droplet and jet flows.[41] An especially intriguing feature of their analysis is that, differently from previous works, here the flow rate and the electric current are not assumed, but instead derived given the applied bias. This study leads to predicting a non-linear dependence of $I$ on $V$, which is physically expectable due to concomitant presence of convective redistribution of charges and ohmic conduction as described above (Equation (2.15b)). The non-linear dependence of $I$ on the applied bias has been found in a number of experiments.[49,289] For instance, a recent study has shown that for specified $F$, $Q$ and $\chi$, the current scales as $FQ^{0.5}\chi^{0.4}$ for a wide variety of polymer solutions.[296] In the model, predicted current values are of the order of hundreds of nA for conductivity of 10 mS m$^{-1}$ and bias of tens of kV,[41] which, as mentioned in Section 2.1.2, all are pretty realistic values for electrospinning experiments. In future, it may be interesting to study hybrid approaches to combine such valuable predictions with models focusing instead on the terminal jet path and radius as in Ref. 46. This may help to shed more light on the interplay between the different parameters and to better link the resulting diameter of electrospun nanofibers to primary parameters more directly controlled by experimentalists.

Other important effects that have not been fully rationalized so far come from the interactions of the fluid transport and of the generated instabilities with liquid–solid or gas–solid interfaces that exist both at the spinneret (as studied by Hohman and coworkers[48] and by Feng[283]) and at the collector, and in non-uniform cross-sectional properties of the jet. In fact, that the conditions of velocity, density, *etc.* in different points of the jet cross-section are uniform is certainly a drastic though generally adopted approximation. In this respect, I. Greenfeld *et al.* have recently proposed an appealing theoretical model and performed three-dimensional random walk simulations of highly entangled semi-dilute polymer solutions under the very high longitudinal acceleration typical of electrospinning.[72] To start, it is quite obvious that, due to such acceleration, each element of liquid within the electrified jet should also exhibit a non-zero radial component ($v_r$) of its velocity, in addition to the longitudinal, axial component (the latter is what we have previously introduced as $v_j$). In particular, the axial velocity may take a quadratically increasing dependence on the $z$-coordinate, with the radial component depending instead linearly on both $z$ and $r$ within the liquid tube given by the jet, *i.e*:[72,297]

$$v_j = v_{j0}\left(1+\frac{z}{z_0}\right)^2 \tag{2.19a}$$

$$v_r = -\frac{r}{2}\frac{\partial v_j}{\partial z} = -v_{j0}\left(1+\frac{z}{z_0}\right)\frac{r}{z_0} \tag{2.19b}$$

Here, $v_{j0}$ is the initial velocity at the beginning of the jet. $z_0$, estimated in the range from hundreds of micrometers to millimeters, is a system-related constant giving the length-scale of the longitudinal acceleration, which depends on the flow rate, applied bias and solution rheology and conductivity characteristics. The minus sign in the right member of Equation (2.19b) indicates of course a radial velocity directed towards the center of the jet, thus supporting the overall accelerating mass transport, which leads to stretching and eventually to nanoscale fiber diameters. Combining the previous two equations and integrating between the initial condition and a generic point of coordinates ($r$, $z$) gives a hyperbolic profile, $r(z)$, for the accelerating jet: $r(z) \propto [1 + (z/z_0)]^{-1}$.

To model the entangled system, the authors depict the polymer network as made of stretching macromolecules given by topological knots or beads, linked by flexible sub-chain segments, similarly to the Rouse model[276] (Figure 2.26). Stretching is associated to leading knot-to-knot distances well beyond their value of equilibrium, which is the state occurring at the beginning of the jet (*i.e.* at the spinneret). As physically expected, such initial, average knot-to-knot distance is inversely proportional to the polymer concentration, $c_p$, and is independent of the number of monomers within a polymer coil (or *MW*). Gaussian coils are assumed to result from the reasonably good solvents used in electrospinning.

The model predicts the remarkable longitudinal stretching taking place along the jet, which results in the concomitant transversal contraction of the polymer network, as well as in macromolecules ordering and longitudinal orientation. This drives the polymer network from its initial equilibrium to an almost fully stretched state by a continuous cross-over. In addition, since the radial contraction of the polymer network in turn leads to an increase of polymer concentration close to the axis of jet, this prediction can be effectively tested by means of X-ray phase-contrast imaging probing the local density in the electrified solution. Indeed, in-line measurements performed at the 32-ID Beamline (Advanced Photon Source) of the Argonne National Laboratory (Figure 2.27)[72,298] have allowed researchers to find the radiation transmission ($T^*$), whose dependence on the jet coordinates (Figure 2.28) can be summarized as follows:

$$T^*(r,z) = g_F(r,z)\exp[-\alpha(r,z)d(r,z)]. \tag{2.20}$$

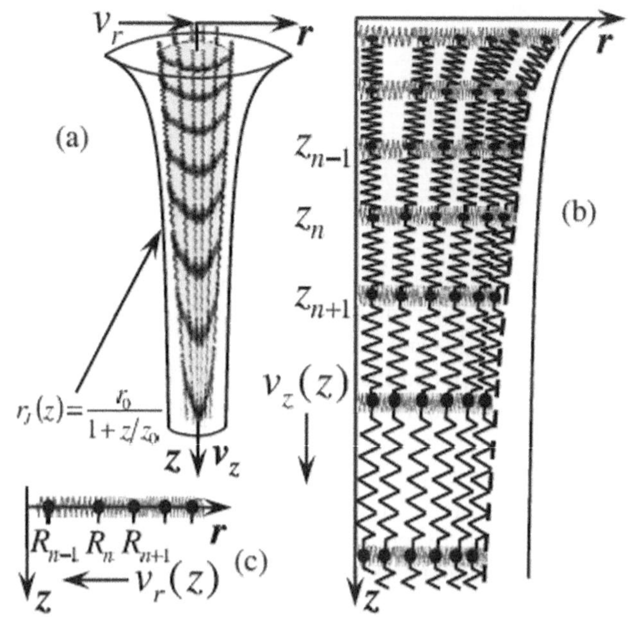


**Figure 2.26** (a) Illustration of polymer network stretching in an electrospun jet, with a cylindrical coordinate system ($r$, $z$). (b) The stretching of the polymer network can be described considering the axial component of the velocity ($v_z$), and various topological knots or beads, linked by flexible sub-chain segments. (c) Same concept along the radial direction, allowing the polymer network contraction to be described leading to a reduction of the network diameter faster than the narrowing of the jet. Reproduced with permission from Ref. 72, *Phys. Rev. E*, 2011, **84**, 041806. Doi: 10.1103/PhysRevE.84.041806. Copyright © 2011, American Physical Society.

In the previous expression, $g_F$ indicates a form factor that describes possible effects due to radiation scattering within the jet, $d$ is the jet thickness and $\alpha$ is the absorption coefficient at the probed point. One way to compare experimental transmission with the developed model passes through assuming the same form factor for measurements and for wave propagation simulations carried out for a homogeneous solution (the latter providing a transmission, $T^{*,sim}$)§§. This assumption allows the variation, $\Delta\alpha$, of the absorption coefficient of the electrospun jet with respect to the solution at rest to be obtained:

$$\Delta\alpha(r,z) = -\ln\left[T^*(r,z)/T^{*,sim}(r,z)\right]/d(r,z) \tag{2.21}$$

§§ Refinement of the form factor is given in Ref. 298.

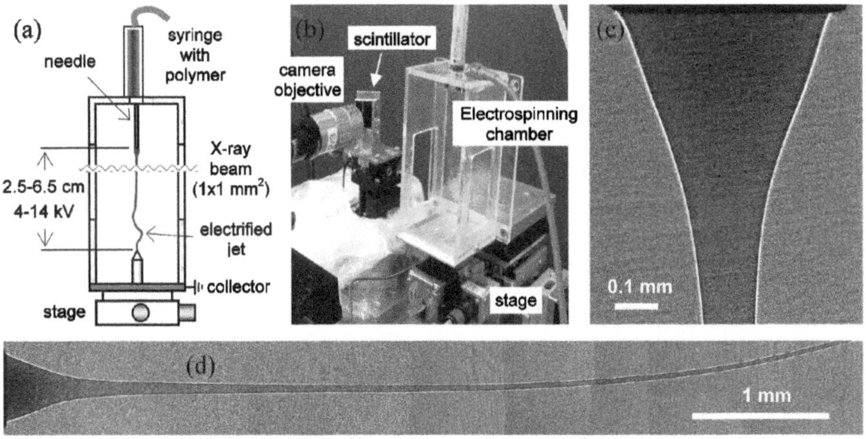


**Figure 2.27** (a) X-ray measurements to probe the interior of an electrospun jet. The X-ray beam, generated from an electron storage ring, penetrates the jet and is then converted to a digital image *via* a fast scintillator crystal. Finally, a charge-coupled device (CCD) camera is timed to capture the first pulse after shutter opening. The integration of this set-up with the electrospinning chamber placed in the position of the sample stage is photographed in (b). (c) Representative X-ray image of an electrospun jet composed of a solution of 5 wt.% PEO in water. (d): Same case, presenting a jet section over a length of 6 mm, reconstructed from 12 sequential images. Reproduced with permission from Ref. 298, *Macromolecules*, 2012, **45**, 3616–3626. Doi: 10.1021/ma300237j. Copyright © 2012, American Chemical Society.

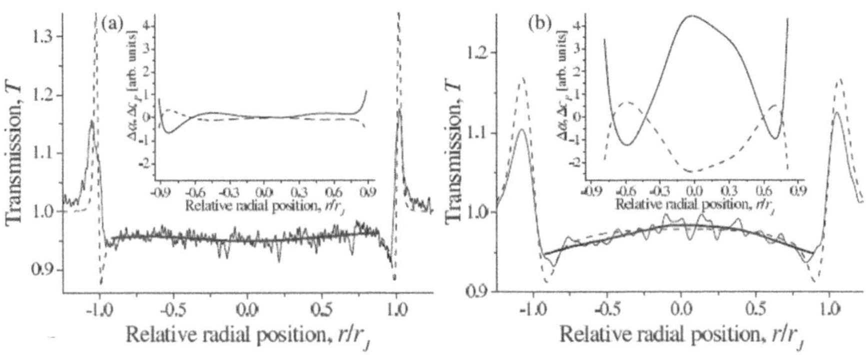


**Figure 2.28** X-ray transmission measurements across an electrospun jet of a solution of 3 wt.% PEO in water, at two different $z$ positions: 0.02 mm (a) and 0.5 mm (b). In other words, the initial and the straight regions of the jet are probed. The thin continuous lines and the thick lines are experimental and smoothed data, respectively, whereas the dashed lines are from simulations. Insets show the calculated $\Delta\alpha$ and $\Delta c_p$. Reproduced with permission from Ref. 72, *Phys. Rev. E*, 2011, **84**, 041806. Doi: 10.1103/PhysRevE.84.041806. Copyright © 2011, American Physical Society.

**Table 2.1** Comparison of characteristic features of main models of electrospinning.

| | *Electrospinning stage* | *Solution rheology* | *Solvent evaporation* | *Polymer conformation* |
|---|---|---|---|---|
| Spivak-Dzenis<br>Refs 271–273 | Straight jets | Non-Newtonian (Power-law) | | |
| Reneker and Yarin model<br>Refs 35, 65, 71, 275 | Straight jets and bending instabilities | Linear viscoelasticity (Maxwellian) | × | |
| Electrohydrodynamic models<br>Refs 46, 48, 57, 58 | Straight jets and whipping instabilities | Newtonian | | |
| Feng<br>Refs 283, 290 | Straight jets | Non-Newtonian | | |
| Carroll and Joo<br>Refs 291–293 | | | | |
| Zussman and coworkers<br>Ref. 72 | Straight jets | Linear viscoelasticity | | × |

Finally, the corresponding variation of the polymer concentration in the solution can be straightforwardly obtained, being related to $\Delta\alpha$ by the following equation:

$$\Delta c_p(r,z) = \Delta\alpha(r,z) / \left[\alpha_p^* - \alpha_s^* \left(\delta_s / \delta_p\right)\right], \tag{2.22}$$

where $\alpha_p^*$ and $\alpha_s^*$ indicate the X-ray mass absorption coefficient of the polymer and of the solvent, respectively $\{\alpha(r, z) = \alpha_p^* c_p(r, z) + \alpha_s^* c_s(r, z)\}$. Here, $c_p$ and $c_s$ are the mass concentration of the polymer and of the solvent, respectively, and $\delta_p$ and $\delta_s$ are the corresponding densities of the two species. It is very promising that the obtained experimental findings are in good agreement with results from theory, for electrospun jets of both PEO and poly(methyl methacrylate) (PMMA) semi-dilute solutions, which indeed exhibits the predicted increase in polymer concentration at the jet axis, at distances of the order of 0.5–1.0 mm from the spinneret. Furthermore, it has been confirmed that this effect is stronger at higher jet strain rates (*e.g.* higher bias and/or lower flow rate), whereas at lower strain rates rapid solvent evaporation has a dominant effect. Such ordered non-uniform polymer network conformations may persist, at least partially, into the structure of the solid nanofibers. Future developments could consider enriching the variety of polymers investigated by real-time X-ray phase-contrast imaging during electrospinning, thus aiming at unveiling how the increase of density at the jet axis could depend on specific steric groups and conformational behavior of different macromolecular compounds.

To conclude this section, we summarize some relevant features of the main models proposed for electrospinning in Table 2.1. A direct comparison is often difficult because the different approaches aim to describe different stages of the jet and may be largely complementary in better fitting the experimental behavior of electrified flows for different electrospinning phases or polymer–solution systems. In any case, wide, cross-cutting testing campaigns on the enormous variety of systems that can be effectively electrospun are still generally missing for the wide majority of these models. In fact, the richness of solution phenomenologies and jet aspects taken into account by the different theoretical pictures proposed to date is certainly a particularly intriguing issue and promising validating-research field for experimentalists.

## References

1. N. Tucker, J. J. Stanger, M. P. Staiger, H. Razzaq and K. Hofman, *J. Engin. Fibers Fabr.*, 2012, **7**(3), 50–131.
2. W. Gilbert, *De Magnete*, 1600, English translation by P. F. Mottelay, Dover Publications, New York, 1991.

3. L. D. Landau, E. M. Lifshitz and L. P. Pitaevskii, *Electrodynamics of Continuous Media*, Butterworth-Heinemann, Oxford, 2nd edn, 1984.
4. J. F. Cooley, US Pat. No. 692,**631**, 1902.
5. W. J. Morton, US Pat. No. 705,**691**, 1902.
6. J. Zeleny, *Phys. Rev.*, 1914, **3**, 69–91.
7. J. Zeleny, *Proc. Camb. Philos. Soc.*, 1915, **18**, 71–79.
8. J. Zeleny, *Phys. Rev.*, 1917, **10**, 1–6.
9. A. Formhals, US Pat. No. 1975504, 1934.
10. A. Formhals, US Pat. No. 2160962, 1939.
11. A. Formhals, US Pat. No. 2187306, 1940.
12. A. Formhals, US Pat. No. 2323025, 1943.
13. A. Formhals, US Pat. No. 2349950,1944 .
14. C. L. Norton, US Pat. No. 2,048,651, 1936.
15. H. L. Simons, US Pat. No. 3,280,229, 1966.
16. P. K. Baumgarten, *J. Colloid Interface Sci.*, 1971, **36**, 71–79.
17. J. Doshi and D. H. Reneker, *J. Electrostatics*, 1995, **35**, 151–160.
18. D. H. Reneker and I. Chun, *Nanotechnology*, 1996, **7**, 216–223.
19. A. G. MacDiarmid, *Rev. Mod. Phys.*, 2001, **73**, 701–712.
20. D. H. Reneker and H. Fong (ed.), *Polymeric Nanofibers*, American Chemical Society, Washington, DC, 2006.
21. S. Ramakrishna, K. Fujihara, W.-E. Teo, T.-C. Lim and Z. Ma, *An Introduction to Electrospinning and Nanofibers*, World Scientific Publishing Co., Singapore, 2005.
22. J. H. Wendorff, S. Agarwal and A. Greiner, *Electrospinning. Materials, Processing and Applications*, Wiley-VCH, Weinheim, 2012.
23. J.-H. He, Y. Liu, L.-F. Mo, Y.-Q. Wan and L. Xu, *Electrospun Nanofibers and Their Applications*, iSmithers, Shawbury, UK, 2008.
24. D. Li and Y. Xia, *Adv. Mater.*, 2004, **16**, 1151–1170.
25. A. Greiner and J. H. Wendorff, *Angew. Chem. Int. Ed.*, 2007, **46**, 5670–5703.
26. I. Chronakis, *J. Mater. Process. Technol.*, 2005, **167**, 283–293.
27. C. Burger, B. S. Hsiao and B. Chu, *Annu. Rev. Mater. Res.*, 2006, **36**, 333–368.
28. C. Feng, K. C. Khulbe and T. Matsuura, *J. Appl. Polym. Sci.*, 2010, **115**, 756–776.
29. A. L. Andrady, *Science and Technology of Polymer Nanofibers*, John Wiley & Sons, Inc., Hoboken, 2008.
30. V. E. Kalayci, P. K. Patra, Y. K. Kim, S. C. Ugbolue and S. B. Warner, *Polymer*, 2005, **46**, 7191–7200.
31. A. G. Bailey, *Electrostatic Spraying of Liquids*, Wiley, New York, 1998.
32. J. S. Chang, A. J. Kelly and J. M. Crowley, *Handbook of Electrostatic Processes*, Dekker, New York, 1995.
33. S. A. Theron, E. Zussman and A. L. Yarin, *Polymer*, 2004, **45**, 2017–2030.

34. F. Kremer and A. Schönhals (ed.), *Broadband Dielectric Spectroscopy*, Springer, Berlin and , Heidelberg, 2003.
35. D. H. Reneker, A. L. Yarin, H. Fong and S. Koombhongse, *J. Appl. Phys.*, 2000, **87**, 4531–4547.
36. G. I. Taylor, *Proc. R. Soc. London, Ser. A*, 1964, **280**, 383–397.
37. G. I. Taylor and A. D. McEwan, *J. Fluid. Mech.*, 1965, **22**, 1–15.
38. G. I. Taylor, *Proc. R. Soc. London, Ser. A*, 1966, **291**, 145–158.
39. G. I. Taylor, *Proc. R. Soc. London, Ser. A*, 1969, **313**, 453–475.
40. A. L. Yarin, S. Koombhongse and D. H. Reneker, *J. Appl. Phys.*, 2001, **90**, 4836–4846.
41. S. N. Reznik, A. L. Yarin, A. Theron and E. Zussman, *J. Fluid Mech.*, 2004, **516**, 349–377.
42. D. Lukáš, A. Sarkar, L. Martinová, K. Vodsed'álková, D. Lubasová, J. Chaloupek, P. Pokorný, P. Mikeš, J. Chvojka and M. Komárek, *Textile Progress*, 2009, **4**, 59–140.
43. P. K. Bhattacharjee and G. C. Rutledge, in *Comprehensive Biomaterials*, ed. P. Ducheyne, K. E. Healy, D. W. Hutmacher, D. W. Grainger and C. J. Kirkpatrick, Elsevier, Oxford, 2011.
44. S. B. Warner, A. Buer, S. C. Ugbolue, G. C. Rutledge and M. Y. Shin, in *National Textile Center Annual Report*, November 1998, 83–90.
45. L. M. Bellan, H. G. Craighead and J. P. Hinestroza, *J. Appl. Phys.*, 2007, **102**, 094308.
46. S. V. Fridrikh, J. H. Yu, M. P. Brenner and G. C. Rutledge, *Phys. Rev. Lett.*, 2003, **90**, 144502.
47. S. Tripatanasuwan and D. H. Reneker, *Polymer*, 2009, **50**, 1835–1837.
48. M. M. Hohman, M. Shin, G. Rutledge and M. P. Brenner, *Phys. Fluids*, 2001, **13**, 2221–2236.
49. J. M. Deitzel, J. Kleinmeyer, D. Harris and N. C. Beck Tan, *Polymer*, 2001, **42**, 261–272.
50. D. H. Reneker and A. L. Yarin, *Polymer*, 2008, **49**, 2387–2425.
51. S. Chiu-Webster and J. R. Lister, *J. Fluid Mech.*, 2006, **569**, 89–111.
52. Y. Dzenis, *Science*, 2004, **304**, 1917–1919.
53. J. M. Deitzel, J. D. Kleinmeyer, J. K. Hirvonen and N. C. Beck Tan, *Polymer*, 2001, **42**, 8163–8170.
54. J. Kameoka and H. G. Craighead, *Appl. Phys. Lett.*, 2003, **83**, 371–373.
55. M. M. Bergshoef and G. J. Vancso, *Adv. Mater.*, 1999, **11**, 1362–1365.
56. C. Huang, S. Chen, C. Lai, D. H. Reneker, H. Qiu, Y. Ye and H. Hou, *Nanotechnology*, 2006, **17**, 1558–1563.
57. M. Shin, M. M. Hohman, M. P. Brenner and G. C. Rutledge, *Appl. Phys. Lett.*, 2001, **78**, 1149–1151.
58. M. M. Hohman, M. Shin, G. Rutledge and M. P. Brenner, *Phys. Fluids*, 2001, **13**, 2201–2220.
59. T. M. Squires and S. R. Quake, *Rev. Mod. Phys.*, 2005, **77**, 977–1026.
60. S. A. Berger, L. Talbot and L.-S. Yao, *Ann. Rev. Fluid Mech.*, 1983, **15**, 461–512.

61. T. Han, D. H. Reneker and A. L. Yarin, *Polymer*, 2007, **48**, 6064–6076.
62. S. Koombhongse, W. X. Liu and D. H. Reneker, *J. Polym. Sci., Part B: Polym. Phys.*, 2001, **39**, 2598–2606.
63. A. J. Guenthner, S. Khomobhongse, W. Liu, P. Dayal, D. H. Reneker and T. Kyu, *Macromol. Theory Simul.*, 2006, **15**, 87–93.
64. A. Arinstein and E. Zussman, *Phys. Rev. E*, 2010, **76**, 056303.
65. A. L. Yarin, S. Koombhongse and D. H. Reneker, *J. Appl. Phys.*, 2001, **89**, 3018–3026.
66. X.-F. Wu, Y. Salkovskiy and Y. A. Dzenis, *Appl. Phys. Lett.*, 2011, **98**, 223108.
67. L. Pauchard, J.-P. Hulin and C. Allain, *Europhys. News*, 2005, **36**, 9–11.
68. C.-L. Pai, M. C. Boyce and G. C. Rutledge, *Macromolecules*, 2009, **42**, 2102–2114.
69. H. Chang and A. S. Lodge, *Rheol. Acta*, 1972, **11**, 127–129.
70. P.-G. De Gennes, *J. Chem. Phys.*, 1974, **60**, 5030–5042.
71. C. J. Thompson, G. G. Chase, A. L. Yarin and D. H. Reneker, *Polymer*, 2007, **48**, 6913–6922.
72. I. Greenfeld, A. Arinstein, K. Fezzaa, M. H. Rafailovich and E. Zussman, *Phys. Rev. E*, 2011, **84**, 041806.
73. D. E. Smith and S. Chu, *Science*, 1998, **281**, 1335–1340.
74. L. Larrondo and R. St. J. Manley, *J. Polym. Sci.: Polym. Phys. Ed.*, 1981, **19**, 921–932.
75. H. Fong and D. H. Reneker, *J. Polym. Sci., Part B: Polym. Phys.*, 1999, **37**, 3488–3493.
76. P. K. Kahol and N. J. Pinto, *Synth. Met.*, 2004, **140**, 269–272.
77. W. Liang and C. R. Martin, *J. Am. Chem. Soc.*, 1990, **112**, 9666–9668.
78. Z. Cai, J. Lei, W. Liang, V. Menon and C. R. Martin, *Chem. Mater.*, 1991, **3**, 960–966.
79. C.-L. Pai, M. C. Boyce and G. C. Rutledge, *Polymer*, 2011, **52**, 2295–2301.
80. Y. Liu, H. Antaya and C. Pellerin, *J. Phys. Chem. B*, 2010, **114**, 2373–2378.
81. S. F. Fennessey and R. J. Farris, *Polymer*, 2004, **45**, 4217–4225.
82. L. M. Bellan and H. G. Craighead, *Polymer*, 2008, **49**, 3125–3129.
83. J. S. Stephens, S. Frisk, S. Megelski, J. F. Rabolt and D. B. Chase, *Appl. Spectrosc.*, 2001, **55**, 1287–1290.
84. M. Richard-Lacroix and C. Pellerin, *Macromolecules*, 2012, **45**, 1946–1953.
85. M. V. Kakade, S. Givens, K. Gardner, K. H. Lee, D. B. Chase and J. F. Rabolt, *J. Am. Chem. Soc.*, 2007, **129**, 2777–2782.
86. S. Pagliara, M. S. Vitiello, A. Camposeo, A. Polini, R. Cingolani, G. Scamarcio and D. Pisignano, *J. Phys. Chem. C*, 2011, **115**, 20399–20405.
87. R. Jaeger, H. Schönherr and G. J. Vancso, *Macromolecules*, 1996, **29**, 7634–7636.

88. R. Jaeger, M. M. Bergshoeft, C. Martín i Batlle, H. Schlinherr and G. J. Vancso, *Macromol. Symp.*, 1998, **127**, 141–150.
89. H.-M. Liem, P. Etchegoin, K. S. Whitehead and D. D. C. Bradley, *Adv. Funct. Mater.*, 2003, **13**, 66–72.
90. M. C. Gather and D. D. C. Bradley, *Adv. Funct. Mater.*, 2007, **17**, 479–485.
91. R. Dersch, T. Liu, A. K. Schaper, A. Greiner and J. H. Wendorff, *J. Polym. Sci., Part A: Pol. Chem.*, 2003, **41**, 545–553.
92. M. Bognitzki, W. Czado, T. Frese, A. Schaper, M. Hellwig, M. Steinhart, A. Greiner and J. H. Wendorff, *Adv. Mater.*, 2001, **13**, 70–72.
93. S. Megelski, J. S. Stephens, D. B. Chase and J. F. Rabolt, *Macromolecules*, 2002, **35**, 8456–8466.
94. C. L. Casper, J. S. Stephens, N. G. Tassi, D. B. Chase and J. F. Rabolt, *Macromolecules*, 2004, **37**, 573–578.
95. H. Hou, Z. Jun, A. Reuning, A. Schaper, J. H. Wendorff and A. Greiner, *Macromolecules*, 2002, **35**, 2429–2431.
96. S. A. Theron, A. L. Yarin, E. Zussman and E. Kroll, *Polymer*, 2005, **46**, 2889–2899.
97. G. Kim, Y. S. Cho and W. D. Kim, *Eur. Polym. J.*, 2006, **42**, 2031–2038.
98. S. Kidoaki, I. K. Kwon and T. Matsuda, *Biomaterials*, 2005, **26**, 37–46.
99. C. H. Kim, K. S. Khil, H. Y. Kim, H. U. Lee and K. Y. Jahng, *J. Biomed. Mater. Res., Part B*, 2006, **78B**, 283–290.
100. B. Duan, X. Y. Yuan, Y. Zhu, Y. Y. Zhang, X. L. Li, Y. Zhang and K. D. Yao, *Eur. Polym. J.*, 2006, **42**, 2013–2022.
101. B. Ding, E. Kimura, T. Sato, S. Fujita and S. Shiratori, *Polymer*, 2004, **45**, 1895–1902.
102. Y. Hong, K. Fujimoto, R. Hashizume, G. Ryotaro, S. Jianjun, J. John, K. Tobota and W. R. Wagner, *Biomacromolecules*, 2008, **9**, 1200–1207.
103. A. Holzmeister, M. Rudisile, A. Greiner and J. H. Wendorff, *Eur. Polym. J.*, 2007, **43**, 4859–4867.
104. S. Neubert, D. Pliszka, A. Góra, A. Jaworek, E. Wintermantel and S. Ramakrishna, *J. Appl. Polym. Sci.*, 2012, **125**, 820–827.
105. I. C. Um, D. F. Fang, B. Hsiao, A. Okamoto and B. Chu, *Biomacromolecules*, 2004, **5**, 1428–1436.
106. J. Ji, K. Ghosh, B. Li, J. C. Sokolov, R. A. F. Clark and M. H. Rafailovich, *Macromol. Biosci.*, 2006, **6**, 811–817.
107. S. H. Kim, S. Kim, S. Nair and E. Moore, *Macromolecules*, 2005, **38**, 3719–3723.
108. P. Gupta, S. R. Trenor, T. E. Long and G. L. Wilkes, *Macromolecules*, 2004, **37**, 9211–9218.
109. H. Penchev, D. Paneva, N. Manolova and I. Rashkov, *Macromol. Biosci.*, 2009, **9**, 884–894.
110. X. Xu, J.-F. Zhang and Y. Fan, *Biomacromolecules*, 2010, **11**, 2283–2289.
111. L. M. Bellan, G. W. Coates and H. G. Craighead, *Macromol. Rapid Commun.*, 2006, **27**, 511–515.

112. J. D. Schiffman and C. L. Schauer, *Biomacromolecules*, 2007, **8**, 2665–2667.
113. J. Lyons, C. Li and F. Ko, *Polymer*, 2004, **45**, 7597–7603.
114. L. Larrondo and R. St. J. Manley, *J. Polym. Sci.: Polym. Phys. Ed.*, 1981, **19**, 909–920.
115. L. Larrondo and R. St. J. Manley, *J. Polym. Sci.: Polym. Phys. Ed.*, 1981, **19**, 933–940.
116. T. D. Brown, P. D. Dalton and D. W. Hutmacher, *Adv. Mater.*, 2011, **23**, 5651–5657.
117. P. D. Dalton, K. Klinkhammer, J. Salber, D. Klee and M. Möller, *Biomacromolecules*, 2006, **7**, 686–690.
118. O. O. Dosunmu, G. G. Chase, W. Kataphinan and D. H. Reneker, *Nanotechnology*, 2006, **17**, 1123–1127.
119. A. L. Andrady, D. S. Ensor and R. J. Newsome, *US Pat.*, 7,134,857, 2006.
120. C. C. Liao, S. S. Hou, C. C. Wang and C. Y. Chen, *Polymer*, 2010, **51**, 2887–2896.
121. C.-C. Liao, C.-C. Wang and C.-Y. Chen, *Polymer*, 2011, **52**, 4303–4318.
122. M. Li, Y.-Z. Long, D. Yang, J. Sun, H. Yin, Z. Zhao, W. Kong, X. Jiang and Z. Fan, *J. Mater. Chem.*, 2011, **21**, 13159–13162.
123. K. Sarkar, C. Gomez, S. Zambrano, M. Ramirez, E. de Hoyos, H. Vasquez and K. Lozano, *Mater. Today*, 2010, **13**, 12–14.
124. Z. McEachin and K. Lozano, *J. Appl. Polym. Sci.*, 2012, **126**, 473–479.
125. M. R. Badrossamay, H. A. McIlwee, J. A. Goss and K. K. Parker, *Nano Lett.*, 2010, **10**, 2257–2261.
126. H. Pan, L. Li, L. Hu and X. Cui, *Polymer*, 2006, **47**, 4901–4904.
127. X. Cui, L. Li, J. Xu and F. Xu, *J. Appl. Polym. Sci.*, 2010, **116**, 3676–3681.
128. S. Sarkar, S. Deevi and G. Tepper, *Macromol. Rapid Commun.*, 2007, **28**, 1034–1039.
129. R. Kessick, J. Fenn and G. Tepper, *Polymer*, 2004, **45**, 2981–2984.
130. J.-H. He, Y. Wu and N. Pang, *Int. J. Nonlinear Sci. Num. Simul.*, 2005, **6**, 243–248.
131. E. Sanders, R. Kloefkorn, G. Bowlin, D. Simpson and G. Wnek, *Macromolecules*, 2003, **36**, 3803–3805.
132. H. Qi, P. Hu, J. Xu and A. Wang, *Biomacromolecules*, 2006, **7**, 2327–2330.
133. X. Xu, L. Yang, X. Xu, X. Wang, X. Chen, Q. Liang, J. Zeng and X. Jing, *J. Control. Release*, 2005, **108**, 33–42.
134. J. C. Sy, A. S. Klemm and V. P. Shastri, *Adv. Mater.*, 2009, **21**, 1814–1819.
135. A. V. Bazilevsky, A. L. Yarin and C. M. Megaridis, *Langmuir*, 2007, **23**, 2311–2314.
136. X. H. Li, C. L. Shao and Y. C. Liu, *Langmuir*, 2007, **23**, 10920–10923.

137. C. Kim, Y. I. Jeong, B. T. N. Ngoc, K. S. Yang, M. Kojima, Y. A. Kim, M. Endo and J.-W. Lee, *Small*, 2007, **3**, 91–95.
138. A. L. Yarin, *Polym. Adv. Technol.*, 2011, **22**, 310–317.
139. A. L. Yarin, E. Zussman, J. H. Wendorff and A. Greiner, *J. Mater. Chem*, 2007, **17**, 2585–2599.
140. Y. Dror, W. Salalha, R. Avrahami, E. Zussman, A. L. Yarin, R. Dersch, A. Greiner and J. H. Wendorff, *Small*, 2007, **3**, 1064–1073.
141. S. Sinha-Ray, D. D. Pelot, Z. P. Zhou, A. Rahman, X.-F. Wu and A. L. Yarin, *J. Mater. Chem.*, 2012, **22**, 9138–9146.
142. M. Chen, M. Dong, R. Havelund, V. R. Regina, R. L. Meyer, F. Besenbacher and P. Kingshott, *Chem. Mater.*, 2010, **22**, 4214–4221.
143. E. A. Buyuktanir, M. W. Frey and J. L. West, *Polymer*, 2010, **51**, 4823–4830.
144. S. N. Reznik, A. L. Yarin, E. Zussman and L. Bercovici, *Phys. Fluids*, 2006, **18**, 062101.
145. E. Zussman, A. L. Yarin, A. V. Bazilevsky, R. Avrahami and M. Feldman, *Adv. Mater.*, 2006, **18**, 348–353.
146. D. Sun, C. Chang, S. Li and L. Lin, *Nano Lett.*, 2006, **6**, 839–842.
147. C. Chang, K. Limkrailassiri and L. Lin, *Appl. Phys. Lett.*, 2008, **93**, 123111.
148. M. Rinaldi, F. Ruggeri, L. Lozzi and S. Santucci, *J. Vac. Sci. Technol. B*, 2009, **27**, 1829–1833.
149. G. Zheng, W. Li, X. Wang, D. Wu, D. Sun and L. Lin, *J. Phys. D: Appl. Phys.*, 2010, **43**, 415501.
150. H.-Y. Kim, M. Lee, K. J. Park, S. Kim and L. Mahadevan, *Nano Lett.*, 2010, **10**, 2138–2140.
151. C. Chang, V. H. Tran, J. Wang, Y.-K. Fuh and L. Lin, *Nano Lett.*, 2010, **10**, 726–731.
152. G. S. Bisht, G. Canton, A. Mirsepassi, L. Kulinsky, S. Oh, D. Dunn-Rankin and M. J. Madou, *Nano Lett.*, 2011, **11**, 1831–1837.
153. F.-L. Zhou, P. L. Hubbard, S. J. Eichhorn and G. J. M. Parker, *Polymer*, 2011, **52**, 3603–3610.
154. D. Chen, S. Lei and Y. Chen, *Sensors*, 2011, **11**, 6509–6516.
155. C. Hellmann, J. Belardi, R. Dersch, A. Greiner, J. H. Wendorff and S. Bahnmueller, *Polymer*, 2009, **50**, 1197–1205.
156. H. Wang, G. Zheng, W. Li, X. Wang and D. Sun, *Appl. Phys. A.*, 2011, **102**, 457–461.
157. J. Kameoka, R. Orth, Y. Yang, D. Czaplewski, R. Mathers, G. W. Coates and H. G. Craighead, *Nanotechnology*, 2003, **14**, 1124–1129.
158. S.-Y. Min, T. S. Kim, J. H. Cho and T.-W. Lee, presented at the *2nd International Conference of Electrospinning*, Jeju, Korea 2012.
159. Y. Srivastava, M. Marquez and T. Thorsen, *J. Appl. Polymer Sci.*, 2007, **106**, 3171–3178.

160. S. Madhugiri, A. Dalton, J. Gutierrez, J. P. Ferraris and K. J. Balkus, *J. Am. Chem. Soc.*, 2003, **125**, 14531–14538.
161. Z. Sun, E. Zussman, A. L. Yarin, J. H. Wendorff and A. Greiner, *Adv. Mater.*, 2003, **15**, 1929–1932.
162. I. G. Loscertales, A. Barrero, M. Marquez, R. Spretz, R. Velarde-Ortiz and G. Larsen, *J. Am. Chem. Soc.*, 2004, **126**, 5376–5737.
163. D. Li and Y. Xia, *Nano Lett.*, 2004, **4**, 933–938.
164. J. H. Yu, S. V. Fridrikh and G. C. Rutledge, *Adv. Mater.*, 2004, **16**, 1562–1566.
165. Y. Zhang, Z.-M. Huang, X. Xu, C. T. Lim and S. Ramakrishna, *Chem. Mater.*, 2004, **16**, 3406–3409.
166. I. G. Loscertales, A. Barrero, I. Guerrero, R. Cortijo, M. Marquez and A. M. Gañán-Calvo, *Science*, 2002, **295**, 1695–1698.
167. D. Li, A. Babel, S. A. Jenekhe and Y. Xia, *Adv. Mater.*, 2004, **16**, 2062–2066.
168. D. Li, J. T. McCann and Y. Xia, *Small*, 2005, **1**, 83–86.
169. J. T. McCann, D. Li and Y. Xia, *J. Mater. Chem.*, 2005, **15**, 735–738.
170. Y. Z. Zhang, J. Venugopal, Z.-M. Huang, C. T. Lim and S. Ramakrishna, *Biomacromolecules*, 2005, **6**, 2583–2589.
171. Y. Z. Zhang, X. Wang, Y. Feng, J. Li, C. T. Lim and S. Ramakrishna, *Biomacromolecules*, 2006, **7**, 1049–1057.
172. F. Yi and D. A. LaVan, *Macromol. Biosci.*, 2008, **8**, 803–806.
173. S. W. Lee, G. D. Moon and U. Jeong, *J. Mater. Chem.*, 2009, **19**, 743–748.
174. S. W. Lee, H. J. Lee, J. H. Choi, W. G. Koh, J. M. Myoung, J. H. Hur, J. J. Park, J. H. Cho and U. Jeong, *Nano Lett.*, 2010, **10**, 347–351.
175. C. S. Reddy, A. Arinstein, R. Avrahami and E. Zussman, *J. Mater. Chem.*, 2009, **19**, 7198–7201.
176. H. Chen, N. Wang, J. Di, Y. Zhao, Y. Song and L. Jiang, *Langmuir*, 2010, **26**, 11291–11296.
177. Y. Zhao, X. Cao and L. Jiang, *J. Am. Chem. Soc.*, 2007, **129**, 764–765.
178. J. Di, Y. Zhao and J. Yu, *J. Mater. Chem.*, 2011, **21**, 8511–8520.
179. V. Kalra, J. H. Lee, J. H. Park, M. Marquez and Y. L. Joo, *Small*, 2009, **20**, 2323–2332.
180. M. Ma, V. Krikorian, J. H. Yu, E. L. Thomas and G. C. Rutledge, *Nano Lett.*, 2006, **6**, 2969–2972.
181. V. Kalra, S. Mendez, J. H. Lee, H. Nguyen, M. Marquez and Y. L. Joo, *Adv. Mater.*, 2006, **18**, 3299–3303.
182. H. Chen, N. Wang, J. Di, Y. Zhao, Y. Song and L. Jiang, *Langmuir*, 2010, **26**, 11291–11296.
183. B. Carlberg, T. Wang and J. Liu, *Langmuir*, 2010, **26**, 2235–2239.
184. D. Zhang and J. Chang, *Nano Lett.*, 2008, **8**, 3283–3287.
185. C. Del Gaudio, A. Bianco and M. Grigioni, *Ann. Ist. Super. Sanità*, 2007, **44**, 178–186.

186. X. Wang, K. Zhang, M. Zhu, H. Yu, Z. Zhou, Y. Chen and B. S. Hsiao, *Polymer*, 2008, **49**, 2755–2761.
187. P. D. Dalton, D. Klee and M. Möller, *Polymer*, 2005, **46**, 611–614.
188. G. Chang and J. Shen, *Macromol. Rapid Comm.*, 2010, **31**, 2151–2154.
189. L.-Q. Liu, M. Eder, I. Burgert, D. Tasis, M. Prato and H. Daniel Wagner, *Appl. Phys. Lett.*, 2007, **90**, 083108.
190. W. E. Teo, R. Gopal, R. Ramaseshan, K. Fujihara and S. Ramakrishna, *Polymer*, 2007, **48**, 3400–3405.
191. U. Ali, Y. Zhou, X. Wang and T. Lin, *J. Text. I.*, 2011, **103**, 80–88.
192. B. K. Gu, M. K. Shin, K. W. Sohn, S. I. Kim, S. J. Kim, S.-K. Kim, H. Lee and J. S. Park, *Appl. Phys. Lett.*, 2007, **90**, 263902.
193. W. E. Teo and S. Ramakrishna, *Nanotechnology*, 2006, **17**, R89–R106.
194. S. Pagliara, A. Camposeo, A. Polini, R. Cingolani and D. Pisignano, *Lab Chip*, 2009, **9**, 2851–2856.
195. T. Tamura and H. Kawakami, *Nano Lett.*, 2010, **10**, 1324–1328.
196. P. Anzenbacher, Jr. and M. A. Palacios, *Nat. Chem.*, 2009, **1**, 80–86.
197. P. Anzenbacher Jr., F. Li and M. A. Palacios, *Angew. Chem. Int. Ed.*, 2012, **51**, 2345–2348.
198. D. Li, Y. Wang and Y. Xia, *Nano Lett.*, 2003, **3**, 1167–1171.
199. D. Li, Y. Wang and Y. Xia, *Adv. Mater.*, 2004, **16**, 361–366.
200. L. Liu and Y. Dzenis, *Nanotechnology*, 2008, **19**, 355307.
201. J. Xie, M. R. MacEwan, W. Z. Ray, W. Liu, D. Y. Siewe and Y. Xia, *ACS Nano*, 2010, **4**, 5027–5036.
202. B. Sundaray, V. Subramanian and T. S. Natarajan, *Appl. Phys. Lett.*, 2004, **84**, 1222–1224.
203. A. Theron, E. Zussman and A. L. Yarin, *Nanotechnology*, 2001, **12**, 384–390.
204. E. Zussman, A. Theron and A. L. Yarin, *Appl. Phys. Lett.*, 2003, **82**, 973–975.
205. J. Kameoka, D. Czaplewski, H. Liu and H. G. Craighead, *J. Mater. Chem.*, 2004, **14**, 1503–1505.
206. P. Katta, M. Alessandro, R. D. Ramsier and G. G. Chase, *Nano Lett.*, 2004, **4**, 2215–2218.
207. P. Kiselev and J. Rosell-Llompart, *J. Appl. Polym. Sci.*, 2012, **125**, 2433–2441.
208. D. Y. Yang, B. Lu, Y. Zhao and X. Y. Jiang, *Adv. Mater.*, 2007, **19**, 3702–3706.
209. D. Yang, J. Zhang, J. Zhang and J. Nie, *J. Appl. Polym. Sci.*, 2008, **110**, 3368–3372.
210. Y. Liu, X. Zhang, Y. Xia and H. Yang, *Adv. Mater.*, 2010, **22**, 2454–2457.
211. A. Jafari, J.-H. Jeon and I.-K. Oh, *Macromol. Rapid Commun.*, 2011, **32**, 921–926.

212. C. Ayres, G. L. Bowlin, S. C. Henderson, L. Taylor, J. Shultz, J. Alexander, T. A. Telemeco and D. G. Simpson, *Biomaterials*, 2006, **27**, 5524–5534.
213. J. M. Corey, D. Y. Lin, K. B. Mycek, Q. Chen, S. Samuel, E. L. Feldman and D. C. Martin, *J. Biomed. Mater. Res., Part A*, 2007, **83A**, 636–645.
214. X. Lu, C. Wang and Y. Wei, *Small*, 2009, **5**, 2349–2370.
215. X. M. Sui, C. L. Shao and Y. C. Liu, *Appl. Phys. Lett.*, 2005, **87**, 113115.
216. H. Sun, H. Zhang, J. Zhang, H. Wei, J. Ju, M. Li and B. Yang, *J. Mater. Chem.*, 2009, **19**, 6740–6744.
217. V. Vohra, A. Devaux, L.-Q. Dieu, G. Scavia, M. Catellani, G. Calzaferri and C. Botta, *Adv. Mater.*, 2009, **21**, 1146–1150.
218. V. Vohra, A. Devaux, L.-Q. Dieu, G. Scavia, M. Catellani, G. Calzaferri and C. Botta, *ACS Nano*, 2010, **4**, 1409–1416.
219. S. Schlecht, S. Tan, M. Yosef, R. Dersch, J. H. Wendorff, Z. Jia and A. Schaper, *Chem. Mater.*, 2005, **17**, 809–814.
220. H. Liu, J. B. Edel, L. M. Bellan and H. G. Craighead, *Small*, 2006, **2**, 495–499.
221. C. Erisken, D. M. Kalyon and H. Wang, *Nanotechnology*, 2008, **19**, 165302.
222. Y. Xin, Z. H. Huang, L. Peng and D. J. Wang, *J. Appl. Phys.*, 2009, **105**, 086106.
223. Q. B. Yang, D. M. Li, Y. L. Hong, Z. Y. Li, C. Wang, S. L. Qiu and Y. Wei, *Synth. Met.*, 2003, **137**, 973–974.
224. V. V. Ramanan, K. C. Hribar, J. S. Katz and J. A. Burdick, *Nanotechnology*, 2011, **22**, 494009.
225. C.-L. Zhang, K.-P. Lv, H.-P. Cong and S.-H. Yu, *Small*, 2012, **8**, 648–653.
226. M. Bashouti, W. Salalha, M. Brumer, E. Zussman and E. Lifshitz, *ChemPhysChem*, 2006, **7**, 102–106.
227. A. Stoiljkovic, M. Ishaque, W. Justus, L. Hamel, E. Klimov, W. Heckmann, B. Eckhardt, J. H. Wendorff and A. Grinwer, *Polymer*, 2007, **48**, 3974–3981.
228. F. Di Benedetto, A. Camposeo, L. Persano, A. M. Laera, E. Piscopiello, R. Cingolani, L. Tapfer and D. Pisignano, *Nanoscale*, 2011, **3**, 4234–4239.
229. L. Persano, A. Camposeo, F. Di Benedetto, R. Stabile, A. M. Laera, E. Piscopiello, L. Tapfer and D. Pisignano, *Adv. Mater*, 2012, **24**, 5320–5326.
230. D. Li, T. Herricks and Y. Xia, *Appl. Phys. Lett.*, 2003, **83**, 4586–4588.
231. W. K. Son, J. H. Youk, T. S. Lee and W. H. Park, *Macromol. Rapid. Commun.*, 2004, **25**, 1632–1637.
232. K. H. Hong, J. L. Park, I. H. Sul, J. H. Youk and T. J. Kang, *J. Polym. Sci. Part B: Polym. Phys.*, 2006, **44**, 2468–2474.
233. M. Graeser, M. Bognitzki, W. Massa, C. Pietzonka, A. Greiner and J. H. Wendorff, *Adv. Mater.*, 2007, **19**, 4244–4247.

234. X. Lu, Y. Zhao and C. Wang, *Adv. Mater.*, 2005, **17**, 2485–2488.
235. C. Wang, E. Yan, Z. Sun, Z. Jiang, Y. Tong, Y. Xin and Z. Huang, *Macromol. Mater. Eng.*, 2007, **292**, 949–955.
236. F. Ko, Y. Gogotsi, A. Ali, N. Naguib, H. Ye, G. Yang, C. Li and P. Willis, *Adv. Mater.*, 2003, **15**, 1161–1165.
237. H. Hou, J. J. Ge, J. Zeng, Q. Li, D. H. Reneker, A. Greiner and S. Z. D. Cheng, *Chem. Mater.*, 2005, **17**, 967–973.
238. Y. Dror, W. Salalha, R. L. Khalfin, Y. Cohen, A. L. Yarin and E. Zussman, *Langmuir*, 2003, **19**, 7012–7020.
239. Q. Bao, H. Zhang, J.-X. Yang, S. Wang, D. Y. Tang, R. Jose, S. Ramakrishna, C. T. Lim and K. P. Loh, *Adv. Funct. Mater.*, 2010, **20**, 1–10.
240. D. M. Luzhansky, presented at *INTC 2003: International Nonwovens Technical Conference*, 2003.
241. WIPO PATENTSCOPE, full-text keywords: (Polymer) and (Nanofibers) and (Electrospinning). http://patentscope.wipo.int/search/en/search.jsf (accessed November 2012). The World Intellectual Property Organization (WIPO) bears no responsibility for the integrity or accuracy of the data contained herein, in particular due, but not limited, to any deletion, manipulation, or reformatting of data that may have occurred beyond its control.
242. W. Dayong, H. Wang and J. Cao, International Patent Application No. PCT/CN2010/000922, 2010.
243. A. M. dos Santos, J. Dierck, M. Troch, M. Podevijn and E. Schacht, *Macromol. Mater. Eng.*, 2011, **296**, 637–644.
244. J. S. Varabhas, G. G. Chase and D. H. Reneker, *Polymer*, 2008, **49**, 4226–4229.
245. T. H. Grafe and K. M. Graham, presented at *INTC 2002: International Nonwovens Technical Conference*, 2002.
246. Inovenso Ltd, Industrial Machines, http://www.inovenso.com/products/industrial-nanofiber-machines/(accessed November, 2012).
247. Y. Filatov, A. Budyka and V. Kirichenko, *Electrospinning of Micro- and Nanofibers: Fundamentals and Applications in Separation and Filtration Processes*, Begell House, Inc., New York, 2007.
248. C. J. Luo, S. D. Stoyanov, E. Stride, E. Pelanb and M. Edirisinghe, *Chem. Soc. Rev.*, 2012, **41**, 4708–4735.
249. A. Varesano, F. Rombaldoni, G. Mazzuchetti, C. Tonina and R. Comotto, *Polym. Int.*, 2010, **59**, 1606–1615.
250. A. Varesano, R. A. Carletto and G. Mazzuchetti, *J. Mater. Process. Technol.*, 2009, **209**, 5178–5185.
251. A. L. Yarin and E. Zussman, *Polymer*, 2004, **45**, 2977–2980.
252. J. C. Almekinders and C. Jones, *J. Aerosol. Sci.*, 1999, **30**, 969–71.
253. O. Jirsák, F. Sanetrnik, D. Lukáš, V. Kotek, L. Martinova and J. Chaloupek, International Patent Publication No. WO 2005/024101, 2005.

254. Elmarco, NS Production Lines 1600, http://www.elmarco.cz/products/ns-organic-production-lines-1600/(accessed November, 2012).
255. S. Petrik and M. Maly, *MRS Proceedings*, 2009, **1240**, 1240-WW03-07.
256. Dienes Apparatebau, Nanofiber Spinning Installations, http://www.dienes.net/docs/Dienes%20Article%20-%20Nanofiber%20Spinning%202010-10%20en.pdf (accessed November, 2012).
257. IME Technologies, Climate-Controlled Electrospinning, http://www.imetechnologies.nl/uploads/images/ime_electrospinning_brochure_2012.pdf (accessed November, 2012).
258. D. A. Saville, *Annu. Rev. Fluid Mech.*, 1997, **29**, 27–64.
259. G. I. Taylor, *Proc. R. Soc. London, Ser. A*, 1966, **291**, 159–166.
260. D. A. Saville, *Phys. Fluids*, 1970, **13**, 2987–94.
261. D. A. Saville, *J. Fluid Mech.* 1971, **48**, 815–827.
262. D. A. Saville, *Phys. Fluids*, 1971, **14**, 1095–99.
263. A. M. Gañán-Calvo, J. Davila and A. Barrero, *J. Aerosol. Sci.*, 1997, **28**, 249–275.
264. A. M. Gañán-Calvo, *Phys. Rev. Lett.*, 1997, **79**, 217–220.
265. A. M. Gañán-Calvo, *J. Fluid Mech.*, 1997, **335**, 165–188.
266. A. M. Gañán-Calvo, *J. Aerosol. Sci.*, 1999, **30**, 863–872.
267. A. M. Gañán-Calvo, J. M. López-Herrera and P. Riesco-Chueca, *J. Fluid Mech.*, 2006, **566**, 421–445.
268. A. M. Gañán-Calvo, R. Gonzáles-Prieto, P. Riesco-Chueca, M. A. Herrada and M. Flores-Mosquera, *Nat. Phys.*, 2007, **3**, 737–742.
269. A. M. Gañán-Calvo, *Phys. Rev. Lett.*, 2007, **98**, 134503.
270. A. M. Gañán-Calvo and J. M. Montanero, *Phys. Rev. E*, 2009, **79**, 066305.
271. A. F. Spivak and Y. A. Dzenis, *Appl. Phys. Lett.*, 1998, **73**, 3067–3069.
272. A. F. Spivak and Y. A. Dzenis, *J. Appl. Mech.*, 1999, **66**, 1026–1028.
273. A. F. Spivak, Y. A. Dzenis and D. H. Reneker, *Mech. Res. Commun.*, 2000, **27**, 37–42.
274. R. B. Bird, R. C. Armstrong and O. Hassager, Dynamics of Polymeric Liquids, Wiley, New York, 2nd edn, 1987, vol. **1**.
275. D. H. Reneker, A. L. Yarin, E. Zussman and H. Xu, *Adv. Appl. Mech.*, 2007, **41**, 42–195.
276. P. E. Rouse, *J. Chem. Phys*, 1953, **21**, 1272–1280.
277. A. L. Yarin, *J. Non-Newtonian Fluid Mech.*, 1990, **37**, 113–138.
278. A. L. Yarin, *Free Liquid Jets and Films: Hydrodynamics and Rheology*, Longman Scientific & Technical and Wiley & Sons, Harlow, New York, 1993.
279. A. Ziabicki, *Fundamentals of Fibre Formation*, Wiley, London, 1976.
280. G. C. Rutledge and S. V. Fridrikh, *Adv. Drug Del. Rev.*, 2007, **59**, 1384–1391.
281. J. Eggers, *Rev. Mod. Phys.*, 1997, **69**, 865–929.
282. E. J. Hinch, *Perturbation Methods*, Cambridge University Press, New York, 1991.

283. J. J. Feng, *Phys. Fluids*, 2002, **14**, 3912–3926.
284. J. W. S. Rayleigh, *Philos. Mag.*, 1882, **14**, 184–186.
285. F. Savart, *Annal. Chim.*, 1833, **53**, 337–374, plates in vol. 54.
286. J. Plateau, *Acad. Sci. Bruxelles Mém.*, 1849, **23**, 5–150.
287. J. W. S. Rayleigh, *Proc. London Math. Soc.*, 1878, **10**, 4–13.
288. J. J. Kaufman, G. Tao, S. Shabahang, E.-H. Banaei, D. S. Deng, X. Liang, S. G. Johnson, Y. Fink and A. F. Abouraddy, *Nature*, 2012, **487**, 463–467.
289. Y. M. Shin, M. M. Hohmann, M. P. Brenner and G. C. Rutledge, *Polymer*, 2001, **42**, 9955–9967.
290. J. J. Feng, *J. Non-Newtonian Fluid Mech.*, 2003, **116**, 55–70.
291. C. P. Carroll and Y. L. Joo, *Phys. Fluids*, 2006, **18**, 053102.
292. C. P. Carroll and Y. L. Joo, *J. Non-Newtonian Fluid Mech.*, 2008, **153**, 130–148.
293. C. P. Carroll and Y. L. Joo, *Phys. Fluids*, 2009, **21**, 103101.
294. C. J. Angammana and S. H. Jayaram, *Proc. ESA Annual Meeting on Electrostatics*, 2011, E2.
295. C. J. Angammana and S. H. Jayaram, *IEEE Trans. Ind. Appl.*, 2011, **47**, 1109–1117.
296. P. K. Bhattacharjee, T. M. Schneider, M. P. Brenner, G. H. McKinley and G. C. Rutledge, *J. Appl. Phys.*, 2010, **107**, 044306.
297. S. N. Reznik and E. Zussman, *Phys. Rev. E*, 2010, **81**, 026313.
298. I. Greenfeld, K. Fezzaa, M. H. Rafailovich and E. Zussman, *Macromolecules*, 2012, **45**, 3616–3626.

CHAPTER 3

# Other Fabrication Technologies for Polymer Nanofibers

## 3.1 Self-assembly and Polymerization Methods

Self-assembly processes can be observed at very different length-scales in nature and in technology. Systems showing assembling and self-patterning behavior include, besides molecules, various classes of colloidal particles and nanocrystals,[1–3] of which we have already seen a beautiful example in Figure 1.4b, mesoscopic[4–7] and even macroscopic objects.[8–11] Flows, liquid evaporation, sedimentation, electric fields, *etc.* can assist assembling during the formation of structures by individual components. For instance, in the late 1990s Whitesides and coworkers have extensively investigated meso- and macroscale self-assembly, driven by capillary forces in liquid media,[12] at the interface between two fluids,[9,13] and based on object shape-selective recognition.[8] Many studies have been specifically focused on self-assembling soft matter components and systems.[14–17]

The most general definition of self-assembly is likely that of a spontaneous arrangement of physical systems into a stable structure, whose organization involves some kind of inter-component interactions linking individual elements, and which is globally closer to thermodynamic equilibrium than the original state. It is frequently claimed that the fact that self-assembling systems aim to reach thermodynamically stable states has, as a direct consequence, that the resulting arrangements, structures and materials show relatively low densities of defects, and often present self-repairing properties, which make them appealing for technological applications.[18] In particular, a distinctive feature of molecular self-assembly processes is that re-arrangements are driven by relatively weak

RSC Nanoscience & Nanotechnology No. 29
Polymer Nanofibers: Building Blocks for Nanotechnology
By Dario Pisignano

Published by the Royal Society of Chemistry, www.rsc.org

(*i.e.* non-covalent) inter-molecular interactions (Van der Waals, electrostatic, dipolar, hydrogen bonds, *etc.*).[19,20] Furthermore, attractive interactions have to compete with other classes of forces, such as, for example, those due to repulsive local interactions and thermal agitation, whose intensity is basically of their same order of magnitude, a feature that is close to those distinctive of the Soft Matter concept (Section 1.1). This is a basic prerequisite for self-assembly to take place, since the competition between different kinds of inter-object forces guarantees that the achieved supramolecular arrangements are rarely fully irreversible, thus leaving some space for further rearrangements. In other words, this gives a degree of *adjustability* to the overall supramolecular system, which is therefore able to explore adjacent spatial and energetic configurations and to find the most favorable, eventually ordered arrangements*. In particular, many relevant examples of self-assembly processes can be found in biological systems, such as the three-dimensional folding of polypeptides into proteins, a somehow *templated self-assembly* driven by specific sequences of amino acids.[10] Scientists have started mimicking some of these phenomena in a quite effective way. For instance, copolypeptides can be assembled by various routes and hierarchically up to form fibers.[21] At a different scale, native cellulose fibers from plants can be isolated, functionalized and assembled to form new materials such as aerogels, and so on.[21] All these concepts have influenced in depth polymer science through the years, and inspired the development of many bottom-up approaches for realizing one-dimensional or quasi-one-dimensional polymer nanostructures. We here describe several self-assembly strategies to fabricate polymer nanofibers. The last part of this Section (Section 3.1.4) reviews some alternatives that are instead based on polymerization methods, in which monomers are the building blocks and the formation of covalent bonds and of macromolecules directly leads to nanostructures.

### 3.1.1 Self-assembly from Marginal Solvents

A lot of research on self-assembled nanofibers has been focused on polymer precipitation from solvents, whose ability to dissolve a given polymer species is poor and generally strongly temperature-dependent (so-called *marginal solvents*).[22–28] In this case, the solubility is relatively good only at high enough temperature (Section 1.4.3). Consequently, cooling the solution under controlled conditions can lead to the formation of solid, precipitating nanofibers, which often exhibit enhanced microcrystalline order as in the case

* This is just the opposite of the mechanisms (ineffective molecular rearrangements, steric hindrance, *etc.*) that lead to the disordered, glassy state of matter (Section 1.4.1). It is especially intriguing that nowadays nano-manufacturing processes and technologies aim to exploit more and more these two main (and competing) possible phenomenologies exhibited by soft matter, *i.e.* either large or small molecular mobility, either high or low adjustability of the supramolecular state, through either self-assembly or glass transition-related molding approaches, respectively. The distinction between *bottom-up* and *top-down* nanofabrication concepts and methods can be largely, though not completely, drawn on this basis.

of regioregular poly-3-hexylthiophene (P3HT). Such resulting order can be rationalized by considering that, in conditions of poor solubility, the polymer molecules try to minimize the energetically unfavorable solvent-backbone interaction sites ($N_s\phi_p$ or $N_pN_r\phi_s$ according to the Flory–Huggins model, where $\phi_p$ and $\phi_s$ indicate the polymer and the solvent volume fractions, respectively, as described in Section 1.4.3). This so-called *solvophobicity* thus drives the polymer system towards a quite closed configuration, in which, for instance, P3HT chains package with thiophene rings disposed face-to-face.[29–31]

Solvent-based nanofiber assembly methods capitalize important, seminal studies of the precipitation of polythiophene whiskers† or mats from solutions of cyclohexanone, *p*-xylene, methanol, *etc.*, and of the related mechanisms of crystallization and molecular orientation within the resulting nanostructures.[22–24] These works have led to clear indications of the preferential orientation of the $\pi-\pi$ stacking direction along the fiber longitudinal axis. We will come back to this $\pi-\pi$ stacking geometry and to the particular crystalline structure typical of P3AT films and nanofibers in Section 6.1.1 due to the importance of this supramolecular configuration for charge-carrier transport. Here, we want to emphasize once again that the fact that these materials show a clear, various crystallization phenomenology is not at all obvious. As detailed in Section 1.4.1, barriers against molecular rearrangement (*i.e.* against the achievement of energetic minima) defining the energy landscape peculiar of polymeric materials hardly allow them to reach completely crystalline states. Even when crystallization takes place, polycrystalline samples are by far the most frequent product. In fact, the crystallization kinetics of P3ATs is still a matter of investigation. The self-assembly of P3ATs from solutions is found to proceed through isothermal crystallization whose kinetics follows the Avrami equation:[24,33,34]

$$\phi(t) = 1 - e^{-k_C t^n}, \tag{3.1}$$

where $\phi$ indicates the volume fraction of the crystalline phase, $k_C$ indicates the overall process rate and the exponent $n$ is related to the specific crystallization process. Measured $n$ values are in the range of 0.6–1.4[24] and indicate a one-dimensional heterogeneous nucleation and a linear growth of the crystalline structure.[33–35]

Self-assembled fibers of P3HT have typical reported length of 0.2–5 μm, width as small as 15 nm and height well below 10 nm (Figure 3.1).[25–28] The width and height of P3AT fibers also show a weak increment upon increasing the length of the alkyl chain.[25,26] A recent study performs a systematic investigation of nanofibers self-assembled starting from several different solvents and by P3ATs with alkyl chains of different lengths (from 3 to 9 C atoms).[36] All these fibers show a diameter of (20 ± 5) nm, but they exhibit

† This approach is also called *whisker method* in the literature and it is appreciated particularly for the ease of processing and the relatively high amount of obtainable nanofibers, offering good perspectives in view of an industrial application.[25,31,32]

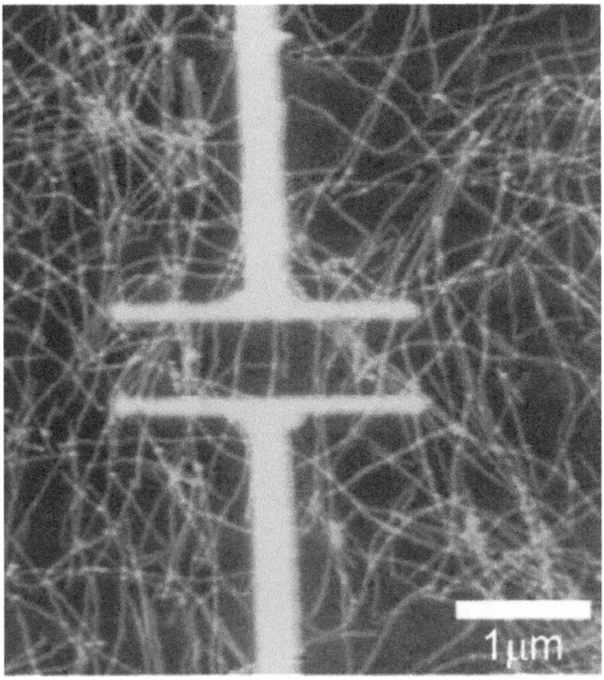


**Figure 3.1** Example of self-assembled nanofibers made of P3HT, between metal electrodes for characterization. Adapted with permission from Ref. 28, *J. Phys. Chem. B.*, 2004, **108**, 19169–19179. Doi: 10.1021/jp047023a. Copyright © 2004, American Chemical Society. Further details on the microcrystalline organization internal to P3HT nanofibers are given in Section 6.1.

various degrees of supramolecular ordering and different crystalline phases. The effect of different solvents on the resulting nanofiber morphology can also be investigated by means of phase-contrast AFM, which is especially informative to distinguish surface fibrillar domains.[37] The fiber diameter is also reported to increase with the polymer *MW* up to a threshold of about 10,000 a.m.u..[38] Below such *MW*, the width of the whiskers corresponds well to the length of extended polymer chains, reaching a maximum value of 13–15 nm at the threshold value. Above 10,000 a.m.u., the width keeps constant around the maximum, losing its *MW*-dependence, thus suggesting a corresponding folding of P3AT molecules within the fibers.[38] Also polythiophene diblock copolymers containing side chains of different polarity, such as hexyl (non-polar) and triethyliene glycol (polar) groups, are found to self-assemble in solution in the form of crystalline fibers, and the molecules can be driven to take well-defined helical structures that in turn assemble in very beautiful super-helices (Figure 3.2).[39] Finally, crystalline-crystalline diblock *co*-P3ATs self-assemble in solution of 1,2-dichlorobenzene, forming nanowires with width around 13 nm and length up to the μm-scale.[40]

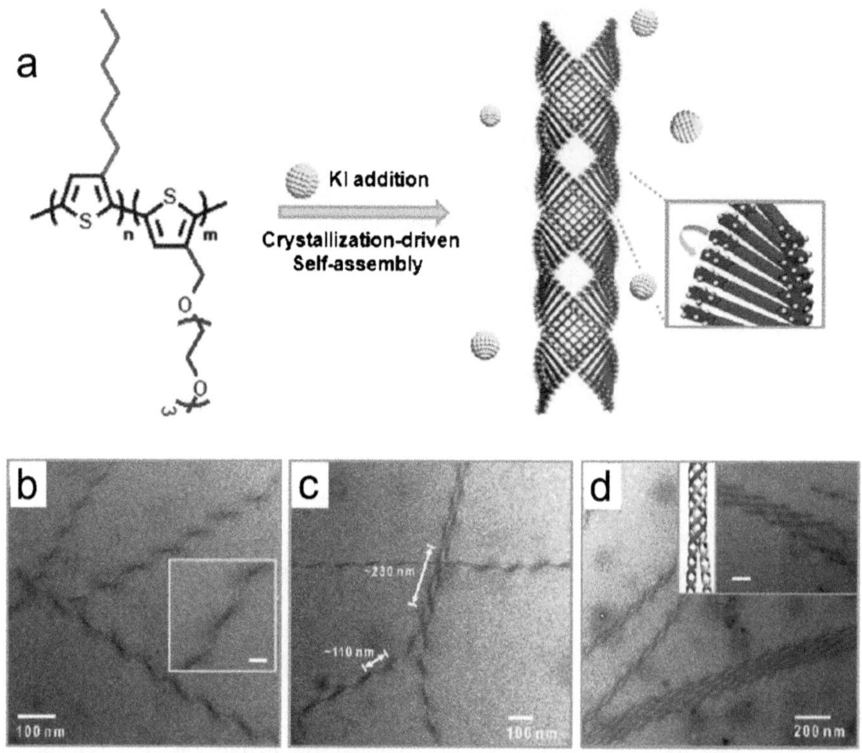


**Figure 3.2** (a) Molecular structure of polythiophene diblock copolymers, and schematics of their assembly into super-helices. Crystallization occurs in the presence of $K^+$ ions. (b) TEM micrographs of the assembled structures of the copolymer, P3HT-*block*-poly(3-triethylene glycol thiophene) with P3HT: poly(3-triethylene glycol thiophene) weight ratio of 2:1, evidencing the formation of regular helices. Inset: magnified image (scale bar = 20 nm). (c, d) TEM images showing the transformation from single- (c) to multiple strands (d). Inset in (d): TEM micrograph (scale bar = 100 nm), and corresponding scheme illustrating the transition from double to quadruple helical structures. Adapted with permission from Ref. 39, *J. Am. Chem. Soc.*, 2011, **133**, 10390–10393. DOI: 10.1021/ja2038547. Copyright © 2011, American Chemical Society.

Other types of elongated polymer nanostructures, in the form of so-called nanobelts, are obtained from poly(benzobisimidazobenzophenanthroline) (BBL) through an approach in solution.[41] For this material, which has the important characteristic of a favored transport of electrons (Sections 6.2–6.3), nanobelts self-assemble upon adding dropwise a (0.2 mg $mL^{-1}$) solution in methanesulfonic acid to a stirring vial of (4:1) chloroform:methanol.[41] Chloroform and methanol serve, respectively, as weakly interacting solvent and deprotonating base for BBL, thus priming the self-assembly process. The resulting nanostructures show a distinctive supramolecular organization compared with typical P3AT nanofibers, as described in Section 6.2.

Furthermore, the BBL nanostructures can be dispersed in many solvents, including water, and can thus be handled and deposited onto solid substrates to realize electronic devices.

In general, however, a point of weakness of solvent-based assembly strategies can be seen in the difficulty of obtaining spatial control, accurate positioning or mutual alignment of the synthetized nanostructures. Typically, dispersed colloidal nanoparticles (both organic and inorganic), nanofibers *etc.* are firstly deposited from their solutions, and then integrated into devices by specifically designed lithographic, printing or transfer methods. Transfer and related printing techniques for device fabrication have recently been reviewed by the groups of Y. Huang, R. G. Nuzzo and J. A. Rogers.[42] Alternative approaches have also been tried in order to address orientation issues for self-assembling polymer nanofibers. In particular, conductive nanofibers of poly(3,3‴-didodecylquarterthiophene) have been demonstrated by self-assembly through anisotropic crystallization in a commercially available nematic liquid crystal (E44), in which the polymer is miscible, thus allowing mutual alignment of the formed nanofibers to be achieved.[32] Other examples of liquid crystal templates are given in Section 3.3.2.

### 3.1.2 Casting Methods

Solvent-based methods to obtain polymer nanofibers have been developed into several techniques in which solution casting, together with the resulting interaction with the solid substrate and solvent evaporation conditions, take some role in favoring the self-assembly of nanostructures. Very narrow distributions in the size of P3HT fibrils are reported by optimizing the processing conditions of casting from toluene solutions.[43] Crystalline microwires are also demonstrated by casting chloroform solutions of P3HT on Si substrates functionalized with octadecyltrichlorosilane, followed by slow solvent evaporation in a closed jar favoring a condition of increased solvent vapor pressure.[44] Similar, elongated nanostructures can be produced by casting toluene solutions of block copolymer molecules alternating P3AT segments together with PS or poly(methyl acrilate),[45,46] and fibrils can be deposited from solvents for oligofluorene-based molecules and chiral oligomers of oligo(thiophene)s and oligo(*para*phenylene vinylene)s as well, with some of these compounds forming helical fibers and the substrate polarity playing an important role in affecting chirality.[47]

Nanofibrous structures can assemble even simply as a consequence of spin-coating processes, in which the rapid solvent evaporation assists the formation and the supramolecular organization of the polymer molecules. This is for instance the behavior of poly[3,4-(4,5-didodecylphenylene)dioxythiophene] deposited from hot (130 °C) solutions of *o*-dichlorobenzene.[48] The resulting fibers have height of 3 nm and width of about 10 nm. Similar nanostructures, with height in the range of 2–15 nm and width from a few nm to about 20 nm, are obtained by casting several different systems, such as poly(phenylene ethylene) (PPE) derivatives from tetrahydrofuran (THF),[49,50] poly(3,3‴-didodecyl quarter thiophene) from toluene,[51] P3HT from chloroform under various conditions of solvent vapor pressure,[52,53] and so on.

Once obtained, polythiopene whiskers can be modified in their internal, supramolecular organization, thus giving rise to a peculiar configuration with the polymer chains perpendicular to the solid substrate underneath, by means of an annealing procedure with carbon disulfide ($CS_2$) vapors.[54] The $CS_2$ molecules, once adsorbed and locally dissolving the pristine, solid polymer structure, are supposed to assist the chains in their transition towards the orientation perpendicular to the substrate.[54] Another recent, related line of research concerns the realization of nanocomposites embedding polythiophene nanowires in an amorphous thermoplastic matrix. The nanowires can be obtained in a very elegant way from blends of polythiophene and PS, exploiting selectively marginal solvents for the polythiophene fraction such as *o*-dichlorobenzene[55] or dichloromethane[‡].[56]

Self-assembled nanotubes of J-aggregates of the amphiphilic cyanine dye 3,3′-bis(2-sulfopropyl)-5,5′,6,6′-tetrachloro-1,1′-dioctylbenzimidacarbocyanine are produced and immobilized on solid surfaces by means of a so-called drop flow, consisting simply in casting a droplet of the organic solution on a tilted substrate and then leaving it to dry slowly in air.[59] Another different method, recently proposed to produce highly ordered and conductive nanofibers, is solvent vapor enhanced drop casting (SVED).[60] SVED consists in exposing a cast polymer solution to an atmosphere saturated with solvent vapor. This allows the self-assembly of the polymer molecules to be driven on the solid surface during the time interval over which solvent evaporation takes place. Processing options available to the operator include the specific choice of the saturating solvent vapor, the surface energy and the wettability properties of the substrate-liquid interface, all of which can affect the rate of solvent evaporation. In turn, the evaporation rate affects the self-assembly and supramolecular organization of the polymer nanofibers that finally lie on the substrate. This technique can be advantageous for some polymers for which other solution-based procedures are not effective in order to produce fibers of controllable diameter. For instance, the cyclopentadithiophene-benzothiadiazole (CDT-BTZ) is reported to form only fibers with diameter of roughly 10 nm by means of standard drop casting or related post-processing methods. On the contrary, SVED allows CDT-BTZ fibers with lateral dimension in the range of 80–600 nm to be obtained, by using dichlorobenzene for both preparing the polymer solutions and saturating the processing atmosphere.[60] Heating slightly above

[‡] More generally, the production of nanofibers by mixing polymer solutions with non-solvents is evolving towards commercialization, exploiting spinning techniques as those described in the previous Chapter (Section 2.4.1). The XanoShear™ technology based on results from the group of O. D. Velev and developed by Xanofi (www.xanofi.com) allows nanofibers to be produced by controlling the precipitation of polymer molecules, taking place when the original solution is mixed with a non-solvent, in the presence of significant shearing of the medium.[57] This technique is reported to yield up to 12 kg of fiber material per hour *via* needleless, sheared wet spinning.[58]

room temperature (*i.e* at 60 °C) helps in reaching saturation conditions without degrading the polymer system, and surface functionalization by means of hexamethyldisilazane (HMDS) allows the affinity of polymer molecules toward the substrate to be enhanced. Of course, using the same solvent for both casting and in the saturated atmosphere slows down the evaporation of the cast fraction, thus providing ample time to the polymer molecules for assembling in a more ordered configuration.[52,53] Similar findings about the importance of solvent evaporation rates are found in experiments on poly(9-9′dioctylfluorene)-PEO copolymers, deposited from either THF or toluene solutions: toluene, evaporating much more slowly, differently from THF enables the formation of well-defined fibrils.[61]

### 3.1.3 Other Self-assembly Examples

Using molecular self-assembly at the interface between two fluids (liquid–liquid or liquid–gas) is another especially effective route to produce polymer nanostructures. For example, nanofibers of regioregular polythiophenes can be obtained by assembly at the water/air interface exploiting Langmuir–Blodgett techniques.[62,63] To this aim, the used polythiophene molecules are amphiphilic, presenting a hydrophilic side that naturally orients towards water and a hydrophobic side at the other part of the molecular backbone axis. This is made possible by alternating hydrophobic (say, $C_{12}H_{25}$) and hydrophilic side groups along the chain. An isothermal compression, performed by the Langmuir–Blodgett method once a molecular monolayer forms on the surface of water, finally leads to the assembly of solid nanostructures presenting the typical $\pi$–$\pi$ stacking. The resulting wires are slightly bigger than those produced by marginal solvents (widths and heights of about 60 nm and 10 nm, respectively).[63] In general, amphiphilic branched molecules can be assembled within nanofibers by exploiting multiple, weak inter-molecular interactions.[64] Nanofibers can also be obtained by self-assembled polyphenylene dendrimers, in which fluorescent chromophores can be attached in various positions.[65]

Another beautiful example is given by the self-assembling triblock copolymer, poly(ferrocenyldimethylsilane-*b*-2-vinylpyridine-*b*-2,5-di-(2′-ethylhexyloxy)-1,4-phenylvinylene) in 2-propanol solutions.[66] This material forms cylindrical micelles, whose fragments obtained by sonication can in turn be subsequently grown into fiber-like micelles by adding additional block polymer as THF solution (Figure 3.3a). The sequential treatment of a solution of micelles of the triblock copolymer with controlled amounts of a diblock copolymer finally leads to multiblock co-micelles with alternating compartments, which constitute nice prototypes of light-emitting polymer nanofiber barcodes (Figure 3.3b).[66] The self-assembly in solution is also demonstrated for the regioregular polythiophene-based block copolymer P3HT-*b*-PDMS, where it leads to the formation of cylindrical micelles (Figure 3.4a–d) with a crystalline core of P3HT and a PDMS corona.[67] These micelles have uniform width of about 19 nm, but are still polydispersed in length, varying from

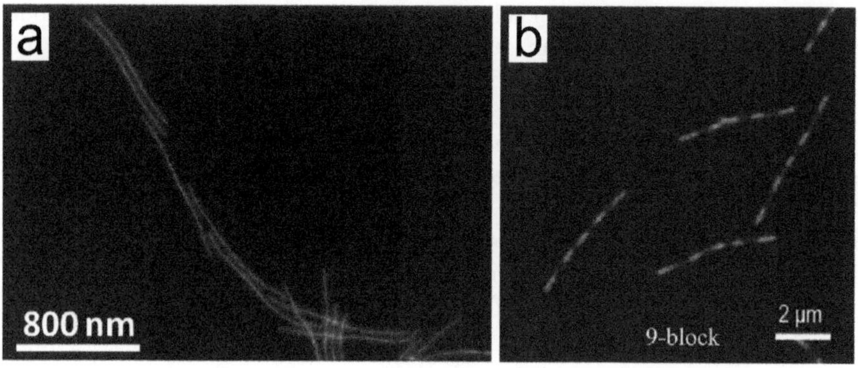


**Figure 3.3** (a) TEM image of cylindrical micelles of the self-assembling triblock copolymer, poly(ferrocenyldimethylsilane-*b*-2-vinylpyridine-*b*-2,5-di-(2′-ethylhexyloxy)-1,4-phenylvinylene). These nanostructures are obtained by adding 1 mg of triblock copolymer as a THF solution (10 mg $mL^{-1}$) to 50 μg of poly(ferrocenyldimethylsilane-*b*-2-vinylpyridine-*b*-2,5-di-(2′-ethylhexyloxy)-1,4-phenylvinylene) seed fragments in 1.0 mL of 2-propanol. (b) Laser confocal fluorescence micrograph of 9-block co-micelles with alternating segments made of the diblock copolymer, poly(ferrocenyldimethylsilane-*b*-2-vinylpyridine), and of the triblock copolymer. Adapted with permission from Ref. 66, *J. Amer. Chem. Soc.*, 2011, **133**, 9095–9103. Doi: 10.1021/ja202662u. Copyright © 2011, American Chemical Society.

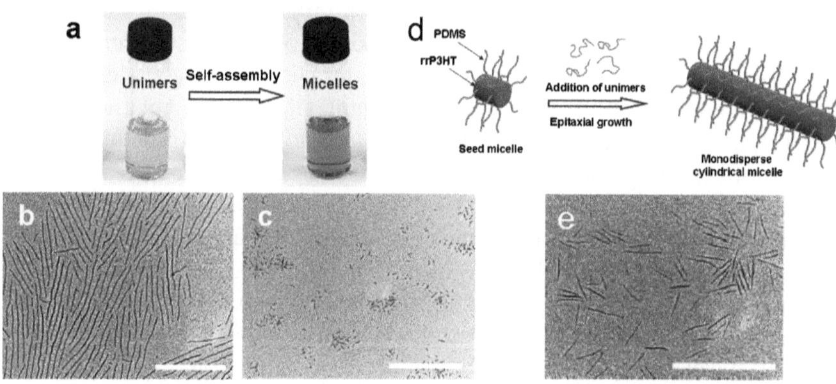


**Figure 3.4** (a) Photographs of P3HT-*b*-PDMS unimers in toluene (left) and of cylindrical micelles in 85% (v/v) ethoxyethane/toluene (right). (b, c) TEM micrographs of (b) P3HT-*b*-PDMS cylindrical micelles formed in ethoxyethane/toluene and (c) seed micelles prepared by sonication of the long cylindrical micelles. Scale bars: 500 nm. (d) Scheme of the formation of monodisperse P3HT-*b*-PDMS cylindrical micelles *via* crystallization-driven self-assembly. Small seed micelles serve as initiators. (e) TEM image of the monodisperse cylindrical micelles of P3HT-*b*-PDMS. Scale bar: 500 nm. Adapted with permission from Ref. 67, *J. Amer. Chem. Soc.*, 2011, **133**, 8842–8845. Doi: 10.1021/ja202408w. Copyright © 2011, American Chemical Society.

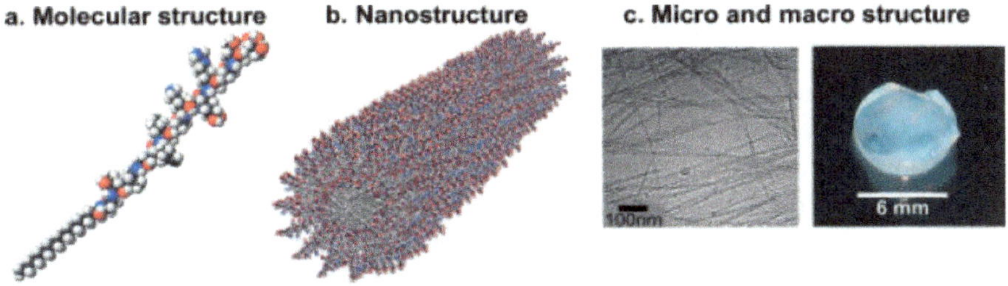


**Figure 3.5** Schematics of the molecular structure (a) and of the resulting fibrous nanostructure (b) of a self-assembling peptide-amphiphile. In molecular graphics different atoms (C, H, O and N) are represented with different colors (black, white, red and blue). (c) TEM micrograph of nanofibers, and photograph of the macroscopic nanofiber sample. Adapted with permission from Ref. 72, *Adv. Mater.*, 2005, **17**, 2612–2617. Doi: 10.1002/adma.200500855. Copyright © 2005, Wiley-VCH Verlag GmbH & Co. KGaA.

400 nm to about 2 μm (Figure 3.4b). However, monodisperse, P3HT/PDMS nanocylinders of controlled length can be prepared (Figure 3.4e), using seed initiators obtained by sonication of the first-generation cylindrical micelles (Figure 3.4c,d).[67] The process is called crystallization-driven self-assembly. We will see other examples of self-assembled nanostructures when reviewing other template methods using micelles (Section 3.3.2), and describe their optical properties in more depth in Section 5.2.

As one may expect due to the wide variety of self-assembly processes that can be found in nature,[16] other approaches have been developed involving biological molecules.[68] Such techniques are utilizable for the production of biochemically active organic nanofibers, or of fiber architectures for tissue engineering (Section 7.7). For instance, the self-assembling properties of peptide amphiphiles[69] are widely used to obtain nanofibers (Figure 3.5).[70–72] Dialkyl chain amphiphiles that include peptides could allow the components of the ECM surrounding living cells to be mimicked, a strategy that is explained in Chapter 7. The enzyme-mediated degradation of networks of peptide amphiphiles nanofibers has also been investigated: gels of fibers incubated with type IV collagenase undergo a weight loss of 50% in one week, and degrade completely within one month, and the nanofibers, imaged by TEM, turn into rounded or multistranded aggregates.[72] Furthermore, it has been found that some peptide nanofibers show extraordinary capabilities of dynamic *reassembly*, reconstituting scaffold architectures, even many times, after deliberated destruction by means of sonication processes.[73] For this class of compounds, the self-assembly is programmable to some extent by molecular design, since the molecular organization depends on the specific composition in terms of the constituent amino acids, and, in addition, the process is affected by a number of solution parameters and especially pH.[68,74]

Finally, new and straightforward assembly approaches promise today significant advantages for the realization of polymer nanofibers. For instance, simply heating water-dispersed PLA particles above $T_g$ leads to the formation of nanofibers.[75] Self-assembly of monomer units through non-covalent intermolecular interactions deserves to be mentioned as well, playing a crucial role in the formation of so-called supramolecular polymers.[76,77] Indeed, the realization of quasi-one-dimensional organic nanostructures can benefit greatly from supramolecular polymerization processes. For instance, linear supramolecular polymers are formed *via* the host guest interaction between the benzo-21-crown-7 molecule and dialkylammonium salt at high concentrations, and this effect has recently been demonstrated as an effective route to reach entanglement and viscosity properties sufficient to process low-molecular-weight systems by electrospinning (Figure 3.6).[78] By a different approach, polydiacetylenes are obtained by electrospinning nanofibers with distributed diacetylene monomers (10,12-pentacosadiynoic acid) in a PMMA matrix, which induces a self-assembly process of the monomers within the fibers, and by the subsequent photo-polymerization of the supramolecular diacetylene structures.[79]

### 3.1.4 Polymerization Methods

Another class of methods to fabricate polymer nanofibers from solutions is not, or not only, based on supramolecular assembly, but instead builds directly on chemical reactions involving the starting molecules. In particular, the oxidative polymerization of PAN[80] allows nanofibers to be obtained in several ways. In this well-known process, aniline is polymerized by means of an oxidant and in the presence of a strong acid dopant.[81] The resulting polymer is insoluble in the aqueous medium, therefore it deposits under various forms. In many cases, polymerization can result in the formation of elongated nanostructures, fibers and bundles with various morphologies. Ammonium persulfate [APS, $(NH_4)_2S_2O_8$] is among the most frequently used oxidants, and even quite diluted monomer solutions ($\cong$ 8 mM) allow PAN nanofibers to be obtained, with the additional advantage of reducing secondary growth processes (and consequently the presence of more branched and less uniform fiber morphologies), which may originate at additional nucleation sites formed on the surface of just-formed nanofibers in presence of excess monomer.[82] The process is quite robust and can be implemented by a large variety of doping agents, such as camphorsulfonic acid ($C_{10}H_{16}O_4S$), hydrochloric acid (HCl), perchloric acid ($HClO_4$), methanesulfonic acid ($CH_3SO_3H$), hydrogensulfated fullerenol with –(O)$SO_3H$ groups[83] and many others. Recently reported methods based on sodium chlorite ($NaClO_2$)[84] and potassium biiodate [$KH(IO_3)_2$][85] as oxidants allow nanofibers to be produced with especially good conduction properties (Section 6.2). Another, more exotic method is based on using recombinant *Coprinus cinereus* peroxidase (CiP) as biocatalyst.[86] In general, catalysts of biological origin can be very advantageous in nano-manufacturing, opening new perspectives for the realization of materials and nanomaterials at room or low temperature, and under environmentally

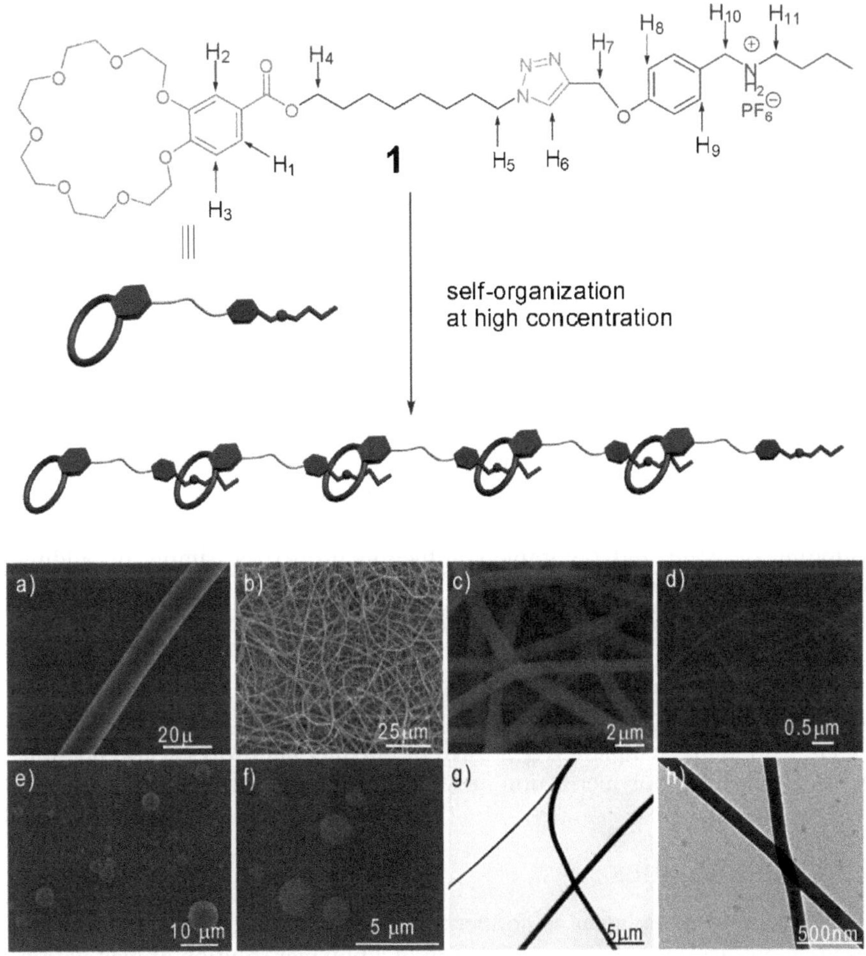


**Figure 3.6** Top scheme: Schematics of the formation of a linear supramolecular polymer from the self-organization of a heteroditopic monomer constituted by benzo-21-crown-7 and dialkylammonium salt. Bottom panels: (a) SEM micrograph of a rod-like fiber, drawn from a chloroform solution. (b–d) SEM micrographs of nanofibers realized by electrospinning the linear supramolecular polymer. Spinning from acetonitrile solutions results instead in deposited droplets (e, f). (g, h) TEM micrographs of the electrospun nanofibers. Adapted with permission from Ref. 78, *Chem. Commun.*, 2011, **47**, 7086–7088. Doi: 10.1039/c1cc11790d.

and user-friendly conditions of pH, pressure, absence of toxic chemicals, *etc.* Firstly, aniline solutions in $CCl_4$ and a CiP solution in camphorsulfonic acid are prepared in parallel. As is well known, peroxidase is an oxidoreductase enzyme, catalyzing the oxidation of many molecules in the presence of hydrogen peroxide, and this can be exploited for polymerization reactions.[87]

The catalytic pathway following the mixing of the two solutions and the subsequent addition of $H_2O_2$ is well schematized in Ref. 86:

$$\mathrm{CiP} + \mathrm{H_2O_2} \rightarrow \mathrm{CiP'} \tag{3.2a}$$

$$\mathrm{CiP'} + \mathrm{aniline} \rightarrow \mathrm{aniline}^{*} + \mathrm{CiP''} \tag{3.2b}$$

$$\mathrm{CiP''} + \mathrm{aniline} \rightarrow \mathrm{aniline}^{*} + \mathrm{CiP} \tag{3.2c}$$

where CiP′ and CiP″ indicate intermediate forms of the enzyme molecule and aniline* stands for the radical species derived from the oxidation of aniline. These free radicals then polymerize to form PAN, which is deposited in the form of disordered mats of fibers with diameter in the range of 50–100 nm and with length reaching a few μm. These fibers are thus achieved in their doped state. As an alternative, they can be produced in the undoped state, thus exhibiting different and tailorable conduction properties, simply by adding 30% $NH_4OH$ to the first PAN solution.[87] The electrical properties of fibers are better described in Section 6.2. Of course, oxidative polymerization is widely used for other conjugated polymers as well, sometimes mediated by the formation of micellar aggregated structures (Section 3.4), and it often allows nanofiber morphologies to be obtained, as in the case of polypyrrole (PPy).[88–90] Finally, template-free growth of polymer nanowires is also possible by electrochemical polymerization on various substrate surfaces.[91,92]

## 3.2 Nanofluidics

The production of polymer nanofibers by nanofluidic methods is inherently based on physical effects. The penetration of a polymer solution or melt within nanochannels can also be significantly affected by non-covalent interactions at the interface with the surrounding walls of the capillary. Indeed, nanofluidics uses the filling of capillaries with transversal size below 100 nm by the polymer in order to template the formation of nanofibers. The motion of the solution or of the melt in the nanochannels can either be driven by spontaneous capillarity, or be assisted by external pumping as explained in the following. After capillary filling, the evaporation of the residual solvent from the possibly open terminations of the nanochannel, the diffusion of the solvent molecules through the material constituting the walls of the capillary or the solidification of the polymer melt upon cooling below $T_g$ finally lead to the deposition of solid nanostructures.

In general, nanofluidic phenomena are involved in the realization of polymer nanofibers in many different ways. Template methods using membranes with pores of size in the order of a few tens of nm to fabricate polymer nanotubules (Section 3.3.1) are a very relevant example.[93–99] In this section, we focus on nanofluidics using lithographically made channels and on similar methods that can be generally classified as soft lithographies (Section 1.2.1).[100] To produce sub-100 nm features, which in turn allow the final capillaries to be obtained, a variety of high-resolution patterning methods can be used, including electron beam lithography (EBL), focused ion-beam (FIB) milling, UV photo-lithography, holography *etc.* (Section 1.2.1). All these techniques are quite expensive, and some of them have relatively low throughput because of their inherently serial character, however these drawbacks can be largely overcome by replicating the initial structures of the so-produced master samples into elastomeric materials, which is the inspiring concept of soft lithographies. Furthermore, as introduced in Section 1.2.1, PFPE compounds[101] can provide valuable alternatives to silicone elastomers in order to produce highly resistant elements, thus opening nanofluidics to the use of polymer solutions made with non-polar organic solvents. Therefore, REM is very often the technique of choice to produce nanochannels, whose shape can then be transferred to polymers by either spontaneous or pressure-driven penetration of the solution or melt into the recessed features of the elastomeric elements. In the first case, the process is basically a nanofluidic variant of the so-called microfluidic lithography or micromolding in capillaries,[102,103] and capillary filling is induced by capillary forces. Typically, capillaries can be assembled onto a substrate simply by (i) cutting the elastomeric mold by a razor blade, to open the edges of the recessed features that will constitute the three polymer walls of the nanochannel, (ii) placing the mold on the substrate under its own weight or by slight, calibrated pressure, (iii) depositing a μL drop of a polymer solution at the edge of the mold in order to make the channels spontaneously filled by capillary action and finally (iv) peeling off the elastomeric replicas after solvent evaporation. Analogously to wetting processes on free solid surfaces (Section 4.2.2), the capillarity process is thermodynamically favored and the liquid fills the channels spontaneously if the free energy variation, $\Delta U_G$, due to an advancement of the polymer front, is $<0$. In the opposite case, one would need external connections coupled to a pump in order to force the polymer solution to fill the nanochannels. This can be carried out by (i) temporarily sealing the elastomeric element with the substrate underneath (generally by conformal contact eventually assisted by surface functionalization, oxygen plasma treatments, *etc.*, to prevent delamination at the substrate/elastomer interface when the external pressure is applied) and (ii) connecting the so-assembled device (substrate + textured elastomer) to syringe infusion pumps for liquid injection, through inlet and outlet holes that are generally opened in the elastomer by means of punches.[104]

In particular, considering spontaneous capillarity and the simplified case of four identical channel walls, the penetration of a polymer solution into the nanochannel (*i.e.*, the motion of the meniscus along the longitudinal coordinate, $z$) can be roughly described by the equation:[102,105]

$$\frac{dz}{dt} = \frac{R_H \sigma_{LV} \cos \theta_W}{4\eta z}, \tag{3.3}$$

where $\theta_W$ indicates the contact angle between the liquid and the surface of the capillary, $\eta$ is the solution viscosity, $\sigma_{LV}$ is the liquid-vapor interfacial free energy per unit area (Section 4.2.2) and $R_H$ is twice the hydraulic radius of the channel, which is given by $A/p$, with $A$ surface-area and $p$ perimeter of the channel cross-section. For instance, for a capillary with rectangular cross-section with base, $w_c$, and height, $h_c$, which is the most common geometry of lithographically made nanochannels, one has $R_H = w_c h_c/(w_c + h_c)$, which is typically of the order of a few tens of nm in nanofluidics. In fact, nanofluidic channels are a very particular environment in which fluids can somehow undergo extreme conditions. For instance, the typical shear rates, $\dot{\gamma}$, in the channels can be estimated as $\sim(dz/dt)/h_c$,[106,107] which is generally $>>1\ s^{-1}$ (up to $10^5\ s^{-1}$) in capillaries with $R_H = 10$–100 nm.[108] Polymer solution penetration velocities range from the order of 100 nm $s^{-1}$ to about 1 mm $s^{-1}$, monotonously decreasing upon moving from the entrance of nanochannels along the capillary length. Integrating Equation (3.3) leads to the Lucas–Washburn equation for the penetration dynamics, which is widely employed to describe the advancement motion of a fluid front within lithographic capillaries:[103,109,110]

$$z = Bt^{0.5} \tag{3.4}$$

where $B = (R_H \sigma_{LV} \cos\theta_W/2\eta)^{1/2}$. However, the previous equation rigorously holds in the case of Newtonian liquids and in the absence of significant interaction between the liquid and the roughness of the wetting walls, which is rarely the case of polymer solutions used for producing nanofibers. In fact, experimental findings for the capillary flow of organic solutions in lithographically made channels often evidence departures from Equation (3.4).[107,108] Readers interested in more details about polymer transport and liquid manipulation in nanofluidic environments are referred to the book edited by J. B. Edel and A. J. deMello in this series.[111]

Regardless of the specific filling dynamics, channels with sub-100 nm cross-section are characterized by Reynolds numbers (Equation (2.3)), which can be as low as $10^{-7}$, indicating a strongly laminar flow. This regime can be useful aiming at achieving molecularly oriented nanofibers. Furthermore, the process is additive, thus minimizing material waste. For example, the nanofluidic approach has been developed for realizing nanofibers of several light-emitting conjugated polymers, such as MEH-PPV, poly[(9,9-dioctylfluorenyl-2,7-diyl)-*co*-(1,4-benzo-{2,1′,3}-thiadiazole)], poly[(9,9-dioctylfluorenylene-2,7-diyl)-*co*-(1,4-diphenylenevinylene-2-methoxy-5-{2 ethylhexyloxy}-benzene)], *etc.*, in which molecular ordering is evidenced by the polarization properties of emitted light (Section 5.2).[108] In another work, the confinement of the liquid volume during solvent evaporation is obtained near the contact point between a plane and a spherical surface made of silica or glass.[112] This allows highly aligned arrays of organic semiconductor nanowires to be realized.

Nanofluidic phenomena are also closely involved in nanofibers assembly through *vertical* penetration of a polymeric fluid or melt into the recessed features of the elastomer. This is a complementary nanofabrication method consisting in the pattern transfer to a deposited polymer film, *via* soft molding. For instance, after placing the elastomeric element on a previously deposited film of the target polymer, the molding process can be performed by increasing the temperature above $T_g$.[113,114] On photo-curable polymers, such as polyurethanes, patterning can be carried out by the simultaneous penetration into features and curing of the material by UV-irradiation through the transparent elastomer.[115] Afterwards, the replica is separated from the cured polymer. By properly choosing the initial master geometry and processing parameters, isolated nanofibers are finally obtained on the substrate. During replica peeling off, one can observe some rearrangement of the shape of the realized features and even intentionally increase their aspect ratio.[114] The polymer structures can eventually detach from the substrate thus resulting in free-standing, flexible nanofibers (Figure 3.7).

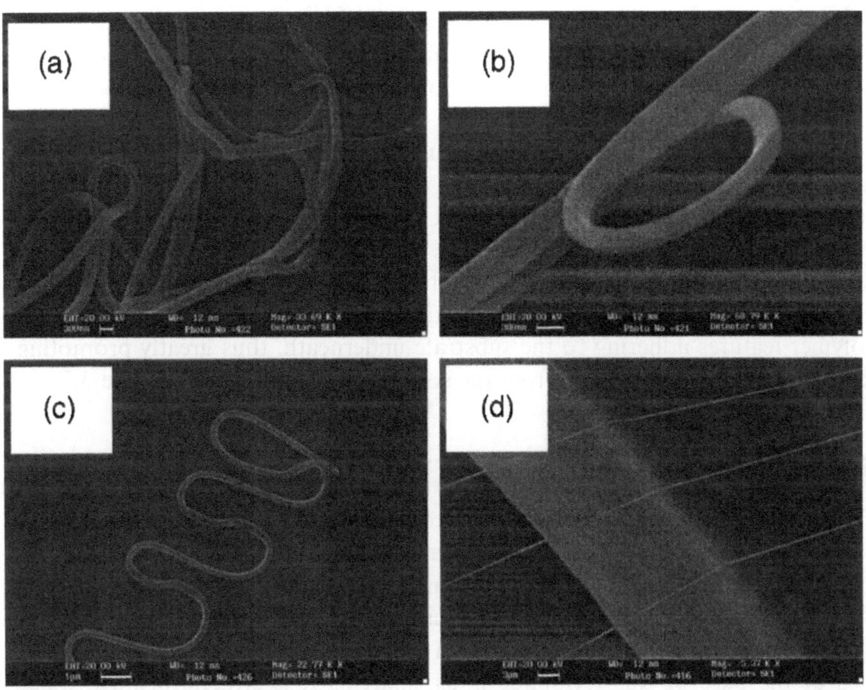


**Figure 3.7** Polyurethane fibers realized by soft molding lithography. (a) Disordered bundle. (b) Knot made by flexible fibers with circular cross-section. (c) Single fiber, diameter ≅ 300 nm. (d) Anchoring of an elastomeric fragment by polyurethane fibers. Reproduced with permission from Ref. 115, *Appl. Phys. Lett.*, 2005, **87**, 123109. Doi: 10.1063/1.2046731. Copyright © 2005, American Institute of Physics.

Equation (3.3) can be used (defining as $z$ the vertical penetration coordinate) to estimate the maximum height, $z_m$, that the polymer can reach during its vertical capillarity rise within the time necessary for complete polymerization. To this aim one needs to know the details of the polymerization kinetics of the filling material. For instance, the initial stage of the polymerization kinetics can often be approximated by an exponential increase of the viscosity:[116]

$$\eta(t) = \eta_0 \exp(k_0 t), \tag{3.5a}$$

where $\eta_0$ is the viscosity at $t = 0$ and $k_0$ is the viscosity-rate kinetic constant [$s^{-1}$]. The solution of Equation (3.3) then becomes:

$$z = \left\{ \frac{R_H \sigma_{LV} \cos\theta_W}{2\eta_0 k_0} [1 - \exp(-k_0 t)] \right\}^{0.5}, \tag{3.5b}$$

which allows the value of $z_m$ to be found, which is reachable by the polymer before becoming fully solid. Estimates on commercial polyurethanes with initial viscosity of $155 \times 10^{-3}$ Pa s and a viscosity-rate kinetic constant of $8.2 \times 10^{-3}$ $s^{-1}$ leads to $z_m$ of the order of millimetres.[115,117] This is far higher than the height values practically achievable in fabricated molds. In other words, the height of the produced nanostructures is basically only limited by the aspect ratio (height/width) of the recessed features of the elastomeric replicas. For example, arrays of vertical nanofibers with 50 nm diameter and 500 nm height have recently been realized by replicating the features of a Si master into a photo-curable polyurethane acrylate and used to assemble highly sensitive, flexible strain-gauge sensors following Pt coating.[118] In addition, high aspect ratio values (together with the absence of polymer bottom residues inter-connecting adjacent polymer nanostructures after patterning) tend to decrease the stability of the polymer features adhering to the substrate underneath, thus greatly promoting the detachment of nanofibers from the solid surface as shown in Figure 3.7.

## 3.3 Template Synthesis

The concept of producing polymer nanostructures by means of external templates has widened enormously, as described later in this section; however, it largely originates from the research on membrane-based synthesis performed by C. R. Martin and coworkers at the Colorado State University (now at the University of Florida) and by a few other groups from the late 1980s to early 1990s.[94–99] The idea at the base of this technique is as smart as it is straightforward. Since membranes can nowadays be produced at industrial scale, and are commercially available (*e.g.* by Nucleopore,[119] Poretics, Alltech,[120] Anotech,[121] Whatman[122], *etc.*), with a very large number of cylindrical nanopores that are highly uniform in size, why not use them as templates for addressing the production of nanostructures made of a material growing in the cavities?

In the pores, the process of growth of solid organic nanostructures can proceed through well-established synthetic methods, such as oxidative polymerization (Section 3.1).[87,95] During the polymerization reaction the template pores have a double effect, with a tight interplay of physical and chemical mechanisms, thus considerably affecting the synthesis and giving rise to new phenomena. On one side, the role of the pores is evidently physical, delimiting the microenvironment in which the polymerization process occurs to a very tiny reaction volume, strongly influencing the diffusional behavior of molecules in the confined condition[123] and setting a maximum value to the achievable diameter of the resulting nanostructures. Some works especially focus on the physical, nanofluidic aspects of template methods, starting from a solution already containing the desired polymer molecules, and driving it through the pores of the membrane.[124] To facilitate infiltration, the membrane can be placed between the solution volume and vacuum (by a water aspirator) until complete solvent evaporation.[93] Besides solutions, polymer melts can be used with good process control, simply by heating thermoplastic materials above their $T_g$ to favor the penetration into the nanochannels because of the drastically reduced viscosity (Section 1.4.1).[125] These latter examples are therefore membrane-based variants of the spontaneous or pressure-driven nanofluidics described in the previous section.

On the other hand, the role played by the template can be chemical, influencing strongly the polymerization reaction dynamics for monomers in solution or in vapor phase.[126,127] Interfacial aspects between different solutions possibly present within the pores and at the pore surfaces are so relevant that some authors include these methods among interfacial polymerization techniques.[128] In fact, regardless of the specific polymerization route, the reactions tend to start just at the walls of the capillaries, and then the growth of the nanostructures proceeds towards the central axis of the pores.[94,119,129–131] "Molecular anchors" on the pore walls, such as, for example, anionic sites for polyheterocyclic compounds, bind the newly generated polymer nanostructure.[99] Consequently, longer polymerization times generally allow thicker walls of the nanostructures to be obtained, and ultimately produce fully solid nanofibers, whereas limiting the growth time is an effective method to obtain hollow nanocylinders with very thin walls. Recently, the thickness of the wall of hollow nanostructures has been studied by paying particular attention to the used solvents and electrolytes, and better highlighting how lower polymerization rates lead more easily to hollow cylinders with thin walls.[132] These structures find several interesting applications once engineered with specific surface functionalization or sealing elements.

### 3.3.1 Hard Templates

The polymerization in template membranes can be induced by using oxidating chemicals (a few examples of oxidative agents promoting the polymerization of PAN are given in Section 3.1, 2 M $FeCl_3$ solutions can be used for obtaining

PPy,[94] *etc.*), but also electrochemically[§]. The oxidating species can be present in an immersion bath together with monomers, or they can be separated by the monomer solution by the template membrane, which leads the reagents to encounter each other and to react within the nanopores.[94,119] Electrochemical methods are very easy to integrate with template synthesis,[120,121,135,136] because membranes and all the pores passing through them can be easily contacted with an anode electrode in a single step. This can be carried out by simply depositing a metal on one of the faces of the membrane. Incidentally, depositing the metal electrodes onto both the sides of the sample allows one straightforwardly to measure the conduction properties of the polymerized nanostructures,[96,119,137] even without removing the template membrane (Section 6.2). The process is schematized in Figure 3.8. Importantly, these nanostructures largely outperform the corresponding bulk materials in terms of electric conductivity,[95] which is related to the enhanced alignment of the molecular chains induced by template synthesis (especially in the outer region of the polymerized nanocylinders, which corresponds to the volume in close proximity of the pore walls). This is also evidenced by means of polarized absorption.[119,131,138]

The most commonly used membranes are made in polycarbonate by particle track-etching or in alumina[139] by electrochemical methods. The membrane material is of great importance, determining the eventual functionalization methods to adopt in order to tune the surface chemistry of the pores. The order and arrangement of the pores also varies depending on the production method and on the membrane material, with membranes of anodic alumina generally outperforming track-etched polycarbonate membranes. The same holds for the surface density of the pores, which ranges in the order of $10^5$ to $10^{11}$ pores per $cm^2$.[97,140] The pore diameter is from the μm-scale down to a few nm. With the appropriate choice of the processing parameters, of the specific synthesis material and of the chemistry of the internal walls of the pores, the resulting, templated nanostructures can show highly controlled shapes and sizes, and can be either solid or hollow as described above. In addition, the produced nanocylinders do not need to remain entrapped within the template membrane, because the latter can be selectively removed in order to collect the synthesized polymer nanostructures, which can then be handled and positioned within devices. For example, methylene chloride can be used to dissolve membranes in polycarbonate while preserving PPy nanostructures.

[§] Many methods allow polymer nanofibers to be fabricated *via* electrochemical growth driven by pre-defined metal features. Such features can be, for instance, gap metal electrodes, and electropolymerization can be performed in the gap region and assisted by a current flowing between the two contacts through an electrolytic solution. In this way, conductive polymer nanowires are fabricated, which bridge the electrodes to realize a nanoelectronic device.[133] In another approach, a substrate embedding pre-fabricated Pt electrodes treated with fluoro-alkanes is used to drive electrochemical polymerization of conductive polymers, and the obtained nanowires can be transferred onto a desired, final target substrate by exploiting suitable adhesion layers.[134]

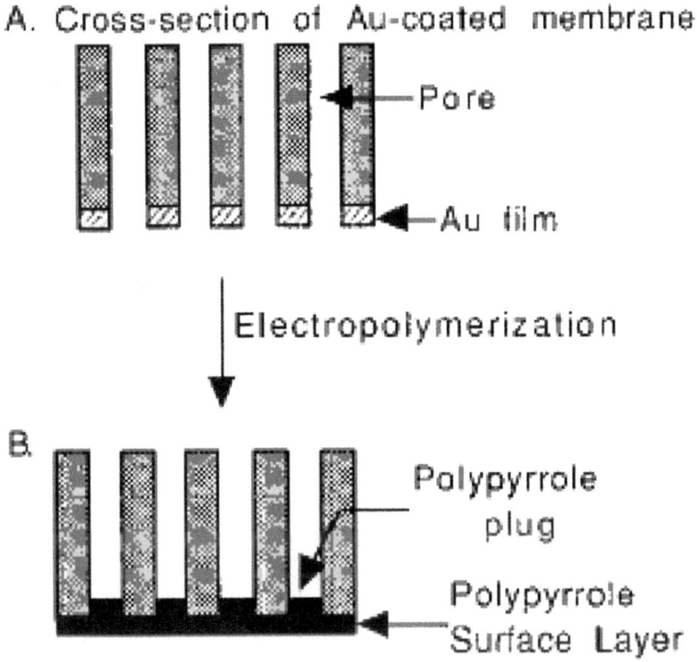


**Figure 3.8** Scheme of the template membrane method allowing PPy nanofibers and nanotubes to be synthesized. The technique consists in using an Au-coated membrane (A), and in electro-polymerizing a film of the conjugated material (B). Nanostructures are then formed in the pores of the membranes by further polymerization, and conduction measurements can be performed by exploiting the bottom Au layer as electrode. Adapted with permission from Ref. 97, *Chem. Mater.*, 1996, **8**, 1739–1746. Doi: 10.1021/cm960166s. Copyright © 1996, American Chemical Society.

Of course, this method is not viable to fabricate very long nanofibers and nanowires. For pores of small tortuosity, as in most experiments, the maximum length of the cylindrical tubules that can be achieved would correspond to the membrane thickness, which can be of the order of tens of μm.[141] However, for many applications this is largely compensated for by the high monodispersity of the template-synthesized wires, by the peculiar capability of this method to produce nanostructures aligned by default in the membrane pores and by the exceptional control achieved on the fiber diameter down to a few nm.[126] The flexibility in terms of the used material is another advantage. Template synthesis has been applied to a large variety of materials, including semiconductors, carbon-based materials and metals.[97] Polymers processed in this way include PS, PMMA, PVDF, poly(2,6-dimethyl-1,4-phenylene oxide), poly(bisphenol A carbonate) and poly(lactic acid), forming cylinders with diameters down to the 10 nm-scale.[93] Also melt-wetting has been initially demonstrated on PS and PMMA, in addition to polytetrafluoroethylene (PTFE), to produce ordered nanotubes.[125] Membrane templates can also be employed to produce biodegradable nanofibers, as is well established for PCL,[142,143] or to cross-link dense arrays of polyacrylate nanofibers, which can be extracted without destroying the templates and used to realize second-generation silica replicas of the original membranes.[144] In addition, a large amount of work is published concerning the application of the method to conjugated polymers.[93,96,99,119,121,123,124,127,129–131,135,136,138,140,141,145–154] A diameter of 3 nm is demonstrated for templated nanostructures of PAN[127,155] and of PAC, then used to obtain carbon wires upon pyrolysis, in a mesoporous aluminosilicate host.[126] This extremely high resolution would not be easy to obtain even in a planar configuration and by means of costly and low-throughput, serial lithography approaches such as EBL. Hollow PAN nanotubules can also be produced as mentioned above,[140] as well as almost monodisperse, conductive nanowires of PPy, which are demonstrated both in polycarbonate[135,137,156] and in alumina[121] membranes. Other systems nanostructured by means of template electrochemical synthesis include copolymers of divinylbenzene and ethylvinylbenzene,[120] and poly(3,4-ethylenedioxythiophene) (PEDOT),[157–159] whose nanotubules growth follows mechanisms that are dependent on the used oxidation potential (Figure 3.9).[136]

Interesting variants of the method have been proposed to better fit with insoluble conjugated polymers. In particular, poly(*p*-phenylenevinylene) (PPV) nanotubes or nanowires can be obtained by membrane-templated chemical vapor deposition (CVD) of the α,α′-dichloro-*p*-xylene monomer precursor of this conjugated polymer.[122] As an alternative, the same kinds of nanostructures can be fabricated by infiltrating the membrane by solutions of the poly(*p*-xylene-α-alkylsulfur) chloride precursor.[141] After filling of the nanochannels by solutions[141] or by vapor deposition[122] and the eventual solvent evaporation, PPV hollow or solid nanostructures of diameter from 30 to 450 nm are achieved through thermal conversion of the precursors at temperatures ranging from 110 °C to 270 °C, under vacuum or inert atmosphere. The resulting PPV nanotubes show peculiar optical properties that will be described in Section 5.2.1. Oxidative CVD has also been used to template a copolymer of 3,4-ethylenedioxythiophene and 3-thiopheneethanol with alumina membranes.[160] One should recall that the process defined by infiltrating a liquid or a gas precursor within the sub-μm pores of hard membranes to realize nanomaterials is also referred to as *nanocasting*. Interested readers can refer to the dedicated book edited by A.-H. Lu, D. Zhao and Y. Wan in this series.[161]

Combining this approach with metal deposition on one side of the membrane and with subsequent electrochemical growth of nickel in the voids of the previously formed PPV nanotubes allows interesting Ni/PPV core/shell nanostructures to be generated (Figure 3.10), which show both light-emission and magnetic properties.[162] Another intriguing microstructure can be produced by filling the pores of a membrane by $SiO_2$ nanoparticles, which in turn serve as templates for the deposition of infiltrating polymers such as polyallylamine hydrochloride and poly(sodium 4-styrenesulfonate) (Figure 3.11a).[163] This allows nanoparticle/polymer fibers to be realized (Figure 3.11b), from which

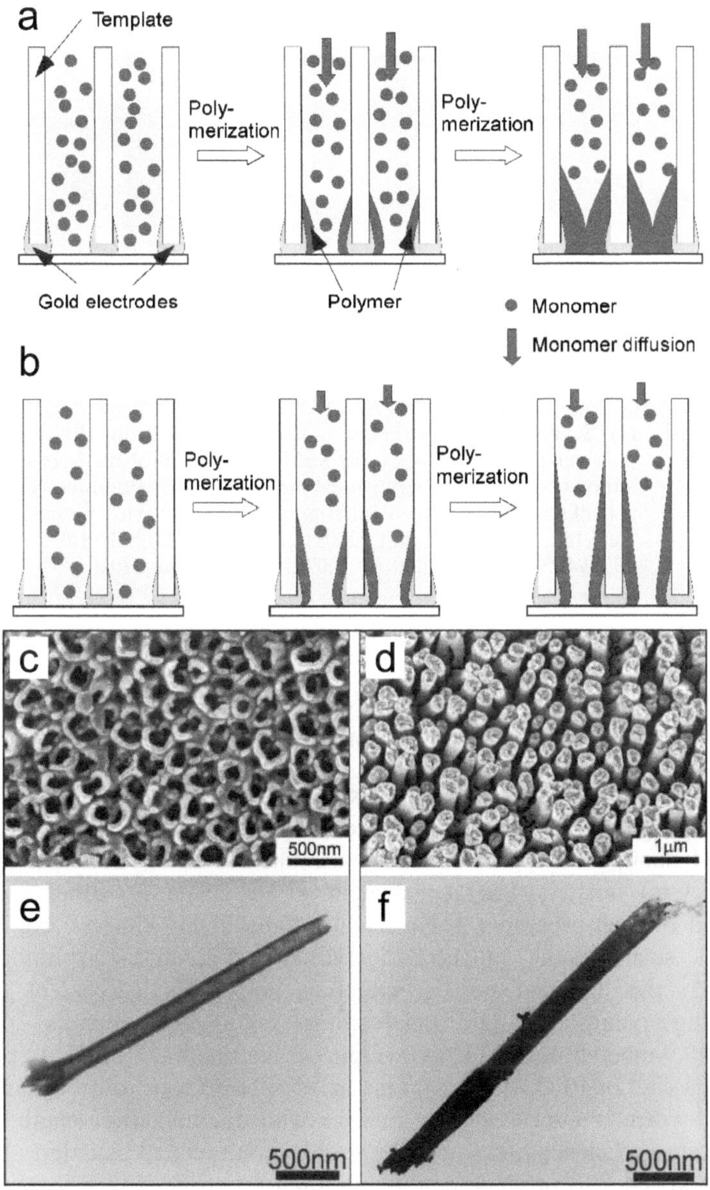


**Figure 3.9** (a, b) Schematics of the growth of template PEDOT nanostructures at high oxidation potential (1.4 V). The two cases shown correspond to conditions of (a) high monomer concentration (slow reaction rates) and (b) low monomer concentration (fast reaction rates), which lead to the formation of solid fibers or hollow nanotubules, respectively. (c–f): SEM and TEM micrographs of nanotubules (c, e, respectively) and solid nanostructures (d, f, respectively), obtained by low oxidation potential (<1.4 V). Adapted with permission from Ref. 136, *Acc. Chem. Res.*, 2008, **41**, 699–707. Doi: 10.1021/ar7002094. Copyright © 2008, American Chemical Society.

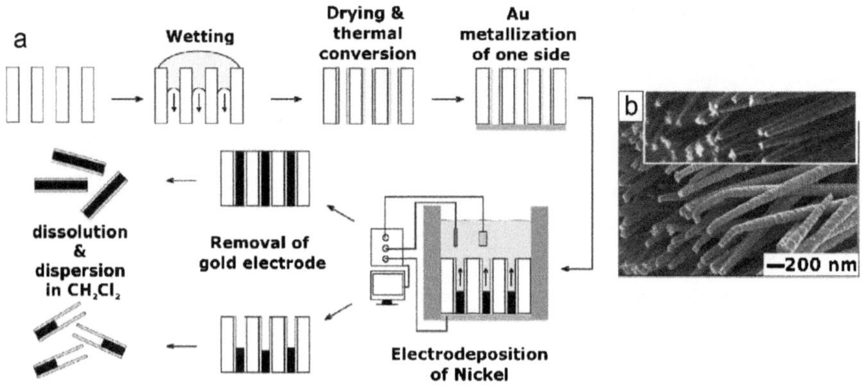


**Figure 3.10** (a) Scheme of the template synthesis for realizing hybrid, coaxial polymer-metal nanowires. (b) SEM micrograph of the hybrid PPV–Ni nanowire array after removal of the templating membrane. The inset is a backscattered electron micrograph. Adapted with permission from Ref. 162, *Nanotechnology*, 2009, **20**, 405601. Doi: 10.1088/0957-4484/20/40/405601. Copyright © 2009, IOP Publishing Ltd.

the silica fraction can be removed to produce filaments of hollow polymeric capsules usable as drug carriers.

Concerning luminescence, waveguiding and other optical properties of membrane-templated polymer nanowires, an especially relevant work has been carried out by the group of G. Redmond at the Tyndall National Institute.[147–154,164,165] Their generally semicrystalline nanostructures include oriented nanotube arrays,[150] Y-shaped junctions (produced by exploiting branched nanopores)[153] and aligned assemblies of nanowires in a nematic liquid crystal host.[152] The nanostructures are made of either solution-processed or melt-processed polymers, including highly emissive conjugated polymers such as the poly[(9,9-dioctylfluorenyl-2,7-diyl)-*co*-(bithiophene)] (F8T2),[147] the hole-transporting material, poly(*N*-vinylcarbazole) (PVK), embedding organo-lanthanide chelate complexes or other dopants[164,165] and the poly(9,9-dioctylfluorene) (PFO) (Figure 3.12a and b).[148–150,154] In particular, in the case of PFO, this technique is reported to lead to an enhanced β-phase,[151] which is associated to a highly extended conjugation length, higher luminescence efficiency and lower threshold for stimulated emission. As with other approaches to produce nanofibers, the formation of this phase is here likely favored by the local conditions of stress and molecular alignment induced by the nanofluidic penetration of PFO solutions within the pores of alumina membranes. The so-obtained nanostructures show a rich variety of photonic properties, such as polarized emission, light confinement and guiding capability, lasing, *etc.* (Sections 5.4, 5.5).

Finally, recent works have opened new routes for the development of more complex nanostructures, such as light-emitting hybrid CdS nanocrystals/oligo(*p*-phenylene vinylene) nanorods,[166] heterojunction nanowires composed

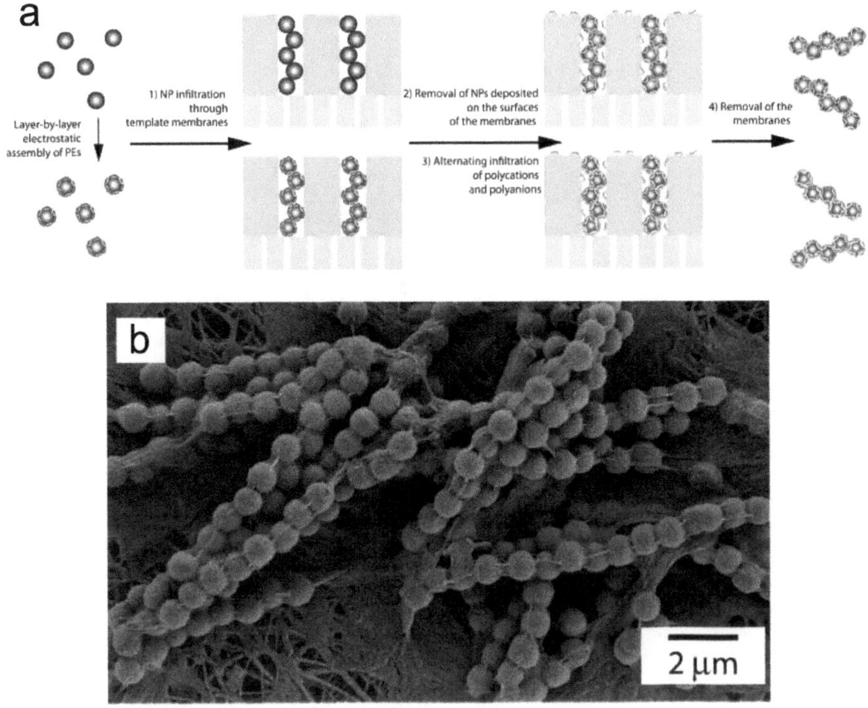


**Figure 3.11** (a) Schematics of templated polyelectrolyte encapsulation of nanoparticles, coated without (top schemes) and with polyelectrolyte multilayer shells (bottom schemes). (b) Resulting microfibers made by silica nanoparticles and template polymers. Adapted with permission from Ref. 163, *Adv. Mater.*, 2011, **23**, 1376–1379. Doi: 10.1002/adma.201004048. Copyright © 2011, Wiley-VCH Verlag GmbH & Co. KGaA.

of a CdS and a PPy segment[167] or color barcode nanowires, which have different code compartments.[168] These nanowires are realized by pressure differentiation,[166] by sequential steps of electrochemical deposition of an inorganic and an organic component[167] or by electrochemical polymerization of segments made of different conjugated polymers within alumina template membranes as reported for poly(3-butylthiophene)/poly(3-methylthiophene)/PEDOT systems.[168] Electrochemical synthesis in anodic alumina templates also allows other beautiful nanostructures to be obtained, such as multi-segmented nanorods alternating Au, other metals and PPy,[169] or Au and a copolymer of 3-hexylthiophene and of a thiophene unit tethered with $C_{60}$ (Figure 3.12c, d).[170] These nanorods are proposed as tools for studying non-ohmic conduction behavior in asymmetric elongated nanostructures embedding metal/conjugated polymer junctions,[169] and photo-induced charge separation and transport at nanointerfaces.[170]

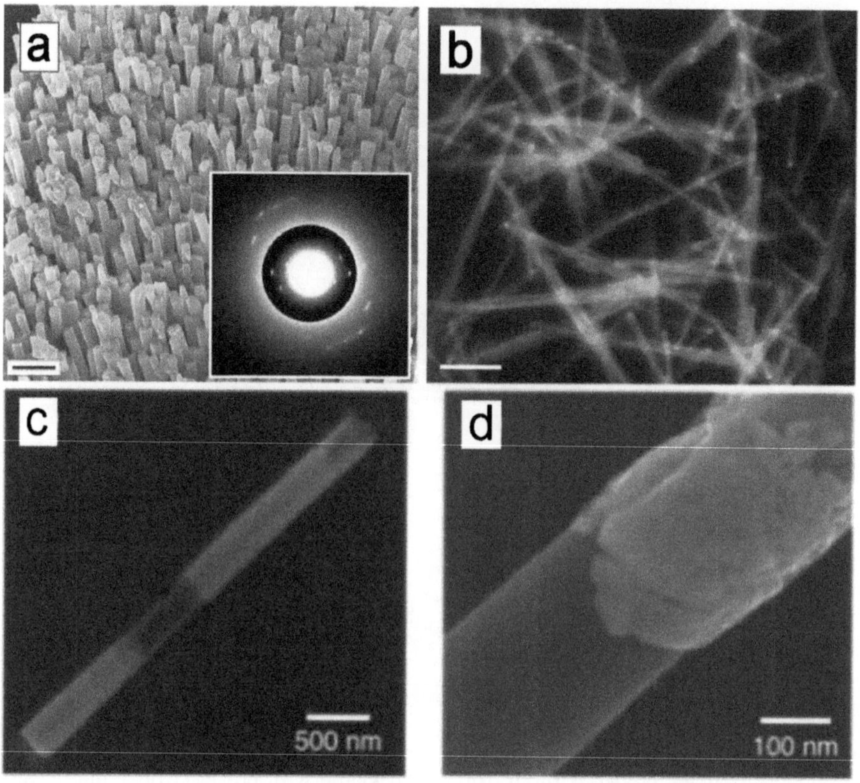


**Figure 3.12** (a) SEM micrograph of an array of PFO nanowires following template removal. Scale bar = 1 μm. Inset: Selected-area electron diffraction data for a single nanowire. (b) Epifluorescence micrograph of a PFO nanowire mat. Scale bar = 5 μm. Adapted with permission from Ref. 149, *Nat. Nanotechnol.*, 2007, **2**, 180–184. Doi: 10.1038/nnano.2007.35. Copyright © 2007, Nature Publishing Group. (c, d) Multisegmented nanorod of Au/copolymer. The copolymer is composed of 3-hexylthiophene and of a thiophene tethered with $C_{60}$. Adapted with permission from Ref. 170, *J. Phys. Chem. Lett.*, 2012, **3**, 478–481. Doi: 10.1021/jz300015e. Copyright © 2012, American Chemical Society.

As described above, polycarbonate and alumina are frequently used as membrane materials for template synthesis, however other options are possible,[171] including glass,[172] other mesoporous (*i.e.* having pore diameter in the range of 2–50 nm) solids[124–127,173,174] and nanostructured surfaces.[175] MEH-PPV filaments down to the single molecule scale,[124] core-shell nanostructures of PPy wrapped in a PPy/PMMA composite cladding[176] and PEDOT nanotubes with different morphologies[175] can be obtained in this way. In general, the methods described so far in this section are referred to as *hard-template* techniques, to distinguish them from the *soft-template* (sometimes called *pseudo-template*)[136] processes that are carried out using templates that have a soft, molecular origin.

## 3.3.2 Soft Templates

Like self-assembly, pseudo-template processes are inherently bottom-up, often very elegant, sometimes sophisticated nanofabrication approaches. A pioneering work based on this concept has demonstrated how the shape of the pores of proteins forming the surface layer of the bacteria, *Sulfolobus acidocaldarius*, and arranged in a two-dimensional crystal, can be transferred to nanostructured titanium oxide, and how the so-obtained periodic pattern can further serve as a mask for patterning a graphite substrate underneath.[177] Subsequently, the concept of templating the formation of nanostructures by means of soft matter has been extended to quasi-one-dimensional, polymer systems. Examples of soft templates used in this way are nematic liquid crystals such as mixtures of 4-(*trans*-4-*n*-propylcyclohexyl)ethoxybenzene and 4-(*trans*-4-*n*-propylcyclohexyl)butoxybenzene or 4-cyano-4′-*n*-hexyl biphenyl, often added with chiral dopants,[178–182] the E44 mixture,[32] phenyl benzoate compounds with oriented smectic A phase,[183] anionic surfactant sodium bis(2-ethylhexyl) sulfosuccinate/*p*-xylene/water mixtures with reverse hexagonal liquid crystalline phase,[184] polymerizable amphiphilic liquid crystal monomers,[185] surfactants [18,186–190] block copolymers,[17,18,191–198] poly(acrylic acid),[199] thiolated cyclodextrins[200] and various types of other molecules and structures of biological derivation including DNA,[201–205] polypeptide-based filaments[206] and amyloid fibers made of bovine insulin.[207–209] We review below some relevant examples.

Liquid crystals are systems characterized by some degree of orientational order and widely used in many modern technologies, with the additional advantage that ordering can be controlled through temperature, electric fields and other methods. Therefore, liquid crystals can be exploited as templates to address in a spatially anisotropic way the assembly of other functional compounds, through several kinds of inter-molecular interactions, which can finally result in the formation of nanofibers.[210] This approach has been initially investigated in depth for aligning films of polyacetylene (PACT),[178,211,212] and widely developed by researchers at the University of Tsukuba and by several other groups.[213,214] In particular, films of helical PACT that present a nanofibrous morphology can be synthesized through polymerization assisted by the catalyst titanium tetra-*n*-butoxide-triethylaluminum dissolved in a chiral nematic liquid crystal.[178] Recently, this technique has been further developed by synthesizing PACT films in the nematic liquid crystal reaction field, which is aligned by a gravity-induced flow.[215] Acetylene is led to polymerize from the gas phase in the liquid crystal matrix, and isolated nanofibers are obtained by ultrasonication of the polymer film in ethanol (Figure 3.13). This procedure, allowing possible problems of fiber entanglements that take place in polymer films obtained by using isotropic solvents to be overcome, is successfully applied for achieving separated nanofibers with lengths up to a few tens of μm.[215] Electrochemical polymerization in nematic liquid crystal fields is also reported for bithiophene,[179] and for 2,5-di(2-furyl)toluene, which forms conjugated polymer nanotubes up to ~1 μm long and with an inner diameter down to

75 nm.[181] Once electrochemically polymerized in liquid crystal electrolyte solutions with different phases (nematic, cholesteric and smectic) in the presence of a magnetic field (Figure 3.14), PEDOT fibrils resemble the original arrangement of the liquid crystals.[182] Several nanofibrous morphologies are also obtained by polymerizing PPy in a cholesteric liquid-crystal electrolyte.[180] Finally, uniaxially aligned arrays of hydrogen-bonded supramolecules nanofibers that present electroactive properties are fabricated by tetrathiafulvalene derivatives assembled in an oriented, smectic A phase of phenyl benzoate liquid crystals.[183] In general, these concepts have fertilized a wide field of research. A rich variety of functional polymers has been synthesized that already incorporate liquid crystalline substituent groups in their molecular structures, thus being controllable by means of external fields, and that show many interesting properties including photo-induced switching, polarized light emission and ferroelectricity.[216] A few of these nanofiber-forming systems and their applications are described later in the book.

Also amphiphilic molecules like surfactants can be used as pseudo-templates to produce polymer nanofibers with relatively high throughput. PPy has been also structured in the form of nanowires and nanoribbons by using lamellar inorganic/organic mesostructures, formed *in situ*, during polymerization, between surfactant cations and oxidizing anions.[188] This last process is schematized in Figure 3.15a. Hydrophobic monomer molecules stay confined along the affine side of micelles, which in these experiments form a lamellar (cetyltrimethylammonium)$_2$S$_2$O$_8$ mesostructure. As a consequence, PPy is formed by oxidative polymerization by the persulfate anions, concomitantly

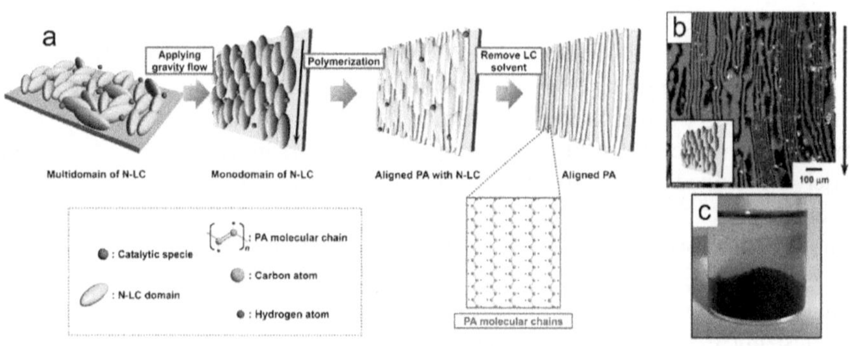


**Figure 3.13** (a) Scheme of the synthesis of aligned polyacetylene (here PA) films in a nematic liquid crystal (here, N-LC) exploiting gravity-induced alignment. (b) Polarized optical micrograph of the aligned liquid crystal at 13 °C. Arrow: gravity direction. (c) Photograph of polyacetylene fibers, mostly precipitated at the bottom of a vial containing an ethanol solution after several hours of ultra-sonication. A minor amount of fibers is still dispersed in the supernatant, hence separated fibers can be easily collected and deposited on solid substrates. Adapted with permission from Ref. 215, *Nanoscale*, 2010, **2**, 509–514. DOI: 10.1039/b9nr00254e.

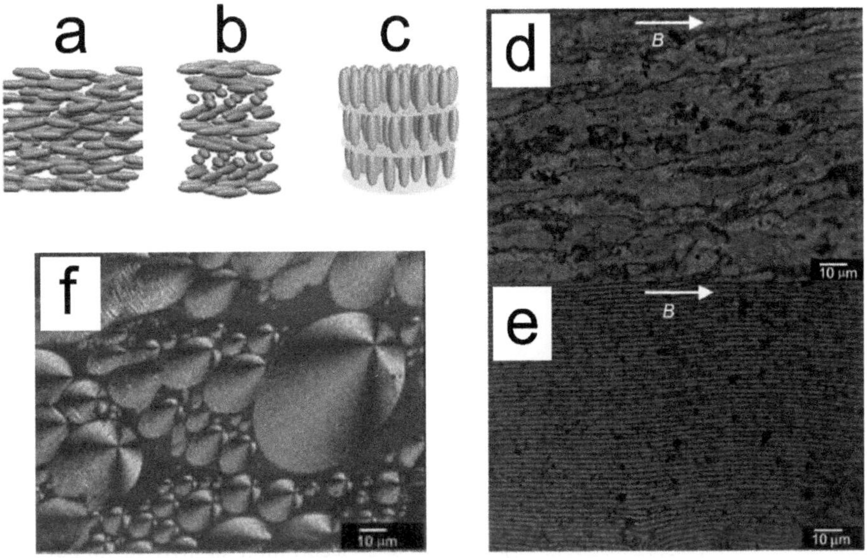


**Figure 3.14** Schematics of the molecular arrangement of (a) nematic, (b) cholesteric and (c) smectic A phases of a templating solution. Corresponding polarized optical micrographs of PEDOT fibrils obtained by electro-polymerization in a nematic (d) and a cholesteric (e) electrolyte, respectively, under a magnetic field of 4 T. (f) Polarized optical micrograph of a PEDOT sample prepared in a smectic A phase under 0 T. Adapted with permission from Ref. 182, *J. Mater. Chem.*, 2010, **20**, 1891–1898. Doi: 10.1039/b920983b.

reducing persulfate to sulfate. After completion of the polymerization process, the template mesostructure degrades spontaneously thus releasing the conjugated polymer nanostructures, and then the possible persistence of excess APS in the solution can ultimately promote the formation of longer and more structurally flexible fibers.[188] Therefore, in these experiments the threshold monomer concentration necessary to obtain a given class of PPy nanostructures (more flexible wires, ribbons, *etc.*) varies depending on the surfactant concentration in the system (Figure 3.15b).[217] Other examples exploiting amphiphilic molecules are described in the next section, because the intrinsic interfacial polymerization mechanisms involved can also be combined with template free approaches.

The most relevant point of strength of block copolymers,[191,192] as soft templates to realize polymer nanofibers and many other kinds of nanostructures, is their peculiar supramolecular organization. Some examples within individual nanofibers have been given in Section 2.3.1 concerning the particular case of nanostructures prepared by coaxial electrospinning. Importantly, since the different blocks [poly(A) or poly(B) or simply A and B as introduced in Section 1.2.1] can have poor or no mutual affinity, phase separation is very frequently observed in such systems, and the resulting

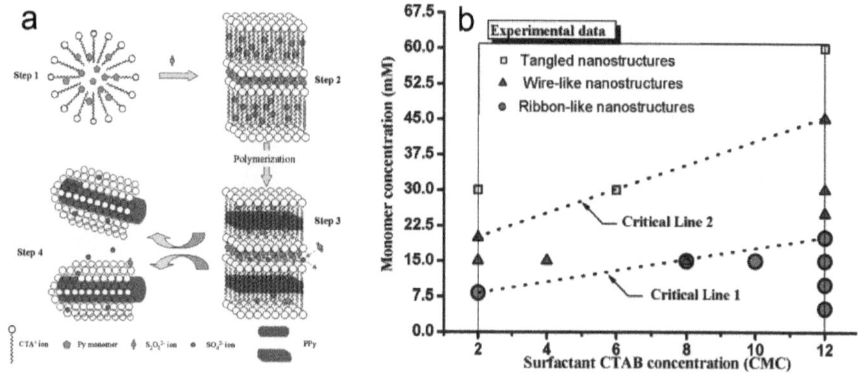


**Figure 3.15** Synthesis of quasi-one-dimensional conjugated polymer nanostructures by oxidative polymerization in lamellar mesostructures. (a) Scheme of the different stages in a typical synthesis of PPy nanoribbons or nanowires in an inorganic/organic mesostructure. Adapted with permission from Ref. 188, *Chem. Commun.*, 2004, 1852–1853. Doi: 10.1039/b405255b. (b) Threshold concentration of pyrrole monomer needed for obtaining a specific kind of nanostructure *vs.* surfactant concentration in the system of pyrrole/cetyltrimethylammonium bromide (CTAB)/APS. For pyrrole concentrations below the Critical Line 1, between Critical Lines 1 and 2 and above Critical Line 2, ribbons, wires and highly entangled wires are mainly obtained, respectively. Adapted with permission from Ref. 217, *J. Phys. Chem. B*, 2006, **110**, 1158–1165. Doi: 10.1021/jp054335k. Copyright © 2006, American Chemical Society.

domains can be exploited in various ways allowing the synthesis of nanostructures to be templated on relatively large areas. The underlying physics of nanoscale and microscale phase separation is beautifully reviewed in Ref. 192. Furthermore, the block copolymer nanostructured patterns can be aligned in many ways, including the use of chemically patterned substrates,[218] electric fields, graphoepitaxy[194] and several other lithographic approaches.[17,18] Especially beautiful examples are given by PS-*b*-PMMA lamellar structures obtained by self-assembled monolayers patterned by extreme UV lithography or EBL,[219,220] by PS-*b*-PDMS domains on PDMS brush substrate surfaces,[221] and by PS-*b*-isoprene cylindrical and lamellar domains in nanofibers electrospun from THF solutions.[222] Another nice and exotic supramolecular assembly in nanofibers is obtained by using amphiphiles (3-*n*-pentadecylphenol), hydrogen-bonded to a block copolymer [PS-*b*-poly(4-vinylpyridine)] electrospun from DMF solutions.[223]

Based on the peculiar supramolecular organization properties of block copolymers, many interesting demonstrations of soft templating by these systems, though mainly targeting inorganic applications, involve or have as side products polymer nanofibers and nanowires. For instance, PS-*b*-poly (2-vinylpyridine) (PS-*b*-P2VP) films form cylindrical domains that are usable as soft templates for metal nanowires (Figure 3.16).[194] This passes through the protonation of the P2VP block, the subsequently increased concentration of metal complexes within these regions, and the selective etching of the polymer component by oxygen plasma. Complex gyroid networks can be produced by poly(4-fluorostyrene)-*b*-poly(D,L-lactide) and used to structure titania in dye-sensitized solar cells.[196,197] Other works are more focused on the realization and characterization of quasi-one-dimensional polymer nanostructures. PS-*b*-P2VP is a particularly versatile molecular system, which is usable to this aim as well. Exploiting monolayer micelle films of P2VP in a PS matrix, electrochemically grown PAN nanofibers can be obtained, after selectively

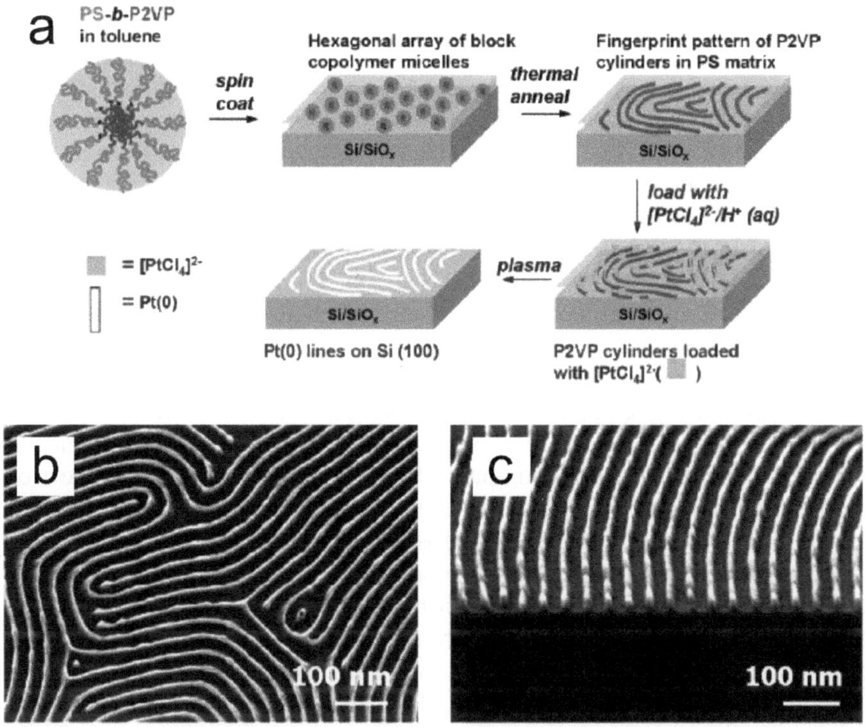


**Figure 3.16** (a) Scheme of the realization process of cylindrical domains of block copolymers, and of their use to assemble and align metallic nanowires. A toluene solution containing polymer micelles is spin-cast onto native oxide-coated Si. A pseudo-hexagonal array of block copolymer micelles is formed. Thermal annealing produces P2VP cylinders embedded in a matrix of PS. Immersion in an acidic solution leads to P2VP swelling due to protonation. Following the addition of anionic metallic salts, these are loaded by the positively charged P2VP. The polymer can then be removed by a plasma treatment. (b, c) Plan view and cross-sectional SEM micrographs of Pt nanowires self-assembled on PS-*b*-P2VP nanostructures. Adapted with permission from Ref. 194, *ACS Nano*, 2008, **2**, 489–501. Doi: 10.1021/nn700341s. Copyright © 2008, American Chemical Society.

removing the P2VP by acetic acid to produce pores to be then re-filled by the conjugated polymer.[193] This concept of selectively removing a block to realize template nanocavities proves to be very flexible both chemically and in terms of experimental geometries. The reactive ion etching (RIE) of the PS component of PS-*b*-PDMS domains on PDMS brush surfaces has been used to achieve a mask through which another etching step can be performed against the exposed region of a PEDOT:poly(styrenesulfonate) (PSS) film, which is underneath the block copolymer layer.[224] In this way, conjugated polymer wires are obtained with diameter as low as 15 nm, a quite impressive resolution considering that it is obtained without exposure-based or mechanical lithographies, and without using hard templates. Furthermore, the resulting pattern shows enhanced sensitivity when employed for gas sensing compared to reference film, due to the very high surface-to-volume ratio. Very dense arrays of PPy nanorods can be also fabricated on indium-tin oxide (ITO) by electrochemical polymerization within a porous diblock copolymer template made by PS-*b*-PMMA.[195] Due to the preliminary deposition of a hydroxy-terminated random copolymer system, PMMA domains orient perpendicularly to the underlying substrate, which finally results in cylindrical volumes of PMMA that can be removed by acetic acid and then re-filled by electropolymerized PPy. Overall, these latter experiments are therefore smart variants of membrane templating, finally allowing roughly aligned and vertically standing, conductive PPy nanorods with good internal molecular orientation to be produced. The so-demonstrated density of nanostructures ($\sim 10^{11}$ per $cm^2$) is comparable to the highest values achievable by hard membrane templates.[195] In addition, one should mention that in principle polymer templates for producing nanofibers can be fabricated by any sort of lithographic, spinning or growth methods. For instance, one can realize large amounts of polymer fibers by electrospinning, and then exploit these as templates for second-generation nanostructures including nanotubes, by coating or functionalization of the fiber external surface (*via* CVD, thermal evaporation, polymerization, wet processing *etc.*). This is the principle at the base of the so-called "tubes by fiber templates" (TUFT) process.[225] Such a method has been used with electrospun PLA to template CVD-grown poly(*p*-xylylene) hollow tubes with inner diameter down to the sub-10 nm scale,[225,226] nanostructured titanium dioxide achieved by sol-gel processing,[227] PAN grown by oxidative polymerization[228] and electrochemically grown PPy and PEDOT (Figure 3.17a,b).[229] Nanofibers doped or coated by ferric tosylate (FeTos) or other ferric oxidants for polymerization are also largely used as templates for the deposition of conductive polymers from the vapor phase.[230–233] As displayed in Figure 3.17c, the latter method requires two sequential steps, in which firstly the oxidant is processed together with a polymer solution to fabricate composite functional nanofibers, and then the monomer in the form of vapor comes into contact with this template and finally polymerizes due to the fiber-embedded oxidant agent. Electrically conductive, Au-impregnated PAC/polythiophene core–shell nanofibers can be produced by similar

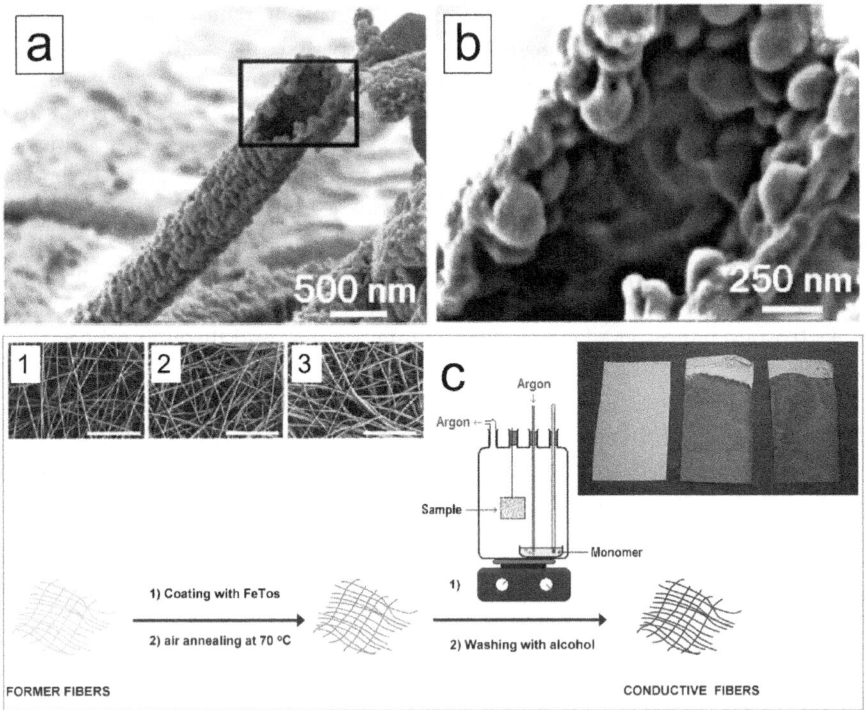


**Figure 3.17** (a) A PEDOT nanotube polymerized around a poly(lactide-*co*-glycolide) (PLAGA) nanofiber, which is then dissolved. (b) Higher-magnification image. Adapted with permission from Ref. 229, *Adv. Mater.*, 2006, **18**, 405–409. Doi: 10.1002/adma.200501726. Copyright © 2006, Wiley-VCH Verlag GmbH & Co. KGaA. (c) Schematics of the production process of hybrid nanostructures made by electrospinning nanofibers, then used as templates for the deposition of a conductive polymer coating through oxidative polymerization from monomers in the vapor phase. Right inset: photographs of as-produced mats of electrospun PAC nanofibers (left sample, sample size $\cong$ 4 $\times$ 2.3 cm$^2$), and after coating with PEDOT (middle) and with PPy (right). Left insets (1–3): Corresponding SEM micrographs. Scale bar = 10 μm. Adapted with permission from Ref. 231, *Chem. Mater.*, 2010, **22**, 2474–2480. DOI: 10.1021/cm902986g. Copyright © 2010, National Research Council Canada.

processing, *i.e.* carrying out sequentially electrospinning and polymerization from the gas phase of the thiophene monomer ($C_4H_4S$).[234] Incidentally, treating preliminarily fabricated nanofibers by vapors or solutions is also an effective route to produce nanocomposite fibers embedding or decorated by semiconductor, metal or organic nanoparticles. As already mentioned in Section 2.3.3, PbS nanoparticles dispersed in nanofibers are straightforwardly obtained by electrospinning PVP together with lead acetate and then exposing samples to $H_2S$ gas at room temperature.[235] Electrospun nanofibers of PMMA doped with an Au-containing salt, once treated by a dilute $NaBH_4$ solution, undergo reduction of the Au-salt component.[236] After washing, electroless-plating can rely on the Au particles on the surface of the fibers to catalyze the reduction of $Au^{3+}$ by hydroxylamine and finally to achieve electrically continuous metal films on the surface of the mats. Hydroxylated PEDOT nanotubes have been fabricated and decorated with spherical and cylindrical substructures by vapor deposition polymerization on electrospun nanofibers.[237] Finally, conjugated polymers such as the poly[2-(3-thienyl) ethanol butoxy carbonyl-methyl urethane] can be assembled onto the surface of electrospun nanofibers by electrostatic interactions, and through processing in solutions.[238]

DNA is especially intriguing as a non-covalent bonding, soft template material directing polymer nanostructures, because of its controllable length, relative stability, self-assembly properties, chemical flexibility and remarkable capability of forming hybrid macromolecules with many synthetic compounds.[201] These properties have made DNA one of the most used programmable materials in nanotechnology.[239–242] In particular, the anionic character of DNA is recognized to be especially important, providing it with high affinity for cationic species, including oxidatively doped polymers. The formation of DNA/polymer hybrids has been described to proceed through sequential stages, in which at first the phosphoric acid groups of DNA dope the polymer, followed by uncoiling of the polymer on DNA.[243] Such features have been exploited for growing PAN assemblies and nanostructures.[202,203,244,245] Conjugated polymer/DNA complexes are formed in solution and, in addition, DNA strands can be immobilized with a high degree of control on solid substrates and then used to drive the position and the orientation of the produced nanowires.[202] The measured diameter of DNA/PPy complex wires can be as small as 5 nm, and AFM evidences a particular assembly process in which single polymer wires twist around each other forming electrically conductive "nanoropes".[204] Boundary regions are also observed, where the nanorope unravels, which can be interpreted as intermediate between segments that have right-handed and left-handed twists (Figure 3.18).[204] With regard to the alignment and orientation of the so-obtained nanostructures, one should underline that DNA, like many other polymer molecules and fibrils,[209,246,247] can be aligned by so-called *molecular combing*, namely disentangling and extending molecular chains by means of a moving air–solution interface.[248] Typically, a droplet of macromolecule solution is deposited onto a substrate and then dragged by a cover slip,[248] a pipette,[249,250] a gentle gas flow,[251] *etc.*, which pulls the liquid and spreads it on the solid surface. The latter is usually chemically modified to selectively anchor one end of the macromolecules,[248] or to tailor the interaction with them,[249,250,252] thus making the polymer more effectively stretchable. The order of magnitude of the force so-exerted on the molecules at the liquid–air interface near the receding meniscus can be estimated as $2\pi\sigma_{LV}\,r$, where $r$ indicates the radius of the cylindrical rod schematizing the

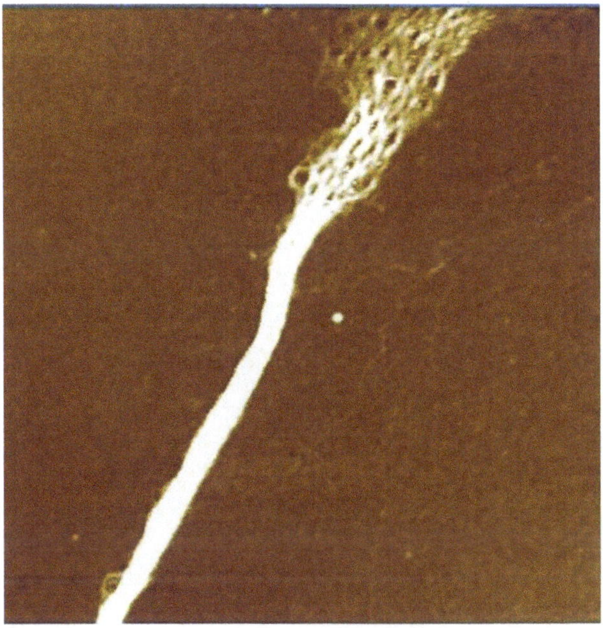

**Figure 3.18** AFM micrograph of PPy–DNA nanoropes on a $SiO_2$/Si surface. The assembling process lasts over 48 h prior to deposition and imaging. Image size ≅ 2 μm × 2 μm. Vertical color scale = 8 nm. Courtesy of Prof. B. R. Horrocks. Adapted with permission from Ref. 204, *Adv. Funct. Mater.*, 2008, **18**, 2444–2454. Doi: 10.1002/adfm.200701336. Copyright © 2008, Wiley-VCH Verlag GmbH & Co. KGaA.

chain, and is in the region of $10^{-9}$ N.[247,248] This, being much larger than entropic forces, is sufficient for stretching and aligning polymers perpendicular to the wetting line, but not for breaking covalent bonds that keep a macromolecule together. Viscous drag in the liquid can also contribute, but to a much lower extent ($\sim 10^{-12}$ N)[247] according to Stokes' law (Section 2.1.1). It is clear that molecular combing relies substantially on nanofluidic phenomena occurring in the reduced volume of the receding liquid film at the meniscus near the wetting line (Section 3.2). This method has been used to align PEDOT/PSS and DNA/PPy complex nanowires, and to embed them in devices comprising metal electrodes.[204,249,253] Also complexes of DNA with poly{3-[(*S*)-5-amino-5-carboxyl-3-oxapentyl]-2,5-thiophenylene hydrochloride}, poly{3-[(*S*)-5-amino-5-methoxycarboxyl-3-oxapentyl]-2,5-thiophenylene hydrochloride}[205] and other polythiophene derivatives[254] can be easily aligned on chemically modified surfaces by molecular combing methods. In general, DNA base pairing and hybridization properties promise great advantages for inter-connects of organic nanowires and nanofibers in nanoscale circuits and devices.[239]

The arrangement of various proteins in the form of amyloid fibers is another effect that can be exploited for pseudo-templating the formation of polymer

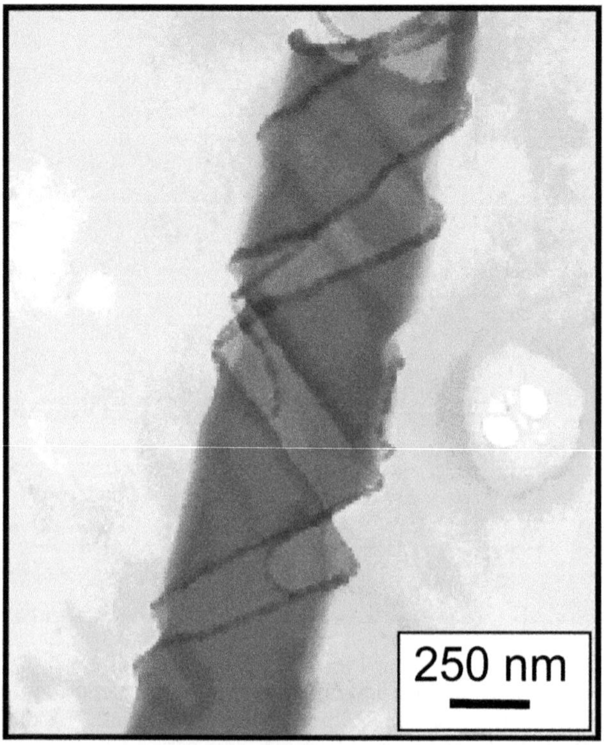


**Figure 3.19** TEM micrograph of PPy strands templated on an open 1,2-bis(10,12-tricosadiynoyl)-*sn*-glycero-3-phosphocholine tubule. The darker gray regions are the formed PPy strands. Adapted with permission from Ref. 260, *Chem. Mater.*, 2000, **12**, 1222–1228. Doi: 10.1021/cm990736z. Copyright © 2000, American Chemical Society.

nanofibers. Amyloid fibers are widely studied aggregates of proteins, with a so-called cross β-sheet architecture and width in the range of 5–13 nm. They are implicated in a number of pathologies including Alzheimer's, Parkinson's and Huntington's diseases,[68,255,256] and can be produced and characterized *in vitro*.[257,258] Researchers at Linköping University have shown how to use them as robust templates, allowing conjugated polymers to be assembled in a controlled fashion.[207,259] For instance, luminescent wires can be fabricated integrating semiconducting conjugated oligoelectrolytes[207,209] with bovine insulin amyloid fibers over lengths of tens of μm, and then aligned by molecular combing approaches as described above. Tape-like nanostructures are also formed by the assembly of biosynthetic triblock peptide copolymers with an amino acid substituted polythiophene derivative, taking place through ionic interactions.[198] Finally, PPy is synthesized with high spatial control on the edges and seams of tubules made of the diacetylenic phospholipid 1,2-bis(10,12-tricosadiynoyl)-*sn*-glycero-3-phosphocholine (Figure 3.19).[260] Other methods rely instead on micelles, driving the self-assembly of molecules or the polymerization of monomers. These approaches are better described in Sections 3.1.3 and 3.4, respectively.

Overall, given the flexibility of template synthesis in terms of obtainable nanostructures (hollow, solid, *etc.*), it is not surprising that with some optimization one can produce nanomaterials addressing the needs of quite varied applications. An important example is the fabrication of hollow, miniaturized capsules of polymer embedding enzyme molecules.[130] To this aim, after partial electrochemical growth of PPy on the metal deposited on one side of the template membrane, extra plugs of PPy are used in the pores, chemical polymerization is left to run for a while to obtain thin walls of PPy in the internal volumes of the pores and the desired enzyme molecules are positioned in the remaining voids before sealing the so-obtained capsules and dissolving the template. The overall process is schematized in Figure 3.20.

Analogously to self-assembled nanofibers (Section 3.1), issues related to manipulating and positioning with good spatial accuracy template-synthesized nanofibers are important in view of the utilization of these nanostructures in devices. Recently, a dielectrophoretic method has been proposed to handle and orient PFO nanowires with their longitudinal axis perpendicular to the edge of an electrode.[154] This allows values as high as 0.89 to be reached for the orientational order parameter, $S$ (Equation (2.10)). Pseudo-template concepts may help in this framework. Liquid crystal matrices[151] and electrochemically etched W wire probes with three-dimensional positioning control[261] may be other convenient tools to manipulate polymer nanowires.

Finally, it should be mentioned that the broad concept of "template" synthesis could be generalized to quite different methods that use, for instance, seeding components (instead of porous membranes) in the monomer solution in order to drive the polymerization process leading to nanofibers. In these techniques, the very early stages of polymerization are crucial in driving the subsequent reaction and somehow determining the final supramolecular organization. This is the case of the so-called "nanofiber seeding" or "seeded-guided polymerization",[262–264] which is something intermediate between the assembling and template techniques. The method consists in seeding a monomer solution with nanofibers made of a different material. These assist the polymerization of the target compound, with an imprinting effect at the nanoscale level that is critical in directing the polymerization towards nanofiber formation (which is not achieved in the absence of the seeding components). In addition, seeding techniques allow one to produce bulk quantities of nanofibers with lengths up to tens of μm. Explored seeds include PAN nanofibers produced by interfacial polymerization (Section 3.4), SWCNTs, nanofibers of $V_2O_5$ and other kinds of fibers.[262–266] Demonstrated nanofibers are made of PAN,[262,266] PPy,[263,265] PEDOT,[264] *etc.* Solvents and processing details have to be optimized for the different materials. Especially beautiful nanostructures have been demonstrated by PPy nanotubes with hollow core as small as 6 nm in diameter, because these tubes spontaneously undergo a redox reaction with noble metals forming

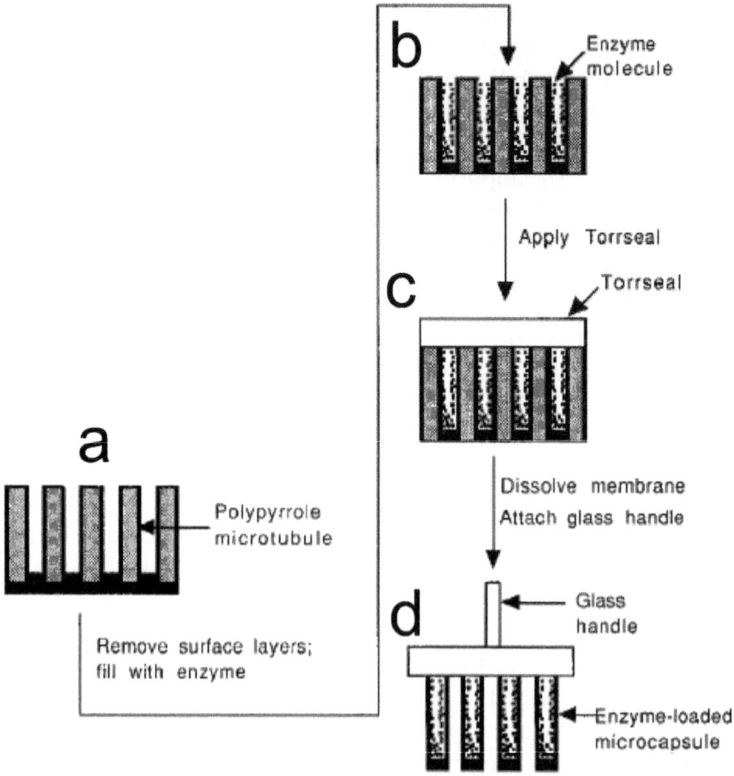


**Figure 3.20** Scheme of the method allowing an array of enzyme-loaded polymeric nanocapsules to be realized. Following the polymerization of a conjugated polymer on the walls of the pores of a template membrane (a), the enzyme is loaded (b), and the system is capped with an epoxy (c) before the membrane is dissolved (d). Adapted with permission from Ref. 97, *Chem. Mater.*, 1996, **8**, 1739–1746. DOI: 10.1021/cm960166s. Copyright © 1996, American Chemical Society.

nanoparticles that decorate the surface of the conjugated polymer.[265] In addition, in the case of Ag, these nanoparticles finally coalesce within the hollow nanochannel along the axis of the nanotube, thus resulting in the formation of a hybrid, core-shell metal/polymer nanostructure.

## 3.4 Interfacial Polymerization

Purely interfacial polymerization is a template-free synthesis method allowing relatively high throughput to be reached (say, grams of nanofibers per run at laboratory scale). The different conditions of interfacial polymerization, such as the structure of the specific monomer and the eventual dopants, the

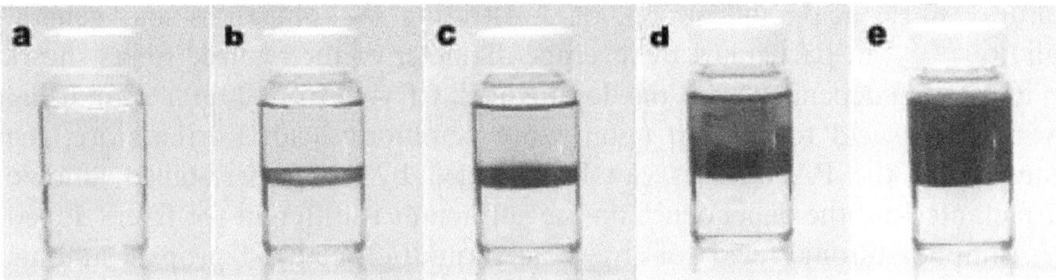


**Figure 3.21** Temporal sequence of photographs of a vial during the interfacial polymerization of aniline. Times: (a) 0 s, (b) 60 s, (c) 90 s, (d) 120 s, (e) 180 s. The aqueous solution of the acid and the oxidant agent fills the volume above the interface. Reproduced with permission from Ref. 269, *Chem. Eur. J.*, 2004, **10**, 1314–1319. Doi: 10.1002/chem.200305211. Copyright © 2004, Wiley-VCH Verlag GmbH & Co. KGaA.

concentration of the reagents and of course the process temperature, all have a remarkable influence on the morphological and functional properties of the obtained nanofibers. A few examples are recalled in the following.

The case of interfacial polymerization of PAN is likely the best known, and the first demonstrated.[128,267–271] As in other kinds of synthetic/assembly and template techniques, oxidative polymerization is indeed very well suited to produce nanostructures, in this case by interfacial reactions. To this aim, instead of a homogeneous solution comprising the monomer, the oxidant agent and the acid dopant as in traditional oxidative polymerization, an interface is used to separate two immiscible liquid systems. A typical example is an aqueous solution comprising the acid and the oxidant component, forming a smooth interface with a solution of aniline in an organic solvent (Figure 3.21).[267,269] In this way, the aniline monomer polymerizes in correspondence of the interface forming nanofibers, and the newly formed nanostructures then gradually diffuse into the aqueous solution. The latter step is particularly important, since the rapid removal of the nanofibers from the reactive interface avoids the onset of secondary growth processes, which can transform fibers in more complex agglomerates as in conventional oxidative polymerization.[80,272] Strategies to suppress secondary growth include using just the amount of reactants that are consumed during the nucleation of nanofibers (which is somehow similar to using dilute monomer solutions in assembly methods)[82] and rapidly mixing the reaction volume.[272] In fact, these studies have a broad and fundamental interest since they lead the nanofibrous morphology to be recognized as intrinsically related to the process of oxidative polymerization of aniline in water.[273] In particular, interfacial polymerization produces PAN fibers that have a diameter in the range of 30–50 nm and length from 0.5 to several μm, promptly usable within devices such as chemical sensors (Section 4.3.2).

The interfacial method, implemented by using camphorsulfonic acid as dopant,[128,267] has been extended to other mineral acids, such as hydrochloric, sulfuric, nitric acid, and so on, demonstrating its robustness and general validity.[268,273] In particular, the average diameter of the realized fibers shows an interesting dependence on the doping acid, varying from 30 nm upon using hydrochloric acid to 120 nm upon using perchloric acid. Furthermore, the diameter of the PAN fibers can be changed by using twin-tailed anionic surfactants, and the dependence on the surfactant is different for fibers doped by camphorsulfonic acid or by 2-acrylamido-2-methyl-1-propanesulfonic acid.[128] Using camphorsulfonic acid, the diameter of the fibers decreases from 48 nm to 28 nm by adding surfactants, whereas using 2-acrylamido-2-methyl-1-propanesulfonic acid the diameter increases from 23 nm to 55 nm by adding surfactants. In another work, the molar ratio of APS to aniline, and the molarity of the $HClO_4$ solution in which APS is dissolved, are demonstrated to affect significantly the nanofiber morphology and conductivity.[274] By introducing initiators such as the *p*-phenylenediamine, which is a dimer of aniline, or the 1,4-benzenediamine, the method can be straightforwardly extended to a variety of substituted PAN derivatives.[275]

We are here in a domain in which interfacial polymerization, soft-templating and self-assembly inter-weave their mechanisms and all contribute to the formation of nanostructures. In this framework, a widely studied phenomenon is the $FeCl_3$ or APS-driven polymerization of pyrrole or aniline at the interface between water and monomer (or monomer/dopant salts) micelles, whose confined volume can hardly be penetrated by the hydrated oxidant molecules, or between hexane and micelles formed by sodium bis(2-ethylhexyl) sulfosuccinate.[186,187,276–284] For example, PAN fibers with diameter down to 76 nm, and nanotubes with outer diameter down to 90 nm, are synthesized by using APS as oxidant and amphiphilic, β-naphthalene sulfonic acid (β-NSA) as dopant, and both the resulting conduction properties and the formation mechanism of the nanostructures are suggested to depend on the relative concentration of the aniline monomer and of NSA.[285,286] For high (≥2) molar ratios of the dopant to the monomer, precipitates of NSA-aniline salt can act as templates driving the formation of fibers, whereas the same role is played by NSA-aniline micelles for low values (0.25, 0.5 *etc.*) of the same molar ratio. Amphiphilic, D-10-camphorsulfonic acid can be exploited to form micelles with the aniline monomer in the same way.[281] Micellar aggregates that can somehow template the formation of nanofibers are also found when polymers, such as PPy, are synthesized.[88,89] Conductive PPy nanofibers that have average diameter between 60 nm and 100 nm are obtained by $FeCl_3$ as oxidant, in the presence of *p*-hydroxyl-azobenzene sulfonic acid as dopant.[282] PPy nanotubes that have diameter down to 90 nm can be fabricated by the so-called method of reverse microemulsion polymerization in an apolar solvent.[186] Here, the cylindrical micelles, formed by aggregation of sodium bis(2-ethylhexyl) sulfosuccinate in apolar solvents upon addition of a $FeCl_3$ aqueous solution, serve as internal templates for the polymerization of PPy as schematized in Figure 3.22a. A similar approach applied to PEDOT leads to ellipsoidal nanoparticles, nanorods or nanotubes, depending on the amount of added

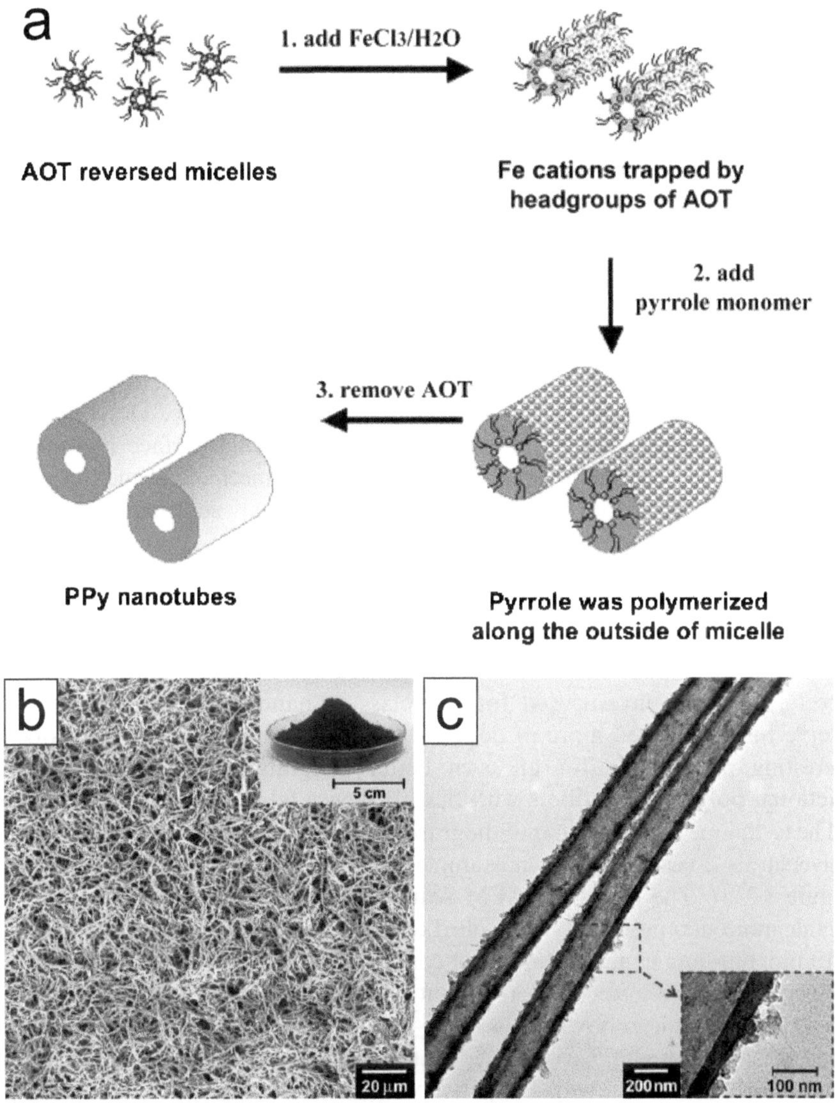


**Figure 3.22** (a) Scheme of PPy nanotube fabrication by reverse microemulsion polymerization. Here, AOT indicates sodium bis(2-ethylhexyl) sulfosuccinate. Adapted with permission from Ref. 186, *Chem. Commun.*, 2003, 720–721. Doi: 10.1039/b211716a. (b) SEM and (c) TEM micrographs of PEDOT nanotubes, realized by chemical oxidation polymerization in reverse (water-in-oil) microemulsions. These nanostructures are obtained using a volume ratio of an aqueous $FeCl_3$ solution to hexane of $2.7 \times 10^{-2}$. The inset in (b) shows a Petri dish filled with PEDOT nanotubes ($\cong$ 3 g) fabricated in a single polymerization run. The inset in (c) shows a magnified image. Adapted with permission from Ref. 187, *Adv. Funct. Mater.*, 2007, **17**, 431–436. DOI: 10.1002/adfm.200600106. Copyright © 2007, Wiley-VCH Verlag GmbH & Co. KGaA.

aqueous solution (Figure 3.22b,c).[187] PAN/PSS nanofiber composites can also be produced by polymerization of aniline at an interface with an aqueous solution of APS and poly(styrene sulfonic acid).[287]

## 3.5 Direct Writing Methods

Many direct writing methods can be used to realize polymer nanostructures, thus extending to the fabrication of nanofibers the concept of serial lithographies (Section 1.2.1). This approach requires scanning serially a solid surface by means of a physical agent (which could vary from an electron, ion or laser beam to a scanning probe tip) localized with very high spatial resolution, and in correspondence of which a physical deposition or a chemical reaction drives the formation of a segment of nanostructure. Drawing and pulling viscous polymeric solutions deposited on a solid surface is another option, usable to obtain suspended fibers. Effective mechanical methods for realizing polymer nanofibers include drawing from solutions by glass micropipettes,[288] or from melts by metal or silica rods.[289] As in nanofluidic methods, once realized the nanostructures can in principle be detached from the underlying substrate by exploiting a properly optimized surface chemistry, reducing the adhesion of the solidified fibers to the substrate, or by transfer techniques based on elastomeric elements. Though handling and patterning aspects are rarely investigated for this class of nanostructures, they would deserve to be explored more in depth in the near future since they can open interesting perspectives for an even better integration of directly written functional polymer nanofibers with flexible device fabrication.

The technique of dip-pen nanolithography (DPN)[290] invented at Northwestern University is a very well-known example of how serial writing strategies work (Figure 3.23a). The tip of an AFM scans the surface and deposits layers of organic molecules in a highly controlled way, with spatial resolution in the scale of 10 nm. μm-long nanofibers made of conjugated polymers can be fabricated in this way.[291] In its electrochemical variant, DPN is used to produce filaments of PEDOT with diameter down to 30 nm.[292] To this aim, a highly doped, tapping-mode AFM tip is preliminarily coated by the monomers, simply by immersion in a chloroform solution. Then, a negative voltage bias is applied between the tip and the target substrate (Si/$SiO_2$) during scans, which leads to electrochemical polymerization tightly localized near the tip, and to deposition of solid polymer features. The absolute value of the needed voltage increases upon increasing the thickness of the $SiO_2$ insulating layer on the substrate, and is between 9 and 15 V for a native oxide layer. In this way, fibers are produced at rates up to 10 nm of length per second.[292] Another mechanism allowing conjugated polymers to be deposited by DPN relies on electrostatic interactions between suitably modified substrate surfaces and the macromolecules delivered by the tip.[293] STM can also be used to induce polymerization with very high spatial resolution, as demonstrated in the realization of nanowires of polydiacetylene.[294] These techniques allow electrochemical polymerization to be combined with

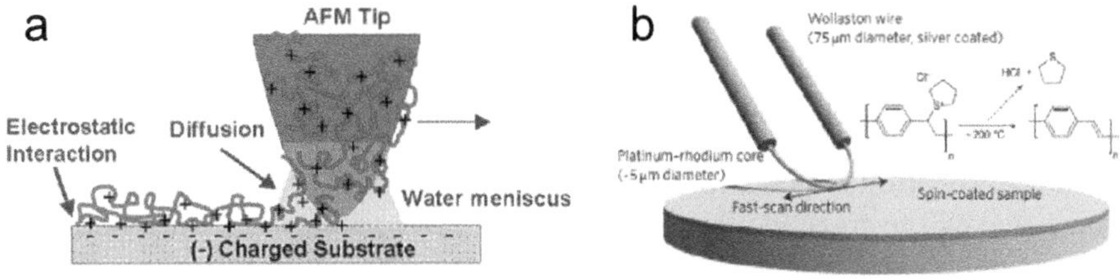


**Figure 3.23** (a) Scheme of DPN for patterning charged conducting polymers. Adapted with permission from Ref. 293, *Adv. Mater.*, 2002, **14**, 1474–1477. Doi: 10.1002/1521-4095(20021016)14:20<1474::AID-ADMA1474>3.0.CO;2-2. Copyright © 2002, Wiley-VCH Verlag GmbH & Co. KGaA. (b) Scheme of thermochemical nanopatterning by a resistively heated wire mounted on an AFM. The wire scan leads to the polymerization of a precursor polymer film, thus allowing different kinds of nanostructures to be fabricated including conjugated polymer nanowires. The inset shows the thermal conversion of poly(*p*-xylene tetrahydrothiophenium chloride) into PPV. Adapted with permission from Ref. 297, *Nat. Nanotechnol.*, 2009, **4**, 664–668. Doi: 10.1038/nnano.2009.254. Copyright © 2009, Nature Publishing Group.

mechanical drawing to stretch the produced nanowires and decrease their diameters.[295] Thermochemical nanopatterning can be carried out by means of a scanning probe as well.[296] This process consists in scanning, on a pre-deposited polymer precursor surface, a thin, resistively heated wire, which is mounted on an AFM and connected to an external temperature control circuit (Figure 3.23b).[297] In this way, the precursor is cross-linked at the passage of wire with very high spatial resolution as demonstrated fabricating sub-30 nm, elongated structures of PPV. Writing speeds are quite good ($\cong$ 100 μm $s^{-1}$).

Electron beam lithography (EBL) can also be used for direct writing on conjugated polymers, allowing various kinds of functional features to be produced in the form of nanofibers. For instance, polyfluorenes generally behave as negative cross-linkable materials (Section 1.2.1) under exposure to radiation beams. Depending on the specific polyfluorene derivative and eventual chemical amplification, sensitivities to electron beams can vary from 2300 μC $cm^{-2}$ down to 4 μC $cm^{-2}$, and lines as narrow as 70 nm have been patterned retaining the optical properties of pristine materials.[298] The group of O. Inganäs has pioneered the field demonstrating the direct EBL patterning of 50 nm wide lines of poly(3-octylthiophene), whose solubility decreases upon electron beam exposure.[299] In these experiments, doses vary from $10^2$ μC $cm^{-2}$ to $10^3$ μC $cm^{-2}$, and the sensitivity is around 200 μC $cm^{-2}$. There is evidence that also MEH-PPV behave as a negative resist, becoming insoluble when exposed by EBL.[300] It should be mentioned that acceleration voltages used for these experiments (50–75 kV) can be significantly higher than values chosen in most patterning processes on conventional electron-beam resists.

Finally, nanofibers can be realized by means of CFL (Section 1.2.1), performed by exploiting two-photon polymerization mechanisms.[301] In these lithographies, two photons are simultaneously absorbed, each having half the energy of the involved optical transition of material.[302] The probability of two-photon absorption depends on the squared intensity of the irradiation light, which makes the method powerful to draw three-dimensional micro-objects because polymerization occurs only within the volume of a spot in which the light is tightly focalized. Combining two-photon polymerization with CFL allows therefore fibers and other organic particles to be produced in a continuous way, and in addition some degree of parallelization can be implemented by using multiple irradiation spots focused in different regions of each microfluidic channel. In particular, helical fibers can be realized by moving the polymerization spot along circumferences in a plane perpendicular to the rectilinear flow of the photo-curable polymer along the microchannel.[301]

However, as anticipated in Section 1.2.1, most writing methods are intrinsically time-consuming due to their serial character, and provide lower throughput than spinning and chemical synthesis. Direct writing can be advantageous when one needs extremely high spatial resolution in positioning the resulting nanostructures. Indeed, because nanofibers can be fabricated directly *in situ* onto previously realized device architectures, metal electrodes, *etc.*, these techniques show unequalled performances in terms of overlay alignment capability and registration accuracy.

## 3.6 Phase Separation

Phase separation in polymer solutions can be driven by temperature variations and exploited to produce nanofibers in the form of inter-connected, highly porous three-dimensional networks. Briefly, the typical process includes the preparation of a polymer solution in a given solvent, phase separation, solvent exchange upon immersion in water, sample freezing and final freeze-drying under vacuum, which finally leads to a foam-like material made of nanofibers.[303] From a practical viewpoint, firstly a vial containing the polymer solution is rapidly cooled in a refrigerator, which determines the gelation of the dissolved polymer within a characteristic time depending on specific compound and on its concentration, on the chosen solvent and of course on temperature. Afterwards the vial is immersed in water, replacing the original solvent, and the gel is then removed from the water, frozen and finally freeze-dried to produce a solid polymer matrix (Figure 3.24). This technique has been used to produce inter-connected, highly porous networks of fibers made of a variety of synthetic polymers such as PLA, PLAGA, poly-DL-lactic acid and biodegradable amphiphilic poly[hydroxyalkyl(meth)acrylate]-graft-poly(L-lactic acid) copolymers.[303–306] The typical range of fiber diameters is 50–500 nm. More recently, the method has been successfully applied to realize three-dimensional architectures of gelatin, showing a nanofibrous structure upon properly choosing the processing parameters.[307] The diameters of the resulting

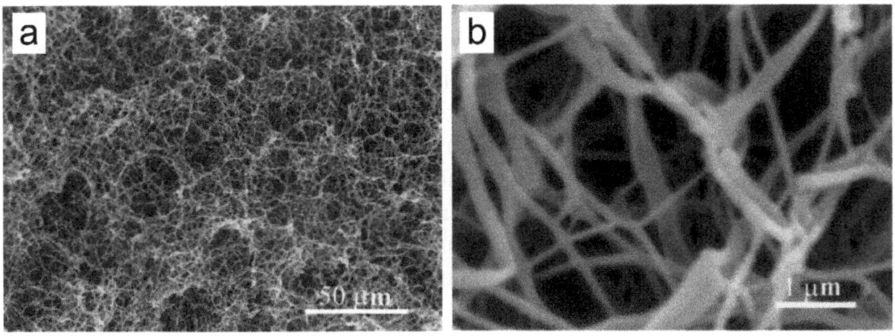


**Figure 3.24** SEM micrographs at different magnifications of a PLA fibrous matrix prepared by phase separation, from 2.5% (wt/v) THF solutions. The gelation temperature is 8 °C. Adapted with permission from Ref. 303, *J. Biomed. Mater. Res.*, 1999, **46**, 60–72. Doi: 10.1002/(SICI)1097-4636(199907)46:1<60::AID-JBM7>3.0.CO;2-H. Copyright © 1999, John Wiley & Sons, Inc.

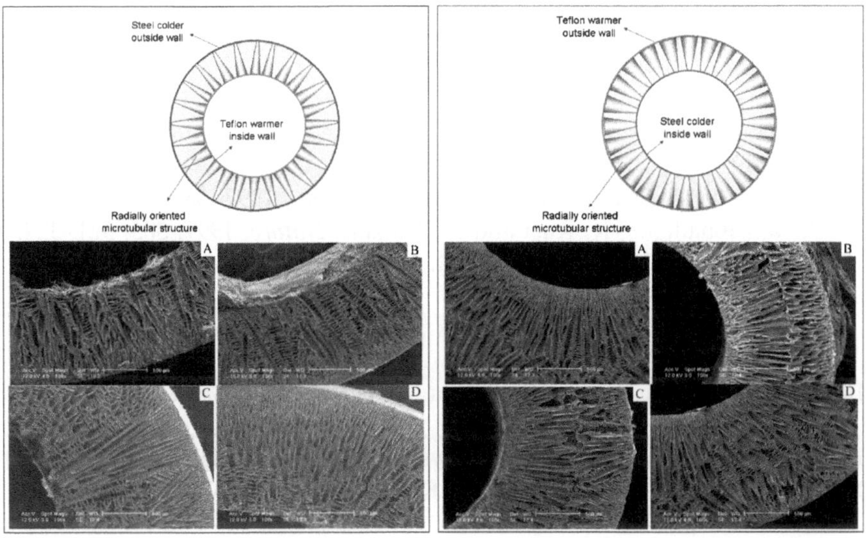


**Figure 3.25** Schematics and SEM micrographs of cross-sections of PLA vessel tubular elements realized under different temperature gradients. The warmer and the colder walls are inverted in the two sets of panels (left and right panels, respectively). The size of the radially oriented voids decreases by moving from the hot to the cold interface with the external template material. In both the investigated cases, PLA benzene solutions at different concentrations are used: (A) 2.5%, (B) 5.0%, (C) 7.5%, (D) 10.0% (weight/volume). Adapted with permission from Ref. 308, *Adv. Funct. Mater.*, 2010, **20**, 2833–2841. Doi: 10.1002/adfm.201000922. Copyright © 2010, Wiley-VCH Verlag GmbH & Co. KGaA.

fibers range roughly from 150 nm to 200 nm, and the lengths are up to about 1–2 μm, for gelatin concentrations varying from 5% to 10%.[307] Of course, polymer samples embedding the nanofibrous structure can be templated at macroscopic scales by external molds confining the processed solution within volumes of the desired shape. More interestingly, by realizing such molds with different materials (steel, Teflon, *etc.*) and exploiting the differences in thermal conductivities of these mold materials, one can control the microphysics of the phase separation process and produce nanofibrous structures oriented along the axial or the radial direction of tubular polymeric samples (Figure 3.25).[308] The characteristics of the so-obtained tubules are especially promising for scaffolds for vascular tissue engineering (Chapter 7).

# References


1. C. A. Mirkin, R. L. Letsinger, R. C. Mucic and J. J. Storhoff, *Nature*, 1996, **382**, 132–188.
2. A. P. Alivisatos, K. P. Johnsson, X. Peng, T. E. Wilson, C. J. Loweth, M. P. Bruchez, Jr. and P. G. Schurtz, *Nature*, 1996, **382**, 609–611.
3. K. Miszta, J. de Graaf, G. Bertoni, D. Dorfs, R. Brescia, S. Marras, L. Ceseracciu, R. Cingolani, R. van Roij, M. Dijkstra and L. Manna, *Nat. Mater.*, 2011, **10**, 872–876.
4. A. S. Dimitrov and K. Nagayama, *Langmuir*, 1996, **12**, 1303–1311.
5. M. Trau, S. Sankaran, D. A. Saville and I. A. Aksay, *Nature*, 1995, **374**, 437–439.
6. S.-R. Yeh, M. Seul and B. I. Shraiman, *Nature*, 1997, **386**, 57–59.
7. A. van Blaaderen, R. Ruel and P. Wiltzius, *Nature*, 1997, **385**, 321–324.
8. A. Terfort, N. Bowden and G. M. Whitesides, *Nature*, 1997, **386**, 162–164.
9. N. Bowden, A. Terfort, J. Carbeck and G. M. Whitesides, *Science*, 1997, **276**, 233–235.
10. G. M. Whitesides and M. Boncheva, *Proc. Natl Acad. Sci. USA*, 2002, **99**, 4769–4774.
11. G. M. Whitesides and B. Grzybowski, *Science*, 2002, **295**, 2418–2421.
12. T. D. Clark, J. Tien, D. C. Duffy, K. Paul and G. M. Whitesides, *J. Am. Chem. Soc.*, 2001, **123**, 7677–7682.
13. N. Bowden, S. R. J. Oliver and G. M. Whitesides, *J. Phys. Chem.*, 2000, **104**, 2714–2724.
14. I. W. Hamley, *Angew. Chem., Int. Ed.*, 2003, **42**, 1692–1712.
15. F. J. M. Hoeben, P. Jonkheijm, E. W. Meijer and A. P. H. J. Schenning, *Chem. Rev.*, 2005, **105**, 1491–1546.
16. M. A. Meyers, P.-Y. Chen, A. Y.-M. Lin and Y. Seki, *Prog. Mater. Sci.*, 2008, **53**, 1–206.
17. I. W. Hamley, *Prog. Polym. Sci.*, 2009, **34**, 1161–1210.
18. Y. H. Kim, D. K. Yoon and H.-T. Jung, *J. Mater. Chem.*, 2009, **19**, 9091–9102.

19. J.-M. Lehn, *Angew. Chem. Int. Ed.*, 1990, **29**, 1304–1319.
20. G. M. Whitesides, J. P. Mathias and C. T. Seto, *Science*, 1991, **254**, 1312–1319.
21. O. Ikkala, R. H. A. Ras, N. Houbenov, J. Ruokolainen, M. Pääkkö, J. Laine, M. Leskelä, L. A. Berglund, T. Lindström, G. ten Brinke, H. Iatrou, N. Hadjichristidis and C. F. J. Faul, *Faraday Discuss.*, 2009, **143**, 95–107.
22. K. J. Ihn, J. Moulton, and P. J. Smith, *Polym. Sci., Part B: Polym. Phys.*, 1993, **31**, 735–742.
23. J. Mårdalen, E. J. Samuelsen and A. Ø. Pedersen, *Synth. Met.*, 1993, **55**, 378–383.
24. S. Malik and A. J. Nandi, *Polym. Sci., Part B: Polym. Phys.*, 2002, **40**, 2073–2085.
25. S. Samitsu, T. Shimomura, S. Heike, T. Hashizume and K. Ito, *Macromolecules*, 2010, **43**, 7891–7894.
26. M. Mas-Torrent, D. den Boer, P. Hadley and A. P. H. J. Schenning, *Nanotechnology*, 2004, **15**, S265–S269.
27. J. A. Merlo and C. D. Frisbie, *J. Polym. Sci. Part B: Polym. Phys.*, 2003, **41**, 2674–2680.
28. J. A. Merlo and C. D. Frisbie, *J. Phys. Chem. B.*, 2004, **108**, 19169–19179.
29. H. Goto, Y. Okamota and E. Yashima, *Macromolecules*, 2002, **35**, 4590–4601.
30. N. Kiriy, E. Jähne, H. J. Adler, M. Schneider, A. Kiriy, G. Gorodyska, S. Minko, D. Jehnichen, P. Simon, A. A. Fokin and M. Stamm, *Nano Lett.*, 2003, **3**, 707–712.
31. S. Samitsu, T. Shimomura and K. Ito, *Thin Solid Films*, 2008, **516**, 2478–2486.
32. S. Samitsu, Y. Takanishi and J. Yamamoto, *Macromolecules*, 2009, **42**, 4366–4368.
33. M. Avrami, *J. Chem. Phys.*, 1939, **7**, 1103–1112.
34. M. Avrami, *J. Chem. Phys.*, 1940, **8**, 212–224.
35. S. Z. D. Cheng and B. Wunderlich, *Macromolecules*, 1988, **21**, 3327–3328.
36. W. D. Oosterbaan, V. Vrindts, S. Berson, S. Guillerez, O. Douhéret, B. Ruttens, J. D'Haen, P. Adriaensens, J. Manca, L. Lutsen and D. Vanderzande, *J. Mater. Chem.*, 2009, **19**, 5424–5435.
37. H. Yang, T. J. Shin, L. Yang, K. Cho, C. Y. Ryu and Z. Bao, *Adv. Funct. Mater.*, 2005, **15**, 671–676.
38. J. Liu, M. Arif, J. Zou, S. I. Khondaker and L. Zhai, *Macromolecules*, 2009, **42**, 9390–9393.
39. E. Lee, B. Hammer, J.-K. Kim, Z. Page, T. Emrick and R. C. Hayward, *J. Am. Chem. Soc.*, 2011, **133**, 10390–10393.
40. P.-T. Wu, G. Ren, C. Li, R. Mezzenga and S. A. Jenekhe, *Macromolecules*, 2009, **42**, 2317–2320.
41. A. L. Briseno, S. C. B. Mannsfeld, P. J. Shamberger, F. S. Ohuchi, Z. Bao, S. A. Jenekhe and Y. Xia, *Chem. Mater.*, 2007, **20**, 4712–4719.

42. A. Carlson, A. M. Bowen, Y. Huang, R. G. Nuzzo and J. A. Rogers, *Adv. Mater.*, 2012, **24**, 5284–5318.
43. R. Zhang, B. Li, M. C. Iovu, M. Jerreries-El, G. Sauvé, J. Cooper, S. Jia, S. Tristram-Nagle, D. M. Smilgies, D. N. Lambeth, R. D. McCullough and T. Kowalewski, *J. Am. Chem. Soc.*, 2006, **128**, 3480–3481.
44. D. Hwan Kim, J. Tark Han, Y. Don Park, Y. Jang, J. H. Cho, M. Hwang and K. Cho, *Adv. Mater.*, 2006, **18**, 719–723.
45. J. Liu, E. Sheina, T. Kowalewski and R. D. McCullough, *Angew. Chem. Int. Ed.*, 2002, **41**, 329–332.
46. G. Sauvé and R. D. McCullough, *Adv. Mater.*, 2007, **19**, 1822–1825.
47. P. Leclère, M. Surin, P. Jonkheijm, O. Henze, A. P. H. J. Schenning, F. Biscarini, A. C. Grimsdale, W. J. Feast, E. W. Meijer, K. Müllen, J. L. Brédas and R. Lazzaroni, *Eur. Polym. J.*, 2004, **40**, 885–892.
48. C. R. G. Grenier, W. Pisula, T. J. Joncheray, K. Müllen and J. R. Reynolds, *Angew. Chem. Int. Ed.*, 2007, **46**, 714–717.
49. P. Samorì, V. Francke, K. Müllen and J. P. Rabe, *Thin Solid Films*, 1998, **336**, 13–15.
50. P. Samorì, V. Francke, T. Magel, K. Müllen and J. P. Rabe, *Opt. Mater.*, 1998, **9**, 390–393.
51. H. Simmonds, N. Bennett, M. Elliott and E. Macdonald, *J. Vac. Sci. Technol. B*, 2009, **27**, 831–835.
52. D. Hwan Kim, Y. Don Park, Y. Jang, S. Kim and K. Cho, *Macromol. Rapid Commun.*, 2005, **26**, 834–839.
53. D. Hwan Kim, Y. Don Park, Y. Jang, S. Kim and K. Cho, *J. Phys. Chem. B*, 2006, **110**, 15763–15768.
54. G. Lu, L. Li and X. Yang, *Adv. Mater.*, 2007, **19**, 3594–3598.
55. G. Lu, H. Tang, Y. Qu, L. Li and X. Yang, *Macromolecules*, 2007, **40**, 6579–6584.
56. L. Qiu, W. H. Lee, X. Wang, J. S. Kim, J. A. Lim, D. Kwak, S. Lee and K. Cho, *Adv. Mater.*, 2009, **21**, 1349–1353.
57. O. D. Velev, S. Smoukov and M. Marquez, US Patent Application No. 2010/0247908, 2010.
58. C. J. Luo, S. D. Stoyanov, E. Stride, E. Pelanb and M. Edirisinghe, *Chem. Soc. Rev.*, 2012, **41**, 4708–4735.
59. D. M. Eisele, J. Knoester, S. Kirstein, J. P. Rabe and D. A. Vanden Bout, *Nat. Nanotechnol.*, 2009, **4**, 658–663.
60. S. Wang, M. Kappl, I. Liebewirth, M. Müller, K. Kirchhoff, W. Pisula and K. Müllen, *Adv. Mater.*, 2012, **24**, 417–420.
61. M. Surin, D. Marsitzky, A. C. Grimsdale, K. Müllen, R. Lazzaroni and P. Leclère, *Adv. Funct. Mater.*, 2004, **14**, 708–715.
62. T. Bjørnholm, D. R. Greve, N. Reitzel, T. Hassenkam, K. Kyaer, P. B. Howes, N. B. Larsen, J. Bøgelund, M. Jayaraman, P. C. Ewbank and R. D. McCullough, *J. Am. Chem. Soc.*, 1998, **120**, 7643–7644.

63. T. Bjørnholm, T. Hassenkam, D. R. Greve, R. D. McCullough, M. Jayaraman, S. M. Savoy, C. E. Jones and J. T. McDevitt, *Adv. Mater.*, 1999, **11**, 1218–1221.
64. M. Ornatska, S. Peleshanko, K. L. Genson, B. Rybak, K. N. Bergman and V. V. Tsukruk, *J. Am. Chem. Soc.*, 2004, **126**, 9675–9684.
65. D. Liu, S. De Feyter, M. Cotlet, U.-M. Wiesler, T. Weil, A. Herrmann, K. Müllen and F. C. De Schryver, *Macromolecules*, 2003, **36**, 8489–8498.
66. F. He, T. Gädt, I. Manners and M. A. Winnik, *J. Amer. Chem. Soc.*, 2011, **133**, 9095–9103.
67. S. K. Patra, R. Ahmed, G. R. Whittell, D. J. Lunn, E. L. Dunphy, M. A. Winnik and I. Manners, *J. Amer. Chem. Soc.*, 2011, **133**, 8842–8845.
68. S. Zhang, *Nat. Biotechnol.*, 2003, **21**, 1171–1178.
69. P. Berndt, G. B. Fields and M. Tirrell, *J. Am. Chem. Soc.*, 1995, **117**, 9515–9522.
70. J. D. Hartgerink, E. Beniash and S. I. Stupp, *Science*, 2001, **294**, 1684–1688.
71. G. A. Silva, C. Czeisler, K. L. Niece, E. Beniash, D. A. Harrington, J. A. Kessler and S. I. Stupp, *Science*, 2004, **303**, 1352–1355.
72. H.-W. Jun, V. Yuwono, S. E. Paramonov and J. D. Hartgerink, *Adv. Mater.*, 2005, **17**, 2612–2617.
73. H. Yokoi, T. Kinoshita and S. Zhang, *Proc. Natl Acad. Sci. USA*, 2005, **102**, 8414–8419.
74. Y. Hong, R. L. Legge, S. Zhang and P. Chen, *Biomacromolecules*, 2003, **4**, 1433–1442.
75. T. Kida, K. Kondo and M. Akashi, *Chem. Commun.*, 2012, **48**, 2319–2321.
76. T. F. A. de Greef and E. W. Meijer, *Nature*, 2008, **453**, 171–173.
77. T. F. A. Greef, M. M. J. Smulders, A. P. H. J. Schenning, R. P. Sijbesma and E. W. Meijer, *Chem. Rev.*, 2009, **109**, 5687–5754.
78. X. Yan, M. Zhou, J. Chen, X. Chi, S. Dong, M. Zhang, X. Ding, Y. Yu, S. Shao and F. Huang, *Chem. Commun.*, 2011, **47**, 7086–7088.
79. S. Kyun Chae, H. Park, J. Yoon, C. Hee Lee, D. June Ahn and J.-M. Kim, *Adv. Mater.*, 2007, **19**, 521–524.
80. D. Li, J. Huang and R. B. Kaner, *Acc. Chem. Res.*, 2009, **42**, 135–145.
81. S. Bhadra, D. Khastgir, N. K. Singha and J. H. Lee, *Prog. Polym. Sci.*, 2009, **34**, 783–810.
82. N.-R. Chiou and A. J. Epstein, *Adv. Mater.*, 2005, **17**, 1679–1683.
83. H. Qiu, M. Wan, B. Matthews and L. Dai, *Macromolecules*, 2001, **34**, 675–677.
84. X. Li and X. Li, *Mater. Lett.*, 2007, **61**, 2011–2014.
85. A. Rahy and D. J. Yang, *Mater. Lett.*, 2008, **62**, 4311–4314.
86. B.-K. Kim, Y. Hwan Kim, K. Won, H. Chang, Y. Choi, K. Kong, B. Whan Rhyu, J.-J. Kim and J.-O. Lee, *Nanotechnology*, 2005, **16**, 1177–1181.

87. S. Kobayashi and H. Higashimura, *Prog. Polym. Sci.*, 2003, **28**, 1015–1048.
88. A. Wu, H. Kolla and S. K. Manohar, *Macromolecules*, 2005, **38**, 7873–7875.
89. W. Zhong, S. Liu, X. Chen, Y. Wang and W. Yang, *Macromolecules*, 2006, **39**, 3224–3230.
90. J. Liu and M. Wan, *J. Mater. Chem.*, 2001, **11**, 404–407.
91. L. Liang, J. Liu, C. F. Windisch, G. J. Exarhos and Y. Lin, *Angew. Chem. Int. Ed.*, 2002, **41**, 3665–3668.
92. S.-J. Choi and S.-M. Park, *J. Electrochem. Soc.*, 2002, **149**, E26–E34.
93. V. N. Cepak and C. R. Martin, *Chem. Mater.*, 1999, **11**, 1363–1367.
94. C. R. Martin, L. S. Van Dyke, Z. Cai and W. Liang, *J. Am. Chem. Soc.*, 1990, **112**, 8976–8977.
95. C. R. Martin, *Science*, 1994, **266**, 1961–1966.
96. C. R. Martin, *Acc. Chem. Res.*, 1995, **28**, 61–68.
97. C. R. Martin, *Chem. Mater.*, 1996, **8**, 1739–1746.
98. J. C. Hulteen and C. R. Martin, *J. Mater. Chem.*, 1997, **7**, 1075–1087.
99. C. R. Martin, *Adv. Mater.*, 1991, **3**, 457–459.
100. Y. Xia and G. M. Whitesides, *Angew. Chem. Int. Ed.*, 1998, **37**, 550–575.
101. J. P. Rolland, E. C. Hagberg, G. M. Denison, K. M. Carter and J. M. DeSimone, *Angew. Chem. Int. Ed.*, 2004, **43**, 5796–5799.
102. E. Kim, Y. Xia and G. M. Whitesides, *Nature*, 1995, **376**, 581–584.
103. E. Kim, Y. Xia and G. M. Whitesides, *J. Am. Chem. Soc.*, 1996, **118**, 5722–5731.
104. J. C. McDonald, S. J. Metallo and G. M. Whitesides, *Anal. Chem.*, 2001, **73**, 5645–5650.
105. D. Myers, *Surfaces, Interfaces and Colloids*, Wiley-VCH Publishers, New York, 1999.
106. N. Srivastava and M. A. Burns, *Anal. Chem.*, 2006, **78**, 1690–1696.
107. S. Girardo, R. Cingolani and D. Pisignano, *Anal. Chem.*, 2007, **79**, 5856–5861.
108. C. De Marco, E. Mele, A. Camposeo, R. Stabile, R. Cingolani and D. Pisignano, *Adv. Mater.*, 2008, **20**, 4158–4162.
109. N. L. Jeon, I. S. Choi, B. Xu and G. M. Whitesides, *Adv. Mater.*, 1999, **11**, 946–950.
110. D. Pisignano, F. Di Benedetto, L. Persano, E. Mele and R. Cingolani, *Synth. Met.*, 2005, **153**, 325–328.
111. J. B. Edel and A. J. deMello (ed.), *Nanofluidics* (*RSC Nanoscience and Nanotechnology Series*), Royal Society of Chemistry, Cambridge, 2009.
112. Z. Wang, R. Bao, X. Zhang, X. Ou, C.-S. Lee, J. C. Chang and X. Zhang, *Angew. Chem. Int. Ed.*, 2011, **50**, 2811–2815.
113. K. Y. Suh, Y. S. Kim and H. H. Lee, *Adv. Mater.*, 2001, **13**, 1386–1389.
114. H. E. Jeong, S. H. Lee, P. Kim and K. Y. Suh, *Nano Lett.*, 2006, **6**, 1508–1513.

115. D. Pisignano, G. Maruccio, E. Mele, L. Persano, F. Di Benedetto and R. Cingolani, *Appl. Phys. Lett.*, 2005, **87**, 123109.
116. A. Y. Malkin, S. G. Kulichikhin, M. L. Kerber, I. Y. Gorbunova and E. A. Murashova, *Polym. Eng. Sci.*, 1997, **37**, 1322–1330.
117. E. Mele, F. Di Benedetto, L. Persano, D. Pisignano and R. Cingolani, *Nanotechnology*, 2005, **16**, 391–395.
118. C. Pang, G.-Y. Lee, T.-i. Kim, S. M. Kim, H. N. Kim, S.-H. Ahn and K.-Y. Suh, *Nat. Mater.*, 2012, **11**, 795–801.
119. Z. Cai, J. Lei, W. Liang, V. Menon and C. R. Martin, *Chem. Mater.*, 1991, **3**, 960–966.
120. C. Liu, M. W. Espenscheid, W.-J. Chen and C. R. Martin, *J. Am. Chem. Soc.*, 1990, **112**, 2458–2459.
121. L. S. Van Dyke and C. R. Martin, *Langmuir*, 1990, **6**, 1123–1132.
122. K. Kim and J.-I. Jin, *Nano Lett.*, 2001, **1**, 631–636.
123. P. Enzel and T. Bein, *Synth. Met.*, 1993, **55**, 1238–1245.
124. T.-Q. Nguyen, J. Wu, V. Doan, B. J. Schwartz and S. H. Tolbert, *Science*, 2000, **288**, 652–656.
125. M. Steinhart, J. H. Wendorff, A. Greiner, R. B. Wehrspohn, K. Nielsch, J. Schilling, J. Choi and U. Gösele, *Science*, 2002, **296**, 1997.
126. C.-G. Wu and T. Bein, *Science*, 1994, **264**, 1757–1759.
127. C.-G. Wu and T. Bein, *Science*, 1994, **266**, 1013–1015.
128. X. Y. Zhang, R. Chan-Yu-King, A. Jose and S. K. Manohar, *Synth. Met.*, 2004, **145**, 23–29.
129. Z. Cai and C. R. Martin, *J. Am. Chem. Soc.*, 1989, **111**, 4138–4139.
130. R. Parthasarathy and C. R. Martin, *Nature*, 1994, **369**, 298–301.
131. R. V. Parthasarathy and C. R. Martin, *Chem. Mater.*, 1994, **6**, 1627–1632.
132. P. M. DiCarmine, A. Fokina and D. S. Seferos, *Chem. Mater.*, 2011, **23**, 3787–3794.
133. K. Ramanathan, M. A. Bangar, M. Yun, W. Chen, A. Mulchandani and N. V. Myung, *Nano Lett.*, 2004, **4**, 1237–1239.
134. Y. Chen and Y. Luo, *Adv. Mater.*, 2009, **21**, 2040–2044.
135. R. M. Penner and C. R. Martin, *J. Electrochem. Soc.*, 1986, **133**, 2206–2207.
136. S. Il Cho and S. Bok Lee, *Acc. Chem. Res.*, 2008, **41**, 699–707.
137. A. Rahman and M. K. Sanyal, *Phys. Rev. B*, 2007, **76**, 045110.
138. W. Liang and C. R. Martin, *J. Am. Chem. Soc.*, 1990, **112**, 9666–9668.
139. C. W. Wang, Z. Wang, M. Li and H. Li, *Chem. Phys. Lett.*, 2001, **341**, 431–434.
140. M. Delvaux, J. Duchet, P.-Y. Stavaux, R. Legras, S. Demoustier-Champagne, *Synth. Met.*, 2000, **113**, 275–280.
141. F. Massuyeau, J. L. Duvail, H. Athalin, J. M. Lorcy, S. Lefrant, J. Wéry and E. Faulques, *Nanotechnology*, 2009, **20**, 155701.
142. S. L. Tao and T. A. Desai, *Nano Lett.*, 2007, **7**, 1463–1468.

143. J. R. Porter, A. Henson and K. C. Popat, *Biomaterials*, 2009, **30**, 780–788.
144. S. Grimm, R. Giesa, K. Sklarek, A. Langner, U. Gösele, H.-W. Schmidt and M. Steinhart, *Nano Lett.*, 2008, **8**, 1954–1959.
145. C. R. Martin, R. Parthasarathy and V. Menon, *Synth. Met.*, 1993, **55**, 1165–1170.
146. J. P. Spatz, B. Lorenz, K. Weishaupt, H. D. Hochheimer, V. P. Menon, R. V. Parthasarathy, C. R. Martin, J. Bechtold and P.-H. Hor, *Phys. Rev. B*, 1994, **50**, 14888–14892.
147. G. A. O'Brien, A. J. Quinn, D. A. Tanner and G. Redmond, *Adv. Mater.*, 2006, **18**, 2379–2383.
148. D. O'Carroll, I. Lieberwirth and G. Redmond, *Small*, 2007, **3**, 1178–1183.
149. D. O'Carroll, I. Lieberwirth and G. Redmond, *Nat. Nanotechnol.*, 2007, **2**, 180–184.
150. S. Moynihan, D. Iacopino, D. O'Carroll, P. Lovera and G. Redmond, *Chem. Mater.*, 2008, **20**, 996–1003.
151. D. O'Carroll, D. Iacopino, A. O'Riordan, P. Lovera, É. O'Connor, G. A. O'Brien and G. Redmond, *Adv. Mater.*, 2008, **20**, 42–48.
152. S. Moynihan, P. Lovera, D. O'Carroll, D. Iacopino and G. Redmond, *Adv. Mater.*, 2008, **20**, 2493–2502.
153. D. O'Carroll, D. Iacopino and G. Redmond, *Adv. Mater.*, 2009, **21**, 1160–1165.
154. A. O'Riordan, D. Iacopino, P. Lovera, L. Floyd, K. Reynolds and G. Redmond, *Nanotechnology*, 2011, **22**, 105602.
155. C.-G. Wu and T. Bein, *Chem. Mater.*, 1994, **6**, 1109–1112.
156. A. Rahman, M. K. Sanyal, R. Gangopadhayy, A. De and I. Das, *Phys. Rev. B*, 2006, **73**, 125313.
157. J. L. Duvail, P. Rétho, S. Garreau, G. Louarn, C. Godon and S. Demoustier-Champagne, *Synth. Met.*, 2002, **131**, 123–128.
158. J. L. Duvail, P. Rétho, V. Fernandez, G. Louarn, P. Molinié and O. Chauvet, *J. Phys. Chem. B*, 2004, **108**, 18552–18556.
159. Y. Z. Long, J. L. Duvail, Z. J. Chen, A. Z. Jin and C. Z. Gu, *Polym. Adv. Technol.*, 2009, **20**, 541–544.
160. D. Bhattacharyya, R. Yang and K. K. Gleason, *J. Mater. Chem.*, 2012, **22**, 17147–17152.
161. A.-H. Lu, D. Zhao and Y. Wan (ed.), *Nanocasting – A Versatile Strategy for Creating Nanostructured Porous Materials* (*RSC Nanoscience and Nanotechnology Series*), Royal Society of Chemistry, Cambridge, 2010.
162. J. M. Lorcy, F. Massuyeau, P. Moreau, O. Chauvet, E. Faulques, J. Wery and J. L. Duvail, *Nanotechnology*, 2009, **20**, 405601.
163. D. Kohler, M. Schneider, M. Krüger, C.-M. Lehr, H. Möhwald and D. Wang, *Adv. Mater.*, 2011, **23**, 1376–1379.
164. S. Moynihan, D. Iacopino, D. O'Carroll, H. Doyle, D. A. Tanner and G. Redmond, *Adv. Mater.*, 2007, **19**, 2474–2479.

165. K. Dawson, P. Lovera, D. Iacopino, A. O'Riordan and G. Redmond, *J. Mater. Chem.*, 2011, **21**, 15995–16000.
166. Y. Guo, Y. Li, J. Xu, X. Liu, J. Xu, J. Lv, C. Huang, M. Zhu, S. Cui, L. Jang, H. Liu and S. Wang, *J. Phys. Chem. C*, 2008, **112**, 8223–8228.
167. Y. Guo, Q. Tang, H. Liu, Y. Zhang, Y. Li, W. Hu, S. Wang and D. Zhu, *J. Amer. Chem. Soc.*, 2008, **130**, 9198–9199.
168. D. Hyuk Park, Y. Ki Hong, E. Hei Cho, M. Suk Kim, D.-C. Kim, J. Bang, J. Kim and J. Joo, *ACS Nano*, 2010, **4**, 5155–5162.
169. S. Park, S.-W. Chung and C. A. Mirkin, *J. Amer. Chem. Soc.*, 2004, **126**, 11772–11773.
170. H. Imahori, S. Kitaura, A. Kira, H. Hayashi, M. Nishi, K. Hirao, S. Isoda, M. Tsujimoto, M. Takano, Z. Zhe, Y. Miyato, K. Noda, K. Matsushige, K. Stranius, N. V. Tkachenko, H. Lemmetyinen, L. Qin, S. J. Hurst and C. A. Mirkin, *J. Phys. Chem. Lett.*, 2012, **3**, 478–481.
171. G. A. Ozin, *Adv. Mater.*, 1992, **4**, 612–649.
172. R. J. Tonucci, B. L. Justus, A. J. Campillo and C. E. Ford, *Science*, 1992, **258**, 783–785.
173. C. T. Kresge, M. E. Leonowicz, W. J. Roth, J. C. Vartuli and J. S. Beck, *Nature*, 1992, **359**, 710–712.
174. J. S. Beck, J. C. Vartuli, W. J. Roth, M. E. Leonowicz, C. T. Kresge, K. D. Schmitt, C. T.-W. Chu, D. H. Olson, E. W. Sheppard, S. B. McCullen, J. B. Higgins and J. L. Schlenker, *J. Am. Chem. Soc.*, 1992, **114**, 10834–10843.
175. M. Döbbelin, R. Tena-Zaera, P. M. Carrasco, J.-R. Sarasua, G. Cabañero and D. J. Mecerreyes, *Polym. Sci., Part A: Polym. Chem.*, 2010, **48**, 4648–4653.
176. J. Jang, B. Lim, J. Lee and T. Hyeon, *Chem. Commun.*, 2001, 83–84.
177. K. Douglas, G. Devaud and N. A. Clark, *Science*, 1992, **257**, 642–644.
178. K. Akagi, G. Piao, S. Kaneko, K. Sakamaki, H. Shirakawa and M. Kyotani, *Science*, 1998, **282**, 1683–1686.
179. H. Goto and K. Akagi, *Macromolecules*, 2005, **38**, 1091–1098.
180. H. Goto, *J. Polym. Sci., Part A: Polym. Chem.*, 2007, **45**, 1377–1387.
181. H. Goto and K. Kawabata, *Macromolecules*, 2008, **41**, 4551–4554.
182. H. Goto and S. Nimori, *J. Mater. Chem.*, 2010, **20**, 1891–1898.
183. T. Kitamura, S. Nakaso, N. Mizoshita, Y. Tochigi, T. Shimomura, M. Moriyama, K. Ito and T. Kato, *J. Am. Chem. Soc.*, 2005, **127**, 14769–14775.
184. L. Huang, Z. Wang, H. Wang, X. Cheng, A. Mitra and Y. Yan, *J. Mater. Chem.*, 2002, **12**, 388–391.
185. R. C. Smith, W. M. Fischer and D. L. Gin, *J. Am. Chem. Soc.*, 1997, **119**, 4092–4093.
186. J. Jang and H. Yoon, *Chem. Commun.*, 2003, 720–721.
187. H. Yoon, M. Chang and J. Jang, *Adv. Funct. Mater.*, 2007, **17**, 431–436.
188. X. Zhang, J. Zhang, Z. Liu and C. Robinson, *Chem. Commun.*, 2004, 1852–1853.

189. X. Zhang and S. K. Manohar, *Chem. Commun.*, 2004, 2360–2361.
190. P. Anilkumar and M. Jayakannan, *J. Phys. Chem. B*, 2009, **113**, 11614–11624.
191. F. S. Bates and G. H. Fredrickson, *Annu. Rev. Phys. Chem.*, 1990, **41**, 525–557.
192. F. S. Bates and G. H. Fredrickson, *Phys. Today*, 1999, **52**, 32–38.
193. X. Li, S. Tian, Y. Ping, D. H. Kim and W. Knoll, *Langmuir*, 2005, **21**, 9393–9397.
194. J. Chai and J. M. Buriak, *ACS Nano*, 2008, **2**, 489–501.
195. J. I. Lee, S. Cho, S.-M. Park, J. K. Kim, J. K. Kim, J.-W. Yu, Y. C. Kim and T. P. Russell, *Nano Lett.*, 2008, **8**, 2315–2320.
196. E. J. W. Crossland, M. Kamperman, M. Nedelcu, C Ducati, U. Wiesner, D. M. Smilgies, G. E. S. Toombes, M. A. Hillmyer, S. Ludwigs, U. Steiner and H. J. Snaith, *Nano Lett.*, 2009, **9**, 2807–2812.
197. E. J. W. Crossland, M. Nedelcu, C. Ducati, S. Ludwigs, M. A. Hillmyer, U. Steiner and H. J. Snaith, *Nano Lett.*, 2009, **9**, 2813–2819.
198. F. Li, A. A. Martens, A. Aslund, P. Konradsson, F. A. de Wolf, M. A. Cohen Stuart, E. J. R. Sudhölter, A. T. M. Marcelis and F. A. M. Leermakers, *Soft Matter*, 2009, **5**, 1668–1673.
199. J. M. Liu and S. C. Yang, *J. Chem. Soc., Chem. Commun.*, 1991, 1529–1531.
200. S.-J. Choi and S.-M. Park, *Adv. Mater.*, 2000, **12**, 1547–1549.
201. A. Houlton, A. R. Pike, M. A. Galindo and B. R. Horrocks, *Chem. Commun.*, 2009, 1797–1806.
202. Y. Ma, J. Zhang, G. Zhang and H. He, *J. Am. Chem. Soc.*, 2004, **126**, 7097–7101.
203. Y. Bardavid, J. Ghabboun, D. Porath, A. B. Kotylar and S. Yitzchaik, *Polymer*, 2008, **49**, 2217–2222.
204. S. Pruneanu, S. A. F. Al-Said, L. Dong, T. A. Hollis, M. A. Galindo, N. G. Wright, A. Houlton and B. R. Horrocks, *Adv. Funct. Mater.*, 2008, **18**, 2444–2454.
205. P. Björk, A. Herland, I. G. Scheblykin and O. Inganäs, *Nano Lett.*, 2005, **5**, 1948–1953.
206. J. Wigenius, P. Björk, M. Hamedi and D. Aili, *Macromol. Biosci.*, 2010, **10**, 836–841.
207. A. Herland, P. Björk, K. P. R. Nilsson, J. D. M. Olsson, P. Asberg, P. Konradsson, P. Hammarstrom and O. Inganäs, *Adv. Mater.*, 2005, **17**, 1466–1471.
208. A. Herland, D. Thomsson, O. Mirzov, I. G. Scheblykin and O. Inganäs, *J. Mater. Chem.*, 2008, **18**, 126–132.
209. A. Herland, P. Björk, P. Hania, I. Scheblykin and O. Inganäs, *Small*, 2007, **3**, 318–325.
210. T. Kato, *Science*, 2002, **295**, 2414–2418.
211. K. Araya, A. Mukoh, T. Narahara and H. Shirakawa, *Synth. Met.*, 1986, **14**, 199–206.

212. K. Akagi, S. Katayama, M. Ito, H. Shirakawa and K. Araya, *Synth. Met.*, 1989, **28**, D51–D56.
213. M. Goh, S. Matsushita and K. Akagi, *Chem. Soc. Rev.*, 2010, **39**, 2466–2476.
214. T. Mori, M. Kyotani and K. Akagi, *Chem. Sci.*, 2011, **2**, 1389–1395.
215. M. Kyotani, S. Matsushita, M. Goh, T. Nagai, Y. Matsui and K. Akagi, *Nanoscale*, 2010, **2**, 509–514.
216. K. Akagi, *J. Polym. Sci. A: Polym. Chem.*, 2009, **47**, 2463–2485.
217. X. Zhang, J. Zhang, W. Song and Z. Liu, *J. Phys. Chem. B*, 2006, **110**, 1158–1165.
218. D. O. Shin, B. H. Kim, J.-H. Kang, S.-J. Jeong, S. H. Park, Y.-H. Lee and S. O. Kim, *Macromolecules*, 2009, **42**, 1189–1193.
219. S. O. Kim, H. H. Solak, M. P. Stoykovich, N. J. Ferrier, J. J. de Pablo and P. F. Nealey, *Nature*, 2003, **424**, 411–414.
220. M. P. Stoykovich, M. Müller, S. O. Kim, H. H. Solak, E. W. Edwards, J. J. de Pablo and P. F. Nealey, *Science*, 2005, **308**, 1442–1446.
221. Y. S. Jung and C. A. Ross, *Nano Lett.*, 2007, **7**, 2046–2050.
222. V. Kalra, P. A. Kakad, S. Mendez, T. Ivannikov, M. Kamperman and Y. L. Joo, *Macromolecules*, 2006, **39**, 5453–5457.
223. T. Ruotsalainen, J. Turku, P. Heikkilä, J. Ruokolainenn, A. Nykänen, T. Laitinen, M. Torkkeli, R. Serimaa, G. ten Brinke, A. Harlin and O. Ikkala, *Adv. Mater.*, 2005, **17**, 1048–1052.
224. Y. S. Jung, W. C. Jung, H. L. Tuller and C. A. Ross, *Nano Lett.*, 2008, **8**, 3776–3780.
225. M. Bognitzki, H. Hou, M. Ishaque, T. Frese, M. Hellwig, C. Schwarte, A. Schaper, J. H. Wendorff and A. Greiner, *Adv. Mater.*, 2000, **12**, 637–640.
226. H. Hou, Z. Jun, A. Reuning, A. Schaper, J. H. Wendorff and A. Greiner, *Macromolecules*, 2002, **35**, 2429–2431.
227. R. A. Caruso, J. H. Schattka and A. Greiner, *Adv. Mater.*, 2001, **13**, 1577–1579.
228. H. Dong, S. Prasad, V. Nyame and W. E. Jones, *Chem. Mater.*, 2004, **16**, 371–373.
229. M. R. Abidian, D.-H. Kim and D. C. Martin, *Adv. Mater.*, 2006, **18**, 405–409.
230. S. Nair, E. Hsiao and S. H. Kim, *Chem. Mater.*, 2009, **21**, 115–121.
231. A. Laforgue and L. Robitaille, *Chem. Mater.*, 2010, **22**, 2474–2480.
232. A. Laforgue and L. Robitaille, *Macromolecules*, 2010, **43**, 4194–4200.
233. A. Laforgue, *J. Mater. Chem.*, 2010, **20**, 8233–8235.
234. L. Wang, J. Hu, H. Zhang and T. Zhang, *Chem. Commun.*, 2011, **47**, 6837–6839.
235. X. Lu, Y. Zhao and C. Wang, *Adv. Mater.*, 2005, **17**, 2485–2488.
236. G. Han, B. Guo, L. Zhang and B. Yang, *Adv. Mater.*, 2006, **18**, 1709–1712.

237. O. S. Kwon, S. J. Park, J. S. Lee, E. Park, T. Kim, H.-W. Park, S. A. You, H. Yoon and J. Jang, *Nano Lett.*, 2012, **12**, 2797–2802.
238. X. Wang, Y.-G. Kim, C. Drew, B.-C. Ku, J. Kumar and L. A. Samuelson, *Nano Lett.*, 2004, **4**, 331–334.
239. E. Braun, Y. Eichen, U. Sivan and G. Ben-Yoseph, *Nature*, 1998, **391**, 775–778.
240. N. C. Seeman, *Nature*, 2003, **421**, 427–431.
241. K. Keren, M. Krueger, R. Gilad, G. Ben-Yoseph, U. Sivan and E. Braun, *Science*, 2002, **297**, 72–75.
242. Y. He, Y. Chen, H. Liu, A. E. Ribbe and C. Mao, *J. Am. Chem. Soc.*, 2005, **127**, 12202–12203.
243. A. Dawn and A. K. Nandi, *Macromolecules*, 2005, **38**, 10067–10073.
244. R. Nagarajan, S. Roy, J. Kumar, S. K. Tripathy, T. Dolukhanyan, C. Sung, F. Bruno and L. A. Samuelson, *J. Macromol. Sci., Pure Appl. Chem.*, 2001, **38**, 1519–1537.
245. P. Nickels, W. U. Dittmer, S. Beyer, J. P. Kotthaus and F. C. Simmel, *Nanotechnology*, 2004, **15**, 1524–1529.
246. L. Tskhovrebova and J. Trinick, *J. Mol. Biol.*, 1997, **265**, 100–106.
247. M. J. Sherratt, C. Baldock, J. L. Haston, D. F. Holmes, C. J. P. Jones, C. A. Shuttleworth, T. J. Wess and C. M. Kielty, *J. Mol. Biol.*, 2003, **332**, 183–192.
248. A. Bensimon, A. Simon, A. Chiffaudel, V. Croquette, F. Heslot and D. Bensimon, *Science*, 1994, **265**, 2096–2098.
249. L. Dong, T. Hollis, S. Fishwick, B. A. Connolly, N. G. Wright, B. R. Horrocks and A. Houlton, *Chem. Eur. J.*, 2007, **13**, 822–828.
250. H. Nakao, H. Hayashi, T. Yoshino, S. Sugiyama, K. Otobe and T. Ohtani, *Nano Lett.*, 2002, **2**, 475–479.
251. Z. Deng and C. Mao, *Nano Lett.*, 2003, **3**, 1545–1548.
252. G. Liu and J. Zhao, *Langmuir*, 2006, **22**, 2923–2926.
253. S. Samitsu, T. Shimomura, K. Ito, M. Fujimori, S. Heike and T. Hashizume, *Appl. Phys. Lett.*, 2005, **86**, 233103.
254. P. Björk, D. Thomsson, O. Mirzov, J. Wigenius, O. Inganäs and I. G. Scheblykin, *Small*, 2009, **5**, 96–103.
255. J. D. Sipe and A. S. Cohen, *J. Struct. Bio.*, 2000, **130**, 88–98.
256. J. C. Rochet and P. T. Lansbury, Jr., *Curr. Opin. Struct. Biol.*, 2000, **10**, 60–68.
257. M. Lindgren, K. Sörgjerd and P. Hammarström, *Biophys. J.*, 2005, **88**, 4200–4212.
258. L. J. Domigan, J. P. Healy, S. J. Meade, R. J. Blaikie and J. A. Gerrard, *Biopolymers*, 2011, **97**, 123–133.
259. E. O. Gabrielsson, K. Tybrandt, P. Hammarström, M. Berggren and K. P. R. Nilsson, *Small*, 2010, **6**, 2153–2161.
260. M. Goren, Z. Qi and R. B. Lennox, *Chem. Mater.*, 2000, **12**, 1222–1228.

261. K. Reynolds, J. Komulainen, J. Kivijakola, P. Lovera, D. Iacopino, M. Pudas, J. Vähäkangas, J. Röning and G. Redmond, *Nanotechnology*, 2008, **19**, 485301.
262. X. Zhang, W. J. Goux and S. K. Manohar, *J. Am. Chem. Soc.*, 2004, **126**, 4502–4503.
263. X. Zhang and S. K. Manohar, *J. Am. Chem. Soc.*, 2004, **126**, 12714–12715.
264. X. Zhang, A. G. MacDiarmid and S. K. Manohar, *Chem. Commun.*, 2005, 5328–5330.
265. X. Zhang and S. K. Manohar, *J. Am. Chem. Soc.*, 2005, **127**, 14156–14157.
266. R. C. Y. King and F. Roussel, *Synth. Met.*, 2005, **153**, 337–340.
267. J. Huang, S. Virji, B. H. Weiller and R. B. Kaner, *J. Am. Chem. Soc.*, 2003, **125**, 314–315.
268. J. Huang and R. B. Kaner, *J. Am. Chem. Soc.*, 2004, **126**, 851–855.
269. J. Huang, S. Virji, B. H. Weiller and R. B. Kaner, *Chem. Eur. J.*, 2004, **10**, 1314–1319.
270. J. Huang and R. B. Kaner, *Angew. Chem.*, 2004, **116**, 5941–5945.
271. J. C. Michaelson and A. J. McEvoy, *J. Chem. Soc., Chem. Commun.*, 1994, 79–80.
272. J. Huang and R. B. Kaner, *Angew. Chem. Int. Ed.*, 2004, **43**, 5817–5821.
273. J. Huang and R. B. Kaner, *Chem. Commun.*, 2006, 367–376.
274. C. Su, G. Wang, F. Huang and X. Li, *J. Mater. Sci.*, 2008, **43**, 197–202.
275. H. D. Tran and R. B. Kaner, *Chem. Commun.*, 2006, 3915–3917.
276. B.-J. Kim, S.-G. Oh, M.-G. Han and S.-S. Im, *Synth. Met.*, 2001, **122**, 297–304.
277. H. Qiu and M. Wan, *J. Polym. Sci. Part A: Polym. Chem.*, 2001, **39**, 3485–3497.
278. Y. Yang and M. Wan, *J. Mater. Chem.*, 2002, **12**, 897–901.
279. K. Huang and M. Wan, *Chem. Mater.*, 2002, **14**, 3486–3492.
280. M. Wan, Z. Wei, Z. Zhang, L. Zhang, K. Huang and Y. Yang, *Synth. Met.*, 2003, **135–136**, 175–176.
281. L. Zhang and M. Wan, *Nanotechnology*, 2002, **13**, 750–755.
282. K. Huang, M. Wan, Y. Long, Z. Chen and Y. Wei, *Synth. Met.*, 2005, **155**, 495–500.
283. L. Zhang, Y. Long, Z. Chen and M. Wan, *Adv. Funct. Mater.*, 2004, **14**, 693–698.
284. Z. Zhang, Z. Wei and M. Wan, *Macromolecules*, 2002, **35**, 5937–5942.
285. Z. Wei, Z. Zhang and M. Wan, *Langmuir*, 2002, **18**, 917–921.
286. Y. Long, Z. Chen, J. Shen, Z. Zhang, L. Zhang, K. Huang, M. Wan, A. Jin, C. Gu and J. L. Duvail, *Nanotechnology*, 2006, **17**, 5903–5911.
287. A. R. Hopkins, D. D. Sawall, R. M. Viaalhermosa and R. A. Lipeles, *Thin Solid Films*, 2004, **469–470**, 304–308.
288. A. S. Nain, J. C. Wong, C. Amon and M. Sitti, *Appl. Phys. Lett.*, 2006, **89**, 183105.

289. X. Xing, Y. Wang and B. Li, *Opt. Exp.*, 2008, **16**, 10815–10822.
290. R. D. Piner, J. Zhu, F. Xu, S. H. Hong and C. A. Mirkin, *Science*, 1999, **283**, 661–663.
291. A. Noy, A. E. Miller, J. E. Klare, B. L. Weeks, B. W. Woods and J. J. DeYoreo, *Nano Lett.*, 2002, **2**, 109–112.
292. B. W. Maynor, S. F. Filocamo, M. W. Grinstaff and J. Liu, *J. Am. Chem. Soc.*, 2002, **124**, 522–523.
293. J. H. Lim and C. A. Mirkin, *Adv. Mater.*, 2002, **14**, 1474–1477.
294. Y. Okawa and M. Aono, *Nature*, 2001, **409**, 683–684.
295. H. X. He, C. Z. Li and N. J. Tao, *Appl. Phys. Lett.*, 2001, **78**, 811–813.
296. R. Szoszkiewicz, T. Okada, S. C. Jones, T.-D. Li, W. P. King, S. R. Marder and E. Riedo, *Nano Lett.*, 2007, **7**, 1064–1069.
297. O. Fenwick, L. Bozec, D. Credgington, A. Hammiche, G. M. Lazzerini, Y. R. Silberberg and F. Cacialli, *Nat. Nanotechnol.*, 2009, **4**, 664–668.
298. Y. Doi, A. Saeki, Y. Koizumi, S. Seki, K. Okamoto, T. Kozawa, and S. Tagawa, *J. Vac. Sci. Technol. B*, 2005, **23**, 2051–2055.
299. S. H. M. Persson, P. Dyreklev, and O. Inganäs, *Adv. Mater.*, 1996, **8**, 405–408.
300. J. G. Park, G. T. Kim, J. H. Park, H. Y. Yu, G. McIntosh, V. Kristic, S. H. Jhang, B. Kim, S. H. Lee, S. W. Lee, M. Burghard, S. Roth and Y. W. Park, *Thin Solid Films*, 2001, **393**, 161–167.
301. S. C. Laza, M. Polo, A. A. R. Neves, R. Cingolani, A. Camposeo and D. Pisignano, *Adv. Mater.*, 2012, **24**, 1304–1308.
302. S. Kawata, H.-B. Sun, T. Tanaka and K. Takada, *Nature*, 2001, **412**, 697–698.
303. P. X. Ma and R. Zhang, *J. Biomed. Mater. Res.*, 1999, **46**, 60–72.
304. I. O. Smith, X. H. Liu, L. A. Smith and P. X. Ma, *WIREs Nanomed. Nanobiotechnol.*, 2009, **1**, 226–236.
305. X. Liu and P. X. Ma, *Biomaterials*, 2010, **32**, 259–269.
306. L. A. Smith, X. Liu, J. Hu, P. Wang and P. X. Ma, *Tissue Eng. Part A*, 2009, **15**, 1855–1864.
307. X. Liu and P. X. Ma, *Biomaterials*, 2009, **30**, 4094–4103.
308. H. Ma, J. Hu and P. X. Ma, *Adv. Funct. Mater.*, 2010, **20**, 2833–2841.

CHAPTER 4

# *Structural and Surface Properties of Polymer Nanofibers and Their Applications*

## 4.1 Structural and Mechanical Properties

Individual polymer nanofibers may exhibit improved mechanical properties compared to film or bulk samples. The factors contributing to such improved performances are both (i) related to the possibly enhanced orientation of polymer backbones along the longitudinal axis of nanofibers, and (ii) due to a general expected increase of the strength of filament-shaped materials as the fiber diameters decrease. Indeed, oriented molecular chains (mechanism (i)) would increase the uniaxial tensile strength along the direction of alignment through a structural, bulk-related effect originating in the fiber internal region. On the contrary, flaws present at the surface (mechanism (ii), known as Griffith's criterion in the field of fracture mechanics) weaken the fibers and can ultimately lead to mechanical failure. In fact, cracks may form in correspondence of these flaws and then propagate in the sample. Fibers of smaller radius, having an external surface per unit length ($2\pi r$) lower than thicker fibers, would be less subject to crack formation and hence to failure related to surface flaws effects. A. A. Griffith has summarized these issues in the law:[1]

$$\sigma_F (l_F)^{1/2} \cong C_G, \tag{4.1}$$

RSC Nanoscience & Nanotechnology No. 29
Polymer Nanofibers: Building Blocks for Nanotechnology
By Dario Pisignano

Published by the Royal Society of Chemistry, www.rsc.org

where $\sigma_F$ indicates the stress at failure, $l_F$ is the length of surface flaws and $C_G$ is a constant. According to Griffith, $l_F$ is limited in the case of thin fibers because of the reduced sample radius. For instance, admitting a linear proportionality between $l_F$ and $r$, according to the previous equation the stress at failure would scale as $r^{-1/2}$. The constant, $C_G$, of Equation (4.1) can be derived on the basis of thermodynamic considerations on the crack propagation. This clearly has a well-defined associated energy budget due to the formation of two new solid-air interfaces at the crack, each having its own free energy, $U_G$, while an amount of elastic energy is lost simultaneously. Consequently, $C_G$ has to be related to the fiber Young's modulus, $E$, and to $U_G$. Of course, this holds for brittle fractures such as in glass. Other eventual components of dissipated energy, such as those related to viscoelasticity (Section 1.4.2), should be added to the overall energy balance to obtain refined expressions of $C_G$ for different classes of materials, including polymers. In fact, mechanical models describing films or fibers with sub-µm thicknesses or diameters take into account various surface properties, such as free-energies, tension, elasticity, *etc.*, whose contribution becomes more and more relevant upon decreasing the transversal size.[2–6] In addition, in polymer nanofibers mechanical properties have to be closely related to the other mechanism mentioned above, namely the eventual alignment of molecular backbones along the fiber length and the possibly increased degree of crystallinity. To this aim, polymer nanofibers have been analyzed taking into account their supramolecular structure, which is globally mainly amorphous but can include domains with mutually aligned molecules (Figure 4.1).[7,8] Hence, the orientation correlation length, in the scale of few hundreds of nm, is a good estimate for the fiber diameter below which one expects a remarkable increase of the elastic modulus. In addition, the temperature dependence of the shear modulus below the glass transition temperature, investigated for PS fibers, shows interesting variations influenced by the fiber diameter.[7] Overall, correlating with precision mechanical and microscopic fundamental properties of polymer nanofibers is a particularly challenging and inter-disciplinary task, which requires experimental data coming from very different measurement techniques to be coupled. This will undoubtedly constitute an intriguing research task in the near future. Incidentally, stress concentration points such as surface flaws, but also polymeric backbone ends, entanglement, *etc.*, are related not only to mechanical failure but also to phonon scattering, namely heat transfer. Consequently, nanofiber geometries, with their lower linear density of defects and higher molecular alignment, can also lead to a greatly enhanced thermal conductivity in polymers, which is indeed observed experimentally.[9] These and other thermal aspects are better described in Section 4.3.2.

Experimentally, mechanical properties such as Young's modulus and yield strength can be measured *via* tensile tests. These can be performed by stress-strain experiments on nanofiber samples of different degrees of complexity, ranging from non-wovens to individual fibers. One termination of the sample is clamped in a loading frame, while a controlled displacement is applied on the

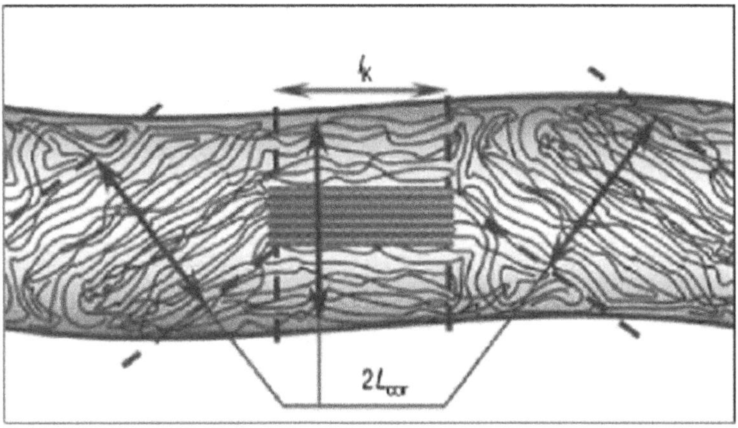


**Figure 4.1** Scheme showing the possible arrangement of crystalline domains and oriented, amorphous matrix domains in a polymer nanofiber. Here, $L_{cor}$ and $l_k$ indicate the orientation correlation length and the length of freely joined straight segments (so-called Kuhn segment) in the macromolecules. Reproduced with permission from Ref. 7, *Nat. Nanotechnol.*, 2007, **2**, 59–62. Doi: 10.1038/nnano.2006.172. Copyright © 2007, Nature Publishing Group.

other termination with a macroscopic, standard configuration, or with the use of microelectromechanical systems.[10–12] The corresponding load is measured during the experiment, which allows the tensile stress and the longitudinal strain to be obtained (Section 1.4.2.1 and Figure 1.12). Some exemplary curves are shown in Figure 4.2a. At low strain (for instance, below about 25%–30% for the curves in Figure 4.2a, respectively), the stress-strain relation is linear, namely the regime is elastic and the Young's modulus is immediately obtained from the ratio, $\sigma/\gamma_L$. Above a threshold value of the longitudinal strain (so-called proportional limit), the stress-strain curve becomes sub-linear, which indicates that the material is undergoing microscopic rearrangements and eventually plastic flow. The yield stress is the value at which the character of the deformation changes from elastic to plastic. The slope of the curve decreases but it is still $d\sigma/d\gamma_L > 0$, and the corresponding regime is named strain hardening. Finally, a maximum value of the measured stress (ultimate tensile strength, which is around 90% and 115% for the curves A and B in Figure 4.2a, respectively) is reached. For larger deformations the sample undergoes the so-called necking regime (in which $d\sigma/d\gamma_L < 0$) and ultimately fracture. The moduli of polymer nanofibers can reach the order of several tens of GPa for highly oriented samples. Maximum elongations can range from a few percent to three to four times the original length, depending of course on the nature of the used polymer. Tensile strengths are mostly in the range of tens of MPa. For example, the Young's modulus and the tensile strength of

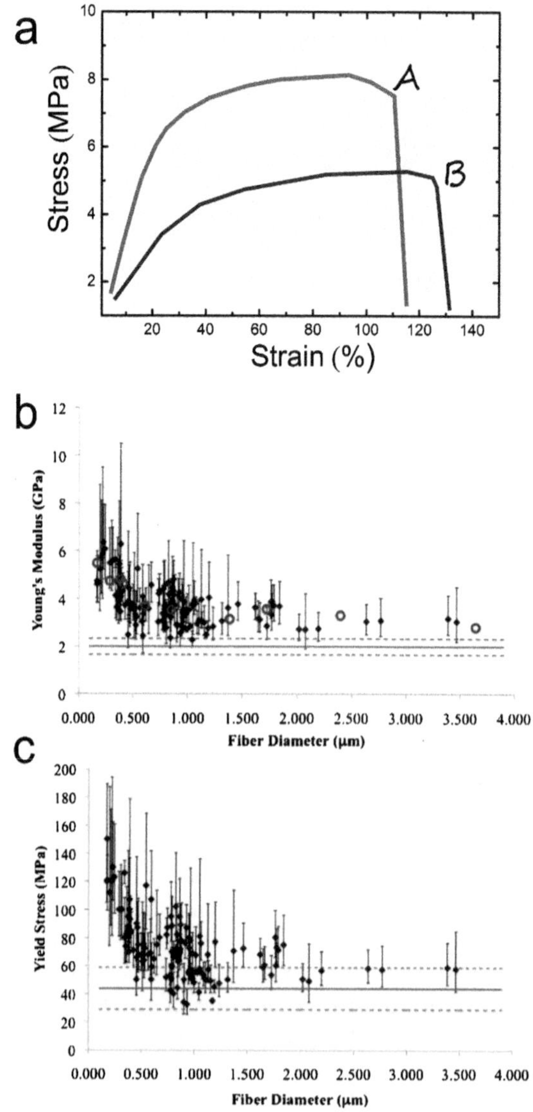


**Figure 4.2** (a) Example of stress-strain curves of nanofiber mats. Two different curves are labeled with A and B. The mechanical properties can change remarkably, for instance upon varying the composition of a blend of polymers. (b, c) Dependence of Young's modulus (b) and yield stress (c) on the diameter of electrospun fibers of poly(trimethyl hexamethylene terephthalamide). Diamonds: data for individual fibers, deformed in uniaxial extension. Open circles in (b) represent values obtained from modeling and measurements of the molecular orientation within fibers. Solid and dashed lines indicate data for bulk films and one standard deviation limit, respectively. Adapted with permission from Ref. 15, *Polymer*, 2011, **52**, 2295–2301. Doi: 10.1016/j.polymer.2011.03.041. Copyright © 2011, Elsevier Ltd.

single electrospun PCL fibers are respectively at least two-fold and an order of magnitude higher than those of a PCL film.[13] Furthermore, PCL and poly(caprolactone-*co*-ethyl ethylene phosphate) exhibit a dependence of $E$ and of the tensile strength on the fiber diameter, with the stiffness increasing upon decreasing the diameter from the μm-scale to hundreds of nm.[13,14]

Analogous experiments have been carried out on single electrospun fibers made by poly(trimethyl hexamethylene terephthalamide), which also show an increase of the Young's modulus and of the yield stress upon decreasing the diameter below about 500 nm (Figure 4.2b,c).[15] Other polymer systems showing an abrupt increase of the elastic or of the shear modulus for diameters below or approaching the scale of 100 nm include fibers made of PS,[16] nylon[7,17] and poly(2-acrylamido-2-methyl-1-propanesulfonic acid).[18] In general, different dopants and fillers embedded in the polymer can determine various and different mechanical properties of the resulting nanofibers, both increasing and decreasing stiffness and strength being possible.[13,16] For example, SWCNTs in spun PVA fibers are demonstrated to lead to an increase of the modulus and of the strength, which is linearly dependent on the volume fraction occupied by nanotubes up to fractions of about 10% and then saturates for higher volume fraction values.[19]

Several other methods are available in order to further improve the mechanical properties of polymer nanofibers. In particular, heating nanofibers above the $T_g$ of the constituent polymer (generally for a few tens of minutes) allows fibers to be locally inter-connected at their junctions, thus leading to a microscopically networked structure that shows increased tensile strength.[20–22] This approach has been tested on a variety of polymers including PLA,[20] poly(etherimide)[21] and PCL.[22] In addition, infiltrating gel solutions during the thermal treatment allows deformation and shrinkage of the pristine sample shape to be prevented.[22]

Another useful method to investigate the mechanical properties of single polymer nanofibers in the regime of small deformations ($\Delta l$) is by measuring the force-displacement curves during a nanoindentation produced using a scanning probe tip (Figure 4.3). The Hertz model[23–25] is often adopted for describing contact mechanics and calculating the compressive modulus, fitting the linear part of the load-deformation curves in the region of small indentation (say, $\Delta l \leq 10$ nm). The relation between the normal load, $F_N$, applied to the body of a polymer nanofiber, and the resulting deformation is then given by:

$$F_N = \frac{4}{3} r^{*1/2} E^* \Delta l^{3/2}, \tag{4.2}$$

where, $r^*$ and $E^*$ respectively indicate the relative radius of curvature and the combined strain modulus of the scanning probe tip and of the fiber, which also takes into account the Poisson ratios, *i.e.*:

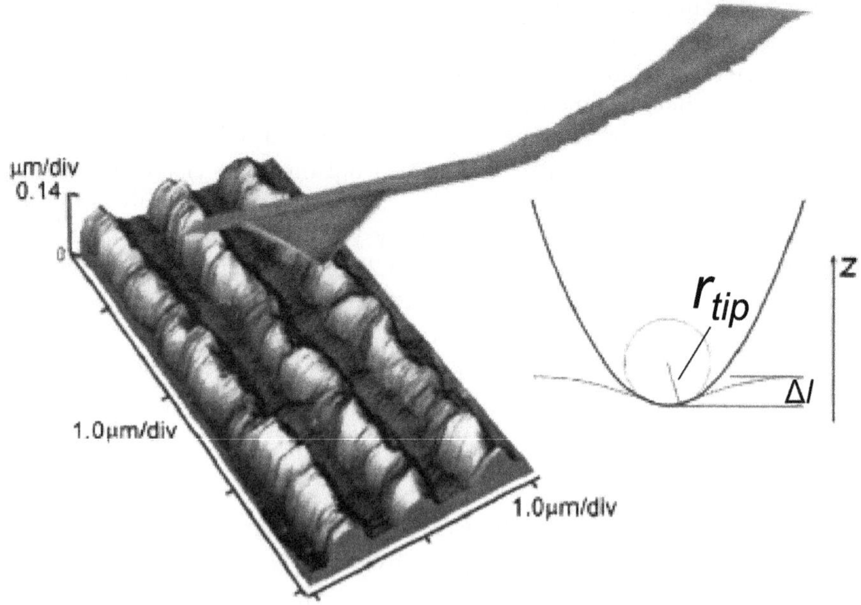


**Figure 4.3** AFM micrograph of aligned polyurethane fibers produced by soft molding (Section 3.2) and lying on silicon, and scheme of a scanning probe tip used for collecting force-displacement curves during elastic nanoindentation and hence fiber compressive modulus. Inset: scheme of the elastic compression of the polymer surface, and of the *z* vertical coordinate system of the AFM system. $r_{tip}$: radius of the contact paraboloid, $\Delta l$: indentation. Adapted with permission from Ref. 26, *Appl. Phys. Lett.*, 2005, **87**, 123109. Doi: 10.1063/1.2046731. Copyright © 2005, American Institute of Physics.

$$\frac{1}{r^*} = \frac{1}{r} + \frac{1}{r_{tip}}, \tag{4.3a}$$

and

$$\frac{1}{E^*} = \frac{(1 - \nu_P^2)}{E} + \frac{\left(1 - \nu_{P,tip}^2\right)}{E_{tip}}. \tag{4.3b}$$

Normally, in force-displacement measurements one has $E_{tip} >> E$ (for example $E_{tip}$ is about 150 GPa for silicon tips). For polymer nanofibers, compressive moduli measured by nanoindentation measurements may range from about 1 to a few tens of MPa depending on the polymer system, degree of internal orientation and stretching, curing and other processing conditions.[26]

In the case of nanofiber yarns or arrays, the number of fibers contributing to the overall mechanical behavior increases enormously, leading to a complex phenomenology where networking due to entanglements and joints between nanofibers, friction effects at nanoscale and the amount of voids, namely the degree of porosity in the sample, all can play an important role. In a mat, fibers deform under an external stress by means of cooperative motions, with bending and stretching combining at microscale. Recently, the mechanical stability of disordered fiber networks has been studied by a lattice-based model.[27,28] In another work, a detailed micromechanical model of the elastic-plastic properties of non-wovens composed of nanofibers has been developed, and validated by electrospun mats of polyamide.[29] Core-shell structures have an influence on the mechanical properties of nanofibers as well. For example, the effective Young's modulus (Section 1.4.2.1), $E_{eff}$, of a mat of core-shell fibers is estimated by the following equation:[30]

$$E_{eff} = (1-P)\left\{\left[1-\left(\frac{r_{in}}{r_{out}}\right)^2\right]E_{out} + \left(\frac{r_{in}}{r_{out}}\right)^2 E_{in}\right\}, \qquad (4.4)$$

where $P$ indicates the porosity of the mat, which is introduced in Section 4.2.1, $r_{in}$ and $r_{out}$ are the radius of the core and of the external shell, respectively, and $E_{in}$ and $E_{out}$ are the corresponding Young's moduli of the inner and of the outer material. For nanofibers with a liquid core[30] or for hollow nanotubes,[31] only the shell parameters matter in determining the modulus of the nanofiber samples.

Of course, a technical application directly related to the mechanical properties of polymer nanofibers is the realization of nanocomposites reinforced by embedding the fibers within a matrix. Aligned nanofibers are of course preferable, allowing several properties of the resulting composites to be enhanced including the work of fracture (J $m^{-2}$), the flexural strength, the elastic modulus, *etc.* Besides orientation, the volume fraction occupied by nanofibers, their aspect ratio ($\sim l/2r$), the degree of uniformity of their dispersion, and the quality of the interface between the fibers and the embedding matrix are important in determining the properties of nanocomposites. A wide literature is available in the field.[32–36] Finally, mechanical properties of mats composed of polymeric or nanocomposite nanofibers are crucial for biomedical applications such as wound dressing, regenerative medicine and tissue nanoengineering,[37–40] where nanofibrous sheets and scaffolds should guarantee elasticity and suturability, or approach the natural structural properties of tissues and organs.

### 4.1.1 Carbon Nanofibers

An especially important and wide field of use of quasi-one-dimensional organic systems, which exploits directly their extraordinary mechanical properties, is based on the realization of carbon nanofibers and on their embedment within composite materials for reinforcement.[41–45] Fibers made of glass and carbon have been increasingly used in reinforced composites within plastic resins and concrete. Today carbon nanomaterials are opening new perspectives for such technologies. In particular, carbon nanofibers are significantly cheaper than carbon nanotubes, and are potentially producible with a much higher throughput. They can also be employed as building blocks for catalytic supports,[46] nanoelectronics,[47–49] energy storage[50–52] and filtration.[53] A possible, hybrid approach is of realizing composite nanofibers embedding carbon nanotubes, or even pure carbon nanotubes fibers.[54] A few examples are given in Section 2.3.3. Catalytic growth processes are also well established, and currently represent the technology of choice for commercializing carbon nanofiber products.[55–57] However, what is especially of interest in the framework of this book is that carbon nanofibers can be also effectively produced starting from polymer nanofibers, and carbonizing the nanomaterial by pyrolysis. To this aim, the typical procedure basically replicates on nanofibers the conventional method by which carbon fibers with diameter in the range of a few μm are made at industrial scale. A few hands-on notes summarizing the process are the following (interested readers can find comprehensive, recent reviews in Refs 45 and 58):

(i) realization of PAC nanofibers by electrospinning; this is typically carried out by DMF solutions, and a convenient set of parameters includes a polymer solution concentration of 6–15% in weight, a needle-collector distance of 10–40 cm and an applied voltage bias of 10–30 kV;[59–61] other techniques can be used for PAC nanofiber production, including polymerization in porous aluminum oxide templates (Section 3.3.1), chemical vapor growth, *etc.*;[45]
(ii) oxidative stabilization of the so-realized nanofibers, performed by moderate heating (at $T$ = 250 °C) in air, over time intervals of a few hours;[59]
(iii) a further heating step at $T$ = 1000 °C or above under inert atmosphere or under vacuum, for carbonization; suitable heating rates are in the region of 5 °C min$^{-1}$.[59] Many variants of tube furnaces and heating elements are commercially available, which allow these temperatures to be reached quite easily at laboratory level.

A variety of nanostructures can be obtained by tailoring the process above and combining it with other nanofabrication, lithographic or growth methods.[62] Other precursor polymers can be used, such as, for instance, PAV or cellulose, though PAC is the most frequently chosen material for realizing the pristine nanofibers. The polymer precursor morphology and the degree of internal molecular alignment within fibers are especially relevant also in this frame, because these properties are likely to affect significantly the mechanical properties of the resulting carbon fibers.[63] Carbon fibers with a hollow, multicore internal structure are producible from electrospun fibers with two phase-separated components, one of which is decomposed during the subsequent thermal treatment.[64] Boron-carbide-carbon nanofibers are obtainable from a poly(norbornenyldecaborane) precursor.[65] One of the most beautiful examples of organic nanofiber structures is displayed in the TEM micrographs in Figure 4.4.[59] Though these hierarchical nanostructures are not made of polymers but are instead composed of carbon, they deserve a place of honor in an inter-disciplinary book on polymer nanofibers, because they are realized by an extremely elegant and instructive nanofabrication process, thus providing both an intriguing source of inspiration for the design of

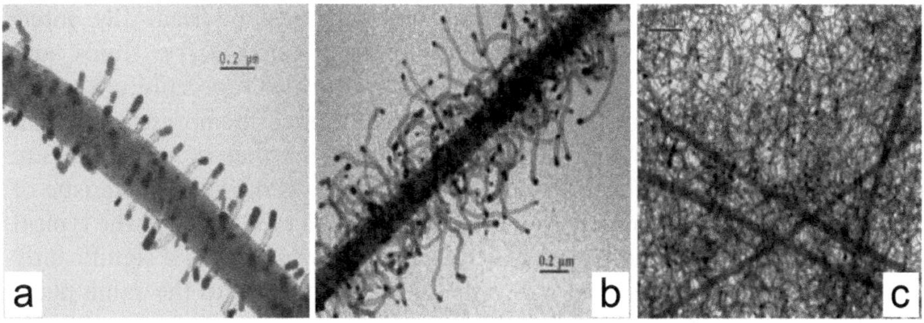


**Figure 4.4** TEM micrographs of hierarchical carbon nanostructures made by electrospinning, carbonization and catalytic growth of lateral carbon nanotubes. The length of the tube is controlled by supplying hexane vapors for different times during the catalytic growth: 3 min (a), 5 min (b) and 20 min (c), respectively. Hexane vapors are carried by an argon flow delivered at the furnace at 600 mL min$^{-1}$. Adapted with permission from Ref. 59, *Adv. Mater.*, 2004, **16**, 69–73. Doi: 10.1002/adma.200306205. Copyright © 2004, Wiley-VCH Verlag GmbH & Co. KGaA.

novel nanosystems and an excellent exemplary case of complementary technologies combined together in a smart way. These nanostructures are fabricated by a sequence of steps including electrospinning of PAC mixed with Fe(acetylacetonate)$_3$ from DMF solutions, annealing, reduction of $Fe^{3+}$ to Fe in $H_2$ atmosphere, carbonization and hexane decomposition at the surface of the formed Fe nanoparticles, which serve as catalysts for the lateral growth of carbon nanotubes from the main body of carbon fibers.[59] In this way, the hexane vapor provides a continuous source of carbon for the nanotube growth, whose length is controlled by means of the time during which hexane is supplied (Figure 4.4).

# 4.2 Porosity and Wettability Properties

Different kinds of "porosity" can be present and measured in polymer nanofiber samples. Porosity can be a bulk property of a sample, namely a collective feature of polymer nanofibers assembly, non-woven, *etc.*, or an internal property of individual nanofibers, or a surface property of either nanofiber-based coatings or of single filaments. These features impact on many properties of nanomaterials, ranging from fluid diffusion into and out of non-wovens or single nanofibers, to wettability, ability of drug delivery, of effectively exchanging bio-chemical signals, and so on.

## 4.2.1 Porosity

### *4.2.1.1 Porosity of Nanofibers Assemblies*

For non-woven mats, porosity is generally given by three-dimensional, unfilled voids present between nanofibers. The pores are delimited by the fiber bodies, entanglements and joints. These voids are generally hydraulically interconnected, thus resulting in continuous paths available for fluids and molecules flowing within the nanofiber mats, which is especially critical in determining the filtration performances of nanofiber membranes (Section 4.3.1), and for transporting gases, nutrients, drugs, *etc.* when these samples are used as biological scaffolds (Sections 7.1.1 and 7.7). Another characteristic of disordered assemblies of polymer nanofibers has to be recognized in the typical size and shape of pores in the plane of deposition, which can be significantly different from those measured along the direction normal to the same plane (*i.e.* in the cross-section of the mat). Of course, a simple way to estimate this kind of porosity, $P$ (%), could be based on simple geometrical considerations, assisted by a reasonable topological sketch of the nanofiber ensemble. A densely packed, uniaxially aligned array of uniform nanofibers, each of radius $r$ and length $l$ (as schematized in Figure 4.5a) would have a porosity given by:

$$P = 1 - \frac{V_{fiber}}{V_{cell}}, \tag{4.5}$$

where $V_{fiber} = \pi r^2 l$ and $V_{cell}$ is the unit cell volume ($= 2\sqrt{3}r^2 l$). The resulting porosity is therefore quite low, about 9%. Another model describes a topologically networked assembly of nanofibers in a cubic lattice, where wire segments of uniform length ($l$) constitute the sides of the lattice cells (Figure 4.5b). In this case, considering that there are three side segments for each lattice unit cell, one has:[66]

$$P = 1 - \frac{3V_{fiber}}{V_{cell}}, \tag{4.6}$$

where $V_{fiber}$ is the volume of each side cylinder ($= \pi r^2 l$) and $V_{cell} = l^3$. This expression allows for a very high porosity (99.9% for fibers with an aspect ratio

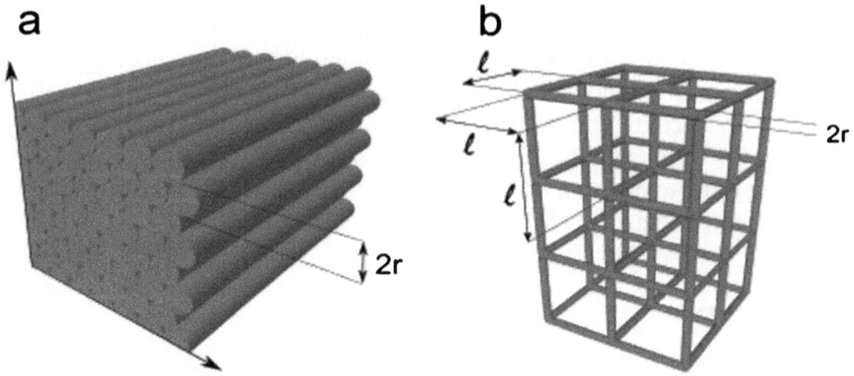


**Figure 4.5** (a) Schematics of a uniaxially aligned array of uniform nanofibers, and relevant geometrical parameters. (b) Schematics of a cubic fiber network.

of 100). Real samples consist of partially or fully disordered assemblies of fibers, and have porosities intermediate between the values obtained by Equations (4.5) and (4.6).

However, such geometrical evaluations can be quite inaccurate in describing real samples. On one hand, SEM and other tools for morphological investigations rarely allow with the needed precision the exact packing of nanofibers in the deepest regions of non-wovens to be determined. The size of individual nanofibers in an ensemble often presents a relevant dispersion in terms of radius and length. On the other hand, topological models to describe assemblies are generally oversimplified. For these reasons, other methods are often preferred to determine porosity. For example, one can use the following formula:[66,67]

$$P = \frac{\delta_0 - \delta}{\delta_0}, \tag{4.7}$$

where $\delta_0$ is the density of the polymer of which nanofibers are made (also called skeletal density), and $\delta$ is an apparent density, extracted by weighting and considering the geometrical dimensions of the sample. Since crystalline and amorphous components of polymer samples have different values of $\delta_0$, the skeletal density can be estimated as:[66]

$$\delta_0 = \frac{1}{\frac{1-\phi}{\delta_a} + \frac{\phi}{\delta_c}}, \tag{4.8}$$

where $\phi$ is the crystallized fraction of the polymer, and $\delta_a$ and $\delta_c$ indicate the density of the material in its amorphous and crystalline phase, respectively. The previous equations allow an estimate of porosity to be obtained easily. For instance, for electrospun non-woven mats, $P$ values up to 80–90% are not unusual. Values of $P$ as high as >98% can be reached by nanofiber networks realized by phase-separation (Section 3.6).[66] Furthermore porosity, together with the characteristics of diffusional flows in polymer nanofiber samples, can be evaluated by numerical simulations, particularly using Monte-Carlo methods.[68–71] Some of these concepts and how they can impact on applications, and especially on filtration, will be explained in more depth in Section 4.3.1. Finally, a few experimental techniques have been developed that allow not only the overall resulting porosity but also the pore size and distribution in nanofiber assemblies and non-woven mats to be measured.[72]

*Mercury intrusion porosimetry* uses Hg, which is a highly non-wetting liquid, and determines the pressure that is required to push it through the sample. Besides the well-known toxicity of mercury, an issue with soft polymer samples, including polymer nanofiber assemblies, is given by the possible distortions in pores due to the high applied pressure.[72] In addition, the method is destructive for samples. *Liquid extrusion porosimetry* is more suitable for soft polymeric samples. Here, a wetting liquid is pushed through a nanofiber mat, thus extruding an equivalent volume of liquid, previously infiltrated into the mat, out from the porous sample. An even more effective, and non-destructive, technique is *capillary flow porometry*,[72–74] in which a chemically inert gas is led to flow through a dry sample of polymer nanofibers, and then through the same mat, preliminarily wet with a known liquid. The volumetric flow rate of the gas upon changing the wetting conditions of the mat is measured *vs.* pressure. The typical obtained data curves are shown in Figure 4.6. The pore diameter can be finally calculated as $4\cos\theta_W\sigma_s/\Delta p$, where $\sigma_s$ indicates the surface tension of the wetting liquid, $\theta_W$ is the contact angle of the liquid and $\Delta p$ is the measured differential pressure.[73,74] In particular, when a distribution of pores is present in the sample, as is practically always the case with polymer nanofiber mats, various aspects of the pore size distribution can be inferred from these experiments. The accessed range of pores is evidenced in Figure 4.6. For example, the onset of the gas flow, corresponding to lower values of $\Delta p$, allows the size of the largest pores present in the ensemble (so-called *bubble point* shown in Figure 4.6) to be determined, which are the first capillary paths through which a flow activates upon increasing the driving pressure. *Vice versa*, the flow through the pores having the smallest capillary size is activated only at higher values of pressure gradients. This condition is achieved at the differential pressure in which the flow data measured under wet conditions become almost superimposed to those collected under dry conditions, as shown in Figure 4.6. The mean pore diameter can be estimated by the intercept of the curve collected after wetting with that given by half data measured in dry conditions. Physically reasonable values of the pore size in non-woven mats

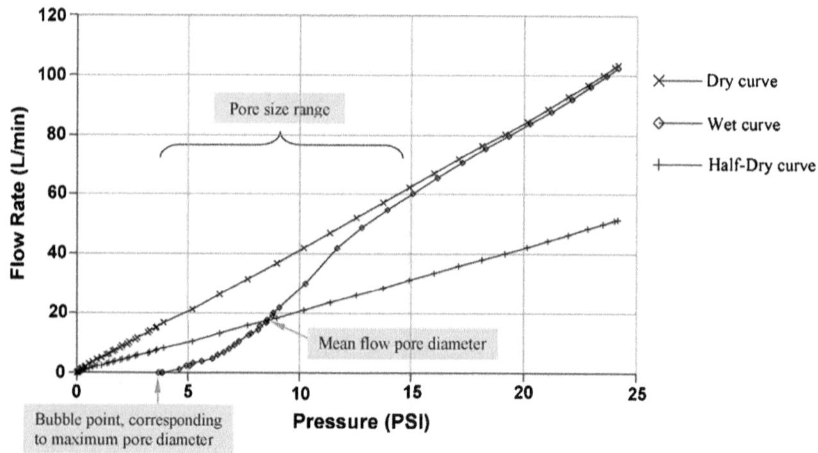


**Figure 4.6** Typical curves obtained by capillary flow porometry on a nanofiber non-woven, under dry (crosses) and wet (diamonds) conditions. A curve corresponding to half the curve measured in dry conditions is also shown (plus signs). Reproduced with permission from Ref. 73, *J. Membr. Sci.*, 2006, **286**, 104–114. Doi: 10.1016/j.memsci.2006.09.020. Copyright © 2006, Elsevier B.V.

made of polymer nanofibers are from one to two orders of magnitude larger than the fiber radius. Indeed, thinner fibers generate a reduced amount of free space in the sample because they are able to pack more closely, which results in an apparent density closer to $\delta_0$ and in smaller average pore size. The resulting dependence of the pore size on fiber diameter can be generally described by roughly linear laws. Of course, similar considerations apply to the cross-sectional shape of individual nanofibers composing a membrane: geometries allowing filaments to pack more tightly lead to smaller average pore size.

### 4.2.1.2 Internal Porosity of Single Nanofibers

A similar concept of porosity, but applied at nanoscale, namely to the internal structure of each nanofiber, accounts for diffusional events of gases or other molecules into, within or out of single fibers. A significant internal porosity can be present even in nanofibers whose external surface is smooth and continuous when observed, for instance, by SEM or AFM. Aspects related to free volume (Section 1.4.1) and to how this is affected by sample processing and ageing are especially relevant in this regard. For instance, the guidelines reported in Section 2.1.4[75,76] help in predicting the resulting porosity and surface morphology of fibers realized by solvent-assisted fabrication methods and when phase-separation[77] between immiscible components is likely to occur, such as in electrospinning. In addition, annealing procedures can change fiber internal porosity and surface morphology considerably.[75]

In general, the fluid diffusion at single-fiber level enables the use of these polymer nanostructures in chemical and biological sensors, where the agent to be detected gradually penetrates into the polymer and causes a measurable variation of a physical property, such as the optical (Section 5.2) or electrical characteristics (Section 6.2). An effective method to produce polymer nanofibers with controlled internal porosity is performing electrospinning directly onto either refrigerated collectors or baths of fluids, including liquid nitrogen.[78] This process would prime a sudden phase separation between the solvent and the polymer component of the electrospun solution. The subsequent removal of the solvent fraction (so-called *freeze-drying*) allows polymer nanofibers whose porosity extends from the surface well into the body of the nanostructures be to obtained. Another approach is electrospinning in a high-humidity environment, where water diffusion into the jet, favored when water-miscible solvents are used, such as DMF, can lead to the formation of internal pores.[75]

### 4.2.1.3 Surface Porosity of Single Nanofibers

In a different way, the surface porosity of individual nanofibers is defined by the topography which can be present on the surface of each polymer filament. Depending on their nature and on the used fabrication method, surface features can extend from the nm-scale to hundreds of nm, thus possibly being much larger than internal voids and free volumes eventually present within the polymer nanofibers. In other words, surface roughness may or may not match with the eventual internal porosity of the deep regions of each polymer filament. When dislocated on length-scales of many tens or hundreds of nm, surface pores are potentially suitable for photonic applications in the visible and in the near infrared spectral range. In addition to other usages that one can easily envisage for controlled drug release, fibers with porous surfaces can be templates for the growth of second-generation nanostructures as described in Section 3.3.2.

Surface features can be directly transferred onto the surface of nanofibers by means of templates (Section 3.3) or of top-down techniques such as imprinting (Section 5.6), or can be the result of various processes of phase separation between two or more solvents or polymeric components during fiber formation or processing. For instance, a controllable surface porosity can be obtained by electrospinning from solvents that have low boiling point and high vapor pressure, due to the rapid evaporation from solvent-rich regions.[79] Using dichloromethane (boiling point $\cong$ 313 K) and other highly volatile solvents, this has been demonstrated in electrospun fibers of PLA, polycarbonate, PMMA and poly(vinylcarbazole) (Figure 4.7).[79,80] Regions rich in solvent can finally result in sub-100 nm surface pores. Other studies have identified the conditions concerning the buckling, the drying and the phase separation times in electrospun jets, which are at the base of the formation of surface corrugations (Section 2.1.4).[75,76] Finally, as mentioned in Section 2.2, ambient humidity, which is normally detrimental to the electrospinning process, can induce the creation of surface features.[80,81] The resulting pores have typical size of tens or hundreds of nm, increasing with humidity and polymer $MW$, and are due to the condensation of aqueous vapor onto the electrospun jet while the latter is being rapidly cooled because of solvent evaporation.

### 4.2.2 Wettability

Coatings made by polymer nanofibers, like many other classes of surfaces based on inorganic, organic or hybrid micro- and nanostructures, can allow the degree of water adhesion and spreading to be controlled. The mechanisms by which topographical corrugations can induce remarkable variations of wettability properties compared to untextured samples have been widely investigated in the last decade with the aim of realizing artificial, super-hydrophobic or super-oleophobic self-cleaning surfaces. This strategy is largely inspired by natural systems such as plants and insects, and is of enormous interest for textiles, anti-fog windows, illuminating and automotive engineering and biomedical sciences. Polymer nanofibers can be very useful and exhibit peculiar features in this framework. For instance, super-hydrophobic surfaces based on polymer nanofibers are nowadays well-established.[82]

The wettability properties of a given surface are summarized by the contact angle, $\theta_W$, which roughly defines the shape of a deposited liquid drop. Contact

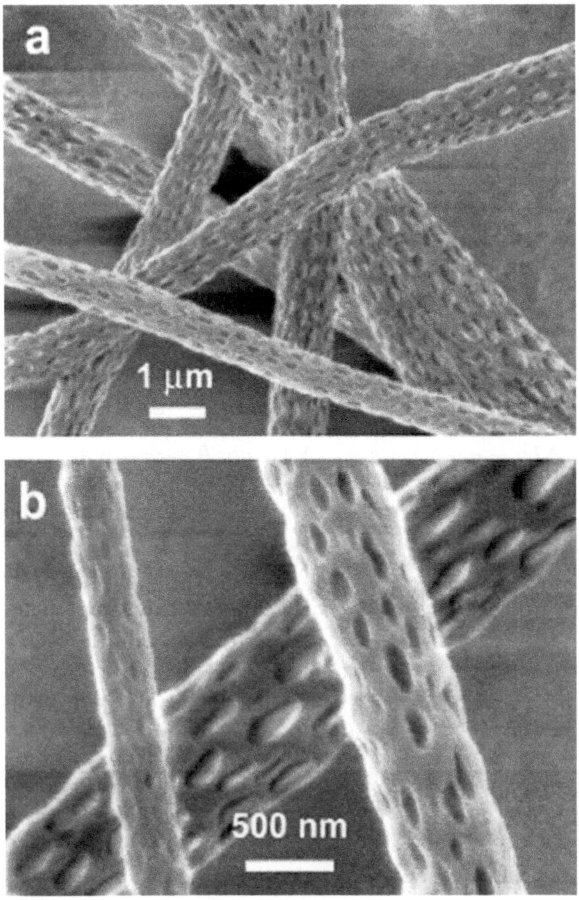


**Figure 4.7** Porous electrospun fibers made of PLA using dichloromethane as solvent. Reproduced with permission from Ref. 79, *Adv. Mater.*, 2001, **13**, 70–72. Doi: 10.1002/1521-4095(200101)13:1<70:: AID-ADMA70>3.0.CO;2-H. 

angle measurements can be performed very easily by gently delivering a minute droplet (with typical volume in the µL range) onto the surface, and lateral optical imaging. In turn, the contact angle is first of all influenced by the various interfacial tensions, or interfacial free energies per unit area, involved near the contact line. For a smooth planar surface, one has:

$$\sigma_{LV} \cos\theta_Y = \sigma_{SV} - \sigma_{SL} \tag{4.9}$$

where $\sigma_{LV}$, $\sigma_{SV}$ and $\sigma_{SL}$ (N m$^{-1}$) represent the interfacial tensions at the boundaries between liquid (*L*), vapor (*V*) and solid (*S*), respectively, and $\theta_Y$ indicates the contact angle on such untextured surface. $\sigma_{LV}$ is the surface tension at the interface between the liquid volume and the external free volume when vapor is present, which is what we also symbolize with $\sigma_s$ in this book (Section 2.1.2). $\sigma_{LV}\cos\theta_Y$ represents therefore the component of the liquid–vapor interfacial tension along the solid plane. A liquid is able effectively to wet the smooth substrate if $\sigma_{SV} > \sigma_{LS}$, namely if $\theta_Y$ is less than 90°. Otherwise the drop remains compact and $\theta_Y$ is above 90°. The previous, well-known relation (Young's equation)[83] is straightforwardly obtained by imposing the equilibrium condition on the projections of the different forces at the contact lines onto the horizontal solid plane. Equivalently, it is highly instructive to derive Equation (4.9) by zeroing the overall free energy variation resulting from slightly moving the contact line on the solid. Such small displacement determines a (positive or negative) change, $\Delta A_{SL}$, of the solid–liquid interfacial area. Taking into account also the other interfaces created or destroyed upon moving the contact line (Figure 4.8a), the free energy variation is given by:

$$\Delta U_G = \sigma_{SL}\Delta A_{SL} + \sigma_{LV}\Delta A_{SL}\cos\theta_Y - \sigma_{SV}\Delta A_{SL}. \tag{4.10}$$

Imposing $\Delta U_G = 0$, thus seeking the equilibrium condition, then immediately leads to Equation (4.9). Such derivation is certainly simplified, not taking into account the variation of the droplet shape due to the total liquid volume conservation,[84] however it is very intuitive, and, what matters most in this context, it can be immediately extended to describe the macroscopically observed wetting behavior on nanostructured surfaces, including coatings, mats and membranes made of polymer nanofibers.

There are two classical models that can be potentially used to explain the effect of the surface roughness on wettability. These two models basically differ

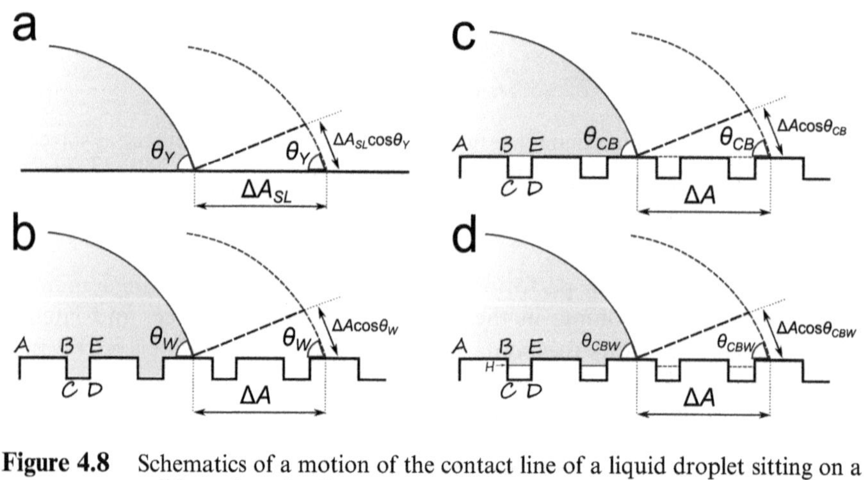


**Figure 4.8** Schematics of a motion of the contact line of a liquid droplet sitting on a solid surface, leading to a corresponding free energy variation for a smooth surface (Young model, a), complete penetration of the liquid into the recessed features of a pattern (Wenzel model, b), and for heterogeneous wetting regimes with absence of penetration (Cassie–Baxter model, c) and for partial penetration (d). In the Wenzel model, $r_W = (AB + BC + CD + DE)/(AB + CD)$ according to the letters in (b). In the Cassie–Baxter model, $\phi = (AB)/(AB + BE)$ according to the letters in (c).

in the degree of penetration of the liquid phase into the recessed features of the solid surface pattern. The complete penetration of the liquid drop into these features, described as homogeneous wetting regime (Figure 4.8b), leads to a variation of the free energy following a displacement of the contact line which is:

$$\Delta U_G = r_W \sigma_{SL} \Delta A + \sigma_{LV} \Delta A \cos\theta_{We} - r_W \Delta \sigma_{SV} A, \tag{4.11}$$

where $\theta_{We}$ is the liquid-solid contact angle on the textured surface and $r_W \geq 1$ is the surface roughness ratio, namely the overall area of the structured surface projected on the horizontal plane of the solid. The condition, $\Delta U_G = 0$, leads to the following Wenzel equation relating $\theta_{We}$ with the Young contact angle:[85]

$$\cos\theta_{We} = r_W \cos\theta_Y. \tag{4.12}$$

Considering water as fluid of interest, the roughness as described by the Wenzel model would therefore enhance by the factor $r_W$ the hydrophilicity ($\theta_Y < 90°$) or hydrophobicity ($\theta_Y > 90°$) of the smooth surface.

In the heterogeneous wetting regime, air can instead be trapped inside the features underneath the liquid drop, which ideally stands on the top of the nanostructures. This situation (schematized in Figure 4.8c) is described by the Cassie–Baxter model.[86] The free energy variation following the contact line displacement is:

$$\begin{aligned}\Delta U_G = {} & \sigma_{SL} \phi \Delta A + \sigma_{LV}(1-\phi)\Delta A \\ & + \sigma_{LV} \Delta A \cos\theta_{CB} - \sigma_{SV} \phi \Delta A.\end{aligned} \tag{4.13}$$

where $\theta_{CB}$ is the Cassie–Baxter contact angle, and $\phi$ is the fraction of the projected solid surface that is wet by the liquid. Consequently imposing $\Delta U_G = 0$ leads to:

$$\cos\theta_{CB} = \phi(\cos\theta_Y + 1) - 1. \tag{4.14}$$

This allows for a richer phenomenology than the Wenzel equation. For example, $\theta_{CB}$ values larger than 90° (*i.e.* hydrophobic or oleophobic surfaces) can be obtained even for materials whose Young contact angle, $\theta_Y$, is below 90°, depending on the value of the $\phi$ parameter. In real physical systems, including surfaces realized by polymer nanofibers, situations fully described by either the pure Wenzel or the Cassie–Baxter model are quite rare. It is reasonable frequently to expect intermediate situations, in which the liquid phase wets partially the surface roughness (Figure 4.8d). Therefore, drops on nanofiber surfaces have to be described by intermediate states, with the fluid partially penetrating into voids between fibers and leaving air pockets beneath to some extent. In this case one has the following expression for $\Delta U_G$:

$$\Delta U_G = \sigma_{SL} r_{W,w} \phi \Delta A + \sigma_{LV}(1-\phi)\Delta A + \sigma_{LV}\Delta A \cos\theta_{WCB} - \sigma_{SV} r_{W,w} \phi \Delta A, \tag{4.15}$$

ultimately leading to the generalized equation for the contact angle, $\theta_{WCB}$:

$$\cos\theta_{WCB} = r_{W,w}\phi\cos\theta_Y + \phi - 1. \tag{4.16}$$

Here, $r_{W,w}$ indicates the roughness ratio of the wet area. Equation (4.16) contains Equations (4.12) and (4.14) as particular cases. When $\phi = 1$ and $r_{W,w} = r_W$, Equation (4.16) turns into the Wenzel equation (Equation (4.12)). On the contrary, when the liquid drop lies on the top of the rough surface, without sinking into the features, one has $r_{W,w} = 1$ and Equation (4.16) reduces to the simplified Cassie–Baxter equation (Equation (4.14)). Equation (4.16) also allows a roughness with two different characteristic length-scales to be described, which can be present on the solid.[87] This is very often the case for polymer nanofiber surfaces. In fact, nanofiber layers and coatings can exhibit two different length-scales of porosity, namely a larger porosity due to the network composed of nanofiber bodies, and a much smaller surface porosity provided by the roughness on each single fiber (Section 4.2.1.3). These two length-scales can be described respectively by the parameters $\phi$ and $r_{W,w}$ entering into Equation (4.16).

Coming more specifically to nanofiber non-woven membranes, what can one specialize of these analyses? This issue has been addressed in depth by researchers at M.I.T. with the aim of designing super-oleophobic surfaces, which are very rare.[88] First of all, it can be easily realized that some threshold values of pressure (or equivalently of the height of a liquid column) should exist, allowing a droplet to be forced to pass from a metastable Cassie–Baxter state to a Wenzel-like state. This threshold pressure is a sort of breakthrough value required to drive the liquid, initially standing on the top layer of nanofibers, to come into contact with the other layer underneath. Such transition can be described by simple geometrical considerations on nanofiber mats. Indeed, given the fiber radius ($r$) and the distance ($\Lambda$) between the bodies of two fibers in the top layer, a state in which the droplet is partially suspended can be described by the height, $h_2$, of the air pocket between the liquid and the second layer of nanofibers, and by the local equilibrium contact angle, $\theta_W$, defined by a single fiber and the bottom surface of the droplet.[88] The relation, $h_2 = r\,(1 - \cos\theta_W)$ holds in an initial, Cassie–Baxter state. When the system undergoes a possible wetting onset, namely a transition to a Wenzel-like state, the sagging of the droplet between two adjacent fibers in the top layer determines a shift, $\delta\theta$, of the local contact angle. In turn, $\delta\theta \cong \sin^{-1}(\Lambda/2\,r_C)$, approximated as $\cong \Lambda/2\,r_C$ for small angular shifts, where $r_C$ is the radius of curvature of the sagging droplet ($r_C \cong p/2\sigma_{LV}$ according to the Laplace equation, Equation (2.2b)). Finally, the sagging amplitude, $h_1$, is roughly $r_C$ (1

– cos $\delta\theta$). The most important step of this analysis is recognizing a so-called robustness parameter, $H^*$, given by the ratio of $h_2$ to $h_1$, that tells how much the Cassie–Baxter state is stable on nanofibers.[88] High values of $H^*$ mean that the droplets prefer to stay suspended on air pockets, with only weak potentiality for sagging into a Wenzel-state. The robustness parameter can be written as follows, where one again uses the approximation of small angle shifts:

$$\begin{aligned} H^* = \frac{h_2}{h_1} &\cong \frac{r(1-\cos\theta_W)}{r_C(1-\cos\delta\theta)} \cong \frac{r(1-\cos\theta_W)}{r_C(\delta\theta)^2/2} \\ &\cong \frac{2r(1-\cos\theta_W)}{r_C(\Lambda/2r_C)^2} = \frac{8rr_C(1-\cos\theta_W)}{\Lambda^2}. \end{aligned} \tag{4.17}$$

This equation highlights how smaller fiber radius, highly wettable materials composing the surface of nanofibers and larger inter-fiber distances are all factors that contribute to lower $H^*$ and hence to reduce the stability of the Cassie–Baxter state. In particular, for mats of equal porosity (*i.e.* with roughly the same $r/\Lambda$ ratios), smaller fibers lead to a greater robustness. These principles can be used to design polymer nanofiber surfaces with highly controlled wettability properties. Exploiting blend nanofibers of PMMA and hydrophobic materials such as polyhedral oligomeric silsesquioxanes with perfluoro-alkyl groups, surfaces that are simultaneously super-hydrophobic and super-oleophilic have been demonstrated, potentially suitable for effectively separating water and alkanes. With a proper calibration of the composition of the polymeric blends, the re-entrant surfaces of nanofiber layers favor instead super-oleophobicity (allowing contact angles up to about 140° to be achieved for decane, hexadecane, *etc.*).[88]

For liquids impacting at some velocity on nanofiber surfaces, the physical scenario can change considerably. The impact of drops has been studied in depth in view of the possible development of spray cooling methods for surfaces coated by nanofibers.[89,90] A significant penetration of liquids into nanofiber non-wovens is found, regardless of the specific wettability of the nanostructures, provided that incoming droplets are fast enough ($\sim$1.5–3.0 m s$^{-1}$).[90]

Wetting phenomena on single suspended nanofibers are different as well. Both numerical[91] and analytical approaches are possible to study the problem, once one has formulated a suitable expression to describe the interaction between a molecule of the liquid and a molecule of the fiber. An option could be, for example, through a London–Van der Waals, power fractional law form such as $\Phi \sim (l_0/r_x)^6$ for $r_x > l_0$, where $r_x$ is the inter-molecular distance and $l_0$ stands for an impenetrable, repulsive core of the order of a few Å.[92] In general, tiny droplets can arrange on free-standing fibers by non-axisymmetrical (*i.e.* clam-shell) or by symmetrical (barrel) configurations. Humidity from air can of course condense on suspended polymer fiber structures. A nice example comes from natural fibers such as spider silk filaments, which exhibit unique properties of directional water collection by virtue of their structure.[93] The configurations of droplets on arrays of flexible suspended fibers, which are frequently present both in nature such as for example in the feathers of birds and in artificial materials such as in wetting textiles, have been also investigated.[94] The final configuration taken by drops and their level of spreading are found to be governed by the elasticity of fibers, their radius and spacing, and by the liquid volume,[94] according to a physical picture that could likely be useful to also depict polymer nanofiber materials.

The situations described so far are basically determined by static wettability properties. In other words, we have seen examples in which, once the surfaces have been produced, their wettability behavior is stable over time. An intriguing variety of polymer nanofibers is instead based on dopants allowing wettability to be controlled in real-time, by means of suitable external stimuli. For instance, upon irradiation with photons at a specific wavelength, $\lambda$, photochromic molecules can change various physico-chemical properties related to the degree of electronic conjugation such as, for instance, absorption/emission, refractive index and dipole moments.[95,96] The latter directly impacts on surface energy and ultimately on wettability. In addition, the transformation can be reversible following an irradiation at a different wavelength. Photo-switchable nanofibers embed these molecules and undergo a reversible transformation between two states as well. A class of photo-switchable nanofibers has been realized by electrospinning optically inert, transparent polymers such as PMMA, doped with 1′,3′-dihydro-1′,3′,3′-trimethyl-6-nitrospiro [2H-1-benzopyran-2,2′-(2H)-indole] (6-$NO_2$-BIPS).[97] The polymer matrix is needed because of the low molecular weight of the photo-chromic molecules, making them unsuitable to be electrospun directly. The photo-chromic properties of 6-$NO_2$-BIPS are based on the transformation of the non-polar, colorless form of this molecular system, which is the so-called spiropyran, into a highly polar, isomeric and colored form (*trans*-merocyanine, Figure 4.9). Merocyanines are characterized by an extended $\pi$-electron delocalization, achieved *via* the photochemical cleavage of the C–O bond in the molecular ring by UV exposure ($\lambda \leq$ 400 nm), and absorb light at visible wavelengths. Furthermore, these molecules revert back to the spiropyran form photo-chemically, by irradiation at green wavelengths. Through this mechanism and alternating exposures at

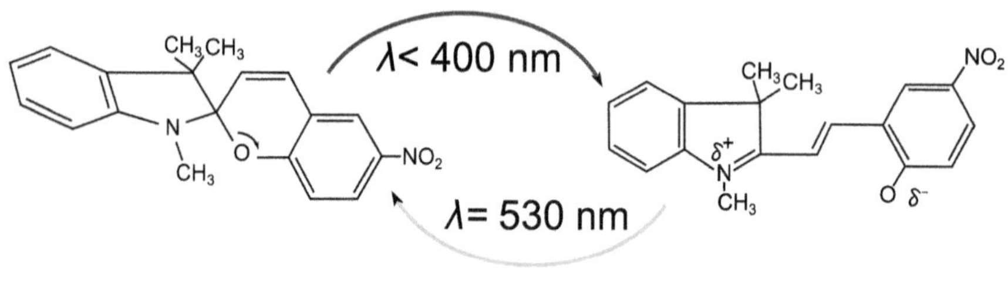


**Figure 4.9** Scheme displaying the conversion of spiropyran (left) and merocyanine (right) following irradiation at different wavelengths.

different wavelengths, polymer nanofiber surfaces can be driven across reversible water contact angle variations. For example, the contact angle is found to vary from 108° (before any irradiation) to 91° when the more hydrophilic merocyanine state is induced by UV light.[97] Such wettability change is significantly enhanced compared to films made of the same material, because of the much higher surface available in nanofiber samples to interact with the liquids. These results evidence the high potential of polymer nanofibers for use in externally controllable, smart surfaces and coatings. In particular, photo-switchable fibers may find application within microfluidic devices, where controlled wettability properties can be exploited in integrated valves and mixing elements.

## 4.3 Applications

### 4.3.1 Filtration, Catalysis and Energy Storage

*Filtration.* Filtration is traditionally the leading, and today the most developed, industrial sector involved in the application of polymer nanofibers.[98,99] The most striking and clear feature of polymer nanofiber mats and non-wovens, namely the extremely high surface-to-volume ratio, together with their highly porous and lightweight nature, make these systems ideal for filtration and purification of gases and liquids. Furthermore, production volumes needed for these applications are generally compatible with the throughput of polymer nanofiber fabrication methods, particularly of electrospinning. The idea of nanofiber-based filters dates back to the years 1937–1938. I. V Petryanov-Sokolov, N. D. Rosenblum and N. A. Fuchs at the Karpov Institute have incorporated electrospun fibers within filtering media, known today as Petryanov filters®.[100,101] These results have given impulse to industrial-scale production, initially mostly aimed at realizing gas masks. The so-produced material has been called *BF* ("Battlefield Filter") and throughputs as high as $2 \times 10^7$ m$^2$ of electrospun fibers per year have been reported by the 1960s.[99] For a long time, the scientific research by the Karpov Institute and the related industrial activities were not allowed to circulate outside the USSR due to military reasons.[99] More recently, however, a dedicated volume has been published, describing in depth the production of these nanofibrous media and their aerosol filtration performances.[102]

Companies involved in the production of filtering elements based on miniaturized fibers include Freudenberg Group[103] (Viledon®, www.freudenberg-filter.com), Donaldson (Ultraweb®, www.donaldson.com),[104] Hollingswarth & Vose (NANOWEB®, www.hollingsworth-vose.com), KX Industries (www.kxtech.com), Toray (www.toray.com), Esfil Tehno (www.esfiltehno.ee), *etc.* Filtration applications are at the base of a huge increase of the worldwide demand for nanofiber products, whose market is expected to grow significantly in the next years. Even filters against bacteriological and chemical agents have rapidly become up-to-date in recent years as security and anti-terrorism have been urgently included among governmental priorities. In all cases, air filter materials have to meet stringent quality standards due to their sanitary importance and widespread use in vehicles, aircrafts, hospitals, clean rooms, *etc.*, and polymer nanofiber-based materials can be very useful in this regard. According to United States standards set by the Department of Energy, so-called high-efficiency particulate air (HEPA) filters have to be able to remove a percentage as high as 99.97% of particles larger than 300 nm.[105] The classification of high-efficiency air filters according to European Union standards is based on overall retention efficiencies at the most penetrating particle size ranging from 85% (E10 filter class) to 99.999995% (U17 class).[106] The efficiency of nanofiber-based filters is very high, and tailorable by the material and flow characteristics as is better detailed below. In addition, nanofiber materials are generally quite easy to clean, which is useful to extend the lifetime of filtration units.

To serve as effective filtering elements, nanofiber membranes should first of all support significant flow of liquids or gases through them. In this way, particles or contaminants would be removed from the fluid flow. A useful phenomenological equation describing the volumetric flow rate ($Q$) of a fluid moving across a porous sample, along the $x$-direction, and in a regime of low Reynolds number, is the following Darcy's law:

$$Q = -\frac{K_P A}{\eta}\frac{\partial p}{\partial x}, \tag{4.18a}$$

where $A$ is the involved cross-sectional area of the medium, and $K_P$ indicates the sample permeability or permeation coefficient [m$^2$]. Dividing by $A$, in terms of the fluid velocity the previous equation reads:

$$v_x = -\frac{K_P}{\eta}\frac{\partial p}{\partial x}. \tag{4.18b}$$

An intuitive interpretation of Darcy's law is that of a fluidodynamic equivalent of Ohm's law for electric conduction, with the volumetric flow of the fluid, the $K_P/\eta$ ratio, and pressure taking the place of electric current, conductivity and applied voltage bias, respectively. The pressure gradient, $\partial p/\partial x$, across the porous material can often be approximated as $\Delta p/\Delta x$, where $\Delta p$ and $\Delta x$ indicate the overall pressure drop and the thickness of the sample, respectively. An apparent flow resistance, $R_F$ [m$^{-1}$], can be defined as:

$$R_F = \frac{\Delta x}{K_P}, \tag{4.19}$$

which is measurable through the product, $A\Delta p/\eta\, Q$.[107] The minus sign in Darcy's law indicates of course that the flow moves towards regions with lower pressure values. Equation (4.18a) also tells us that $Q$ increases linearly upon increasing the cross-sectional area of pores (in analogy with electric current increasing within wires of larger cross-section). Therefore, given the approximately linear dependence of the pore size on the radius, $r$, of nanofibers composing a polymeric sample (Section 4.2.1.1), one concludes that an increasing, roughly quadratic behavior is expected for the flow rate with increasing $r$. In addition, as is evident from Equation (4.19) and as expectable from physical intuition, thicker and less permeable samples have higher resistance against flow. For gases flowing through complex assemblies of polymer nanofibers such as electrospun PAN and PAC non-wovens, the $K_P$ coefficient is in the range of $10^{-14}$–$10^{-13}$ $m^2$.[71,98] Consequently, according to Equation (4.19), a nanofiber non-woven with a thickness of 10 μm, which corresponds to a surface mass of about 2 g $m^{-2}$ considering a porosity of 85% (Equation 4.7), would have a flow resistance of the order of $10^8$–$10^9$ $m^{-1}$.

Of course, Darcy's law works in the framework of the so-called continuum approximation, in which the flowing fluid is treated neglecting finite molecular dimensions. On the other hand, the flow of fluids across membranes made of polymer nanofibers with very thin (~1–100 nm) diameters could give rise to peculiar features. An important parameter in this respect is the Knudsen number (Kn), which is the ratio between the microscopic and macroscopic length-scales involved in the flow process, namely:[108]

$$\mathrm{Kn} = \frac{\lambda_F}{2r_P}, \tag{4.20}$$

where $\lambda_F$ is the mean free path of flowing molecules. The denominator of Equation (4.20) has to contain some relevant characteristic length that is typical of the material or physical system under study. Here, $r_P$ is the pore radius. As described in physics textbooks, for an ideal gas at atmospheric pressure and room temperature, $\lambda_F$ is of the order of a few tens of nm (66 nm for air in normal ambient conditions). Therefore, in systems with mesoscopic capillaries or pores, which may be the case of polymer materials made of nanofibers, there is the possibility to reach regimes in which Kn $\geq$ 1. Indeed, since $r_P \propto r$, the Knudsen number is inversely proportional not only to $r_P$ but also to $r$ according to Equation (4.20), and Kn larger than unity can be obtained at room temperature by nanofibers with diameter around 30–40 nm*. The fiber radius, $r$, is often considered directly in the denominator of Equation (4.20) as characteristic material length.

In regimes where Kn is comparable to or larger than unity, the continuum approximation fails, namely the finite size of molecules becomes important and has to be taken into account for correctly describing flow conditions. The

*In some cases, however, using free-standing media or membranes made of fibers with diameter in the range of tens of nm in filtering elements can be difficult due to possible mechanical instability, high flow resistance and safety issues related to very small particles. A viable approach is using nanofiber membranes to coat more robust supports and filtering media composed of fibers with larger size.

resulting fluid transport, in which molecules collide frequently with the pore walls, is known as Knudsen diffusion.[69,109] An associated phenomenon to be considered for Kn $\geq$ 0.1 (namely for filtering media composed of nanofibers with diameters roughly below 400 nm),[99] is the increase of the relative velocity of fluid molecules at the pore walls (*i.e.* the system passes from so-called no-slip to slip conditions), as shown in Figure 4.10. In particular, slip conditions determine a reduced drag force exerted on particles near the fiber surface, namely an ultimately decreased pressure drop across the filtering material. All these issues are very interesting from the point of view of basic studies, but also for applications. In fact, understanding in depth the collisional mechanisms and the diffusional regimes undergone by fluid particles while moving through a fibrous medium allows a unique underlying phenomenology to be unveiled, characteristic not only of filtration but also of thermal dissipation properties. These aspects are better described in Section 4.3.3.

In general, the filtration of a fluid by means of a porous nanomaterial, namely the removal of particles of contaminants, particulate, dusts, *etc.* from a flow of gas or liquid, is due to at least three different mechanisms by which the particles came into contact with the surface of the solid walls of the pores, where they can remain entrapped or adsorbed. These mechanisms, widely investigated by aerosol science and technology and schematized in Figure 4.11, are:

(i) *interception*, in which contaminant particles follow a streamline of the flow of fluid across the porous material (these streamlines tend to bend, deflecting to avoid the solid walls in ordinary, Kn < 1, flow conditions), and stick on the surface of pores when approaching the solid walls at a distance below the particle size;
(ii) *impaction*, in which contaminant particles are not able to follow a streamline of the flow due to their large size and inertia, therefore colliding onto the obstacles represented by the solid fraction of the porous material (for instance, impinging directly on a fiber);
(iii) *diffusion* due to Brownian motion; the random walk of a particle due to collisions with molecules of the surrounding fluid medium is described,

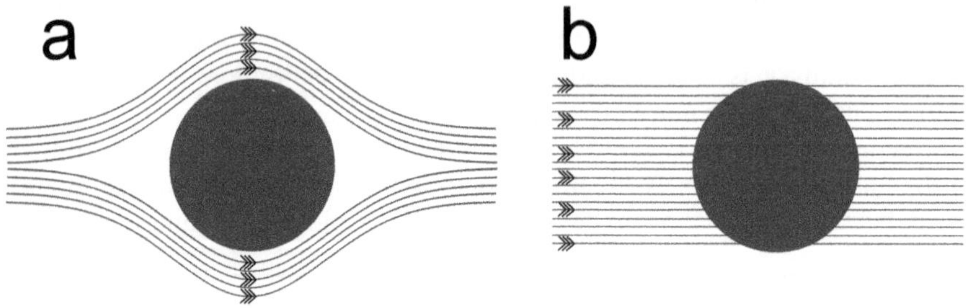


**Figure 4.10** Schemes of the flow lines of a fluid about the body of a polymer nanofiber (here shown in cross-section) for small Kn values (a) and for high Kn values leading to slip conditions (b).

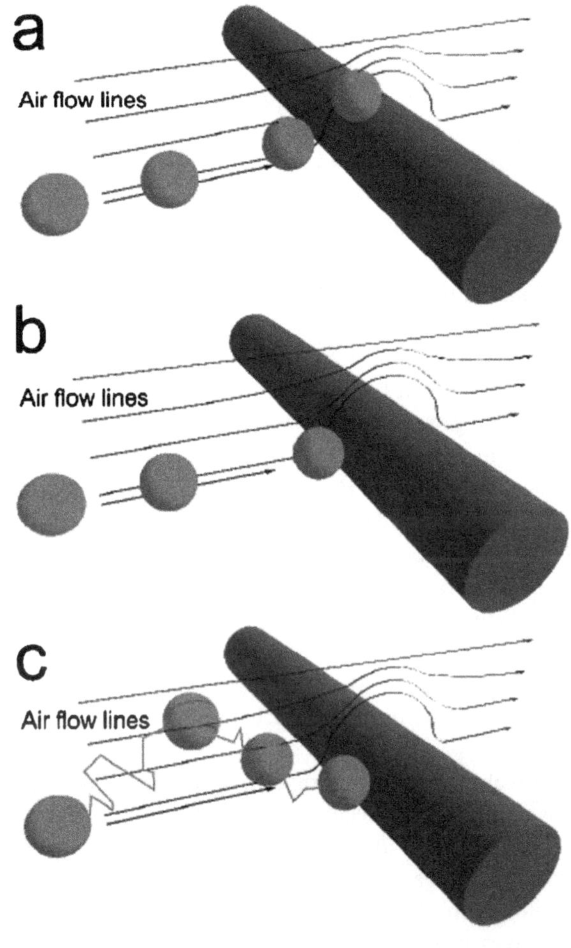


**Figure 4.11** Schemes of the flow lines and particle behavior near a polymer nanofiber, showing the different mechanisms of filtration: interception (a), impaction (b) and diffusion due to Brownian motion (c).

along each spatial dimension ($x$), by a Gaussian law, $p(x) = (1/4\pi Dt)^{1/2} \times \exp[-(x^2/4Dt)]$, where $p$ indicates the probability of finding the particle at a given $x$-coordinate, $D$ is the diffusion coefficient, $t$ is time and the origin of coordinates is set in the initial ($t = 0$) particle position. This Gaussian is a solution of the diffusion equation, $\partial p/\partial t = D\partial^2 p/\partial t^2$. Hence, the broadening of the probability distribution ($<\Delta x^2> = 2Dt$ for a motion in one dimension, and $<\Delta d^2> = 6Dt$ in the three-dimensional case) is linear with time, and depends on the particle diffusion coefficient. From the Stokes–Einstein law ($D = k_BT/6\pi R_P\eta$), one clearly sees that diffusion is favored for particles with smaller radius, $R_P$, and at higher temperatures.

Depending on the size of the considered particles and fibers, and on the flow conditions, the efficiencies of the different filtration mechanisms can be calculated.[110] In particular, diffusion becomes more important and ultimately the dominant mechanism for contaminant particles of progressively smaller size (in the range of 100 nm and below). Furthermore, exotic mechanisms can be triggered when nanofibers with small diameter are involved. As discussed above, it is expected that increasing Kn, as would be for air flowing across polymeric nanofibrous materials, leads to a condition in which streamlines no longer deflect to avoid the solid walls of pores, but instead keep rectilinear paths.[111] Simulations show that, in these conditions, the filtration efficiency increases and the smallest captured particles decrease in size upon reducing the fiber diameter below 100 nm and that, in addition, increasing temperature improves efficiency, which can be related to an enhanced Brownian motion as explained above.[111] The possibility of using polymer nanofibers in coalescence filters deserves to be mentioned as well.[112,113] Adding these nanomaterials to water-in-oil emulsions has been demonstrated to improve the coalescence efficiency, namely the ability to separate tiny water droplets from the oil continuous phase. This is especially important for the removal of water traces from fuels used in vehicles and aircraft. In other applications, such as for cleaning polluted water, the superior capability of molecular and droplet capture and retention exhibited by polymer nanofibers can be very useful. Indeed, highly oil- or chemical-sorbent materials based on polymer nanofibers have been engineered, as in the patented Eminus™ products[114] by Milliken (www.milliken.com).

Today, other novel concepts are rapidly emerging for filtering elements based on polymer nanofibers. Developing new and more efficient technologies for water decontamination, desalinization and purification is critically important due to the increasing scarcity of drinking water and freshwater resources. Nanofiltration by polymer nanofiber non-wovens can open interesting perspectives in this regard, given that good mechanical stability is achieved in the filtering material to support water flux. Rejections of 60–70% and 80% are obtained for NaCl and $MgSO_4$, starting from 2000 ppm concentrations, in water filtered by composite membranes based on electrospun PVDF nanofibers.[115,116] Elements using nanofibers for bioseparation can be manufactured by various methods,[117] and integrated within microfluidic chips.[118] Furthermore, new-generation building blocks for waste-water treatment and decontamination are made by living bacteria incorporated in the inner part of core-shell or hollow fibers.[119,120] Here the idea is of exploiting the mass transport of pollutants and contaminated fluids from a surrounding water environment into the fibers, where the metabolic activity of encapsulated bacteria would result in bioremediation of toxic elements. Overall, these systems would therefore combine the intrinsic filtration capability of polymer nanofiber mats with bio-induced decontamination, thus possibly removing relatively large particles as well as molecular contaminants (given that specific bacteria species are available, able to degrade the desired target pollutants and that can be safely and viably embedded in the fibers.). These systems and some strategies for encapsulation of living cells in nanofibers are described in Section 7.7.5.

*Catalysis.* Catalysis is another very broad and important field of application where permeability properties of polymer nanofibers play a key role. Indeed, jointly with the very high surface-to-volume ratio, permeability is crucial to allow catalytic agents (carried by polymer nanofibers) to come into contact with reactants (usually present in the fluid environment external to nanofiber samples). Another advantage of polymer nanofibers is that they are usable as catalytic elements not only in membranes and mats but also as individual, dispersed filaments, and that they can be easily recovered from the liquid medium. This can be done, for instance, by means of magnetic fields if fibers have been loaded by magnetic nanoparticles during preparation. In heterogeneous catalysis†, this allows the immediate recovery of the catalyst from the reaction environment, which is usually one of the most difficult tasks related to catalytic applications. Furthermore, this allows nanofibers in the liquid medium to be intimately dispersed during use, thus overcoming most of the diffusional and mat permeability issues that could limit reaction rates. Such a strategy can be implemented with several different classes of magnetic nanoparticles.[121–123]

In general, various catalysis working schemes can be designed. In homogenous catalysis, polymer nanofibers are used to deliver catalytic agents, following similar mechanisms as in drug delivery (Section 7.4). Low-molar mass catalysts and enzymes can be incorporated as dopants within or at the surface of polymer nanofibers by blending, by core-shell architectures, by microencapsulation of liquid reservoirs within fibers or by reversible surface-linking, and then released into the surrounding liquid where reactants are present. Both the porosity of individual fibers and the porosity and permeability at the level of the assembly of nanofibers affect the release kinetics, which has to be designed in order to guarantee a sustained and uniform presence of the catalyst in the environment. To this aim, several strategies have been proposed as reviewed in Section 7.4. As anticipated above, heterogeneous catalysis can be implemented by means of polymer nanofibers as well. In this case fibers directly serve as catalyst and reagents have to diffuse to the surface of or even into the fibers. Catalytic agents, differently from homogeneous catalysis, have to be immobilized permanently on the polymer component. Both heterogeneous catalyst complexes and nanoparticles (Fe, Pd, Pt) are viable options. Particles can be loaded or generated within nanofibers in various ways (see, for instance, Sections 2.3.3 and 4.1.1) and there is often no need to link them chemically to the polymer component since, if fibers are stable in the liquid/gaseous environment, the

†As known from chemistry textbooks, in *homogeneous* catalysis the catalyst is in the same phase as reactants, whereas in *heterogeneous* catalysis the catalyst and the reactants are in different phases (typically in solid and fluid phase, respectively).

diffusional escape of particles is quite difficult due to their relatively large size. Instead, molecular catalysts have been immobilized, once covalently bound to a polymer, within polymer nanofibers made of the same polymeric system. Styrene oligomers and polymers have been used in these experiments together with proline as catalyst.[124] An important requisite is that the immobilization of the catalyst, while necessarily increasing the level of steric hindrance in the nanoscale environment surrounding the molecule, does not preclude the activation of those rotational and translational degrees of freedom that enable catalytic action. In most of these processes, at the end of the desired reaction products have to diffuse effectively out of the nanofiber material, thus avoiding the reaction kinetics being altered. *Catalytic growth* is another possibility, in which secondary nanostructures can be obtained at the surface of nanofibers as a product of one or more delivered feeding reactants. Here, the reaction is catalyzed by chemical species, ions, nanoparticles, *etc*., preliminarily loaded in the polymer nanofibers and seeding the growth of daughter structures at specific points, as exemplified by Fe-catalyzed hexane decomposition and growth of carbon nanotubes (Section 4.1.1).[59]

*Energy.* Broadly related to the very high surface-to-volume ratio and tailorable permeability performances of polymer nanostructures, the use of nanofibers within electrochemical devices for energy storage applications is getting a lot of interest. These devices, including fuel cells, batteries and super-capacitors,[125] are becoming more and more important in modern technologies. When changing their redox state, conductive polymer nanostructures (Chapter 6) can exhibit a considerably increased switching speed compared to bulk materials due to the reduced diffusion distance that has to be covered by counterions.[126] For this reason, polymer nanofibers showing fast charge/discharge properties can be appealing for realizing batteries or capacitors, capable of delivering high power density.

The use of nanofibers is increasing in proton exchange membranes for fuel cells as well. These devices are able to convert directly and efficiently chemical energy into electrical energy, and generate a low amount of emitted pollutants.[127] Composite polymer electrolyte membranes made of uniaxially aligned sulfonated polyimide nanofibers and sulfonated polyimide are demonstrated to show high proton conductivity ($\sim 10^{-1}$ S cm$^{-1}$),[128] concomitantly with low gas permeability and long durability. In particular, decreased oxygen permeability and diffusion coefficients ($\sim 9 \times 10^{-13}$ m$^2$ s$^{-1}$) are measured for composites embedding the aligned nanofibers compared to pristine materials.[128] Core-shell nanofibers can encapsulate metal or complex hydrides inside a hydrogen-permeable sheath, which could be at the base of a new approach for hydrogen storage. In this case, a satisfactory dehydrogenation kinetics has to be achieved together with limited release of impurities, which are strongly detrimental for the operation of fuel cells. In this respect, interesting results are reported about the encapsulation of the complex hydride ammonia borane in electrospun PS.[129]

PVDF and poly(vinylidenefluoride-*co*-hexafluoropropylene) electrospun fibers are used to build membranes usable as separators with good results.[130–132] High electrolyte uptake, ionic conductivities of the order of $10^3$ S cm$^{-1}$ and stable charge-discharge behaviors are reported. The new nanofiber-based battery separator (Energain$^{TM}$) by Dupont, improving the performance and lifetime of batteries, is particularly interesting due to the great expectations and foreseen market size of electric vehicles in the near future.[133] Enzymatic biofuel cells based on PPy nanowires have also been reported.[134]

Concerning capacitors, a valuable figure of merit to quantify performances is the specific capacitance, $C_g$ (F g$^{-1}$). For instance, once the charge-discharge process of the capacitor is known, the specific capacitance can be easily estimated as:

$$C_g = \frac{I}{m(\mathrm{d}V/\mathrm{d}t)}, \tag{4.21a}$$

where d*V*/d*t* is the slope of the *V*(*t*) curve and *m* indicates the mass of the active material in a single device electrode. Of course for linear slopes the previous equation can be simplified as:

$$C_g = \frac{It}{m\Delta V}, \tag{4.21b}$$

where *It* (discharge current × time interval) is the total electric charge and $\Delta V$ is the potential bias change during the discharge process. In this framework, the combination of conjugated polymer nanofibers with graphene is opening new, exciting perspectives. For example, vacuum filtration allows one to produce free-standing, composite films of chemically converted graphene and PAN nanofibers obtained by interfacial polymerization (Section 3.4).[135] These films exhibit remarkable flexibility, conductivity of 5.5 S cm$^{-1}$ (one order of magnitude above the value of pure PAN nanofiber films) and they can be embedded within super-capacitor devices that have stable electrochemical capacitance of 210 F g$^{-1}$ at a discharge current density of 0.3 A g$^{-1}$.[135] Other studies have considered graphene nanosheets and PPy nanowires, whose multilayers have a specific capacitance of 165 F g$^{-1}$ at discharge current density of 1 A g$^{-1}$,[136] and chemically modified graphene and PAN nanofiber composites that reach 480 F g$^{-1}$ at 0.1 A g$^{-1}$.[137] Highly dense arrays of PAN nanorods electro-polymerized on ITO films reach more than 3400 F g$^{-1}$ at a current density of 0.5 μA cm$^{-2}$.[138] High power density performances (25 kW kg$^{-1}$ with energy densities keeping stable around 7.5 W h kg$^{-1}$) have been obtained by PEDOT nanotubes.[126] Overall, the field is expanding vigorously due to the concomitant demand for reducing environmental pollution and increasing use of portable electronic and communication devices, and of electric vehicles.

## 4.3.2 Micro- and Nanofluidics, Thermal Management and Sensing

The permeability properties of individual polymer nanofibers and of assemblies composed of them make these nanostructures advantageous also in other application fields where interactions of fluids with polymer surfaces are strongly involved.

*Micro- and nanofluidics.* One of these fields is related to the properties of wettability of polymer nanofibers, which can be exploited to realize smart surfaces, such as those described in Section 4.2.2, but also micro- and nanofluidic systems. Some working principles of micro- and nanofluidics are summarized in Section 3.2. Hollow polymer fibers made by coaxial electrospinning (Section 2.3.1) or by other techniques can be used as disposable capillaries, allowing micro- and nanofluidic processes to be carried out without the need for lithographic fabrication facilities. Carbon tubes[139,140] can be fabricated through these methods as well. It is clear that both spontaneous and pressure-driven capillarity could be used, depending on the wettability features of the internal surface of the hollow tubes. In particular, arrays and bundles of hollow fibers can be inter-connected by external tubing for fluid injection, and the resulting system can support pressure-driven flows of air and liquids, as demonstrated in extensive studies by the Yarin group (Figure 4.12).[139–141]

*Thermal management.* The flow of gases or liquids through polymer nanofiber non-wovens, if efficient, could be used for dissipating heat, for instance for cooling microelectronic components. Yarin and collaborators have explored this possibility as well.[142] Coating heat-transfer surfaces with layers of nanofibers, one can exploit the permeability of such mats to water used for spray cooling. In this approach, the thermal conductivity, $\kappa_T$, of nanofibers is almost irrelevant, since the main role is played by water evaporation, which, due to its significant latent heat, is a very effective way to dissipate heat.[142] It should be mentioned, however, that other recent studies

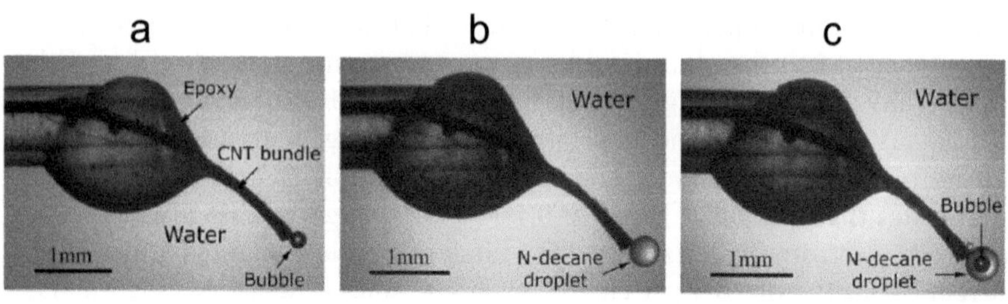


**Figure 4.12** (a) Growing air bubble at the tip of a bundle of carbon tubes immersed in water. An epoxy glue is used to secure a tight connection between the bundle and the end of a glass capillary. (b) Growing an *n*-decane droplet in water. (c) Growing an *n*-decane droplet with a gas bubble growing inside it. Adapted with permission from Ref. 139, *Lab Chip*, 2008, **8**, 152–160. Doi: 10.1039/b711446j.

have highlighted an unexpectedly high thermal conductivity in polymer micro- and nanofibers, due to the enhanced molecular alignment and reduced number of defect sites (polymer chain ends and entanglements per unit volume).[143] Bulk polymers have typical thermal conductivities of the order of $2 \times 10^{-2}$ cal/(K m s) [*i.e.* about $10^{-1}$ W/(K m)], whereas a value of $\kappa_T$ as high as 25 cal/(K m s) [104 W/(K m)], which is larger than a half of the thermal conductivity of pure metals, has been measured by mounting individual polyethylene nanofibers between a heated steel needle and a bi-material AFM cantilever (Figure 4.13).[9] Thermal conductivities exceeding 200 cal/(K m s) [$10^3$ W/(K m)], increasing with the backbone length, are estimated for single polymer chains by molecular dynamics simulations.[144] Overall, the investigation of thermal conductivity properties in polymer nanofiber materials is still an open field, leaving space for different points of view about how to exploit these properties. In other reports, entire new applications and processing methods such as flash welding are proposed, which are based on nanofibers with *low* (bulk-like) thermal conductivity and simultaneously efficient photo-thermal conversion.[145–147]

*Sensing applications.* The high surface-to-volume ratios of nanomaterials also mean great potentiality for sensing, as long as the same materials also have some transducing physical or chemical property that varies significantly following the exposure to either a chemical/biological agent or another stimulus from the surrounding environment. Polymer nanofibers are ideal in this respect because the high surface-to-volume ratio can be effectively combined with complementary properties allowing chemical changes to be

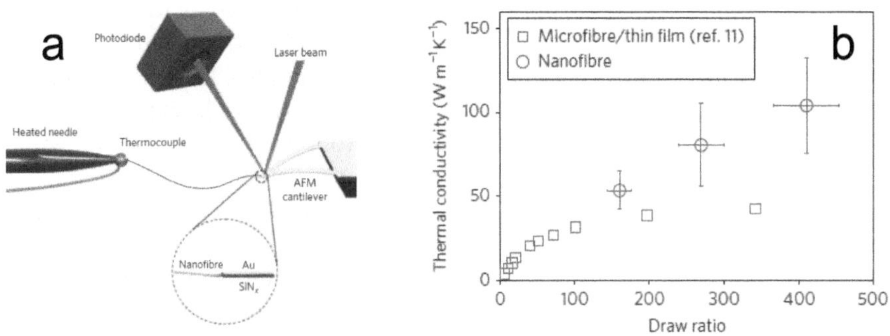


**Figure 4.13** (a) Scheme of the experimental set-up used to measure the thermal conductivity of individual nanofibers. The silicon nitride AFM cantilever is preliminary coated with a 70 nm Au film to realize a bi-material mounting. A laser beam of variable power is focused on the tip of the cantilever, and reflected onto a photo-diode analogously to normal AFM operation. The other ending point of the nanofiber is attached to a micro-thermocouple and a heated steel tip, and the fiber conductance is measured by the deflection of the AFM tip. (b) Circles: thermal conductivities of fiber samples *versus* draw ratios. Squares are data for thin-films and microfibers, from Ref. 143. Adapted with permission from Ref. 9, *Nat. Nanotechnol.*, 2010, **5**, 251–255. Doi: 10.1038/nnano.2010.27. Copyright © 2010, Nature Publishing Group.

detected with high sensitivity. An instructive example is given by the application of light-emitting nanofibers (Section 5.2) for optical sensing. For instance, one can exploit the quenching of the fluorescence of electrospun fibers made of PAA-poly(pyrene methanol) together with cross-linkable polyurethane latex,[148] or of many other light-emitting polymers and blends. Upon exposure to metal ions such as $Fe^{3+}$ and $Hg^{2+}$, and to nitro aromatic compounds such as the explosive 2,4-dinitrotoluene,[149,150] the fluorescence intensity drops down, which can be used to realize highly sensitive sensors. In another study, the fluorescent conjugated polymer, hydrolyzed poly[2-(3-thienyl) ethanol butoxy carbonyl-methyl urethane], has been assembled electrostatically on electrospun nanofibers of cellulose acetate, whose system undergoes quenching upon subsequent exposure to methyl viologen and cytochrome *c*.[151] In these works, the sensing performances are quantified through the Stern–Volmer relation:[148,151–153]

$$\frac{A_{em,0}}{A_{em}} = 1 + K_{SV}[c_Q] \,. \tag{4.22}$$

Here, $A_{em,0}$ and $A_{em}$ are the fluorescence intensities of the unperturbed and of the exposed samples, respectively, $K_{SV}$ ($M^{-1}$) is the so-called Stern–Volmer constant, which describes the association between the quenching and the emitting species, and $c_Q$ (M) indicates the concentration of the quencher that is sensed by the nanofibers.[154] Due to the high surface-to-volume ratios of nanofiber materials, the values of $K_{SV}$ reach the order of $10^6$ $M^{-1}$, *i.e.* are up to three orders of magnitude larger than those obtained with reference films.[148,151] Wearable sensors constituted by arrays of attoL volume-reactors (attoL = attoLitre) corresponding to the nanojunctions between polymer nanofibers loaded with reactants have been also realized, exploiting the fluorescence changes in fluorophore-polyamines following the interaction with metal ions ($Al^{3+}$, $Fe^{3+}$, $Co^{2+}$, $Ni^{2+}$, $Cu^{2+}$, $Zn^{2+}$, $Hg^{2+}$, $Cd^{2+}$, $Ca^{2+}$ and $Mg^{2+}$).[155]

A wide variety of other optical sensors based on polymer nanofibers has been demonstrated, including devices targeting DNA, proteins, volatile organic compounds, *etc.*[152,153,156] Changes of color can also be exploited to sense.[157–159] A very interesting, recently reported application is the colorimetric detection of adulterated gasoline.[160] Fuel adulteration has a tremendous economic impact and is therefore a relevant problem for companies and countries involved in the transformation and commercialization of petroleum combustible derivatives. Polymer nanofibers can be produced with protective matrices, such as PAA and PS, embedding polydiacetylene as schematized in Figure 4.14a and b. When present in gasoline, contaminants such as toluene or methanol dissolve the matrix and induce a *solvatochromic* change, namely a variation of coloration (*i.e.* of the absorption properties) in polydiacetylene. This variation is clearly visible by eye as schematized in Figure 4.14c–f.[160] Humidity sensors can be based on mechanically drawn polymer fibers embedding inorganic (CdSe/ZnS) quantum dots, which respond in a time ($\cong$

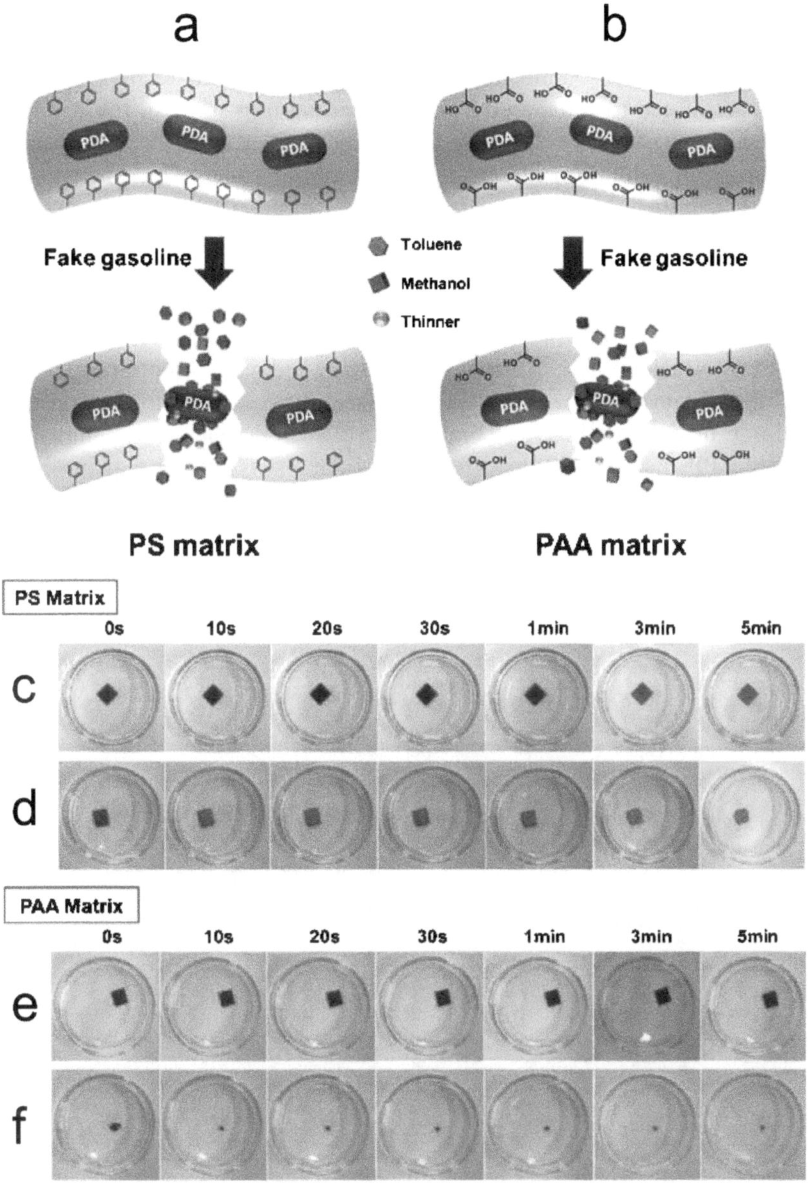


**Figure 4.14** (a, b) Schemes showing the mechanism for the adulterated gasoline-induced chromic change in PS (a) and PAA (b) nanofibers embedding polydiacetylene. Together with toluene and methanol, the thinner is a component of adulterated gasoline, composed of 80% naphtha. (c–f) Polydiacetylene/PS and polydiacetylene/PAA fiber mats in commercial gasoline (c and e) and adulterated gasoline (d and f). A few seconds of exposure of the fiber mats to the analytes are enough to induce a visible color difference. Adapted with permission from Ref. 160, *J. Mater. Chem.*, 2011, **21**, 2648–2655. Doi: 10.1039/c0jm02287j.

90 ms) 1–2 orders of magnitude lower than devices based on thin-films or monolayers (Figure 4.15).[161] In this case, an increase of the emission signal from quantum dots is found upon increasing the ambient relative humidity, which is attributed to water molecules passivating trap states at the surface of nanoparticles. Besides the desired reversibility in their response, these sensors present a good resolution in sensing the humidity value, which is estimated around 1%. Other devices, based on the variation of the amount of light transmitted along nanofibers coupled to fiber tapers, allow $NO_2$ and $NH_3$ to be detected at concentrations down to the order of 0.1–1 ppm.[162]

Finally, another class of sensing devices relies on the electronic properties of polymer nanofibers (Chapter 6). Chemical sensors based on PAN nanofibers made by interfacial polymerization have been tested with 100 ppm of HCl and $NH_3$ vapors, causing respectively an increase and a decrease of many orders of magnitude in the conductivity (Figure 4.16).[163,164] Similar devices have been realized by PPy nanowires fabricated by EBL on resists and subsequent lift-off, and used to detect 240 ppm $NH_3$.[165] Other sensors can be realized by field-effect transistors (FETs, Section 6.3). For instance, the hole mobility in PAN nanofibers is found to decrease from 2.0 to 1.5 $cm^2\ V\ s^{-1}$ upon exposure to 10 ppm $NH_3$, which determines easily detectable variations in the transistor

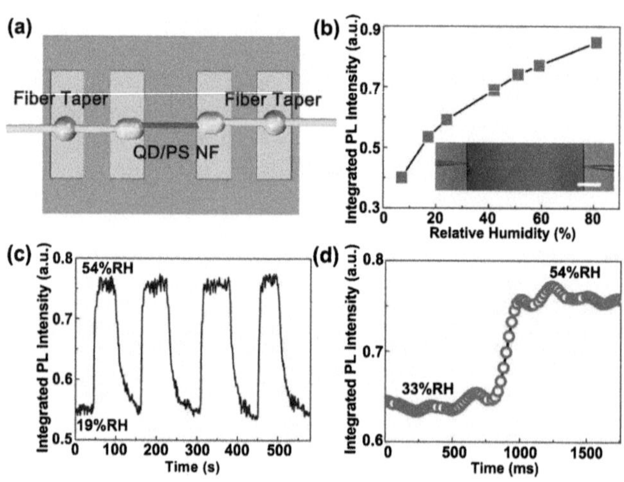


**Figure 4.15** (a) Scheme of a humidity sensor based on polymer nanofibers embedding CdSe/ZnS quantum dots. (b) Integrated photo-luminescence (PL) intensity of the fiber exposed to ambient relative humidity in the range from 7% to 81%. The inset shows an optical micrograph of the emitting fiber, which is excited by launching 532 nm light from one of the two fiber tapers, and whose emission is collected at the other side. Scale bar = 50 μm. (c) Sensor response to alternately cycled 54% and 19% relative humidity. (d) Zoomed response curve evidencing the fast response time (≅ 90 ms). Reproduced with permission from Ref. 161, *Adv. Mater.*, 2011, **23**, 3770–3774. Doi: 10.1002/adma.201101392. Copyright © 2011, Wiley-VCH Verlag GmbH & Co. KGaA.

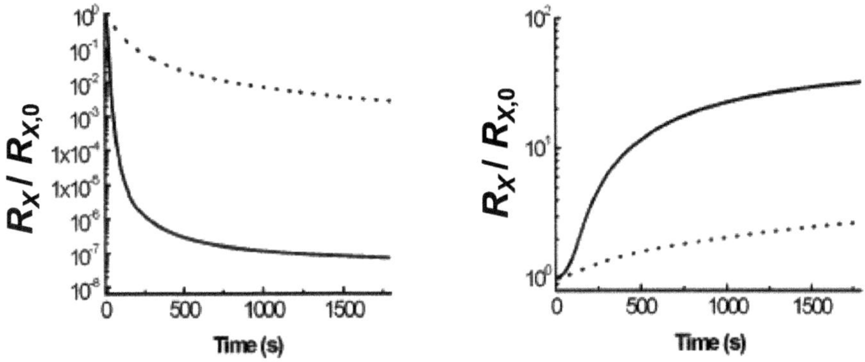


**Figure 4.16** Variation of the relative resistance ($R_X/R_{X,0}$) for an emeraldine base, nanofibrous film (solid line) and for a conventional (dotted line) film, following exposure to 100 ppm HCl vapor in nitrogen (left), and for the same samples after full HCl doping and following exposure to 100 ppm $NH_3$ vapor in nitrogen (right). $R_X \propto \chi^{-1}$. Adapted with permission from Ref. 163, *J. Am. Chem. Soc.*, 2003, **125**, 314–315. Doi: 10.1021/ja028371y. Copyright © 2003, American Chemical Society.

current (in the range of 10–20%).[166] An enhanced sensitivity has been measured for PEDOT:PSS nanowires fabricated by block copolymer assembly and RIE and tested by 325 ppm of ethanol vapors.[167] Composite, $TiO_2$/poly(vinyl acetate) (PVAC) electrospun mats after thermal treatments can result in single-crystal fibrils of titania in the anatase phase.[168] These samples are ideal building blocks of sensors for $NO_2$. The devices show a very high sensitivity, with an increase in the resistance of the nanofibers by more than 8 times when they are exposed to 500 ppb $NO_2$ at room temperature, which would correspond to a detection limit below 1 ppb.[168] Chemical sensors have also been developed by PEDOT nanotubes, and tested with 1000 ppm of various alcohols such as methanol and ethanol.[169] Also in these sensors the response following vapor exposure can be much faster than in thin-films, because of the easier diffusion of gases into the polymer semiconductor. For the same reason, the response of thinner nanofibers is normally faster than that of fibers with larger radius. A mathematical expression has been derived for the temporal behavior of the polymer conductivity ($\chi$), which takes into account the diffusion of molecules into a nanofiber, supposed to have a cylindrical cross-section.[170] Due to the chemical agent ($A$) to be detected and diffusing into the fiber (*i.e.* after the sensor has been exposed to this agent at time, $t = 0$), the $\chi$ variation can be described by:[170]

$$\chi(t)=\chi(t=0)\left[1-\frac{c_{q,A}}{c_B}\left(1-\sum_{n=1}^{\infty}\frac{4}{x_n^2}e^{-x_n^2Dt/r^2}\right)\right]. \tag{4.23}$$

In the previous expression, $c_{q,A}$ indicates the supplied concentration of the conductivity-quenching (namely dedoping) species generated by the diffusing chemical ($c_{q,A} = c_A$ is often assumed as a reasonable simplifying condition), $c_B$ is the initial concentration of the doping species, $D$ is the diffusion coefficient of the dedoping chemical into the cylindrical nanofiber and $x_n$ are solutions given by the equation, $J_0(x_n) = 0$, where $J_0$ is the Bessel function of the first kind of order zero.[170] Equation (4.23) helps to foresee physically expectable numbers for the characteristic times of conductivity variations once the diffusion coefficient is known. As an alternative, fitting experimental temporal curves of conductivity by Equation (4.23) can provide some insight into the diffusional phenomena, allowing to what extent the made assumptions (about the fiber cross-sectional shape, dedoping species, *etc.*) are valid to be understood, or to infer the $D$ values of each specific ($A$+nanofiber) system.[170]

### 4.3.3 Textiles

Dealing with textile applications of fibers, one becomes familiar with the *denier* or *tex* units, which indicate the linear mass density. The denier is the mass expressed in grams and contained in a length, $l$, of 9,000 meters of fiber, or of fiber yarn. For example, a 9,000 meters length of a *B. mori* silk filament with a diameter ($2r$) of about 13 μm has a mass ($\delta\pi r^2 l$, where the density, $\delta$, is about 1.2–1.3 g cm$^{-3}$ for materials such as PMMA, nylon and polycarbonate) around 1.5 g. Hence, this fiber measures 1.5 denier. The same length of a spider silk filament would be almost 14 times lighter, mainly due to the smaller fiber diameter ($2r \cong 3.5$ μm), and would measure about 0.1 denier. In this context, the term *microdenier* describes a filament whose mass is below 1 gram per 9,000 meter length, or, equivalently, a *microfiber* refers to a filament below 1 denier. According to our example above, a spider silk filament is therefore a microdenier fiber. A comparison between electrospun fibers (with diameter in the range from 10 nm to 1 μm)‡ and conventional textile fibers and nanotubes in terms of surface area (m$^2$ g$^{-1}$) and denier is shown on a logarithmic scale in Figure 4.17a.[107] The tex unit corresponds instead to a mass of 1 g per km. The *B. mori* silk filament above would measure about 0.17 tex, and the spider silk filament would measure slightly more than 0.01 text. In tex units, a microfiber would measure 0.11 tex or less. Conventionally, one often calls microfibers filaments below 0.1 tex (1 decitex) when talking in tex units. Artificial microdenier fibers for the textile industry have been commercialized since 1989 in the United States by Dupont.[171] Thin artificial fibers are typically realized by extrusion methods such as melt-spinning through very fine holes and by so-called "islands at sea" approaches. This technology produces multicomponent fibers where many thin melt filaments are embedded in another polymer fiber of larger radius, as in the cross-sectional scheme in the right inset of

‡As described in Chapter 3, several other technologies for the production of polymer nanofibers basically cover the same range as electrospinning in terms of fiber diameters.

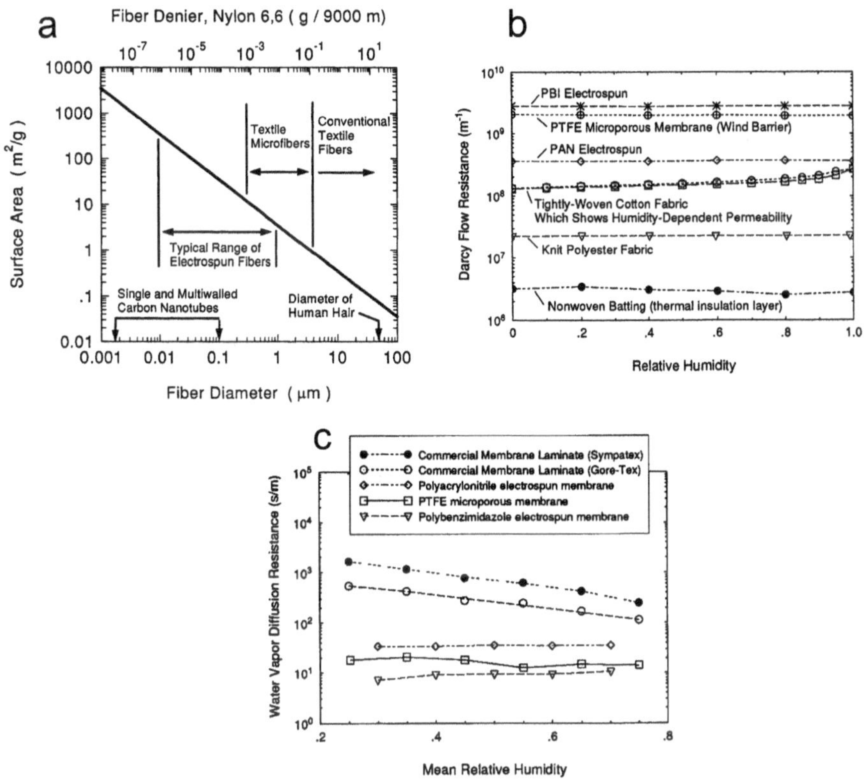


**Figure 4.17** (a) Logarithmic plot of the surface area *vs.* fibers diameter (bottom axis) and denier (top axis). Ranges typical of various classes of fibers are highlighted, including carbon nanotubes, electrospun fibers and conventional textile fibers. The linearly decreasing curve in the logarithmic scale clearly evidences the $r^{-1}$-dependence of fiber surface area. (b) Dependence on humidity of the Darcy flow resistance, $R_F$ (Equation (4.19)), for various textile fabrics and membranes. Hydrophilic natural fibers such as cotton show a humidity-dependent $R_F$. (c) Dependence on humidity of the water vapor diffusion resistance for various textile membranes including electrospun non-wovens. Adapted with permission from Ref. 107, *Colloids and Surfaces A: Physicochem. Eng. Aspects*, 2001, **187–188**, 469–481. Doi: 10.1016/S0927-7757(01)00616-1. Copyright © 2001, Elsevier Science B.V.

Figure 2.11b but with a high number (up to, say, $10^3$) of core fibers. Hills Inc. (www.hillsinc.net) is another company that is recently developing extruded fibers with diameters below 1 μm.

A lot of claims are frequently made in technical literature about the use of polymer nanofibers to confer new functionalities for smart clothing. Envisaged applications include thermal-insulation enhanced compared to conventional microfibers, integration of nanofiber-based plastic electronics or sensors, clothes finished by fluorescent features made by light-emitting polymer fibers, embedded piezoelectric nanofiber-based elements to harvest energy from body movements, higher resistance against wind and controllable wettability, photochromic, electrochromic, antibacterial or self-cleaning properties enabled by polymer nanofiber non-wovens eventually doped by specific functional molecules or nanoparticles.

In particular, the thermal insulation ability is closely related to the properties of the flow of gases or liquids through pores in assemblies of polymer fibers (Sections 4.3.1 and 4.3.2). Here, these aspects deserve an additional explanation. The microscale walls of a porous medium can dissipate heat, towards air or towards a liquid flowing through it, by means of the microscopic collisions with the particles of the fluid, which occur at molecular level. For example, more intense flows, due to a higher collisional rate of impinging particles, extract more energy from a heat-dissipating material, as anyone can experience on a hot summer day by increasing the rotational speed of a fan. However, when the flow has to pass through a porous network like that composed of the textile fibers of a cloth, another physical consideration has to be made. In order to transfer effectively energy to the surrounding fluid (say, air), after a collision with the solid walls of the porous medium each particle of the fluid has to *thermalize* (namely, cooling by repeated collisions with *other* particles of the fluid), before the subsequent collisions with the porous walls. Such a thermalization process due to thermal agitation and mutual collisions of particles is at the base of the normal thermal conductivity of air, which is about $6 \times 10^{-3}$ cal/(K m s) [*i.e.* $2.5 \times 10^{-2}$ W/(K m)] at room temperature. This means, in other words, that heat dissipation is effective for fluid transport regimes in which the mean free path, $\lambda_F$, of flowing molecules is significantly smaller than the distance travelled between two consecutive collisions with the walls of the porous medium (whose distance is practically given by the pore size), namely for small Knudsen numbers (Equation (4.20)). *Vice versa*, in the case of Knudsen diffusion, as that which can take place for gases flowing through polymer nanofiber materials and coatings, thermal insulation properties could be strongly enhanced, because the particles of the gas are not able to thermalize within their surrounding microenvironment of fluid molecules between consecutive collisions with porous walls. Of course, analogous considerations hold for thermal insulation against heat coming from an external environment at higher temperature than the body. Finally, another important property of textile fabrics is wind resistivity, which is clearly related to the Darcy flow resistance, $R_F$ (Equation (4.19)). To achieve high insulation against wind, $R_F$ should be as high as possible. In clothes it is also desirable to have a flow resistance almost independent from external environmental variables, particularly humidity. The flow resistance for different textile fabrics and membranes composed of either natural or synthetic fibers is shown in Figure 4.17b, which allows the higher $R_F$ values of electrospun non-wovens compared to other materials to be appreciated.[107] In addition, differently from hydrophilic natural fibers such as cotton, $R_F$ of PAN or poly(benzimidazole) (PBI) electrospun membranes is practically humidity independent. Also, electrospun non-wovens are surprisingly very "breathable", namely they show a resistance against water vapor diffusion that is comparatively lower than so-called breathable commercial membrane laminates (Figure 4.17c). The combination of wind resistivity and effective water vapor diffusion clearly makes electrospun membranes very promising candidates as coatings to produce protective cloths. Self-cleaning properties can also be obtained, by means of nanofiber-based super-hydrophobic or super-oleophobic surfaces such as those described in Section 4.2.2, photocatalytic schemes,[172] or both.

Yet, notwithstanding these advantages, research is still required in order to make functional polymer nanofibers fully compatible with the textile industry and existing cloth production methods and, of course, to reach the extremely large throughput needed for these applications. Nanofiber coatings have to adhere permanently or be stably anchored on textile fabrics, and resist repeated use, friction and washing. The investigation of the tribological properties of nanofibers surfaces is still in its infancy. In this regard, a pioneering study has highlighted that the wear resistance of these coatings

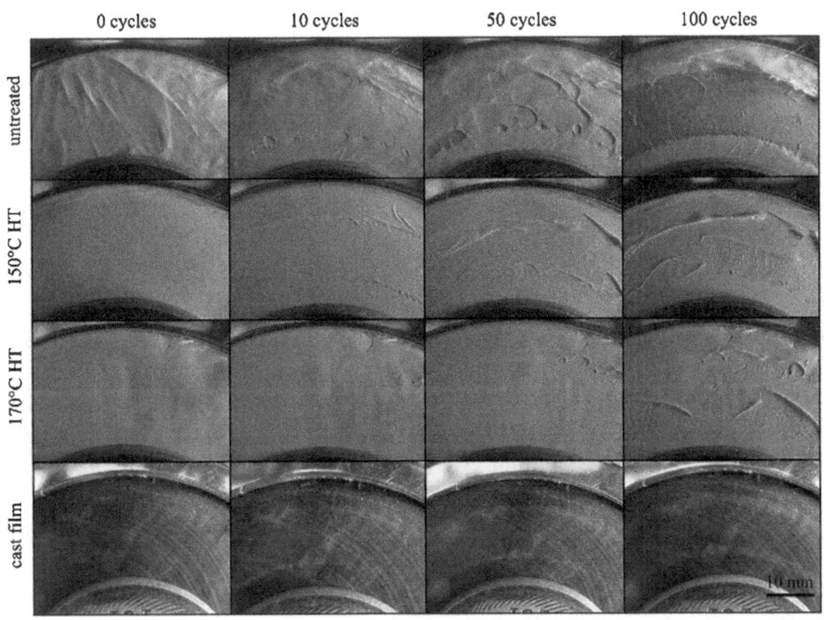


**Figure 4.18** Study of the abrasive wear paths for poly(trimethyl hexamethylene terephthalamide) nanofiber mats at 0, 10, 50 and 100 wear cycles under 100 g applied load. From top to bottom rows, shown samples are as-spun nanofibers, fibers heat-treated at 150 °C and at 170 °C and cast films on a brown Kapton® polyimide support. Scale bar = 1 cm. The polymer $T_g$ is about 150 °C. Reproduced with permission from Ref. 173, *Polymer*, 2012, **53**, 3017–3025. Doi: 10.1016/j.polymer.2012.04.039. Copyright © 2012, Elsevier Ltd.

improves significantly with increasing the temperature of post-deposition annealing, which is consistent with the formation of a continuous and more stable network due to joints between partially melt nanofibers (Figure 4.18).[173] In addition, concerning electrospinning, when aiming at depositing nanofibers directly onto textile substrates to realize functional coatings, one should carefully take into account the effect that non-conducting textiles, placed between the spinneret and the metal counterelectrode, can have on the stability of the electrified jet. These aspects have only recently begun to be investigated in the related literature.[174]

## References

1. A. A. Griffith, *Phil. Trans. R. Soc. Lond. A*, 1921, **221**, 189–235.
2. R. E. Miller and V. B. Shenoy, *Nanotechnology*, 2000, **11**, 139–147.
3. P. Sharma, S. Ganti and N. Bhate, *Appl. Phys. Lett.*, 2003, **82**, 535–537.
4. S. Cuenot, C. Frétigny, S. Demoustier-Champagne and B. Nysten, *Phys. Rev. B*, 2004, **69**, 165410.
5. Y. Wang, M. Rafailovich, J. Sokolov, D. Gersappe, T. Araki, Y. Zou, A. D. L. Kilcoyne, H. Ade, G. Marom and A. Lustiger, *Phys. Rev. Lett.*, 2006, **96**, 028303.
6. S. Buell, K. J. Van Vliet and G. C. Rutledge, *Macromolecules*, 2009, **42**, 4887–4895.
7. A. Arinstein, M. Burman, O. Gendelman and E. Zussman, *Nat. Nanotechnol.*, 2007, **2**, 59–62.
8. A. Arinstein and E. Zussman, *J. Polym. Sci., Part B: Polym. Phys.*, 2011, **49**, 691–707.
9. S. Shen, A. Henry, J. Tong, R. Zheng and G. Chen, *Nat. Nanotechnol.*, 2010, **5**, 251–255.
10. M. Naraghi, I. Chasiotis, H. Kahn, Y. Wen and Y. Dzenis, *Rev. Sci. Instrum.*, 2007, **78**, 085108.
11. M. Naraghi and I. Chasiotis, in *Major Accomplishments in Composite Materials and Sandwich Structures: An Anthology of ONR Sponsored Research*, ed. I. M. Daniel, E. E. Gdoutos and Y. D. S. Rajapakse, Springer, Dordrecht, 2009.
12. N. T. Hiep and B.-T. Lee, *J. Mater. Sci.: Mater. Med.*, 2010, **21**, 1969–1978.
13. S. Y. Chew, T. C. Hufnagel, C. T. Lim and K. W. Leong, *Nanotechnology*, 2006, **17**, 3880–3891.
14. S.-C. Wong, A. Baji and S. Leng, *Polymer*, 2008, **49**, 4713–4722.
15. C.-L. Pai, M. C. Boyce and G. C. Rutledge, *Polymer*, 2011, **52**, 2295–2301.
16. Y. Ji, B. Li, S. Ge, J. C. Sokolov and M. H. Rafailovich, *Langmuir*, 2006, **22**, 1321–1328.
17. M. Burman, A. Arinstein and E. Zussman, *Appl. Phys. Lett.*, 2008, **93**, 193118.

18. M. K. Shin, S. I. Kim, S. J. Kim, S.-K. Kim, H. Lee and G. M. Spinks, *Appl. Phys. Lett.*, 2006, **89**, 231929.
19. F. M. Blighe, K. Young, J. J. Vilatela, A. H. Windle, I. A. Kinloch, L. Deng, R. J. Young and J. N. Coleman, *Adv. Funct. Mater.*, 2011, **21**, 364–371.
20. Y. You, S. Won Lee, S. Jin Lee and W. Ho Park, *Mater. Lett.*, 2006, **60**, 1331–1333.
21. S.-S. Choi, S. G. Lee, C. W. Joo, S. S. Im and S. H. Kim, *J. Mater. Sci.*, 2004, **39**, 1511–1513.
22. S. Jin Lee, S. Heang Oh, J. Liu, S. Soker, A. Atala and J. J. Yoo, *Biomaterials*, 2008, **29**, 1422–1430.
23. H. Hertz, *J. Reine Angew. Math.*, 1882, **92**, 156–171.
24. K. R. Shull, *Mater. Sci. Eng. R*, 2002, **36**, 1–45.
25. S. Tan, R. L. Sherman Jr. and W. T. Ford, *Langmuir*, 2004, **20**, 7015–7020.
26. D. Pisignano, G. Maruccio, E. Mele, L. Persano, F. Di Benedetto and R. Cingolani, *Appl. Phys. Lett.*, 2005, **87**, 123109.
27. C. P. Broedersz, X. Mao, T. C. Lubensky and F. C. MacKintosh, *Nat. Phys.*, 2011, **7**, 983–988.
28. E. van der Giessen, *Nat. Phys.*, 2011, **7**, 923–924.
29. M. N. Silberstein, C.-L. Pai, G. C. Rutledge and M. C. Boyce, *J. Mech. Phys. Solids*, 2012, **60**, 295–318.
30. C. S. Reddy, A. Arinstein, R. Avrahami and E. Zussman, *J. Mater. Chem.*, 2009, **19**, 7198–7201.
31. D. Li and Y. Xia, *Nano Lett.*, 2004, **4**, 933–938.
32. M. M. Bergshoef and G. J. Vancso, *Adv. Mater.*, 1999, **11**, 1362–1365.
33. J.-S. Kim and D. H. Reneker, *Polym. Compos.*, 1999, **20**, 124–131.
34. H. Fong, *Polymer*, 2004, **45**, 2427–2432.
35. M. Tian, Y. Gao, Y. Liu, Y. Liao, R. Xu, N. E. Hedin and H. Fong, *Polymer*, 2007, **48**, 2720–2728.
36. S. Lin, Q. Cai, J. Ji, G. Sui, Y. Yu, X. Yang, Q. Mab, Y. Wei and X. Deng, *Compos. Sci. Technol.*, 2008, **68**, 3322–3329.
37. Y. Hong, K. Fujimoto, R. Hashizume, G. Ryotaro, S. Jianjun, J. John, K. Tobota and W. R. Wagner, *Biomacromolecules*, 2008, **9**, 1200–1207.
38. S. Kidoaki, I. K. Kwon and T. Matsuda, *Biomaterials*, 2005, **26**, 37–46.
39. B. Duan, X. Y. Yuan, Y. Zhu, Y. Y. Zhang, X. L. Li, Y. Zhang and K. D. Yao, *Eur. Polym. J.*, 2006, **42**, 2013–2022.
40. B. Ding, E. Kimura, T. Sato, S. Fujita and S. Shiratori, *Polymer*, 2004, **45**, 1895–1902.
41. W.-H. Zhong, J. Li, L. R. Xu, J. A. Michel, L. M. Sullivan and C. M. Lukehart, *J. Nanosci. Nanotechnol.*, 2004, **4**, 794–802.
42. L. R. Xu, L. Li, C. M. Lukehart and H. Kuai, *J. Nanosci. Nanotechnol.*, 2007, **7**, 1–3.
43. M. P. Manoharan, A. Sharma, A. V. Desai, M. A. Haque, C. E. Bakis and K. W. Wang, *Nanotechnology*, 2009, **20**, 295701.

44. A. J. Rodriguez, M. E. Guzman, C.-S. Lim and B. Minaie, *Carbon*, 2011, **49**, 937–948.
45. S. K. Nataraj, K. S. Yang and T. M. Aminabhavi, *Prog. Polym. Sci.*, 2012, **37**, 487–513.
46. C. H. Kim, B.-H. Kim and K. S. Yang, *Carbon*, 2012, **50**, 2472–2481.
47. J. J. Santiago-Avilés and Y. Wang, *Carbon*, 2003, **1**, 2653–2689.
48. Y. Wang and J. J. Santiago-Avilés, *Appl. Phys. Lett.*, 2006, **89**, 123119.
49. F. Agend, N. Naderi and R. Fareghi-Alamdari, *J. Appl. Polym. Sci.*, 2007, **106**, 255–259.
50. C. Kim and K. S. Yang, *Appl. Phys. Lett.*, 2003, **83**, 1216–1218.
51. E. J. Ra, E. Raymundo-Piñero, Y. H. Lee and F. Béguin, *Carbon*, 2009, **47**, 2984–2992.
52. C. Kim, B. T. N. Ngoc, K. S. Yang, M. Kojima, Y. A. Kim, Y. J. Kim, M. Endo and S. C. Yang, *Adv. Mater.*, 2007, **19**, 2341–2346.
53. X. Song, Z. Wang, Z. Li and C. Wang, *J. Colloid Interf. Sci.*, 2008, **327**, 388–392.
54. W. Lu, M. Zu, J.-H. Byun, B.-S. Kim and T.-W. Chou, *Adv. Mater.*, 2012, **24**, 1805–1833.
55. P. Li, T.-J. Zhao, J.-H. Zhou, Z.-J. Sui, Y.-C. Dai and W.-K. Yuan, *Carbon*, 2005, **43**, 2701–2710.
56. M. H. Al-Saleh and U. Sundararaj, *Carbon*, 2009, **47**, 2–22.
57. N. Roy, R. Sengupta and A. K. Bhowmick, *Prog. Polym. Sci.*, 2012, **37**, 781–819.
58. M. Inagaki, Y. Yang and F. Kang, *Adv. Mater.*, 2012, **24**, 2547–2566.
59. H. Hou and D. H. Reneker, *Adv. Mater.*, 2004, **16**, 69–73.
60. S. Y. Gu, J. Ren and G. J. Vancso, *Eur. Polym. J.*, 2005, **41**, 2559–2568.
61. J. Liu, Z. Yue and H. Fong, *Small*, 2009, **5**, 536–542.
62. F. Ko, Y. Gogotsi, A. Ali, N. Naguib, H. Ye, G. Yang, C. Li and P. Willis, *Adv. Mater.*, 2003, **15**, 1161–1165.
63. E. Zussman, X. Chen, W. Ding, L. Calabri, D. A. Dikin, J. P. Quintana and R. S. Ruoff, *Carbon*, 2005, **43**, 2175–2185.
64. C. Kim, Y. I. Jeong, B. T. N. Ngoc, K. S. Yang, M. Kojima, Y. A. Kim, M. Endo and J.-W. Lee, *Small*, 2007, **3**, 91–95.
65. D. T. Welna, J. D. Bender, X. Wei, L. G. Sneddon and H. R. Allcock, *Adv. Mater.*, 2005, **17**, 859–862.
66. P. X. Ma and R. Zhang, *J. Biomed. Mater. Res.*, 1999, **46**, 60–72.
67. A. Laforgue and L. Robitaille, *Chem. Mater.*, 2010, **22**, 2474–2480.
68. M. M. Tomadakis and T. J. Robertson, *J. Compos. Mater.*, 2005, **39**, 163–188.
69. M. M. Tomadakis and S. V. Sotirchos, *AlCHE J.*, 1991, **37**, 74–86.
70. M. M. Tomadakis and S. V. Sotirchos, *J. Chem. Phys.*, 1993, **99**, 9820–9827.
71. D. Hussain, F. Loyal, A. Greiner and J. H. Wendorff, *Polymer*, 2010, **51**, 3989–3997.
72. A. Jena and K. Gupta, *Int. Nonwovens J.*, 2005, **14**, 25–30.

73. D. Li, M. W. Frey and Y. L. Joo, *J. Membr. Sci.*, 2006, **286**, 104–114.
74. L. Ghasemi-Mobarakeh, M. P. Prabhakaran, M. Morshed, M. H. Nasr-Esfahani and S. Ramakrishna, *Biomaterials*, 2008, **29**, 4532–4539.
75. C.-L. Pai, M. C. Boyce and G. C. Rutledge, *Macromolecules*, 2009, **42**, 2102–2114.
76. P. K. Bhattacharjee and G. C. Rutledge, in *Comprehensive Biomaterials*, ed. P. Ducheyne, K. E. Healy, D. W. Hutmacher, D. W. Grainger and C. J. Kirkpatrick, Elsevier, Oxford, 2011.
77. Y. Z. Zhang, Y. Feng, Z.-M. Huang, S. Ramakrishna and C. T. Lim, *Nanotechnology*, 2006, **17**, 901–908.
78. J. T. McCann, M. Marquez and Y. Xia, *J. Am. Chem. Soc.*, 2006, **128**, 1436–1437.
79. M. Bognitzki, W. Czado, T. Frese, A. Schaper, M. Hellwig, M. Steinhart, A. Greiner and J. H. Wendorff, *Adv. Mater.*, 2001, **13**, 70–72.
80. S. Megelski, J. S. Stephens, D. B. Chase and J. F. Rabolt, *Macromolecules*, 2002, **35**, 8456–8466.
81. C. L. Casper, J. S. Stephens, N. G. Tassi, D. B. Chase and J. F. Rabolt, *Macromolecules*, 2004, **37**, 573–578.
82. L. Jiang, Y. Zhao and J. Zhai, *Angew. Chem. Int. Ed.*, 2004, **43**, 4338–4341.
83. T. Young, *Phil. Trans. R. Soc. Lond.*, 1805, **95**, 65–87.
84. P. Roura and J. Fort, *J. Colloid Interf. Sci.*, 2004, **272**, 420–429.
85. R. N. Wenzel, *Ind. Eng. Chem.*, 1936, **28**, 988–994.
86. A. B. D. Cassie and S. Baxter, *Trans. Faraday. Soc.*, 1944, **40**, 546–551.
87. V. Zorba, L. Persano, D. Pisignano, A. Athanassiou, E. Stratakis, R. Cingolani, P. Tzanetakis and C. Fotakis, *Nanotechnology*, 2006, **17**, 3234–3238.
88. A. Tuteja, W. Choi, M. Ma, J. M. Mabry, S. A. Mazzella, G. C. Rutledge, G. H. McKinley and R. E. Cohen, *Science*, 2007, **318**, 1618–1622.
89. A. Lembach, H. B. Tan, I. V. Roisman, T. Gambaryan-Roisman, Y. Zhang, C. Tropea and A. L. Yarin, *Langmuir*, 2010, **26**, 9516–9523.
90. R. P. Sahu, S. Sinha-Ray, A. L. Yarin and B. Pourdeyhimi, *Soft Matter*, 2012, **8**, 3957–3970.
91. G. McHale and M. I. Newton, *Colloids Surf. A*, 2002, **206**, 79–86.
92. G. O. Berim and E. Ruckenstein, *J. Phys. Chem. B*, 2005, **109**, 12515–12524.
93. Y. Zheng, H. Bai, Z. Huang, X. Tian, F.-Q. Nie, Y. Zhao, J. Zhai and L. Jiang, *Nature*, **463**, 640–643.
94. C. Duprat, S. Protière, A. Y. Beebe and H. A. Stone, *Nature*, 2012, **482**, 510–513.
95. H. Bouas-Laurent and H. Dürr, *Pure Appl. Chem.*, 2001, **73**, 639–665.
96. M.-M. Russew and S. Hecht, *Adv. Mater.*, 2010, **22**, 3348–3360.
97. F. Di Benedetto, E. Mele, A. Camposeo, A. Athanassiou, R. Cingolani and D. Pisignano, *Adv. Mater.*, 2008, **20**, 314–318.

98. R. S. Barhate, C. K. Loong and S. Ramakrishna, *J. Membr. Sci.*, 2006, **283**, 209–218.
99. R. S. Barhate and S. Ramakrishna, *J. Membr. Sci.*, 2007, **296**, 1–8.
100. B. F. Sadovsky, in *Aerosols: Science and Technology*, ed. I. Agranovski, Wiley-VCH Verlag GmbH & Co. KGaA, Weinheim, 2010.
101. http://www.electrospinning.ru/histor-en.html
102. Y. Filatov, A. Budyka and V. Kirichenko, *Electrospinning of Micro- and Nanofibers: Fundamentals and Applications in Separation and Filtration Processes*, Begell House, Inc., New York, 2007.
103. D. Groitzsch and E. Fahrbach, *US Pat.* 4,618,524, 1986.
104. T. H. Grafe and K. M. Graham, presented at *Nonwovens in Filtration – Fifth International Conference*, 2003.
105. DOE Technical Standard. Specification for HEPA Filters Used by DOE Contractors DOE-STD-3020-2005, US Department of Energy, Washington, D.C. 20585, 2005.
106. EN 1822:2009.
107. P. Gibson, H. Schreuder-Gibson and D. Rivin, *Colloids and Surfaces A: Physicochem. Eng. Aspects*, 2001, **187–188**, 469–481.
108. T. M. Squires and S. R. Quake, *Rev. Mod. Phys.*, 2005, **77**, 977–1026.
109. V. N. Burganos and S. V. Sotirchos, *Chem. Eng. Sci.*, 1989, **44**, 2451–2462.
110. T. H. Grafe, M. Gogins, M. Barris, J. Schaefer and R. Canepa, presented at *Filtration 2001 International Conference and Exposition of the INDA (Association of the Nowovens Fabric Industry)*, 2001.
111. B. Maze, H. Vahedi Tafreshi, Q. Wang and B. Pourdeyhimi, *J. Aerosol. Sci.*, 2007, **38**, 550–571.
112. C. Shin and G. G. Chase, *AIChE J.*, 2004, **50**, 343–350.
113. C. Shin, G. G. Chase and D. H. Reneker, *AIChE J.*, 2005, **51**, 3109–3113.
114. http://www.milliken.com/en-us/pands/Documents/Eminus%20US.pdf (accessed November 2012).
115. S. Ramakrishna, R. Jose, P. S. Archana, A. S. Nair, R. Balamurugan, J. Venugopal and W. E. Teo, *J. Mater. Sci.*, 2010, **45**, 6283–6312.
116. S. Kaur, S. Sundarrajan, R. Gopal and S. Ramakrishna, *J. Appl. Polym. Sci.*, 2012, **124**, E205–E215.
117. T. J. Menkhaus, L. Zhang and H. Fong, *Applications of Electrospun Nanofiber Membranes for Bioseparations*, Nova Science Publishers Inc., New York, 2010.
118. L. Matlock-Colangelo, D. Cho, C. L. Pitner, M. W. Frey and A. J. Baeumner, *Lab Chip*, 2012, **12**, 1696–1701.
119. S. Klein, J. Kuhn, R. Avrahami, S. Tarre, M. Beliavski, M. Green and E. Zussman, *Biomacromolecules*, 2009, **10**, 1751–1756.
120. A. Nardi, R. Avrahami, E. Zussman, J. S. Rokem and C. L. Greenblatt, *J. Environm. Protect.*, 2012, **3**, 164–168.
121. C. W. Lim and I. S. Lee, *Nano Today*, 2010, **5**, 412–434.

122. S. Shylesh, V. Schünemann and W. R. Thiel, *Angew. Chem. Int. Ed.*, 2010, **49**, 3428–3459.
123. C.-J. Jia and F. Schüth, *Phys. Chem. Chem. Phys.*, 2011, **13**, 2457–2487.
124. M. Stasiak, C. Röben, N. Rosenberger, F. Schleth, A. Studer, A. Greiner and J. H. Wendorff, *Polymer*, 2007, **48**, 5208–5218.
125. M. Winter and R. J. Brodd, *Chem. Rev.*, 2004, **104**, 4245–4269.
126. S. Il Cho and S. Bok Lee, *Acc. Chem. Res.*, 2008, **41**, 699–707.
127. B. C. H. Steele and A. Heinzel, *Nature*, 2001, **414**, 345–352.
128. T. Tamura and H. Kawakami, *Nano Lett.*, 2010, **10**, 1324–1328.
129. Z. Kurban, A. Lovell, S. M. Bennington, D. W. K. Jenkins, K. R. Ryan, M. O. Jones, N. T. Skipper and W. I. F. David, *J. Phys. Chem. C*, 2010, **114**, 21201–21213.
130. J. R. Kim, S. W. Choi, S. M. Jo, W. S. Lee and B. C. Kim, *Electrochim. Acta*, 2004, **50**, 69–75.
131. S. S. Choi, Y. S. Lee, C. W. Joo, S. G. Lee, J. K. Park and K.-S. Han, *Electrochim. Acta*, 2004, **50**, 339–343.
132. J. R. Kim, S. W. Choi, S. M. Jo, W. S. Lee and B. C. Kim, *J. Electr. Chem.*, 2005, **152**, A295–A300.
133. http://www2.dupont.com/energain/en-us/eg/ua/battery-separators.html (accessed November 2012).
134. J. Kim, S. I. Kim and K.-H. Yoo, *Biosens. Bioelectron.*, 2009, **25**, 350–355.
135. Q. Wu, Y. Xu, Z. Yao, A. Liu and G. Shi, *ACS Nano*, 2010, **4**, 1963–1970.
136. S. Biswas and L. T. Drzal, *Chem. Mater.*, 2010, **22**, 5667–5671.
137. K. Zhang, L. Li Zhang, X. S. Zhao and J. Wu, *Chem. Mater.*, 2010, **22**, 1392–1401.
138. B. K. Kuila, B. Nandan, M. Böhme, A. Janke and M. Stamm, *Chem. Commun.*, 2009, 5749–5751.
139. A. V. Bazilevsky, A. L. Yarin and C. M. Megaridis, *Lab Chip*, 2008, **8**, 152–160.
140. R. Srikar, A. L. Yarin and C. M. Megaridis, *Nanotechnology*, 2009, **20**, 275706.
141. S. Sinha-Ray and A. L. Yarin, *J. Appl. Phys.*, 2010, **107**, 0294903.
142. R. Srikar, T. Gambaryan-Roisman, C. Steffes, P. Stephan, C. Tropea and A. L. Yarin, *Int. J. Heat Mass Transf.*, 2009, **52**, 5814–5826.
143. C. L. Choy, Y. W. Wong, G. W. Yang and T. Kanamoto, *J. Polym. Sci. Part B: Polym. Phys.*, 1999, **37**, 3359–3367.
144. A. Henry and G. Chen, *Phys. Rev. Lett.*, 2008, **101**, 235502.
145. J. X. Huang and R. B. Kaner, *Nat. Mater.*, 2004, **3**, 783–786.
146. D. Li and Y. Xia, *Nat. Mater.*, 2004, **3**, 753–754.
147. C. O. Baker, B. Shedd, P. C. Innis, P. G. Whitten, G. M. Spinks, G. C. Wallace and R. B. Kaner, *Adv. Mater.*, 2008, **20**, 155–158.
148. X. Wang, C. Drew, S.-H. Lee, K. J. Senecal, J. Kumar and L. A. Samuelson, *Nano Lett.*, 2002, **2**, 1273–1275.

149. W.-E. Lee, C.-J. Oh, I.-K. Kang and G. Kwak, *Macromol. Chem. Phys.*, 2010, **211**, 1900–1908.
150. Y. Long, H. Chen, Y. Yang, H. Wang, Y. Yang, N. Li, K. Li, J. Pei and F. Liu, *Macromolecules*, 2009, **42**, 6501–6509.
151. X. Wang, Y.-G. Kim, C. Drew, B.-C. Ku, J. Kumar and L. A. Samuelson, *Nano Lett.*, 2004, **4**, 331–334.
152. B. W. Davis, N. Niamnont, R. Dillon, C. J. Bardeen, M. Sukwattanasinitt and Q. Cheng, *Langmuir*, 2011, **27**, 6401–6408.
153. B. W. Davis, N. Niamnont, C. D. Hare, M. Sukwattanasinitt and Q. Cheng, *ACS Appl. Mater. Interfaces*, 2010, **2**, 1798–1803.
154. J. R. Lakowicz, *Principles of Fluorescence Spectroscopy*, Springer, New York, 3rd edn, 2006.
155. P. Anzenbacher Jr., F. Li and M. A. Palacios, *Angew. Chem. Int. Ed.*, 2012, **51**, 2345–2348.
156. C.-C. Kuo, C. T. Wang and W.-C. Chen, *Macromol. Mater. Eng.*, 2008, **293**, 999–1008.
157. S. Kyun Chae, H. Park, J. Yoon, C. Hee Lee, D. June Ahn and J.-M. Kim, *Adv. Mater.*, 2007, **19**, 521–524.
158. J. Yoon, S. K. Chae and J.-M. Kim, *J. Am. Chem. Soc.*, 2007, **129**, 3038–3039.
159. J. Yoon, Y.-S. Jung and J.-M. Kim, *Adv. Funct. Mater.*, 2009, **19**, 209–214.
160. J. Lee, S. Balakrishnan, J. Cho, S.-H. Jeon and J.-M. Kim, *J. Mater. Chem.*, 2011, **21**, 2648–2655.
161. C. Meng, Y. Xiao, P. Wang, L. Zhang, Y. Liu and L. Tong, *Adv. Mater.*, 2011, **23**, 3770–3774.
162. F. Gu, L. Zhang, X. Yin and L. M. Tong, *Nano Lett.*, 2008, **8**, 2757–2761.
163. J. Huang, S. Virji, B. H. Weiller and R. B. Kaner, *J. Am. Chem. Soc.*, 2003, **125**, 314–315.
164. J. Huang, S. Virji, B. H. Weiller and R. B. Kaner, *Chem. Eur. J.*, 2004, **10**, 1314–1319.
165. B. Dong, D. Zhong, L. Chi and H. Fuchs, *Adv. Mater.*, 2005, **17**, 2736–2741.
166. D. Chen, S. Lei and Y. Chen, *Sensors*, 2011, **11**, 6509–6516.
167. Y. S. Jung, W. C. Jung, H. L. Tuller and C. A. Ross, *Nano Lett.*, 2008, **8**, 3776–3780.
168. I.-D. Kim, A. Rothschild, B. Hong Lee, D. Young Kim, S. Mu Jo and H. L. Tuller, *Nano Lett.*, 2006, **6**, 2009–2013.
169. H. Yoon, M. Chang and J. Jang, *Adv. Funct. Mater.*, 2007, **17**, 431–436.
170. H. Liu, J. Kameoka, D. A. Czaplewski and H. G. Craighead, *Nano Lett.*, 2004, **4**, 671–675.
171. J. A. Kent (ed.), *Kent and Riegel's Handbook of Industrial Chemistry and Biotechnology*, Springer Science+Business Media, LLC, New York, 11th edn, 2007.

172. N. M. Bedford and A. J. Steckl, *Appl. Mater. Interfaces*, 2010, **2**, 2448–2455.
173. M. M. Mannarino and G. C. Rutledge, *Polymer*, 2012, **53**, 3017–3025.
174. A. Varesano, F. Rombaldoni, G. Mazzuchetti, C. Tonina and R. Comotto, *Polym. Int.*, 2010, **59**, 1606–1615.

CHAPTER 5

# *Optical Properties of Polymer Nanofibers and Their Applications*

## 5.1 Organic Light-emitting Materials

Since the discovery of electroluminescence in PPV,[1] tremendous progress has been made in the field of organic-based, ultimately flexible optoelectronics. Research has allowed the unique properties of processability, spectral tunability and chemical variety of light-emitting conjugated materials within organic light-emitting diodes (OLEDs) and displays to be fully exploited. These devices are nowadays largely commercialized within cell- and smartphones, notebooks, color screens, *etc.*

The fundamental molecular properties at the base of such extraordinary scientific and technological success are quite simple to sketch, although the physico-chemical scenario affecting the optical properties is made complex by the molecular disorder and by the interplay of various electronic excitations. The latter have been, and are still today, the subject of huge research efforts. As in most soft matter (Section 1.1), in organic light-emitting materials inter-molecular interactions are mainly determined by relatively weak, non-covalent (*i.e.* Van der Waals, hydrogen) bonds. This feature in turn results in a frequently small overlap between orbitals belonging to different molecules. Consequently, the optoelectronic properties of conjugated materials in the solid state remain to a large extent (though not completely as explained in the following) related to the characteristics of individual molecules. The structural disorder shown by most organic semiconductors in the solid state further limits inter-molecular coupling

RSC Nanoscience & Nanotechnology No. 29
Polymer Nanofibers: Building Blocks for Nanotechnology
By Dario Pisignano

Published by the Royal Society of Chemistry, www.rsc.org

and can also contribute, together with the relatively low material dielectric constant (Section 1.1), to localization of excitons. This means that, instead of using excitonic confinement and quantum effects as in inorganic semiconductors, the strategy to achieve a fine control on the electronic band-gap structure of organic light-emitting materials passes through chemical synthesis and molecular engineering. The design of conjugation length, functional groups, *etc.* is aimed to fabricate molecules emitting in the desired spectral region, with high luminescence efficiency and controllable supramolecular assembly properties.

In this framework, one-dimensional nanostructures, such as polymer nanofibers, could be strategic for developing the next generation of optoelectronic devices based on organic semiconductors and for pushing miniaturization of these components. Indeed, being a practical way to better control and ultimately drive the supramolecular assembly and the orientation of polymer backbones, fiber-shaped nanostructures allow one to play directly with physical mechanisms at the base of optoelectronic properties of conjugated materials. Expectable achievements that are described in this chapter include well-defined polarization of emitted light, low threshold for optical gain, measurable effects on energy transfer and migration phenomena, and so on. The effects on charge-carrier transport properties will be described in detail in the next chapter.

### 5.1.1 Excitations and Electronic Transitions

In most cases Frenkel excitons (Section 1.1) properly describe elementary excitations in conjugated materials. These excitons are almost localized on single polymeric chains,[2] and their binding energy is quite large, often 0.5 eV or above. However, the proximity between molecules in the solid state makes possible the formation of other excited species, in which electrons can be transferred from one molecule to another nearby, thus forming a charge-transfer state. Of course, such processes are disfavored by a low intermolecular coupling. For these reasons, molecular aggregation and supramolecular packing states, as induced by solution processing or by nanostructure fabrication methods, play a pivotal role in affecting the optical spectra.

The presence of a charge at a molecular site promptly induces a localized, structural deformation over a number of atoms in the surrounding molecular environment. The related arrangement and relaxation processes occur on timescales of, say, $10^{-13}$ s.[3] This is what a physicist would call a *polarization field*. The resulting physical system, thereby composed of the charge-carrier (electron or hole) and the neighbor, structurally deformed organic chains is mobile and is termed *polaron* ($P^-$ or $P^+$, respectively). In other words, a $P^-$ ($P^+$) polaron species has spin 1/2 and is primarily generated by an electron (hole),[4] coming from a donor molecule, from an electrical injection event or from exciton-dissociation processes following photo-excitation. The energy levels of this configuration vary from the pristine state, due to the presence of inter-gap levels pulled out respectively from the conduction (for $P^-$ species) and from the valence (for $P^+$ species) bands of the organic compound. How does a polaron behave and interact with other excited species in the organic material? For instance, two oppositely charged polarons can mutually interact Coulombically and, if their attractive interaction is strong enough (*i.e.* with characteristic energy larger than the thermal energy, $k_BT$), a stably coupled, excited complex can be formed, which is a *polaron pair* or a *charge-transfer exciton*. The dynamics of charge transfer excitons is particularly important for organic photovoltaic devices, especially when the electron and the hole are localized across the interface between two materials.[5]

In general, an excitation of the organic compound, such as that induced by an absorbed photon, causes an electronic transition from the highest occupied molecular orbital (HOMO) to the lowest unoccupied molecular orbital (LUMO) level*,†. Large intra-molecular electronic delocalizations, *i.e.* spatially broadened orbitals, generally occur with low degrees of molecular torsion and determine longer conjugation lengths, reduced HOMO-LUMO gaps and red-shifted optical spectra. These concepts have great relevance for analyzing the absorption and emission properties of polymer nanostructures, including active polymer nanofibers, in which molecular conformations at nanoscale can vary considerably compared to bulk materials.

Excitons may be in general in singlet or triplet states, depending on the total spin of the electron-hole pair. When generated by optical absorption, the excitons are initially in the singlet state. The most practical way to schematize the various optical transitions involving singlet excitons is plotting the system energy *vs.* a generalized nuclear configuration coordinate, $X$, as in Figure 5.1. $X$ tells us if the molecular skeleton composed of nuclei is distorted and moves during a transition or a relaxation process. In very fast optical transitions that connect different levels ($S_n$), the nuclei, much heavier than electrons, are not able to move as rapidly as is needed to follow the new electronic configuration. The underlying concept, the so-called adiabatic approximation, is one of the most useful and successful in solid-state physics and quantum chemistry, and photon absorption and emission processes are hence represented by vertical arrows in Figure 5.1 (Franck–Condon transitions). If the vibronic gap (*i.e.* the energy difference between two consecutive horizontal vibrational levels in the ground electronic sheet, $S_0$) is significantly larger than the available thermal energy, before excitation the molecule lies in its electro-vibrational ground state. Upon optical absorption, the system is promoted to the lowest-energy or to an excited vibrational level in the excited electronic sheet (say, $S_1$), which produces an overall absorption spectrum composed of sub-bands (replicas) peaked at different energies, depending on the specific energy gap involved.

---

*In principle the HOMO and LUMO notations refer to one-electron energy levels, disregarding the electron-hole Coulombic interaction, whereas in practice they are often used as synonyms of valence and conduction band, respectively.

†In analogy with the band theory of crystalline solids, the differences between the vacuum level and the HOMO and the LUMO energy level are called *ionization potential* and *electron affinity*, respectively.

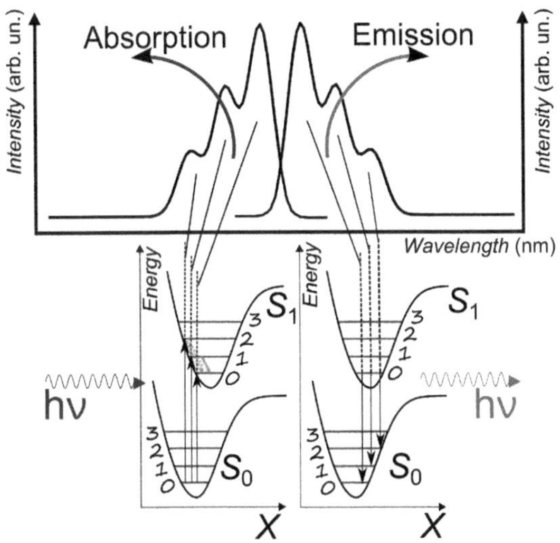


**Figure 5.1** Scheme showing the vibronic progressions in absorption and emission spectra of a conjugated molecule. Here the ground and the first excited electronic sheets are involved in the transitions. Both Franck–Condon (vertical) excitations and de-excitations connecting $S_0$ and $S_1$ and vibrational decays dissipating energy within $S_1$ (small, diagonal arrows in the left $S_1$ scheme) are displayed.

This typically featured shape of absorption spectra in conjugated polymers is called *vibronic progression.*

Once excited to, for instance, the fourth vibrational level internal to the electronic sheet $S_1$, the system firstly loses a part of its energy by relaxing vibrationally, *i.e.* by decaying non-radiatively towards the lowest-energy vibrational state within $S_1$. This relaxation involves the activation of nuclear motions (the transition is not vertical in Figure 5.1) namely of phonons. The associated relaxation process is called *internal conversion*. Afterwards, the system decays radiatively (vertically in Figure 5.1) to the ground electronic level $S_0$, and it again can land on different vibrational levels, which ultimately determines the release of photons of different energy, depending on the specific energy gap‡. This effect generates a vibronic progression also in the emission spectrum. In summary, sequential vibronic replicas are highlighted by spectral peaks at transitions between radiatively coupled states that are characterized by different vibrational levels. In addition, the vibrationally dissipated energy leads to a correspondingly decreased energy of the emission peak with respect to absorption (*Stokes shift*). The Stokes shift has a lot of implications for the experimental investigation and use of light-emitting polymer nanostructures and nanofibers. First of all, the spectral separation between absorption and

‡In many molecular systems, other transitions (*i.e.* singlet-triplet) can also occur, including *intersystem crossing* and radiative singlet-triplet coupling (phosphorescence).

emission can be exploited to filter the excitation photons, thereby isolating the light emitted by the polymer sample. This mechanism is ubiquitously applied in optical microscopy methods for biology and materials sciences, by means of long-wavelength pass filters positioned along the path of collected light. Another motivation for preferring light-emitting polymer systems that have larger Stokes shifts is their reduced self-absorption of emitted photons, because of the smaller overlap between emission and absorption spectra. As better described in the following, this leads to lower optical losses in active polymer nanofiber waveguides (Section 5.4) and in organic solid-state optical amplifiers, in which lasing action can therefore be more easily achieved (Section 5.5). Finally, this would also result in higher extraction efficiencies for photons emitted by electroluminescent devices.

In looking at the scheme in Figure 5.1, a mirror symmetry can be appreciated between the absorption and the fluorescence spectra. In fact, the relative intensity of the different vibronic features are related to the transition-associated variations of the generalized nuclear configuration coordinate, $X$, *i.e.*, in other words, to the degree of structural relaxation undergone by the molecule. The overall supramolecular packing, possible aggregation states and interactions between the polymer species and solvent all affect the intensity of vibronic replicas, and the Stokes shift. Light-emitting nanofibers are an ideal tool in this respect, allowing the assembly of macromolecules to be changed and polymer chromophores to be forced into aligned geometries. This allows one to investigate and utilize the photonic properties of conjugated polymers under stretched, strongly anisotropic spatial conditions.

In OLEDs, electroluminescence is excited by the injection of charges of opposite sign from opposite electrodes, between which the active material is sandwiched.[6] Onto the anode (*i.e.* the hole-injecting electrode, which is generally optically transparent), one or more charge-transporting layers are sequentially deposited including the light-emitting material. The device is completed by a cathode (*i.e.* the electron-injecting electrode[§]) and eventually encapsulated. Film-based architectures (realized by sequential evaporations and/or spin-coating steps) are at the base of today's OLED technology. More complex geometries and shapes taken by the active materials, and particularly light-emitting polymer nanofibers, need very careful and specific cues in order to be embedded within electroluminescent devices (Section 5.3.1). The process of injection of charge carriers can be effectively schematized analogously to the commonly adopted description of inorganic semiconductors. The Fermi levels of the electrode metals and of the organics, which are initially different, align at the equilibrium. A barrier against injection of electrons is formed whenever the Fermi energy of the active layer is smaller than the work function of the metal. The height of the resulting barrier is given by the difference between the metal work function and the electron affinity of the semiconductor material. For

[§]Optically transparent cathode electrodes can also be used for achieving top or bidirectional light extraction.[7,8]

hole injection, the barrier height is instead given by the difference between the ionization potential of the semiconductor and the metal work function. On the contrary, when no barriers are formed the contacts are ohmic. The metal work function is therefore critical to determine the injection properties of electrodes for negative or positive charge carriers. Calcium (2.9 eV), magnesium (3.7 eV) and aluminium (4.0–4.3 eV) can be employed to realize cathodes. For the sake of comparison, one may recall that gold has a work function of 4.7–5.1 eV, and copper is also around 4.7 eV. For anodes, thin-films ($\cong$ 100 nm) of ITO are the most widely used option, because of its low electric resistivity ($2–4 \times 10^{-4}\ \Omega\ cm^{-1}$) and high optical transparency ($\geq$90% of transmission at 550 nm). However, a great variety of cathodes and anodes have been proposed for OLEDs, including graphene-based[9] and fully polymeric injecting electrodes. Several treatments, including oxygen plasmas, can also be carried out on electrode surfaces to further favor charge injection. In addition, injection is strongly affected by the morphology, oxidation state and contaminations possibly present at the metallo-organic interfaces, all issues becoming in principle even more critical when using complex device shapes such as those needed to contact polymer nanofibers. Furthermore, hole- and electron-injecting layers can be inserted, favoring charge carriers in overcoming the energy barriers at the electrode interfaces, and hole- and electron-transporting layers sandwiching the active organics can serve to define radiative recombination sites. A very well-known hole-transporting polymer is PEDOT, doped by PSS. This is also widely used in nanofibers, suggesting that intriguing electrically excitable nanofiber architectures could be fabricated coupling PEDOT with light-emitting polymers.

Once the device has been built, applying a forward bias to the contacts results in the injection of holes and electrons, whose radiative recombination in the organic layers generates photons. While many photons are waveguided or re-adsorbed in the organic and metal layers, a fraction is out-coupled to external free space. To enhance external emission efficiency, multilayer or nanocorrugated OLEDs can be fabricated, allowing internal waveguiding to decrease and surface plasmon effects to be tailored at the metallic cathode.[10–12]

### 5.1.2 Energy Transfer Mechanisms

Two or more molecules in solution or in the solid state can be involved in processes in which an excitation is transferred from a *donor* to an *acceptor*. If the excitation finally concurs to emit an output photon, which is the most interesting case when light-emitting acceptors are involved, the overall energy transfer process can be schematized as follows:

$$[D]^* + [A] \rightarrow [Y] \rightarrow [D] + [A] + h_P\nu\,, \tag{5.1}$$

where $[D]^*$, $[Y]$, $[D]$ and $[A]$ stand for the excited donor, the intermediate excited system (possibly given by a $[D]/[A]^*$ complex), the ground state of the donor and that of the acceptor, respectively, and $h_P\nu$ is the energy of the ultimately emitted photon, which is related to the energy levels of the acceptor. Due to the resulting down-conversion of energy, transfer mechanisms are very effective for controlling the overall emission spectrum of a blend or of an interface of different light-emitting molecular species, and also to increase the luminescence efficiency (because of the reduced spectral overlap between the emission of the acceptor and the absorption of the donor, the latter constituting the larger fraction of total molecules in most practical applications). Analogously to films, in light-emitting polymer nanofibers direct experimental evidence of transfer phenomena is given by intense, low-energy peaks in the fluorescence spectrum of a blend sample. In addition, measuring the time-dependence of the intensities of the donor and of the acceptor fluorescence allows the energy transfer kinetics and the formation and decay of excited intermediates to be monitored.

The most often cited mechanism of energy transfer between two organic species (at an interface between two molecular layers, within a blend or between different chromophore segments at intra-molecular level) is the so-called Förster energy transfer,[13,14] better known as Fluorescence Resonant Energy Transfer (FRET) in biology and biotechnology.[15,16] FRET consists in the transfer of energy from an excited donor to an acceptor molecule *via* resonant, near field dipole–dipole interaction. In this case the transfer rate ($K_{FT}$) from the donor to the acceptor at distance, $d$, is given by:[13,14]

$$K_{FT} = \frac{1}{\tau_D}\left(\frac{d_{FT}}{d}\right)^6, \tag{5.2}$$

where $\tau_D$ indicates the intrinsic fluorescence lifetime of the donor, namely the reciprocal of the fluorescence decay rate ($k_D$) of the $D$ species in the absence of $A$. The latter rate is given by $k_D = k_{D,R} + k_{D,NR}$ (where $k_{D,R} = 1/\tau_{D,R}$ and $k_{D,NR} = 1/\tau_{D,NR}$ are the donor radiative and non-radiative decay rate, respectively). In Equation (5.2), $d_{FT}$ is the Förster radius, expressed by the following expression:

$$d_{FT}^6 = 0.5291\frac{\tilde{\kappa}^2\eta_{PL,D}}{n^4 N_{AV}}\int_0^{+\infty} A_{em,D}(\tilde{\nu})A_{abs,A}(\tilde{\nu})\frac{d\tilde{\nu}}{\tilde{\nu}^4}\,. \tag{5.3}$$

Here, $\eta_{PL,D}$ indicates the luminescence efficiency of the donor, $n$ is the refractive index, $A_{em,D}$ is the normalized emission spectrum of the donor in the absence of acceptor species, $A_{abs,D}$ corresponds to the absorption spectrum of the acceptor (more precisely, the molar decadic extinction coefficient of the acceptor), $\tilde{\nu}$ is the frequency wavenumber and $\tilde{\kappa}^2$ is a factor related to the orientation of the donor and the acceptor dipole moments. $\tilde{\kappa}^2$ is of the order of unity, and an average value of $\tilde{\kappa}^2 = 2/3$ is used most of the time, corresponding to dipole moments that are randomly distributed over all orientations.[17] While such an assumption is generally valid for samples prepared by vacuum thermal deposition, or by processing solutions of low viscosity where Brownian motion is efficient enough to determine randomly oriented ensembles of dipoles, it could be not rigorously valid when orientational anisotropies are present in polymer samples in the solid state, such as in organic nanostructured materials and particularly in light-emitting polymer nanofibers. Furthermore, concerning the dependence of the transfer rate on the donor–acceptor distance, some models go beyond the original point–dipole model, which gives the characteristic $d^{-6}$ dependence of $K_{FT}$, and lead instead to $d^{-2}$, $d^{-3}$ or $d^{-4}$ laws, depending on the particular considered geometry.[18] These alternative descriptions are especially useful when point–surface interactions are analyzed, such as those between an organic chromophore and a much bigger nano- or microcrystal as in organic–inorganic nanocomposite systems.[18,19] What is general is that the efficiency of FRET strongly depends on both the donor–acceptor relative distance and on the overlap integral between the donor emission and the acceptor absorption spectra. For what is usually defined a "good" FRET couple (*i.e.* a combination of donor and acceptor species with significant $A_{em,D}/A_{abs,A}$ overlap), $d_{FT}$ is of roughly 1–2 nm. This is at the base of the great success of this approach, allowing both the resulting color of a multicomponent organic material, as in OLEDs, to be finely tuned and highly sensitive fluorescent tools for biodiagnostics to be developed. We describe in Section 5.2.3 how FRET is used for tailoring the emission of active polymer nanofibers. Other mechanisms of energy transfer are also possible, although to date basically unexplored in polymer nanofibers, such as so-called Dexter transfer.[20] These processes are relevant only at very short donor–acceptor distances (≤1 nm). Readers interested in more details about energy transfer processes in organic materials can refer to the outstanding book edited by R. Farchioni and G. Grosso.[2]

As mentioned above, the occurrence of energy transfer phenomena in a complex macromolecular system can be highlighted by the shift of the emission spectrum. More effective energy transfers between sub-units, prolonged excitonic migrations, longer average conjugation lengths, *etc.* lead to a decrease of the energy available for the emission of photons, *i.e.* a red-shift of the luminescence wavelength. Emissions at higher photon energies correspond instead to prompt radiative decays, relatively faster than the energy transfer in conjugated materials. Oxidative defects and, more generally, conformational disorder, are further reasons for a reduced conjugation length. In general, an important role in determining how excitations are transferred at intra-molecular and inter-molecular scale, ultimately affecting optical properties, is played by all the factors that influence the conformations adopted by molecules and the conformational sub-units into which the conjugation length is broken.[21–28] These factors include the solvents with which solutions are prepared, and other parameters and events belonging to the sample processing history, such as the used solution concentration, eventual phase separation phenomena, spatial confinement, *etc.* Tightly coiled arrangements and torsions of the polymer molecules, generally caused by poor solvents for a given compound, can determine shorter conjugation length, namely relatively blue-shifted emission wavelengths. This is, for instance, the well-established case of MEH-PPV, for which solutions with chlorobenzene, which is a better solvent for the polymer, result in relatively red-shifted emission compared to solutions with THF, because of the different configurations taken by the polymer chains.[29] Solid-state samples prepared by these solutions inherit the characteristics of molecular packing, and the resulting spectral features. We describe these aspects in more depth in the case of polymer nanofibers in Section 5.2.1.

Today, using mesoporous, optically transparent membranes and template methods (Section 3.3.1), scientists are able effectively to tune some energy transfer mechanisms. For instance, infiltrating solutions of MEH-PPV in mesoporous silica presenting hexagonally arrayed channels allows one to separate single polymer chains within individual pores that have a diameter of about 2 nm.[25] In this way, a composite sample is obtained, comprising the silica matrix, a part of molecules that remain on the external surface of the membrane and other molecules that orient within the channels. In particular, in the channels the transition dipoles align parallel to the longitudinal axis of the pores, as evidenced by the polarization of emitted light, and this arrangement suppresses energy transfer along single polymer chains. This is a sort of anticipation of the properties of polarized emission shown by many polymer nanofibers, which similarly arise from the alignment of the transition dipoles of the chromophores along the longitudinal axis of the nanostructures. In addition, when molecules become oriented in the pores of template membranes, inter-molecular transfer is also inhibited because each polymer chain is separated from neighbor molecules by the inter-pore distance.

The most interesting result on these samples is found by means of polarized pump-and-probe spectroscopy.[25] To describe this properly, we recall here a couple of very general definitions about optical anisotropy properties. When analyzing optically a geometrically asymmetric, elongated sample (such as a polymer chain segregated in a nanochannel, or more generally a polymer nanofiber), one can easily distinguish two main axes. In the following, we call these directions $Z$ (the longitudinal axis of the nanostructure) and $X$ (one of the axes perpendicular to $Z$, more precisely the axis whose direction is normal both to $Z$ and to the direction of incident or emitted photons). The direction of light propagation is named $Y$, and supposed to be normal to the sample plan. This geometry, schematized in Figure 5.2, is used when dealing with the polarization properties of light interacting with fiber-shaped nanostructures in the rest of the book. Of course, both the polarization of the incident light (*i.e.* the "incident polarization") and that of the optical signal (transmitted or emitted) collected after the sample (*i.e.* the "analyzed polarization") can be easily measured by means of optical polarizers positioned along the path of photons. Therefore, the polarized absorption or emission spectra can be

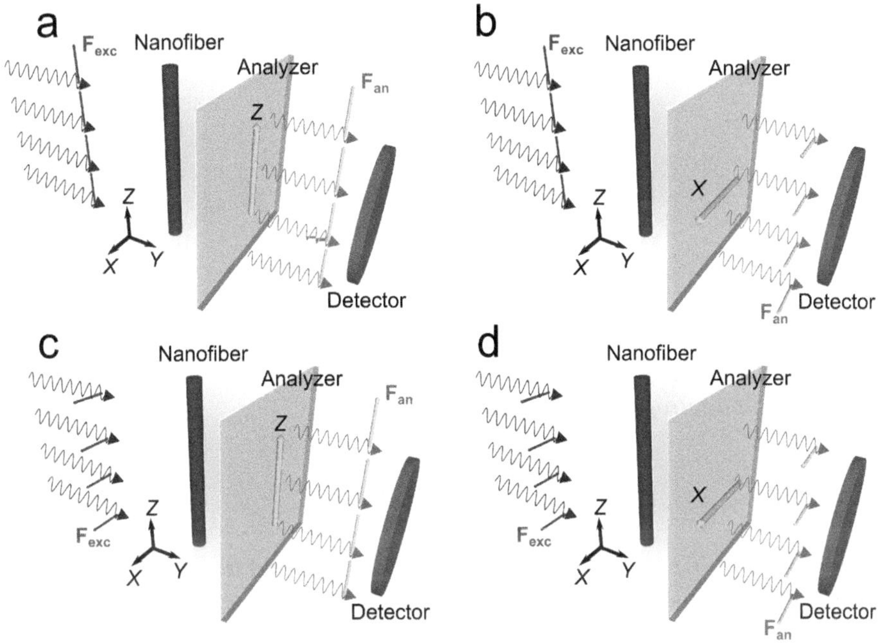


**Figure 5.2** Schemes showing the four possible polarized excitation/detection configurations allowing the polarization of light from polymer nanofibers to be studied. *X*, *Y* and *Z* indicate the directions of axes perpendicular to the nanofiber, the direction of light propagation and fiber longitudinal axis, respectively. $\mathbf{F_{exc}}$ and $\mathbf{F_{an}}$ indicate the incident and the analyzed electric field, respectively. Considering linearly polarized incident light, the excitation component is usually polarized along *Z* (a, c) or along *X* (c, d). Analogously, one can collect the *Z*-polarized (a, c) or the *X*-polarized (b, d) component of the output light, selected by means of the analyzing polarizer along the detection path. According to the notation "incident polarization|analyzed polarization", the four schemes correspond to *Z*|*Z* (a), *Z*|*X* (b), *X*|*Z* (c) and *X*|*X* (d), respectively.

collected by means of four polarization geometries, which we indicate as *Z*|*Z*, *Z*|*X*, *X*|*Z*, *X*|*X*, *i.e.* using the notation "incident polarization|analyzed polarization". For each geometry, one could measure the (absorption or emission) intensity, *e.g.* $I_{Z|Z}$, $I_{Z|X}$, $I_{X|Z}$ and $I_{X|X}$, measured at a specific light wavelength or calculated by simply integrating the area under each corresponding acquired spectrum**. Once this operation is performed, one can finally determine the *polarization* (or *dichroic*) *ratios*:

**Importantly, before plotting spectra and calculating intensity one has to take carefully into account and calibrate the spectral responses (say, $S_Z$ and $S_X$) of the collecting optical system. These can be quite different for light polarized along the two orthogonal directions, *Z* and *X*. The spectral responses, $S_Z$ and $S_X$, have to be used to correct the measured spectra that are polarized along *Z* ($I_{Z|Z}$,$I_{X|Z}$) and $X\left(I_{Z|X},I_{X|X}\right)$ , respectively.

$$\rho_{P,Z} = \frac{I_{Z|Z}}{I_{Z|X}}, \tag{5.4a}$$

and

$$\rho_{P,X} = \frac{I_{X|X}}{I_{X|Z}}. \tag{5.4b}$$

Conversely, the *depolarization ratio*, $I_{Z|X}/I_{Z|Z}$, indicates how much the incident excitation, whose polarization is directed along the longitudinal axis of the nanostructure, is transferred to light polarized along $X$. Finally, the *polarization anisotropy* can be also defined:

$$r_Z(\lambda,t) = \frac{I_{Z|Z} - I_{Z|X}}{I_{Z|Z} + 2I_{Z|X}}, \tag{5.5}$$

where we consider the particular case of $Z$-polarized excitation. Modern microscopy techniques allow complete bidimensional maps of the intensity emitted by luminescent nanostructures *vs.* the angles of polarization of the emitted and of the incident light to be obtained, which proves to be a powerful method for a full assessment of depolarization phenomena.[30] Another straightforward, though less informative, method consists in using unpolarized incoming light and just analyzing the polarization state of the transmitted or emitted signal. This corresponds to measuring polarization ratio values that are averaged over all the directions of possible polarization of incoming light.

Of course, the achievable polarization anisotropy is limited by the degree of orientational order of polymer backbones (Equations (2.7a) and (2.7b)), and by the degree of alignment of the optical transition dipole moments (Equation (2.8)).[31,32] For instance, polarized Raman microscopy is an especially powerful tool to evaluate the order of molecules within single nanofibers. By this method values of the orientational order parameter (Section 2.1.5) up to 0.75 have been found for poly(ethylene terephthalate) (PET) electrospun nanofibers.[33] In addition, interesting correlations can be drawn between the orientational order of molecules and fiber fabrication parameters, used solvents, *etc.* The investigation of electrospun poly[(9,9-dioctylfluorenyl-2,7-diyl)-*co*-(1,4-benzo-{2,1′-3}-thiadiazole)] nanofibers shows that more ordered molecular chains, without nanoscale aggregates, are obtained when the nanostructures are produced from THF solutions rather than from mixtures of THF and dimethyl sulfoxide (DMSO).[34]

As highlighted in Equation (5.5), the polarization anisotropy depends on the light wavelength, and can vary over time, and thus is a very sensitive probe of energy excitation migrations in the nanostructures. For example, for MEH-PPV/silica membrane composite systems one finds values of about 4.5 and below 1 for $\rho_{P,Z}$ and $\rho_{P,X}$, respectively.[25] In addition, the temporal dynamics of

$r_Z$ at 590 nm indicates that a rapid depolarization occurs in randomly oriented molecules outside the nanochannels, followed by a much slower (~250 ps) energy migration to oriented chromophores in the pores, which is evidenced by a recovery of the polarization anisotropy.[25] This intra-chain energy migration toward aligned chromophores has some elegant analogies with photo-synthetic reaction centers in terms of directional flow of harvested energy,[25] thus constituting a very beautiful example of how one-dimensional organic nanostructures can be used to design and control energy excitations and transfer mechanism. Polarized and low-threshold *Amplified Spontaneous Emission* (ASE) is also obtained in conjugated polymers that show optical gain and whose chains are aligned within these nanostructures.[26] These aspects lead us to the next section, introducing stimulated emission properties of organic materials.

### 5.1.3 Stimulated Emission and Lasing from Organics

Here we recall a few important physical concepts and technological issues related to the properties of stimulated emission and lasing action shown by a few classes of light-emitting polymer nanofibers. These fibers are better presented in Section 5.5. A complete, definitive description of the operation principles of laser materials can be found in the book by O. Svelto.[35] Furthermore, for readers interested in more details on organic lasers, excellent specialized reviews exist in the literature.[36–39]

The description of stimulated emission would need at least a few introductory lines based on quantum mechanisms[††]. Let us consider two states of a light-emitting system, having energies, $E_a$ and $E_b$, respectively (with $E_a > E_b$), and a population density of $N_a$ and $N_b$, respectively. The coefficient of optical gain, $\tilde{G}$, in the material (resulting from an above-unity ratio of the stimulated emission to the absorption intensity) is:[35]

$$\tilde{G} = \sigma_{a-b}(N_a - N_b)\ , \tag{5.6}$$

---

[††]Stimulated emission and absorption phenomena are described by quantum mechanics on the basis of the interaction between an incident sinusoidal perturbation like an electromagnetic wave, $A\cos(\omega t)$, and a physical (atomic, molecular, *etc.*) system. The state of the system can be written as a superposition, $\sum_k c_k|\Psi_k\rangle$, of eigenstates of the unperturbed Hamiltonian, and in the initial state one assumes $c_k(t=0) = 1$ for $k = a$, and $c_k(t=0) = 0$ for $k \neq a$, so the system is initially in the eigenstate, $|\Psi_a\rangle$. The expressions for the absorption (if the energy of the final state, $|\Psi_b\rangle$, is higher than that of the initial state, *i.e.* $E_b > E_a$) and the stimulated emission (if $E_b < E_a$) rates, $W_{a\to b} = |c_{a\to b}(t)|^2/t$ ($s^{-1}$), are given by (Fermi's golden rule):[35,40] $W_{a\to b} = \frac{2\pi}{\hbar_P}|\langle\Psi_b|H'|\Psi_a\rangle|^2\delta(E_b - E_a \mp \hbar_P\omega)$, where one calculates the square of the matrix element of the perturbation ($H'$), which connects the initial and the final state, $\delta$ is the Dirac delta-function expressing the conservation of the total energy and $\hbar_P\omega$ is the energy of the absorbed or emitted photon.

where $\sigma_{a\text{-}b}$ indicates the transition cross-section ($m^2$). $\sigma_{a\text{-}b}$ depends on the characteristics of the given material and it is defined as the ratio of the transition rate ($W_{a\rightarrow b}$) to the incident photon flux, *i.e.* $W_{a\rightarrow b}/(I_{\phi,exc}/h\nu)$, where $I_{\phi,exc}$ (photons $cm^{-2}$ $s^{-1}$) indicates the rate of photons impinging per sample unit area. Transition cross-sections are commonly reported in units of $cm^2$, and stimulated emission cross-sections, together with light amplification spectral regions from the molecular excited states, can be directly obtained by pump-probe experiments. In this technique, one measures the differential transmission of a probe pulse intensity, as induced by optically pumping the sample.[41] Reasonable values of gain cross-sections for organic conjugated compounds are of the order of $10^{-16}$–$10^{-15}$ $cm^2$.[42,43] Equation (5.6) shows that a condition of *population inversion* ($N_a > N_b$) determines positive $\tilde{G}$ coefficients, allowing the system to amplify the radiation intensity, $A_{em}$:

$$A_{em}(\lambda) = A_{em,0}(\lambda)e^{[\tilde{G}(\lambda) - \alpha(\lambda)]d}, \tag{5.7}$$

where $A_{em,0}$ indicates the light intensity at the initial point of the path travelled by photons in the material, $d$ is the total travelled length and $\alpha$ is the loss coefficient due to diffraction, scattering, etc. (Section 5.4). In a laser device, a resonator is added to assist stimulated emission in a medium having three or more involved energy levels, by means of a proper *feedback* mechanism that provides a way to select resonant wavelengths. Solid-state lasers based on organic semiconductor materials, though still limited to optical excitation schemes, have attracted increasing interest. Optical gain *via* stimulated emission has been found from a variety of solution-processable conjugated compounds including many light-emitting polymers.[36–39,44] The lasing condition for conjugated compounds is often fulfilled by a four-level scheme, based on the vibronic progressions of the $S_0$ and the $S_1$ states.[37] The population inversion is indeed accomplished by the photo-excitation from the ground vibrational level of $S_0$ to a high-energy sub-level of $S_1$, and the subsequent non-radiative decay to the lowest-energy sub-level of $S_1$ by internal conversion as described in Section 5.1.1.[37] However, for using organics as active media of lasers, other conditions have to be fulfilled, hence the realization of these devices generally requires a lot of effort from both chemistry and nanofabrication. Nanofiber architectures can be strategic in this respect for many reasons, as described in Section 5.5.

Finding and characterizing gaining compounds is the first key task. Basic prerequisites of these materials are a high luminescence efficiency and a large separation between the spectral regions of gain and absorption (both intrinsic and photo-induced). Large Stokes shifts, hence low self-absorption, help considerably in achieving lasing action. In order to assess the eventual gain of a conjugated material, one can check if it shows ASE.[35] ASE occurs by population inversion as well, and results in a very intense emission at a particular angle around a preferential axis of the active sample. This phenomenon is generally highlighted by a relevant line-narrowing (or gain-narrowing), namely by the collapse of the PL spectral broadening at high excitation fluences. ASE describes the amplification of the radiation, initially emitted spontaneously by the molecules. In this case, amplification takes place because of the population inversion (induced directly by the external optical excitation) on the chromophores that emitted photons meet before emerging from the active material. Importantly, ASE is normally assisted by light waveguiding effects within the solid-state sample (for instance, the slab delimited by the substrate/organics/air interfaces in thin-films or the cylindrical waveguide constituted by a polymer nanofiber). Waveguiding is in turn critically affected by the sample geometry and refractive indexes of the involved materials, and is normally found for sample thicknesses above a cut-off threshold.[44] Another distinctive feature of ASE is that line-narrowing is observed even without the feedback provided by external or integrated cavities. For this reason, this phenomenon is often called *mirrorless lasing*. Among the conjugated polymers exhibiting ASE, one should mention MEH-PPV,[44,45] together with a number of other PPV-derivatives, polythiophenes,[46] poly-fluorenes,[47] *etc.* Significant line-narrowing is usually observed at photon excitation densities of the order of $10^{17}$–$10^{18}$ cm$^{-3}$ (*i.e.* 1–10 μJ cm$^{-2}$ in pulsed optical excitation).

## 5.2 Light-emitting Nanofibers

Among other possible application fields of polymeric, miniaturized light-emitting sources and nanostructures, lab-on-a-chip and so-called micro-total analysis systems[48–50] have attracted enormous interest, allowing many of the complex functions traditionally performed in analytic laboratories to be miniaturized in tiny devices with embedded fluidic microchannels. This has clear advantages in terms of reduction of waste and costs, and of increase of portability for both bedside applications and assistance to campaigns of diagnosis and prevention in emerging countries. Many microfluidic devices are based on optical diagnostic methods, which allow excellent precision and sensitivity, making use of chromophores flowing in microchannels. However, as a matter of fact, despite frequent claims about the unequalled portability and cheapness of microfluidic chips, most of these devices still have to be connected to external, often bulky and quite expensive light sources for the excitation of chromophores. The external sources of photons can be lamps, light-emitting diodes and even lasers when monochromaticity, very high brightness and directionality are required to improve analytic and diagnostic performances. On the contrary, using miniaturized, ultimately electrically pumped light sources made of polymer nanofibers embedded or coupled to microchannels would guarantee compact integration in lab-on-a-chip devices and greatly enhance the actual portability of microfluidic architectures.

For example, electrospinning allows light-emitting polymer nanofibers to be realized in many ways. Blending and coaxial methods have been widely used, exploiting the favorable spinning behavior of some optically inert polymers. Fluorescent dye-loading in the mineral oil or transparent polymer core of fibers produced by co-electrospinning is a first straightforward option.[51] A point of strength of core-shell methods is that the external layer may provide some protection against Oxygen, moisture *etc.*, thus improving the photo-stability of light-emitting nanofibers[‡‡]. On the other hand, organic interfaces between the internal and external layers of core-shell nanofibers could induce the formation of defects or the onset of potentially undesired energy transfer processes, therefore the availability of alternative approaches is certainly desirable. A possibility consists in using highly processable polymers, such as PS, PEO, PVP, PMMA, *etc.*, or molecular sieve materials.[54] These can be doped by dyes, fluorescent oligomers or polymers uniformly distributed in these matrices, and electrospun by single or dual-syringe spinnerets. This allows light-emitting nanofibers to be realized without relying on core-shell architectures, as demonstrated both in the visible[54–58] and in the near-infrared[59] spectral range.

By directly processing conjugated light-emitting polymers and producing active nanofibers composed of them, one can open even more intriguing perspectives. This can be a challenging task, particularly by electrospinning due to the often limited solubility, insufficient amount of molecular entanglements and quite low molar mass of many conjugated polymers (*i.e.* $10^3$–$10^4$).[60,61] However, some methods have been developed to overcome these drawbacks. Electrospun, fully conjugated polymer nanofibers have been reported in the 2000s,[60,62–64] although electrospinning of light-emitting polymers is still largely sub-optimal if compared to results, size control, reliability and throughput achieved with more processable, plastic polymers. For example, electrospinning light-emitting polymers from solutions prepared by a single solvent often leads to fibers rich of beads and defects, resulting from axisymmetric instabilities of the jet (Section 2.1.3). A linear density of beads of about 10 $mm^{-1}$ is found along poly[(9,9-dioctylfluorenyl-2,7-diyl)-*co*-(1,4-benzo-{2,1′-3}-thiadiazole)] nanofibers spun from THF solutions.[34] A good

‡‡One has to bear in mind that, regardless of the particular application of light-emitting polymer nanofibers, a critically important issue is the photostability of the realized nanomaterial, especially in air. Oxygen is known to favor considerably quenching of the fluorescence of conjugated materials,[52] reducing the quantum yield and the emission lifetime. Oxygen incorporation within nanofibers can be strongly enhanced through photo-oxidation paths activated under exposure at intense photon beams, and by the high temperatures needed by some nanofabrication approaches (oxygen molecules in a polymer matrix at room temperature have diffusion coefficients of the order of $10^{-8}$–$10^{-7}$ $cm^2$ $s^{-1}$).[53] Last but not least, the shelf-life of light-emitting polymer nanofibers should be taken into account when designing experiments and applications. Fortunately, this is generally good, of the order of many months under ambient storage conditions in darkness, which is in line with shelf-life performances of organic semiconductor materials commercialized in the form of powder or flakes. The storage of active nanofibers can be further improved by implementing suitable packaging and transport conditions, possibly under vacuum or inert atmosphere, upon hermetic sealing.

strategy for electrospinning pure light-emitting polymers consists in using high-*MW* materials,[60] or side-chain substitutions as in polyester side-chain grafted poly(aryleneethynylene)s.[64] Another approach uses suitable mixtures of solvents and non-solvents (at ~5–20% v/v) for each particular polymer.[62,65] Of course, these mixtures can affect significantly not only the resulting nanofiber morphology, but also the optical properties of the nanostructures. For this reason, the eventual use of solvent mixtures should be carefully considered and tested in the desired target application of the produced light-emitting nanofibers.

### 5.2.1 Molecular Aggregation and Orientation in Light-emitting Polymer Nanofibers

Interesting information on the molecular arrangement within nanofibers comes from the spectral features of the PL emission. For instance, nanofibers composed of MEH-PPV or MEH-PPV/P3HT blends within PVP shells are fluorescent, and their absorption spectrum is slightly *red-shifted* and broadened compared to reference films.[66] A red-shift is also found in the PL spectra of poly[2,5-(2′-ethylhexyloxy)]-1,4-phenylenevinylene,[63] P3HT[67] and polyester side-chain grafted poly(aryleneethynylene)[64] nanofibers. Attributing precisely the origin of the observed spectral shifts is often difficult, needing the combination of data collected by complementary spectroscopic and microscopic characterizations. As introduced in Section 5.1.2, red-shifts can indicate more extended conjugation, in turn attributable to stretched molecular conformations along the fiber longitudinal axis. A higher density related to a more effective molecular packing at nanoscale can also contribute to increasing the apparent conjugation length.[68] We have already seen that when poor solvents are used to deposit samples in the solid state, coiled arrangements and torsions of the polymer molecules can reduce the conjugation length, thus leading to blue-shifted emission. However, in some nanofibers systems other, competing effects have been reported as well. For instance, the nanoscale aggregation phenomena induced by the need of polymer molecules to minimize the interaction with the poor solvent can lead to stacking conformations, which corresponds to reduced torsional motions, and consequently to red-shifted emission.[63] In disordered mats, non-wovens, *etc.* the inhomogeneous environment at microscales, due to the random orientation of fibers that enhances light-scattering effects, can also lead to red-shifted optical emission. Finally, internal waveguiding mechanisms in the fibers (Section 5.4), being more effective at low wavelengths, can decrease the relative intensity of low-wavelength light that is externally coupled out from the fibers into free space. For some molecular systems, supramolecular packing with the formation of J-aggregates, as in self-assembled nanotubes of low-molecular-mass cyanines,[69] is another route that leads to narrowed and red-shifted emission spectra, this time in comparison to monomers. Finally, templated, solution-processed wires of F8T2 show a PL spectrum red-shifted from that of THF solutions, and a broadening of the absorption spectrum compared to both solutions and thin-films, which may indicate a significant effect of light-scattering within the disordered assembly of nanowires.[70] The latter mechanism would be especially intriguing, suggesting that the wavelength-scale topological disorder, which is intrinsically present in mats of nanofibers, can affect optical absorption properties. This can be particularly useful in designing and realizing more efficient photo-detectors and photovoltaic devices (Section 5.3.2).

*Vice versa*, other reports on light-emitting polymer nanofibers, realized by nanofluidics (Section 3.2) or electrospinning, evidence a *blue-shift* of the PL peak wavelength up to about 10 nm (~20–40 meV),[71,72] often coupled with better defined spectral vibronic progressions compared to spin-cast films. PPV nanotubes realized by the membrane template method (Section 3.3.1) show a blue-shifted emission by about 100 nm (300 meV), higher luminescence quantum efficiencies and longer PL decays than in reference films, which can be attributed to reduced inter-chain decay channels.[73] In these nanomaterials, prevailing effects should therefore be those suppressing the migration of the excitation towards lower energy chromophores.[74,75] Excitonic migrations that are faster and less spread in space do not allow the system enough time to relax and to emit at higher wavelengths. Moreover, in polymer nanostructures size-dependent effects of optical confinement should be taken into account, which are analogous to the phenomenon of thickness-dependent gain-narrowing and ASE in thin-films (Section 5.1.3). Indeed, thin nanostructures can favor the propagation, and ultimately the amplification if gain media are present, of spectral components having small wavelength.

Finally, other species of light-emitting polymer nanofibers instead present peculiar combinations of spectral features, namely a red-shift of the fluorescence peak wavelength together with better defined spectral vibronic progressions compared to less-ordered molecular systems. Membrane-templated, melt-processed semicrystalline wires of PFO show a red-shifted PL spectrum (by about 16 nm, corresponding to roughly 100 meV) compared with dilute solutions of PFO in THF, together with an enhanced vibronic progression typical of singlet excitonic transitions (Section 5.1.1), which suggests an increased effective conjugation length and a narrowed distribution of emitting chromophores in the fibers (Figure 5.3).[76] In poly[(9,9-dioctylfluorenyl-2,7-diyl)-*co*-(1,4-benzo-{2,1′-3}-thiadiazole)] fibers electrospun from THF, molecular backbones prevalently orient along the length of the nanostructures, determining an excitonic emission at lower energies (shift of about 15 nm, ~60 meV) compared to that from relatively randomly aligned molecules (spun from THF/DMSO mixtures).[34] In other words, the molecular alignment in these polymer fibers leads to a reduced inter-chain separation, and consequently to a relatively increased inter-molecular electronic delocalization, which in turn determines the PL red-shift. Concomitantly, fibers spun from THF can present a more random internal arrangement of the benzothiadiazole and fluorene units composing the molecules. On the

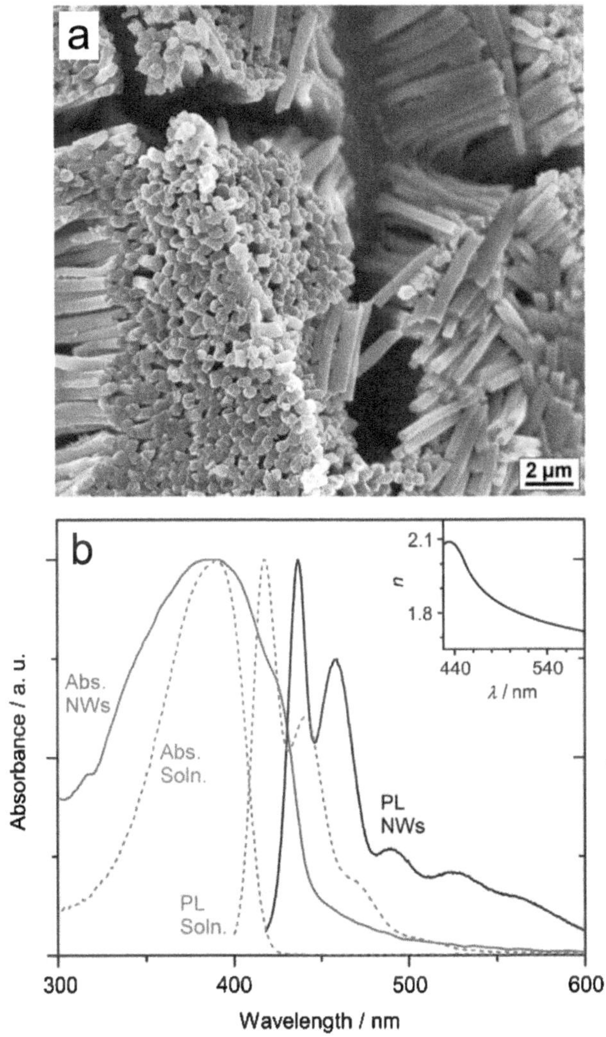


**Figure 5.3** (a) Melt-processed PFO nanowires (here, NWs) following removal of the alumina template membrane. (b) Normalized absorption and PL spectra of a dilute PFO solution (dashed lines) and of PFO nanowires (solid lines). The red-shift of the nanowire PL compared to solution can be clearly appreciated. The inset shows the wavelength-dependence of the refractive index of the PFO nanowires. Adapted with permission from Ref. 76, *Small*, 2007, **3**, 1178–1183. Doi: 10.1002/smll.200600575. Copyright © 2007, Wiley-VCH Verlag GmbH & Co. KGaA.

contrary, when fibers are spun from THF/DMSO, they can be more disordered in terms of backbone alignment along the fiber length, however molecules can have ample time to reach a more energetically favorable configuration, due to the slower evaporation of DMSO. Such a more convenient arrangement can present alternating benzothiadiazole and fluorene units between adjacent molecular chains, which may also be associated to decreased inter-chain energy transfer.[77] Furthermore, possible low-energy sites reached by excitonic migration in nanofibers electrospun from THF/DMSO can be non-emissive, as supported by findings in spin-cast films of poly(9,9-di-*n*-octylfluorene-alt-benzothiadiazole) (F8BT) upon annealing.[77] Overall, these are beautiful examples of the complex and rich phenomenology exhibited by light-emitting polymer nanofibers with respect to excitonic migration and electronic delocalization, and ultimately of emission spectral features. Molecular packing within the fibers, so strictly dependent on the processing conditions, heavily impacts on the degree of electronic delocalization. These examples are therefore especially interesting since the underlying concepts, as described in Section 5.1.2, are quite general and can be applied to various classes of light-emitting nanostructures based on conjugated polymers, regardless of the specific production technique and only on the basis of evidence provided by optical measurements.

### 5.2.2 Polarization Properties

Other information comes from the polarization of emitted light. In conjugated polymers, the emission transition dipole moments are often collinear to the backbone of the active sub-units along the molecular chains. Dipole-chain angular deviations are generally within about 20°.[31,78] For this reason, polymer stretching or flow-induced orientation, achieved during the realization of polymer nanofibers, can provide them with measurable optical anisotropy. Of course, analyzing the polarization state of emitted light (Section 5.1.2) is an indirect but accurate way to probe the alignment of active chromophores. Indeed, an ensemble of transition dipoles aligned along the longitudinal axis of a nanofiber would generate light whose intensity, analyzed by a polarizer positioned along the optical path before the photon detector, can be described by the Malus law:

$$I_{Z|Z}=I_{Z|Z,0}\cos^2\theta_P+I_{Z|Z,1}. \tag{5.8}$$

In the previous equation, $I_{Z|Z}$ indicates the intensity of the emission that is transmitted through the polarizer, $\theta_P$ is the angle between the polarizer axis and the axis of the fiber and $I_{Z|Z,0}$ and $I_{Z|Z,1}$ are the intensity of the polarized component of the emission and of the unpolarized background, respectively. The Malus law describes the absorption of polarized light from aligned transition dipoles as well. In this case, $\theta_P$ indicates the angle between the polarization direction of incoming light (incident normally onto dipoles aligned along the $Z$ direction), and $I_{Z|Z,0}$ and $I_{Z|Z,1}$ are the intensity of the polarized and unpolarized components of light that are transmitted across the sample, respectively. Equation (5.8) usually fits quite well experimental absorption or PL data from polymer nanofibers.

Polarization ratios achieved by polymer nanofibers are usually in the range from 2 to 10. For instance, template-synthetized nanofibers of PACT and PPY show polarized infrared absorption, evidencing a preferential alignment of chains parallel to the fibril axis.[79,80] The effect is more significant by using template membranes with smaller pores. $\rho_{P,Z}$ PL polarization ratios (Equation (5.4a)) are around 2 for light-emitting nanofibers made of various active polymers (MEH-PPV, polyfluorene derivatives, *etc.*) and realized by nanofluidics[71] electrospinning,[81] *etc.* Single PFO nanowires in a nematic liquid crystal host,[82] and uniaxially aligned arrays of these nanowires, oriented by a dielectrophoretic method,[83] exhibit $\rho_{P,Z}$ values between 3 and 4. $\rho_{P,Z}$ and $\rho_{P,X}$ (Equation (5.4b)) PL polarization ratios from poly[(9,9-dioctylfluorenyl-2,7-diyl)-*co*-(1,4-benzo-{2,1′-3}-thiadiazole)] electrospun nanofibers are around 3 and 2, respectively, when a mixture of THF and DMSO is used for spinning, and increase up to about 5 and 4, respectively, when only THF is used as solvent.[34] Analogous findings hold for the infrared absorption dichroic ratios. These results confirm that adding DMSO to the solution, while favoring the electrospinning process, leads to a lower degree of molecular alignment along the longitudinal axis of fibers. Transmission polarization ratios of MEH-PPV nanofibers are up to about 4.5.[81] Aligned arrays of MEH-PPV fibers spun from chloroform/isopropanol and of MEH-PPV/PCL fibers spun from chloroform/DMF show $\rho_{P,Z}$ up to about 7 and 11, respectively.[65] When oriented in mesoporous silica, MEH-PPV filaments reach $\rho_{P,Z} \cong 12$.[84] The PL polarization ratio of uniaxially aligned arrays of PPV[68] or polyfluorene[85] derivative fibers, measured with various conditions of polarization of the excitation light, are from the order of a few units up to 25, depending on the material and processing details. In the cases above, the output light is prevalently polarized along the fiber length, but other polymer species show a quite different behavior. In high-*MW* poly[1-(trimethylsilyl)phenyl-2-phenylacetylene] electrospun from chloroform solutions,[60] self-assembled oligo(*p*-phenylenevinylene)s,[86] and self-assembled nanofibers of poly(3,3‴-didodecylquarterthiophene) aligned in a liquid crystal,[87] optical transition dipoles preferentially orient perpendicularly to the longitudinal axis of the fibers. This can result in an emission intensity, $I_{X|X}$, up to five times higher than $I_{Z|Z}$.

Achieving polarized emission from nanofibers, which can be effectively organized in uniaxially aligned arrays on areas up to many $cm^2$, can be extremely useful to produce macroscopic sources of polarized light. Applications can be ubiquitous, particularly in lab-on-a-chip devices. In microfluidic-based diagnostics, the photo-excitation used to pump chromophores in the chips is generally much more intense than the PL signal to be detected (usually by several orders of magnitude). To remove the excitation light thus making reliably observable the PL from diagnostic chromophores, one usually introduces optical filters along the optical paths. However, this can limit the achievable sensitivity. To tackle these drawbacks, a few methods have been reported in the literature.[88,89] Integrated, polarized light-sources based on light-emitting polymer nanofibers would allow excitation and fluorescence signals to be easily decoupled, by simply using polarization analyzers (with the transmission axis along the $X$ direction defined in Section 5.1.2, namely at $\theta_P = \pi/2$ according to Equation (5.8)), which can remove the $I_{Z|Z,0}$ component (polarized and coming prevalently from the excitation source) and transmit a significant fraction of the useful PL signal (which is normally unpolarized). This has been demonstrated by a uniaxially aligned array of MEH-PPV electrospun nanofibers coupled with a microfluidic channel made of PDMS.[81] The microchannels are defined by a textured PDMS element and a bottom flat substrate of PDMS or quartz, and are realized by soft lithography methods (Section 1.2.1). Then, the microchannel is positioned over the gap of a metal foil collector, commonly used in electrospinning to align and support nanofibers (Section 2.3.2). The bottom of the microfluidic chip is so placed in contact with the suspended light-emitting nanofibers, which are aligned perpendicular to the microchannel longitudinal axis. At the same time, the fibers are separated by the liquid environment by the bottom of the fluidic microchannels, therefore their emission is not affected by local fluctuations (solution concentration, flow rates, pH, *etc.*) in the diagnostic liquid. The overall device is completed by a prism in contact with the suspended array. In this way, nanofibers can be excited by external laser light *via* the evanescent field provided by total internal reflection (in this way external light does not reach the microfluidic chip), the emitted PL from the array is polarized and in turn excites chromophores flowing in the chip and finally the polarization analyzer suppresses the light from the nanofibers letting the desired signal from the chromophores be transmitted for detection. In summary, this architecture results in an increase of the signal-to-noise ratio on the detector by about a factor of two (Figure 5.4).[81] Of course, achieving light emission from nanofibers by electrical injection would be the obvious and more crucial development of this research, opening the way to the realization of really portable microfluidic platforms with integrated photonic excitation. Approaches to produce polymer nanofiber-based OLEDs are described in Section 5.3.1.

## 5.2.3 Light-emission Tunability

Nanofibers with tunable emission color can be obtained by different fabrication methods, but in general they exploit the same FRET mechanism to transfer energy from a high-energy donor to a low-energy acceptor molecule (Section 5.1.2). Varying with the relative concentrations of the two molecules within a blend or a heterostructure allows the desired overall emission spectrum to be obtained. The probability, $P$, of exciton transfer from the donor to the acceptor is given by the ratio, $\frac{K_{FT}}{(1/\tau_D)+K_{FT}}$, where $\tau_D$ is the characteristic time of spontaneous emission of the donor. Examples of polymer-based elongated nanostructures showing FRET include template

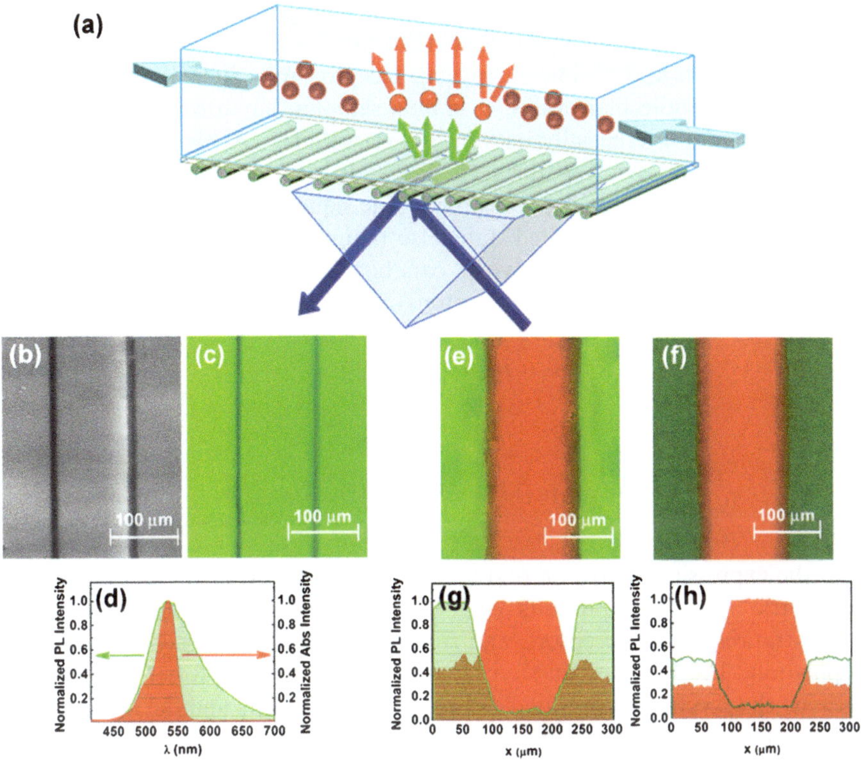


**Figure 5.4** Example of use of an aligned array of light-emitting polymer nanofibers as polarized excitation source coupled to a microfluidic chip. (a) Device scheme, involving an optically transparent microchannel made of PDMS, glass, *etc.*, a transparent substrate separating nanofibers from flowing liquids and a prism for exciting nanofibers by total internal reflection. (b, c) Optical and fluorescence micrographs of the transparent microchannel on the top of the aligned array of light-emitting nanofibers. (d) Nanofiber polarized emission (green spectrum, left vertical scale), matching the absorption of rhodamine chromophores flowing in the microchannels (red spectrum, right scale). (e–h) Rhodamine in the microchannel is excited by the nanofibers (red regions in e, f). The signal from excitation, being polarized, is removed by a polarizer, which leads to an enhanced signal-to-noise ratio. The corresponding fluorescence micrographs (e, f) and PL intensity profiles (g, h) along the microchannel cross-section are shown without (e, g), and with (f, h) the fiber-perpendicular polarizer, respectively. Reproduced with permission from Ref. 81, *Lab Chip*, 2009, **9**, 2851–2856. Doi: 10.1039/b906188f.

synthesized nanotubes of the conjugated polymer, PVK (as donor), used as matrix to host a red emissive organo-lanthanide chelate complex (acceptor),[90] or fluorescent dyes such as Coumarin 6 and Nile red,[91] or electrospun nanofibers of binary blends of conjugated polymers.[92]

In the most frequent case, the acceptor component in the nanofibers interacts with a distribution of donor chromophores. The distance between individual donors and the low-energy gap acceptor unit can be limited not only by the effective molecular radius, as in bulk samples, but also by the nanofiber radius (a condition especially relevant for very thin, ~10 nm fibers). We will come back to this issue in the next pages. In general, the energy transfer dynamics can be described by means of rate equations for the populations, $n_{Do}$ ($n_{Ac}$) of $D$ ($A$) in the excited state, also taking into account that a fraction of molecules of the donor polymer can be directly excited (*i.e.*, by direct absorption of incident excitation photons):

$$\frac{dn_{Do}}{dt} = f_D I_{exc} - (k_D + K_{FT})\, n_{Do}(t)\,, \tag{5.9a}$$

$$\frac{dn_{Ac}}{dt} = f_A(c_A)\, I_{exc} + K_{FT} n_{Do}(t) - k_A n_{Ac}(t)\,. \tag{5.9b}$$

In the expressions above, $f_A(c_A)$ indicates the fraction of the pump photon volumetric rate, $I_{exc}$ (photons cm$^{-3}$ s$^{-1}$), that is directly absorbed by a concentration, $c_A$, of acceptors, and $f_D$ is the fraction absorbed by the donor population. The relative emission intensity from the acceptors (*i.e.* the ratio between the spectrally integrated emission from acceptors and the total sample emission) can be expressed in general as:[93]

$$\phi_{Ac} = \left(1 + \frac{k_{D,R} n_{Do}}{k_{A,R} n_{Ac}}\right)^{-1}, \tag{5.10}$$

where $k_{A,R}$ indicates the acceptor radiative time. The final form of $\phi_{Ac}$ will depend on the exact donor–acceptor interaction (dipole–surface as could be in active nanofibers embedding light-emitting nanocrystals, dipole–dipole which is most likely the case of blend, fully polymer nanofibers, *etc.*). In particular, in order to analyze nanocomposite fibers in which a high-energy polymer donor (such as for example a blue-emitting conjugated polymer) is used to embed low-energy emitting nanocrystals, one should carefully consider the size of the particles and compare it to the typical molecular size of the polymer (say, 1–10 nm),[93] thus choosing the more suitable interaction scheme. Embedded particles much larger than the average molecular size are better described by dipole–surface interaction schemes.

Supposing that a distribution of donors is dislocated along a partially stretched chain of the acceptor, the rate of Förster energy transfer *via* point–point dipole interaction is:

$$K_{FT} = 2\pi\delta \frac{c_A}{c_D} d_{FT}^6 \int_{-l/2}^{+l/2} \int_{r=2a}^{r_F} \frac{1}{\tau_D} \frac{1}{(r^2 + z^2)^3}\, r\, dr\, dz\,, \tag{5.11}$$

where $\delta = 3/(4\pi a^3)$, $d_{FT}$ is the characteristic Förster radius introduced in Equation (5.3), $a$ is the radius of the donor molecules, $l$ is the uniaxially stretched length of the acceptor molecule, $r_F$ is the nanofiber radius and the ratio, $c_A/c_D$, takes into account the relative molar concentration of acceptors and donors. For large uniaxial elongations ($l >> r_F$, $a$), one has:

$$K_{FT} = \frac{\pi^2 c_A \delta d_{FT}^6}{2\tau_D c_D}\left(\frac{1}{8a^3} - \frac{1}{r_F^3}\right). \tag{5.12}$$

The energy transfer rate to the acceptor is suppressed and finely controllable for small fiber diameters ($r_F$ approaching $a$). The resulting color is very effectively tailorable by varying the relative acceptor/donor concentration, $c_A/c_D$. In this respect, polymer nanofibers can exhibit unique properties in terms of emission control.

The importance of tuning color is clear. Illumination and display applications are based on precise quantitative standards to express the color emitted by a material or by a device and received by the human eye. A widely used definition is that provided by the Commission Internationale de l'Eclairage (CIE), associating three coordinates ($X$, $Y$, $Z$) to colors as visualized by humans. These coordinates are given by the normalized projection of the emission spectrum onto three so-called Color Matching Functions, $\overline{x}(\lambda)$, $\overline{y}(\lambda)$ and $\overline{z}(\lambda)$, which are in turn directly related to the perception response of the human eye:[94]

$$X_{CIE} = \frac{\int_0^{+\infty} A_{em}(\lambda)\overline{x}(\lambda)d\lambda}{\Sigma}, \tag{5.13a}$$

$$Y_{CIE} = \frac{\int_0^{+\infty} A_{em}(\lambda)\overline{y}(\lambda)d\lambda}{\Sigma}, \tag{5.13b}$$

$$Z_{CIE} = \frac{\int_0^{+\infty} A_{em}(\lambda)\overline{z}(\lambda)d\lambda}{\Sigma}, \tag{5.13c}$$

where $\Sigma = \sum_{\overline{t}=\overline{x},\overline{y},\overline{z}} \int_0^{+\infty} A_{em}(\lambda)\overline{t}(\lambda)d\lambda$ (*i.e.* $X_{CIE} + Y_{CIE} + Z_{CIE} = 1$). Each color can be therefore represented by its CIE coordinates in the bidimensional space ($X_{CIE}$, $Y_{CIE}$), as in Figure 5.5. In particular, white corresponds to the coordinates (0.33, 0.33), with different tones spreading on a region including greenish, reddish and bluish white. For example, blend nanofibers of the red-emitting conjugated polymer, poly[{2-methoxy-5-(2-ethylhexyloxy)-1,4-(1-cyanovinylenephenylene)}-*co*-{2,5-bis(*N*,*N*′-diphenylamino)-1,4 phenylene}] (acceptor), and the blue-emitting polymer, poly[(9,9-hexylfluorenyl-2,7-diyl)-

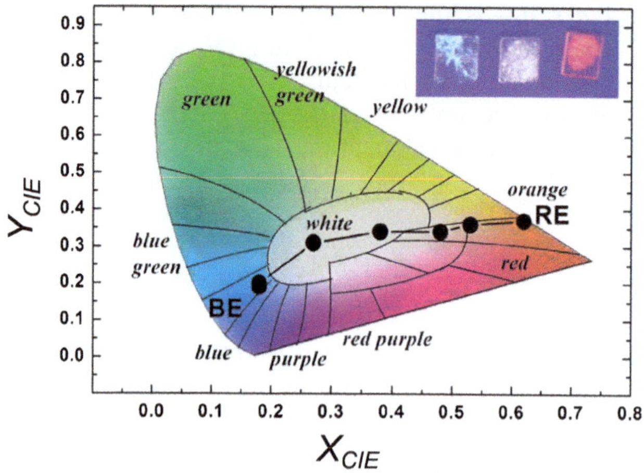


**Figure 5.5** CIE chromaticity coordinates of electrospun, blend light-emitting nanofibers for different acceptor/donor relative concentrations, $c_A/c_D$. From right to left: pure acceptor (red-emitting conjugated polymer, here abbreviated as RE), $c_A/c_D = 1$, $10^{-1}$, $10^{-2}$ (white emission), $10^{-3}$, $10^{-4}$ and 0 (pure blue-emitting conjugated polymer fibers, here BE). Inset: photograph of blue, white and red nanofiber non-wovens deposited onto quartz substrates and emitting under UV excitation. Adapted with permission from Ref. 92, *Appl. Phys. Lett.*, 2009, **94**, 043109. Doi: 10.1063/1.3064139. Copyright © 2009, American Institute of Physics.

alt-*co*-(9-ethyl-3,6-carbazole)] (donor), electrospun from chloroform/DMSO solutions, show a PL emission that covers the whole visible range, including white with CIE coordinates (0.38, 0.34) upon varying the relative concentration ($c_A/c_D$) of the two components from 1 to $10^{-4}$ (Figure 5.5).[92] Other interesting examples are hybrid electrospun nanofibers, such as those made of poly[(9,9-dioctylfluorenyl-2,7-diyl)-*co*-{1,4-benzo-(2,1′-3)-thiadiazole}] or other fluorine-based copolymers blended with PS or PEO and embedding zeolite L crystals, which are aluminosilicate particles having a nanochannel that is ideal for dye-loading.[95,96] White emission is also reported from nanofibers with ZnO nanocrystals dispersed in PVA,[97] and with CdTe nanocrystals dispersed in blends of poly[*p*-xylene-α-(dialkyl-sulfonium halide)] and PVA, having CIE coordinates (0.27, 0.37).[98] Finally, besides electrospun nanofibers, also template-synthesized[91] samples allow well-balanced white emission to be reached, and hetero-epitaxial growth can open a new route for color tuning, at least for nanofibers made of oligomer materials.[99] Importantly, white photo-luminescent nanofibers pumped by inorganic LEDs can be used to realize hybrid devices for illumination, an application that has been developed by researchers at RTI International (Figure 5.6).[100] This technology (Nanofiber Light Improvement Technology, NLITe™) offers an interesting alternative to conventional sources for solid-state lighting, due to its very low cost and power consumption.

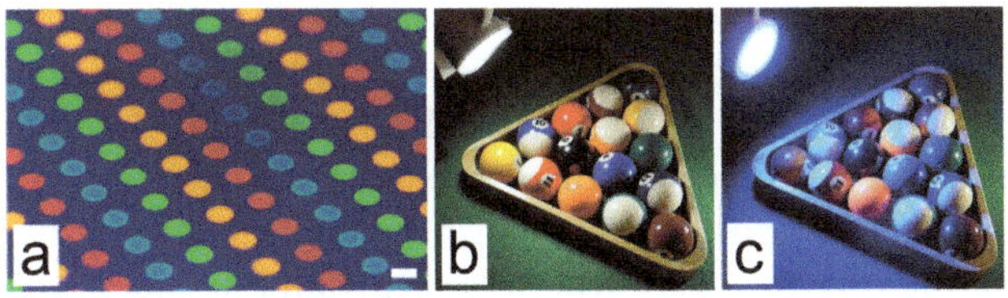


**Figure 5.6** (a) Photograph of NLITe™ photo-luminescent nanofibers emitting different colors. Nanofibers are made of PMMA doped with light-emitting CdSe/ZnS core-shell nanoparticles. Different colors are indicative of different sizes of the embedded nanocrystals. Scale bar: 1 mm. (b, c) Comparison of illumination provided by a NLITe™ source (b) in which white-emitting nanofibers are excited by an inorganic LED (in turn emitting at 450 nm), and by the inorganic LED (used directly, c). Courtesy of Dr. L. Davis. Photos supplied by RTI International.

Findings and applications related to energy transfer mechanisms are not limited to color tuning. FRET from coumarin to rhodamine has been used to highlight the nanojunctions formed at the intersections of perpendicular electrospun nanofibers that can serve as reactors with attoL volume.[101] An efficient energy transfer is observed in electrospun nanofibers made of MEH-PPV/P3HT, which is attributed to the very high effective interfacial area between the donor and the acceptor, due to reduced phase separation length-scale.[58] Another possibility in nanostructured materials, which can be effectively exploited in emissive polymer nanofibers, consists in enhancing the luminescence quantum yield and the emission stability over time, by means of nanoscale metal coatings.[102,103] This effect is related to energy transfer by

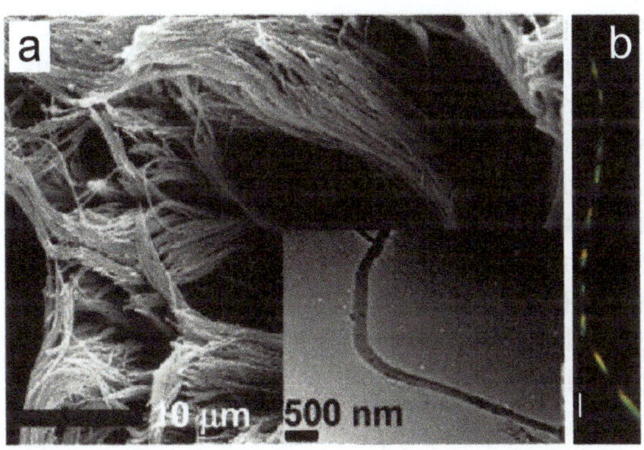


**Figure 5.7** (a) SEM micrograph showing a bundle of flexible light-emitting color barcode nanowires made of poly(3-butylthiophene) and PEDOT. Inset: magnified TEM image of a single strand of the light-emitting color barcode nanowires. (b) Fluorescence, color image of a single barcode nanowire with multiple sequential segments. Scale bar = 2 μm. Adapted with permission from Ref. 104, *ACS Nano*, 2010, **4**, 5155–5162. Doi: 10.1021/nn101096m. Copyright © 2010, American Chemical Society.

surface plasmon resonance coupling. For example, it is used in light-emitting, highly flexible and Cu-coated nanowires obtained by the sequential, template polymerization of different conjugated polymers (Figure 5.7).[104] These and other complex, smart nanostructures are ideal candidates to implement multiple functionalities on single wires, as light-emitting barcodes for biological sensing and for identification systems and as labels for microscale objects that, given the extreme structural flexibility of these nanofibers, can have various and irregular shapes. For these reasons, the development of emissive barcode nanofibers attracted increasing interest in the last few years, with other examples being based on the self-assembly of diblock and triblock copolymers.[105]

## 5.2.4 Other Properties of Light in and from Nanofibers

Another interesting application enabled by the interaction of light with polymer nanofibers consists in manipulating them in a highly controlled way by means of optical tweezers.[106–108] Optical forces that result from scattering of electromagnetic fields incident on a particle, and the shape and refractive index of the trapped object, play a relevant role in the process. Optical tweezers allow a wide variety of micro- and nanosystems to be trapped and studied, including natural and artificial elongated polymeric filaments, and forces in the sub-pN range to be measured.[109–112] In this respect, polymer nanofibers, having a highly anisotropic shape with sub-wavelength diameter and controllable lengths extending well above the μm-scale, are very interesting from the fundamental viewpoint, because they are ideal systems for studying the intermediate regime between the limits described by Rayleigh scattering (*i.e.* scattering particle much smaller than the light wavelength) and by geometrical optics (particle much bigger than the light wavelength). In addition, manipulating polymer nanofibers with extremely high precision would allow photonic circuits and inter-connects to be assembled. In particular, nanofibers can be rotated in a tilted configuration using a highly focused, linear polarized Gaussian beam. A dedicated model for optically trapped linear nanostructures has been developed by O. M. Maragò and coworkers at CNR and University of Messina.[113,114] The mathematical details involve the so-called T-matrix formalism and are beyond the aim of this book. Here, we only recall the expressions of the optical force, $\mathbf{F_N}$, and of the torque, $\mathbf{M}$, exerted on a particle, which are given by the integrals:

$$\mathbf{F_N} = r_S^2 \int_\Omega \hat{\mathbf{r}}_\mathbf{S} \cdot < \mathbf{T_M} > d\Omega \,, \tag{5.14}$$

$$\mathbf{M} = -r_S^3 \int_{\Omega} \hat{\mathbf{r}}_{\mathbf{S}} \cdot <\mathbf{T}_{\mathbf{M}}> \times \hat{\mathbf{r}}_{\mathbf{S}} d\Omega \,. \tag{5.15}$$

In the previous equations, integration is extended over the full solid angle, and $r_S$ indicates the radius of a spherical volume surrounding the center of the particle and to be considered as large as possible (ideally infinite). Finally, an important role is played by $<\mathbf{T}_{\mathbf{M}}>$, which is the time-averaged Maxwell stress tensor in the form of Minkowski:[115,116]

$$<\mathbf{T}_{\mathbf{M}}> = \frac{1}{8\pi} \mathrm{Re}\left[ n_{out}^2 \mathbf{F} \otimes \mathbf{F}^* + \mathbf{B} \otimes \mathbf{B}^* - \frac{1}{2}(n_{out}^2 |\mathbf{F}|^2 + |\mathbf{B}|^2)\mathbf{I} \right] \tag{5.16}$$

where $\otimes$ indicates the dyadic product, **I** is the unit dyadic, **F** is the electric field and **B** is the magnetic field found by expanding both incident and scattered fields as a series of vector harmonics,[117] and $n_{out}$ is the refractive index of the medium surrounding the particle. Now, the optical force and torque on polymer nanofibers can be calculated by Equations (5.14) and (5.15), respectively, by modeling the filament as a linear chain of spheres.[114] The complex hydrodynamics of the trapped polymer fiber, which is suspended in water, is also taken into account. Any orientation and position of the polymer fiber with respect to the incoming trapping beam can be modeled in this way, as exemplified in the schemes in Figure 5.8a and b. Trapping and rotation data, obtained experimentally by a Ti:sapphire laser beam and particle tracking by a back-reflection configuration in an inverted microscope (Figure 5.8c–g) can be accurately described in this way.[114]

The refractive index of polymer nanofibers, together with other optical features, can be changed in real-time by means of photo-chromic dopants embedded in the nanostructures. A class of photo-chromic nanofibers, already introduced in Section 4.2.2, is based on PMMA or other optically inert polymers embedding the low-molar-mass molecule, 6-$NO_2$-BIPS. In these systems, controllable photo-conversion processes can be induced between a spiropyran and a merocyanine form, which exhibit different colors and more generally different degrees of electronic conjugation (Figure 4.9). Several experimental techniques allow one to appreciate the formation of a high-energy, *cis*-merocyanine form, which acts as transient intermediate in the transition kinetics during the conversion of spiropyran, then ultimately leading to the rearrangement to the *trans*-merocyanine in times below the order of 100 ps.[118–123] The cyclic photo-isomerization of these photo-chromic nanofibers has been studied in depth by using the third ($\lambda$ = 355 nm, excitation fluence $\cong$ 240 μJ $cm^{-2}$) and the second ($\lambda$ = 532 nm, excitation fluence $\cong$ 790 μJ $cm^{-2}$) harmonic of a *Q*-switched 0.6 ns-pulsed neodymium-yttrium-aluminum-garnet (Nd:YAG) laser.[124] Exposing the nanofibers to UV light for several intervals, with a total duration of incident pulses ranging from 0 to 0.90 μs, an absorption peak appears at 614 nm, which is the signature of the

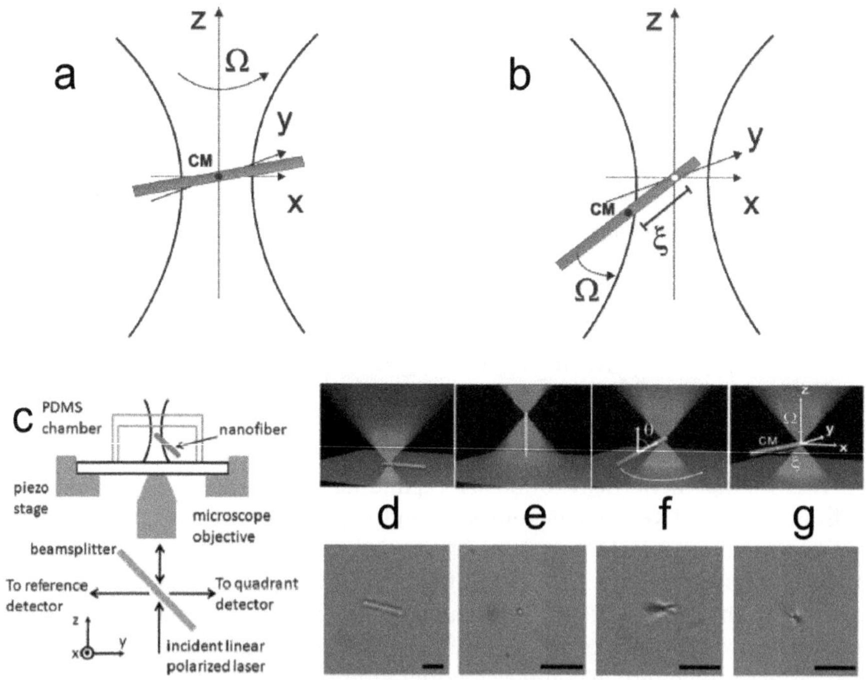


**Figure 5.8** (a) Sketch of a polymer nanofiber trapped and rotated about its center-of-mass (black dot) by means of a focused laser beam. Rotation is in the *xy* plane. Ω indicates the rotation movement. (b) Geometry of the optical trapping of a fiber rotated about a point that is shifted (here by an amount, ξ, towards the edge of the fiber (white dot). (c) Scheme of the optical tweezers set-up, whose detection is based on sample back-scattered light. The arrows indicate the light paths. (d–g) Optical rotation of the polymer fiber, schematized in the top frames and imaged experimentally in the bottom frames, respectively. The fiber longitudinal axis is tilted by an angle ($\theta$) from the optical axis. Scale bar = 2.5 μm. Adapted with permission from Ref. 114, *Opt. Exp.*, 2010, **18**, 822–830. Doi: 10.1364/OE.18.000822. Copyright © 2010, Optical Society of America.

formed merocyanine product (Figure 5.9a). The intensity of the absorption spectrum increases with the UV exposure time in the early stages of irradiation. After 0.54 μs ($\cong$ 0.2 J $cm^{-2}$), the absorption reaches its maximum intensity, which indicates the complete photo-conversion of available molecules to the merocyanine form. Analogously to findings in solution and in bulk samples, the merocyanines in nanofibers can be back-converted to spiropyrans by irradiating with green light (7.7 μs, $\cong$ 10 J $cm^{-2}$).[124] The photo-chromic nanofibers (Figure 5.9b) can undergo photo-induced conversion cycles alternating exposures with different wavelengths (Figure 5.9c), showing some fatigue during cycling (observable mainly as a reduction of the merocyanine signal after each cycle). Similar results concerning cyclic photo-isomerization

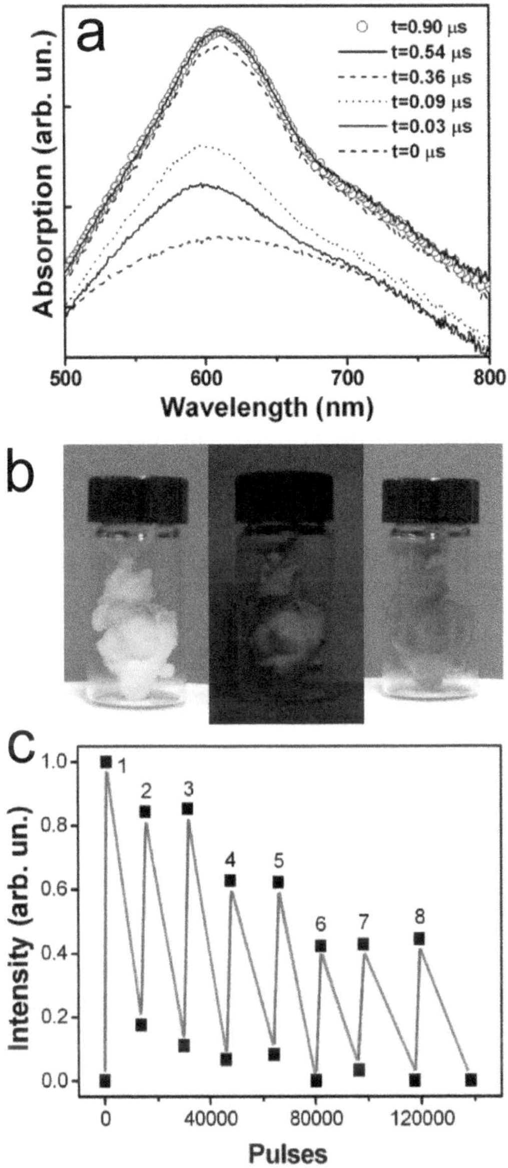


**Figure 5.9** (a) Absorption spectra of photo-chromic nanofibers after irradiation by a 355 nm pulsed laser for 0, 0.03, 0.09, 0.36, 0.54 μs (from bottom to top, lines), and for 0.90 μs (circles). (b) Photographs of the photo-switchable nanofibers in their pristine, spiropyran state (left), during UV-excitation (center) and in the photo-converted, merocyanine form under ambient illumination (right). (c) Absorption intensity measured during cyclic photo-isomerization, carried out by alternatively irradiating the nanofibers by UV and green pulsed laser light. Adapted with permission from Ref. 124, *Adv. Mater.*, 2008, **20**, 314–318. Doi: 10.1002/adma.200700980. Copyright © 2008, Wiley-VCH Verlag GmbH & Co. KGaA.

are obtained by the spirooxazine/merocyanine controlled photo-conversion within nanotubes of a fluorine/anthracene copolymer,[125] fabricated by membrane-templated wetting (Section 3.3.1). PAN nanotubes self-assembled using azobenzenesulfonic acid as dopant show photo-chromic properties as well, evidenced upon UV irradiation.[126]

Electrochromism has also been studied in polymer nanofibers. PEDOT nanotubes have been used in flexible electrochromic devices, exhibiting switching times of 20–30 ms between two states with different coloration, and consequently different reflectivity, following an applied bias of +1 V or –1 V.[127] Faster switching times (~5 ms) have also been observed.[128] Electrochromic nanofibers made of a cross-linked conjugated polymer (following electrospinning of the precursor) have shown a change of color upon applying a bias of 0 V or 1.1 V, within times of the order of seconds,[129] analogously to composite fibers made of polydiacetylene and carbon nanotubes and carrying currents of 10 mA at room temperature.[130]

## 5.3 Polymer Nanofiber-based Optoelectronics

In this section we focus on optoelectronic devices that are based on either the emission or the absorption properties of polymer nanofibers, jointly with transport properties that are needed to lead charge-carriers to either radiatively recombine (as in OLEDs) or separate photo-generated excitons to collect charges (as in photo-detectors or solar cells). These nanofiber-based devices combine photonic and charge-transport features of polymer nanostructures. The charge-carrier transport properties of nanofiber materials are described in detail in Chapter 6.

### 5.3.1 Organic Light-emitting Devices

Polymer nanofibers utilized within OLED architectures can open new routes for the miniaturization of these devices down to the nanoscale, with the additional, appealing possibility of achieving polarized emission. This would be a valuable alternative for fabricating micro- and nanoscale electroluminescent polymer light sources, whose production generally needs quite complex and costly nanofabrication facilities and lithographic techniques.[131–133] A different approach uses template membranes to grow PEDOT nanowires, thus defining areas for the injection of holes and ultimately for light emission that are of the same size as the pores of the membrane.[134] Data storage, miniaturized optical sensors and highly controlled excitation sources for lab-on-a-chip devices are obvious and important targeted applications.

Different electroluminescent device architectures can be developed by means of light-emitting polymer nanofibers. In the simplest case, fibers are just deposited on a hole-injecting layer and capped by a metal cathode. For example, disordered assemblies of PFO nanowires are used within a [glass/ITO/PEDOT:PSS (70 nm)/fibers/Al] prototype structure showing a threshold voltage at 3.4 V.[135] However, it is easily recognized that embedding polymer nanofibers as active media within sandwiched device geometries as in conventional OLEDs would require specific solutions, optimized to avoid short circuits between the two super-imposed metal electrodes and to properly convey charge-carrier through the organic wires. A possible approach consists in depositing nanofibers, made of a light-emitting material and an optically inert, electrically insulating polymer such as PEO, on the usual anode substrates (ITO/PEDOT:PSS).[136] The vertical architecture is completed with a PVK film topped by the device cathode (Ca/Al). Of course, if nanofibers do not cover completely the anode, the two electrodes will sandwich a layer of variable thickness, given by thinner PVK islands and thicker regions where PVK has been deposited onto fibers. Upon applying a voltage bias the current will flow preferentially through PVK regions that have lower electric resistance, as schematized in Figure 5.10a. In order to force current to flow across nanofibers, and hence charge-carriers to recombine radiatively in them and to yield electroluminescence, one has to find a way to render fibers much more conductive than other regions. To this aim, the crucial step consists in annealing samples at 150 °C prior to PVK deposition. This extracts the PEO component, which is phase-separated from the light-emitting polymer, out from nanofibers. PEO fills the empty regions between fibers thus forming an insulating layer, thinner than the fiber diameter. In this way, depositing PVK

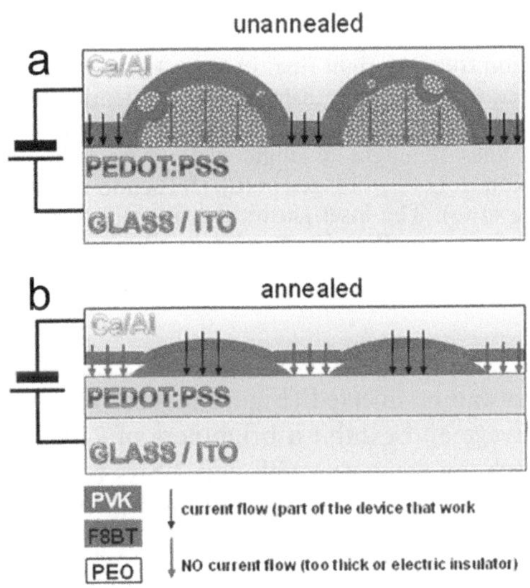


**Figure 5.10** Schemes of the device architectures of unannealed (a) and annealed (b) nanofiber OLEDs based on the light-emitting polymer, poly[(9,9-dioctylfluorenyl-2,7-diyl)-*co*-(1,4-benzo-{2,1′-3}-thiadiazole)] (in this figure, F8BT). Annealing leads to separation of the PEO and the conjugated polymer components, and ultimately to current conduction and electroluminescence from the light-emitting nanofibers. Adapted with permission from Ref. 136, *ACS Nano*, 2011, **5**, 5572–5578. Doi: 10.1021/nn201029c. Copyright © 2011, American Chemical Society.

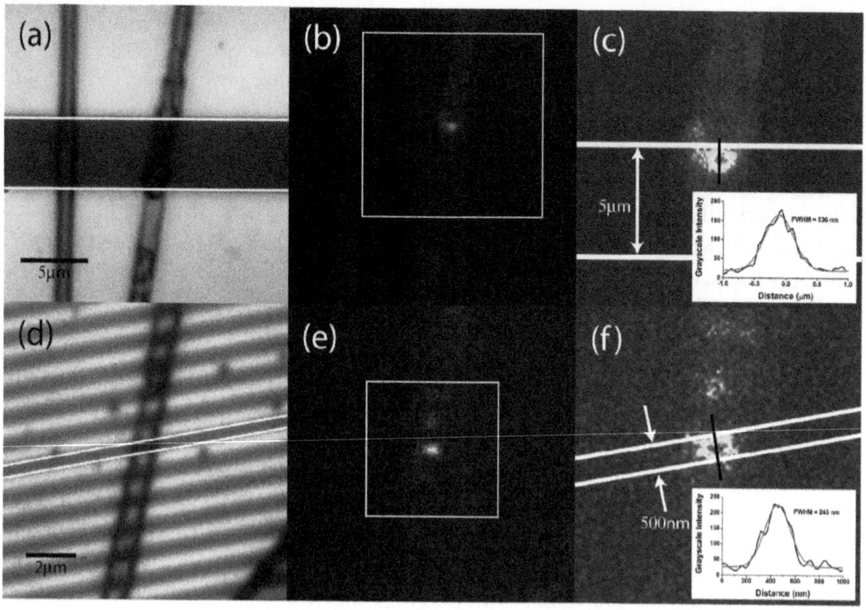


**Figure 5.11** Electroluminescence from an electrospun, light-emitting polymer nanofiber made of ruthenium(II) tris(bipyridine) and PEO deposited on inter-digitated electrodes. (a) Bright field micrograph of two fibers on a 5 μm inter-digitated electrode gap, highlighted by white lines. (b) Emission from the right fiber in (a). (c) Pseudo-color intensity rendition of the square in (b). The inset shows the intensity profile. (d) As in (a), for a single fiber on multiple, 500 nm inter-digitated electrode gaps. Here the white lines highlight a single gap where light emission occurs. (e) Emission from fiber imaged in (d). (f) Pseudo-color intensity rendition of square in (e). The inset shows the intensity profile. Reproduced with permission from Ref. 57, *Nano Lett.*, 2007, **7**, 458–463. Doi: 10.1021/nl062778+. Copyright © 2007, American Chemical Society.

and then the cathode results in effectively contacting the top of fibers, through which charge-carriers can be injected (Figure 5.10b). These OLEDs have 40% of surface fiber coverage and exhibit a brightness of 2300 cd $m^{-2}$ at 6 V.[136]

In another approach, implemented with nanofibers based on ionic transition metal complexes [ruthenium(II) tris(bipyridine)] embedded in PEO, inter-digitated electrodes are used instead of sandwich geometries.[57] Fibers are deposited across the gap between electrodes, and injected carriers recombine yielding luminescence from a point source that has sub-wavelength dimensions ($\sim$0.2 $\times$ 0.3 $\mu m^2$, Figure 5.11).[57]

A different example of miniaturized OLEDs, excellently integrated with nanofiber fabrication technologies and geometries, does not rely on conventional, planar architectures.[137] These devices are realized *via* coaxial electrospinning

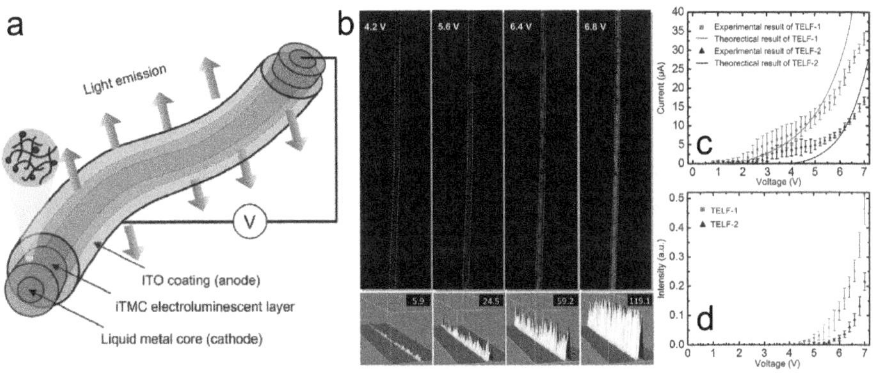


**Figure 5.12** (a) Schematic structure of a metal-core, ITO-shell electroluminescent electrospun fiber. The active medium is an ionic transition-metal complex-based electroluminescent polymer. (b) Electroluminescence of a fiber (TELF-1) upon applying different voltages to the device, under nitrogen atmosphere. Top panels: luminescence micrographs. Bottom panels: corresponding emission intensity profiles. (c) Experimental and theoretical results achieved for the current–voltage characteristics of two different fibers (TELF-1 and TELF-2, which indicate thinner and thicker fiber samples, respectively, obtained by varying electrospinning flow rates). (d) Experimental data for the current–luminance characteristics. Adapted with permission from Ref. 137, *ACS Nano*, 2012, **6**, 622–628. Doi: 10.1021/nn204055t. Copyright © 2011, American Chemical Society.

(Section 2.3.1) of the liquid metal Galinstan[§§] as core, an electroluminescent layer composed of a ruthenium-based ionic transition-metal complex embedded in PEO as shell and an outer coating obtained by ITO evaporation. The electroluminescent fibers show an external electroluminescence efficiency, $\eta_{EL}$, of almost 0.3% and a luminance of 23 cd $m^{-2}$ (Figure 5.12).[137] Future experiments may aim to combine such liquid metal/organic semiconductor core-shell nanofibers with a further ITO shell, produced by co-electrospinning as well. This would allow OLED nanofibers to be extruded fully in continuous electrospinning runs, allowing a much more effective integration of light-emitting devices based on organic semiconductors with technical textiles. Importantly, similar fabrication methods could be applied to the continuous electrospinning of photovoltaic nanofibers as well.

§§The liquid metal Galinstan (composed of Ga, In, and Sn in approximate percentages of 68%, 21% and 10%, respectively) is reported to have $\chi = 3-4 \times 10^4$ S $cm^{-1}$ at room temperature, and a work function around 4.1–4.4 eV, which make it a good cathode material for injecting electrodes into organic semiconductors.[137] For these reasons, and for its liquid phase at room temperature, it may be an excellent choice for use in electrospinning processes and in combination with other techniques for the production of self-contacted, electrically excitable polymer nanostructures.

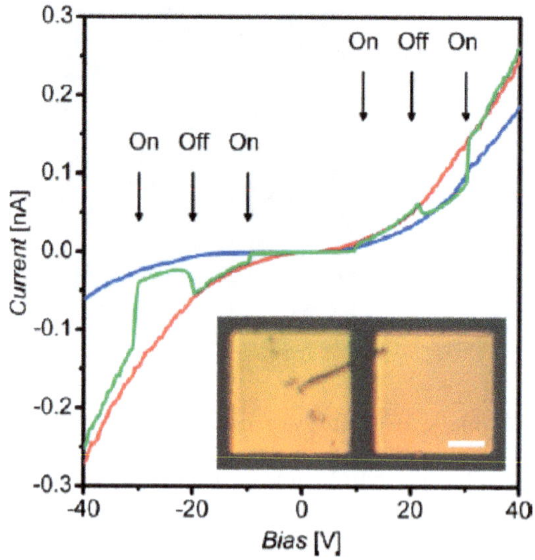


**Figure 5.13** Current–voltage characteristics of a top-contacted F8T2 nanowire, collected in the dark (blue line) and under 405 nm illumination (red line). Reversible switching is measured by stopping the 405 nm illumination (green line). Inset: optical microscopy image of a typical top-contacted device. Scale bar = 10 μm. Adapted with permission from Ref. 70, *Adv. Mater.*, 2006, **18**, 2379–2383. Doi: 10.1002/adma.200601012. Copyright © 2006, Wiley-VCH Verlag GmbH & Co. KGaA.

## 5.3.2 Photo-detectors and Solar Cells

Many examples of photo-detectors have been demonstrated by using polymer nanofibers and nanowires as absorbing and charge-transporting medium. For instance, membrane-templated F8T2 nanowires can be used within two-terminal devices (Figure 5.13), and the current flowing for electric voltage bias up to 40 V between the two electrodes is found to change drastically under illumination at a wavelength of 405 nm (which matches well with the absorption spectrum of the nanomaterial).[70] These devices show a quantum yield of about 0.1%, and a photo-responsivity (amount of photo-generated current per incident light power) of 0.4 mA $W^{-1}$ under an illumination fluence of 40 W $cm^{-2}$.

Templated, multisegmented nanorods alternating Au and a copolymer of 3-hexylthiophene and of a thiophene unit tethered with $C_{60}$ have a dark resistivity of $1.2 \times 10^5$ Ω cm (corresponding to a conductivity, $\chi$ of $8.3 \times 10^{-6}$ S $cm^{-1}$), dropping to $8.4 \times 10^3$ Ω cm ($\chi$ of to $1.1 \times 10^{-4}$ S $cm^{-1}$) under illumination by white light at a fluence of 5 mW $cm^{-2}$.[138] Hybrid nanorods embedding a *p-n* junction of PPy and CdS show an interesting light-dependent rectifying behavior, namely a rectification ratio (*i.e.* the ratio between the current values respectively measured by applying a forward and a reverse bias of 5 V at the rod terminations) increasing from 8 to 13 upon increasing the fluence of white light from 1.12 to 5.76 mW $cm^{-2}$.[139]

Single-bilayer nanotubes of PPV and carbonized PPV can be realized by CVD polymerization in the pores of a membrane, subsequent carbonization at 750 °C for 1 hour under argon and finally a second CVD polymerization step.[140] In two-terminal devices and under illumination by a Xe lamp with fluence of 1.5 W $cm^{-2}$, these nanostructures show a quantum yield of a few percent, and a photo-responsivity around 15 mA $W^{-1}$. Other examples of analogous nanofiber devices, but relying on three-terminal architectures, such as photo-transistors, are described in Section 6.3.1.3.

In addition, organic conjugated materials generally exhibit broad absorption spectra that match well with the spectrum of solar light, which, combined with the relatively good properties of charge transport (Chapter 6), can be exploited for the realization of quite efficient photovoltaic devices. These organic solar cells have attracted increasing interest in the last few years.[141–145] Here, the goal is absorbing sunlight photons to generate electricity, which is done through breaking photo-generated excitons. A photovoltaic cell is therefore basically an asymmetric, diode-like device just as an OLED, but with an inverse applied bias. In this way, the hole and the electron of a photo-generated exciton can be separated, transported and ultimately collected at the electrodes. As usual, efficiency (in terms of amount of collected photo-generated charges per incident photon or of power conversion) and stability (*i.e.* lifetime) have to be optimized to make devices suitable for practical applications. As in OLEDs, various device architectures can be realized. Since absorption intensities of organics are intense, thicknesses in the range of 100 nm are enough to collect light efficiently. The simplest organic solar cell could be built by a single active layer sandwiched between two electrodes, one of which is optically transparent (so through it light comes onto the organic material), exactly like the simplest OLED. In this case, the organic layer has to serve both as photon absorber and as effective transporting layer to transfer separated charges to the contacts. In this respect, *heterojunctions*, namely interfaces between two materials that have different HOMO and LUMO levels and that are deposited sequentially to form a multilayer device, greatly help to dissociate excitons (the latter are much more stable than in bulk inorganic semiconductors due to the large binding energy as introduced in Sections 1.1 and 5.1.1). Excitons migrating at the interface will firstly evolve to a charge transfer state. One material will serve as electron acceptor ($A^-$) and will be in charge of transporting safely (*i.e.* without letting them recombine) electrons to the positively biased electrode, the other material will serve as electron donor ($D^+$) and will transport holes to the negatively biased electrode. The acceptor is the junction material that has the highest electron affinity and ionization potential (a good example is fullerene, $C_{60}$). In multilayer devices, excitons should therefore diffuse at the junction without recombining, to be then broken into separated charges. This could limit the device efficiency because typical diffusion lengths of excitons are small (in the range from a few to about

20 nm), which is the reason why *bulk heterojunctions* are preferably used.[146,147] In bulk heterojunction devices the electron donor and acceptor are intimately mixed in a blend. For example, $C_{60}$ or other acceptors are often blended with conjugated polymers, such as MEH-PPV, P3HT, *etc.* On polymer chains $P^+$ polarons play an important role in the overall charge transport as described in Section 5.1.1. Hole and electron blocking layers can be added near electrodes to improve transport mechanisms. Of course, it is expected that bulk heterojunctions are somehow formed across phase-separated domains, each of which can be richer in one of the components of the blend. For example, ordered core-shell nanorods of P3HT and of the acceptor phenyl-$C_{61}$-butyric acid methyl ester (PCBM) reach a power conversion efficiency of 2.0% upon thermal annealing.[148] Polymer nanofibrous networks, which may form due to phase separation effects, can control the nanoscale morphology of active layers and consequently affect device performances, as studied for P3HT nanowires forming in PCBM upon annealing.[149] In this way, power conversion efficiencies are of 2.7%. Similar efficiency performances are found for devices based on regioregular poly(3-pentylthiophene) or poly(3-butylthiophene), forming beautiful nanofibers in solution or directly in PCBM, with widths of about 15 nm and lengths of many μm (Figure 5.14a,b).[150,151] The nanostructure in the device is found to be critical for the device performances, with a pivotal role played by the network of nanofibers.[152] Sonication can be used to assist the formation of self-assembled nanofibers in the blends, leading to devices that have power conversion efficiency of almost 4.1%.[153]

An annealing-free method to produce ordered nanofibrillar domains of P3HT in PCBM is based on gradually adding a poor solvent, such as hexane, into a P3HT–*o*-dichlorobenzene or P3HT–chloroform solution.[154] The resulting power conversion efficiencies are up to 3.9%. In another approach, P3HT or poly(3-butylthiophene) nanofibers are prepared separately by self-assembly (Section 3.1) and then mixed with PCBM to form the active layer.[155,156] The resulting devices exhibit a power conversion efficiency up to 3.6%, a current density of the order of 10 mA $cm^{-2}$, an open-circuit voltage of about 0.6 V and fill factors around 0.5. Using CdS quantum dots, bound to crystalline P3HT nanowires by means of solvent-assisted grafting and ligand exchange, leads to a further improvement of power conversion efficiency, up to 4.1%.[157]

Important advances are expected from combining nanofiber-based solar cells with textile applications (Section 4.3.3), for realizing coatings generating energy from clothes, curtains, *etc.* Other supports may be envisaged including windows, provided that enough optical transparency is achieved in the photovoltaic fibers. Of course, the often irregular thickness and poor uniformity of nanofibrous mats can be drawbacks limiting performances. P3HT/PCBM nanofibers, obtained by removing the PVP shell of pristine core-shell fibers, have been tested to this aim for the production of so-called *solar cloths*,[158] and found to support a current density of the order of $10^{-6}$ mA $cm^{-2}$, and an open-circuit voltage of 0.12 V.[159] In this respect, one

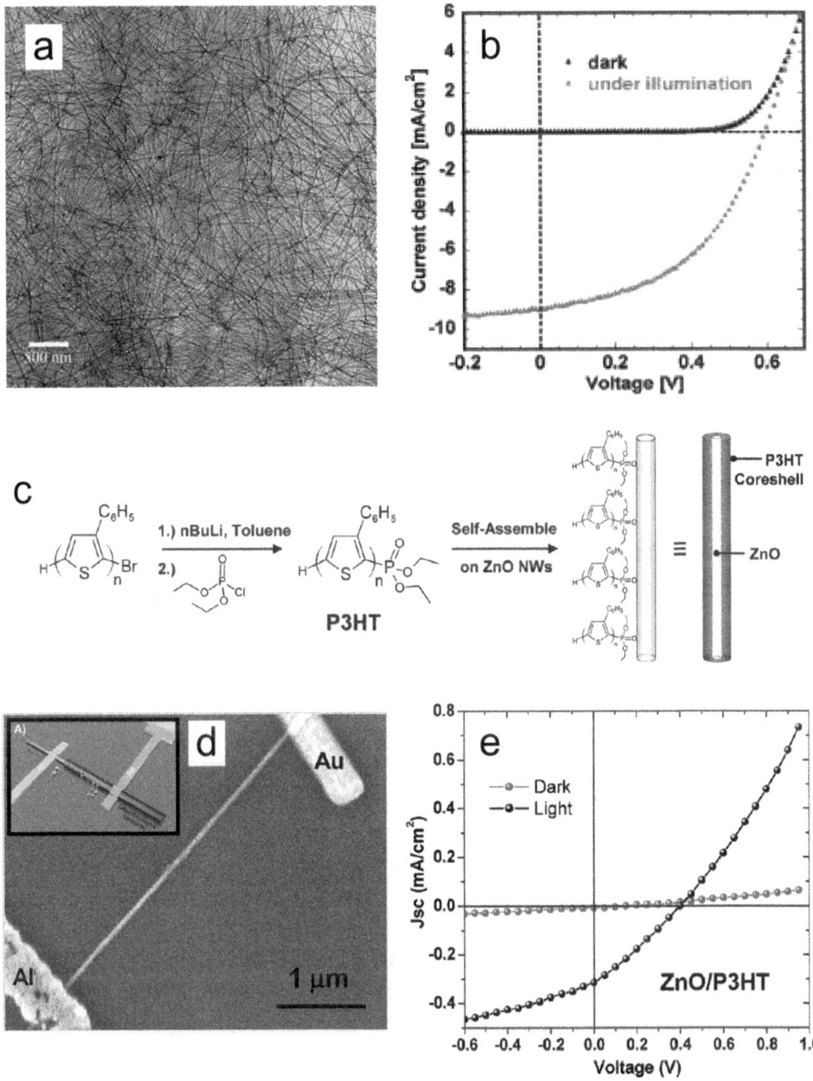


**Figure 5.14** (a) Typical morphology of regioregular poly(3-butylthiophene) nanowires forming upon blending in PCBM, imaged by TEM, and (b) current–voltage curves of a photovoltaic device based on these active layers, in the dark and under illumination at 100 mW cm$^{-2}$. (a, b) Adapted with permission from Ref. 150, *Chem. Mater.*, 2008, **20**, 6199–6207. Doi: 10.1021/cm801324m. Copyright © 2008, American Chemical Society. (c) Synthetic pathway for end-functionalizing P3HT with a phosphonic ester for forming a shell around ZnO nanowires. (d) SEM micrograph of an individual hybrid nanowire solar cell. The inset shows the corresponding device scheme. Contacts are fabricated by EBL. (e) Current–voltage characteristics in the dark and under illumination at 100 mW cm$^{-2}$. (c–e) Adapted with permission from Ref. 164, *Nano Lett.*, 2010, **10**, 334–340. Doi: 10.1021/nl9036752. Copyright © 2010, American Chemical Society.

should also mention that, using optical fibers or drawing methods, various photovoltaic fiber architectures have been proposed, which are based on flexible wire supports and organic materials and have final diameters from hundreds of μm to about 1 mm.[160–163]

Finally, one can realize single nanowire and single nanofiber photovoltaic devices. These sophisticated architectures show lower power conversion performances compared to bulk devices, but are of remarkable and fundamental interest as model systems for studying heterojunctions. For instance, a hybrid solar cell based on end-functionalized polythiophenes grafted onto a ZnO nanowire reaches an efficiency of 0.036% (Figure 5.14c–e).[164] Core-shell approaches have also been applied using P3HT to wrap MWCNTs.[165] A different strategy uses a junction localized at the intermediate point of hybrid CdS/PPy wires, in which the inorganic and the organic components are not in a core-shell geometry but instead form each segment of the overall nanostructure.[166] In these cells, reported efficiencies are of 0.018%.

## 5.4 Waveguiding

Though exhibiting a refractive index ($n$), in the visible and infrared spectral range, that is lower than in most inorganic materials (which results in relatively weaker properties of light confinement), both conjugated polymers and blends of low molar mass chromophores and thermoplastic polymers are largely used to fabricate miniaturized waveguides that allow photons to be confined and driven. For example, in the visible, most conjugated polymers have refractive index in the range of 1.4–2.5, and in many optically inert thermoplastics, such as PS, PMMA, *etc.*, $n$ is around 1.4–1.6. These values can be sufficient to provide some confinement of photons within a solid micro- or nanostructure made of these materials, surrounded by air or deposited on a substrate (*on-chip waveguides*) with lower refractive index. Indeed, these configurations allow effects of internal reflection of light into the organic material. Polymer nanofibers are therefore natural candidates as optical waveguides, offering sub-wavelength light confinement (*i.e.* in lateral dimensions smaller than the light wavelength), high potentiality for pushing miniaturization of optical circuits, very low cost of production, structural flexibility and good performances as both passive (using polymers that do not absorb at the operation wavelength) and active (using light-emitting and eventually gaining polymers and blends) photonic elements.

The waveguiding properties of polymer nanofibers can be characterized in several ways. All of these involve coupling the light coming out from the wires to free space and then to collection optics for observation and measurements. Light can be self-emitted following optical excitation, or escape following the launch into the fiber through external optics. For example, fluorescence micrographs showing bright spots at the terminations of nanofibers, optically excited at a particular location along their length, are the most used qualitative

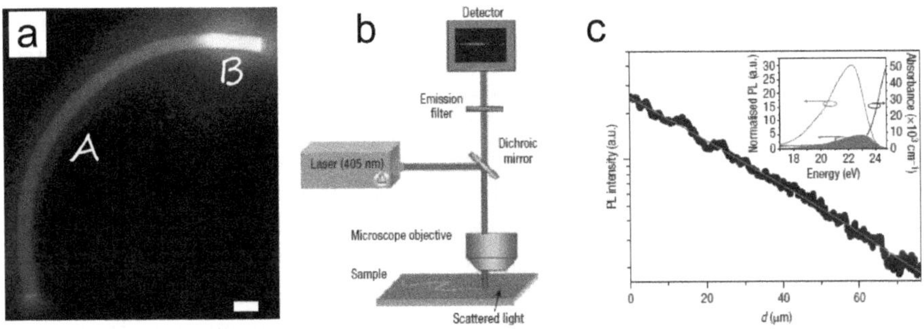


**Figure 5.15** (a) Fluorescence micrograph of an electrospun polymer waveguide. The bright spot (B) is in the region that is directly excited by a laser. The fluorescence can be measured at any point (A) along the longitudinal fiber axis. Note the bright tip at the bottom-left corner of the micrograph, which is not excited directly. Scale bar = 2 μm. (b) Schematics of the experimental set-up used for characterizing polymer nanofiber waveguiding. (c) Dependence of guided light intensity on the distance (*d*) from the directly excited point to the collection point along the nanofiber longitudinal axis. Scattered dots are experimental data, and the superimposed line is the best fit to Equation (5.17). The inset shows the overlap of the polymer nanofiber PL and absorption spectra. (b, c) Adapted with permission from Ref. 62, *Nat. Nanotechnol.*, 2008, **3**, 614–619. Doi: 10.1038/nnano.2008.232. Copyright © 2008, Nature Publishing Group.

tool to demonstrate waveguiding in these nanostructures (Figure 5.15a). This approach can be made quantitative, allowing the loss coefficient, $\alpha$, of the waveguide to be estimated from the Beer–Lambert expression (which is a wavelength-dependent version of Equation (2.20), or, equivalently, a gainless version of Equation (5.7)):

$$A_{em}(\lambda) = A_{em,0}(\lambda)\exp[-\alpha(\lambda)d], \tag{5.17}$$

where $A_{em}$ indicates the light intensity measured at a given point (we label this point with $A$ in Figure 5.15a), which is not excited directly (such as, for instance, the exit tip of the nanofiber), $A_{em,0}$ indicates the light intensity at the excitation point ($B$ in Figure 5.15a) and $d$ indicates the distance between the two points along the fiber longitudinal axis. Most of the time, the loss coefficient is found in units of cm$^{-1}$. For polymer nanofibers and nanowires it can also be conveniently expressed in dB per unit length (for instance in dB μm$^{-1}$), as follows:

$$a(\text{dB}\,\mu\text{m}^{-1}) = \frac{10}{d(\mu\text{m})}\ \text{Log}\,\frac{A_{em,0}}{A_{em}} \tag{5.18}$$

$A_{em}$, $A_{em,0}$ and $d$ are all immediately measurable by fluorescence micrographs of light-emitting nanofibers. Therefore, propagation loss measurements can be performed by simply imaging the PL of a single nanofiber, excited by a tightly focused laser beam. The intensity of the light exiting the tip of the nanofiber can be measured upon varying the distance of the excitation spot from the tip, thus easily obtaining a ($A_{em}$ *vs.* $d$) plot whose fit provides the value of $\alpha$ according to Equation (5.17). A long-pass filter is normally used to cut the exciting laser light, and a spectral selection of detected light also allows the spectral dependence of the loss coefficient to be measured (Figure 5.15b). Another option is based on assuming that the intensity of the scattered light observed from the surface of the nanofiber body is directly proportional to the intensity of light propagating along the fiber axis. Such an assumption is reasonable when scattering centers are uniformly distributed along the fiber***. In this case, the decrease of the scattered light as a function of the position along the nanofiber longitudinal axis, while moving away from the directly excited area, represents the attenuation of the light propagating in the waveguide. The $d$ quantity of Equation (5.17) simply becomes the distance from the excited region to a variable point along the fiber length, which means that a single micrograph can provide all the information needed to obtain a ($A_{em}$ *vs.* $d$) plot and measure $\alpha$ (Figure 5.15c). In most reported cases, the two techniques described above give values of the waveguide losses that are in agreement within experimental errors.[62]

In the last few years, a few groups have applied these methods and explored the potentialities of polymer nanofibers as waveguides†††. Semicrystalline wires, templated in the pores of anodic alumina membranes and produced by several different conjugated polymers (Section 3.3.1), can be effectively used as active waveguides. For instance, PFO wires ($n \cong 1.9$ at 460 nm) with diameter of 200–300 nm and length of about 13 μm show a far-field epifluorescence emission that clearly presents bright spots at the ends of the polymeric fibers,[76] evidencing the waveguiding of emitted light along the wires. The optical loss coefficient is 1060 $cm^{-1}$, which corresponds to 0.46 dB $\mu m^{-1}$.[76] Polymeric nanotubes realized by a wetting template method have loss coefficients of about 480 $cm^{-1}$ (0.21 dB $\mu m^{-1}$) and 300 $cm^{-1}$ (0.13 dB $\mu m^{-1}$), for infrared and red light, respectively.[169] Mechanically drawn polymer microfibers embedding inorganic (CdSe/ZnS) quantum dots or luminescent dyes have loss coefficient of about 50 $cm^{-1}$ ($\sim$0.02 dB $\mu m^{-1}$), and guide light over distances of millimeters.[170,171] Electrospun nanofibers can also work as effective miniaturized waveguides. This has been demonstrated by nanofibers made of a polymeric photo-resist and incorporating luminescent CdSe/ZnS core/shell quantum dots,[172] of rubrene,[56] of conjugated polymers such as poly[(9,9-dioctylfluorenyl-2,7-diyl)-*co*-{1,4-benzo-(2,1′,3)-thiadiazole}][62] and of thermoplastic polymers embedding organic dyes.[55] Reported optical loss coefficients are down to 200–300 $cm^{-1}$ (about 0.10 dB $\mu m^{-1}$).[55,62]

***Scattering centers can be due, for instance, to local agglomerations of the organic material or to nanoscale variations of density and compositional defects.

†††Besides polymer nanofibers, waveguides are also demonstrated by quasi-one-dimensional nanostructures made by compounds of lower molecular weight, such as several classes of oligomers. An example is given by blue-emitting parahexaphenyl fibers grown on mica substrates.[167] Negligible losses are reported for self-assembled nanofibers of thiacyanine, over lengths of more than 200 μm.[168]

The sources of optical losses in polymer nanofibers are generally varied. First of all, one has to bear in mind that different wavelengths are confined with different efficiency in a nanostructure. In the literature, the expression for a cylindrical fiber is typically used to estimate this effect:[76,173,174]

$$\eta_P = 1 - \frac{(2.4e^{-1/\tilde{V}})^2}{\tilde{V}^3}, \tag{5.19}$$

where $\eta_P$ indicates the fractional guided power in the fundamental mode of the waveguide, and $\tilde{V} = (4r/\lambda) \times (n_{in}^2 - n_{out}^2)^{0.5}$. Here, $n_{in}$ and $n_{out}$ respectively indicate the refractive index in the organic material and in the surrounding medium, which is assumed homogeneous. On the basis of the previous formula, the confinement for visible light within polymer fibers that have radius in the scale of 100 nm is calculated to be quite high (>85%).[76,174] However, the equation also tells us that a wavelength dependence, with $\eta_P$ decreasing upon increasing $\lambda$, always exists in the waveguiding performances. This issue can be important, especially for some sensing applications requiring accurate spectral calibration.

As anticipated above and as also evidenced by the previous equation, the surrounding media play a relevant role in determining the effectiveness of internal reflection phenomena. The refractive index of the substrate material, on which the nanostructures are deposited, is demonstrated to be especially important in this respect.[62,76] For instance, optical losses of $5 \times 10^2$ cm$^{-1}$ and $7 \times 10^2$ cm$^{-1}$ are measured for electrospun fibers of the conjugated polymer, poly[(9,9-dioctylfluorenyl-2,7-diyl)-*co*-{1,4-benzo-(2,1′,3)-thiadiazole}], deposited on glass ($n = 1.51$) and silicon ($n = 3.87$), respectively.[62] Evidencing the coupling of initially guided radiation into the underlying substrate, this mechanism can be advantageous for some applications. For example, it can be exploited to eventually transfer a part of waveguided energy to on-chip integrated detectors for sensing and nanophotonics.

Another obvious factor decreasing the intensity of emitted light along the fiber length is *self-absorption*, due to the spectral overlap between the emission and the absorption spectra of the organic material. Differently from Equation (5.19), self-absorption tends to reduce especially the short-wavelength (*i.e.* high-energy) component of the luminescence emission, which is the most overlapping with the long-wavelength tail of absorption spectra (Section 5.1.1). Gradually reducing the intensity of the short-wavelength component of guided light upon increasing the distance ($d$) from the excited (point $B$) to the analyzed (point $A$) region (Equation (5.17)), namely increasing the ($B$ to $A$) optical path available for self-absorption, leads therefore to a more and more red-shifted spectrum. This spectrum is finally shown by the light emitted from the analyzed region ($A$) of the nanofiber (*i.e.*, for instance, from the fiber tip). In other words, for self-absorption, $\alpha$ is higher at lower luminescence wavelengths. Self-absorption is generally quite relevant in active polymer nanofibers. For this reason, negligible losses reported for some organic nanostructures mainly relate to systems with large Stokes shifts.[168]

*Scattering from surface roughness* is another possible phenomenon that reduces the number of photons internally guided along the nanofiber length. Importantly, surface roughness and the scattering thereof can be controlled quite finely by nanofiber surface porosity or patterning (Sections 4.2.1.3 and 5.6). On the other hand, surface scattering is difficult to remove completely, as in most cases the roughness is inherently induced during the fabrication processes. For example, investigating the surface of single, dye-doped PMMA electrospun fibers by means of AFM allows a root-mean-square roughness of about 6 nm to be estimated.[55] This is much smaller than the wavelength of guided light, thus helping to reduce surface scattering. However, a fully smooth surface (with surface corrugation below, say, 1–2 nm) is very difficult to realize in polymer nanostructures, due to effects related to the finite sizes of the macromolecules, molecular aggregations and entanglements, *etc.* Analogously to substrate coupling, surface scattering may even be useful in some respect, allowing to out-couple a part of photons from the body of the fibers into free space. It is also noticed that losses through surface scattering can be related to the orientation of molecular transition dipoles within the nanofibers.[76,175]

Furthermore, *Rayleigh scattering* can occur, induced by irregularities and defects of characteristic size well below the light wavelength, which can be present along the fiber. Also Rayleigh scattering leads to larger optical losses upon decreasing the light wavelength, according to the law:[76,115,176]

$$\alpha = \alpha_0 \lambda^{-4}, \tag{5.20}$$

where the pre-factor, $\alpha_0$, depends on the density, radius and refractive index of the scattering centers.

Distinguishing the different contributions to optical losses is not trivial. An option could be calculating, firstly, the self-absorbed component on the basis of the overlap integral between the absorption and emission spectra of the nanofiber material. The spectra should be rigorously measured in the fiber, where they can be quite different from those of solutions and films due to the different, and often more ordered, supramolecular packing in the nanostructure. Once the self-absorption is estimated, this can be subtracted from the overall optical losses, and if spectral information is available, one can check *a posteriori* if the eventual residual absorption follows the wavelength dependence expected for Rayleigh scattering (Equation (5.20)). Or, *vice versa*, one can estimate at first the contribution of Rayleigh scattering, which however would require a separate estimation of the size and density of scattering defects within the polymer nanofibers (from XRD or TEM data).[76] For instance, a contribution of the order of 25–30% of the overall waveguide losses is attributed to Rayleigh scattering for PFO semicrystalline nanowires.[76]

## 5.5 Lasing

The emission spectra of polymer nanofibers made of compounds with optical gain can show the line-narrowing and intense peaks that are typical of lasing phenomena (Section 5.1.3). As is well known,[35] in order to achieve lasing one needs not only an active medium with sufficiently high stimulated emission cross-section, but also the occurrence of optical feedback nearby (induced by reflection or scattering phenomena). In particular, in polymer nanofibers and nanowires, microcavity effects providing optical feedback can be controlled topologically, on the basis of their dependence on the nanostructure cross-sectional shape,[177] length, ensemble morphologies in aligned or random mats, *etc.*

In general, from assemblies or non-woven mats of nanofibers made of an optically gaining organic material, one measures a large number of lasing (or simply spectrally narrowed) emission modes. This feature can be related to the fiber size dispersion. These structures, with disordered positions and orientations of fibers, are very complex in their resulting photonic properties, and so-called *random lasing* effects could be present as well. Optical paths within closed loops, in which light can travel, can be determined in these ensembles by inter-connects at entanglement points where more nanofibers intersect each other. An extensive work on random lasing in organic nanofibers has been carried out by F. Quochi and coworkers, focusing on epitaxially grown nanostructures of *para*-sexiphenyl.[178–183] In assemblies of nanofibers, random lasing is reported in the blue (around 425 nm), with a very low threshold fluence of 0.5 μJ cm$^{-2}$ (under excitation by the second harmonic of a Ti:sapphire regenerative amplifier, $\lambda$ = 392 nm, pulse duration 150 fs, repetition rate 1 kHz), and can combine with interplaying ASE.[178] Individual nanofibers show random laser effects as well, related to light scattering and partial reflections occurring along the fiber length at points where small cracks are present, which can be modeled by coherent light propagation in one-dimensional random media.[180] Measured threshold fluences for single fiber random lasing action are below 12 μJ cm$^{-2}$.[180] It is likely that, though developed on crystalline oligomer nanostructures, many of these findings can be extended to the lasing behavior of polymer nanofibers.

In other systems the emission is dominated by different lasing mechanisms. Many single polymer nanofibers clearly show the spectral features of a Fabry–Pérot cavity, namely equally spaced modes (in the case of spontaneous emission,[76] or equivalently below the threshold for lasing), followed by a sudden increase of the intensity of one of these modes above the threshold for lasing.[55] For Fabry–Pérot cavities, the light travels back and forth across the fiber length, and photons are reflected in correspondence of the sharp discontinuity in the refractive index at the two terminations of the wire. Analogously to waveguiding (Equation (5.19)), the simplified, prototypical case of a cylindrical tubular structure helps in modeling the problem. For a wire of length $l$ and whose emission or modal distribution has central wavelength $\lambda_c$, the mode spacing, $\Delta\lambda_c$, of the Fabry–Pérot cavity is given by:[184]

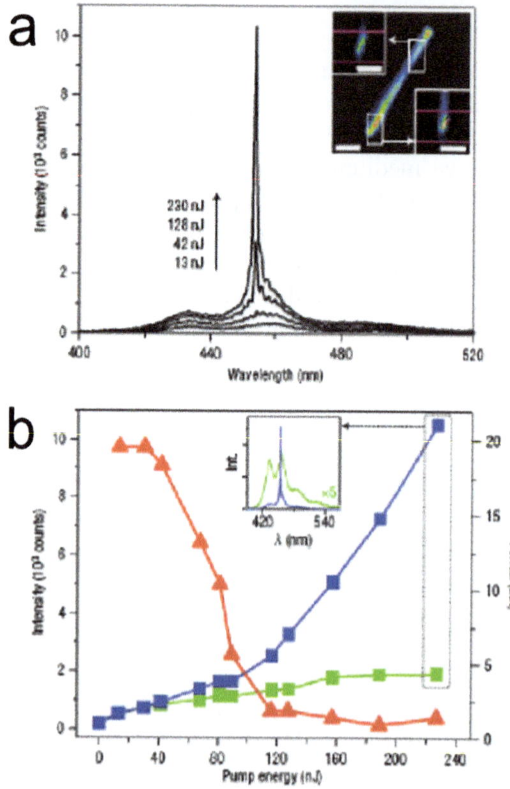


**Figure 5.16** (a) Lasing spectra collected from the tip of a PFO nanowire, for increasing pump energy at room temperature. Inset: emission micrographs of the wire (pump energy = 1.3 nJ). Scale bar: 2 μm. (b) Tip emission intensity (blue squares) and linewidth (red triangles) *vs.* pump energy. The emission from the wire body is also shown (green squares), and it is found to exhibit a much lower, almost linear increase upon increasing the excitation energy. Lines are guides to the eye. Inset: above-threshold spectra, from the wire tip (blue) and body (green, pump energy = 230 nJ). Adapted with permission from Ref. 174, *Nat. Nanotechnol.*, 2007, **2**, 180–184. Doi: 10.1038/nnano.2007.35. Copyright © 2007, Nature Publishing Group.

$$\Delta\lambda_c = \left(\frac{\lambda_c^2}{2l}\right)\left(n - \lambda_c \frac{dn}{d\lambda}\bigg|_{\lambda_c}\right)^{-1}, \tag{5.21}$$

where $dn/d\lambda$ is the refractive index dispersion. Planar end facets, observable in semicrystalline or crystalline nanowires, favor the onset of oscillations in Fabry–Pérot microcavities coinciding with the body of the wires. In addition, data in agreement with Equation (5.21) are collected from amorphous, electrospun nanofibers as well. Indeed, the previous expression correctly predicts the mode

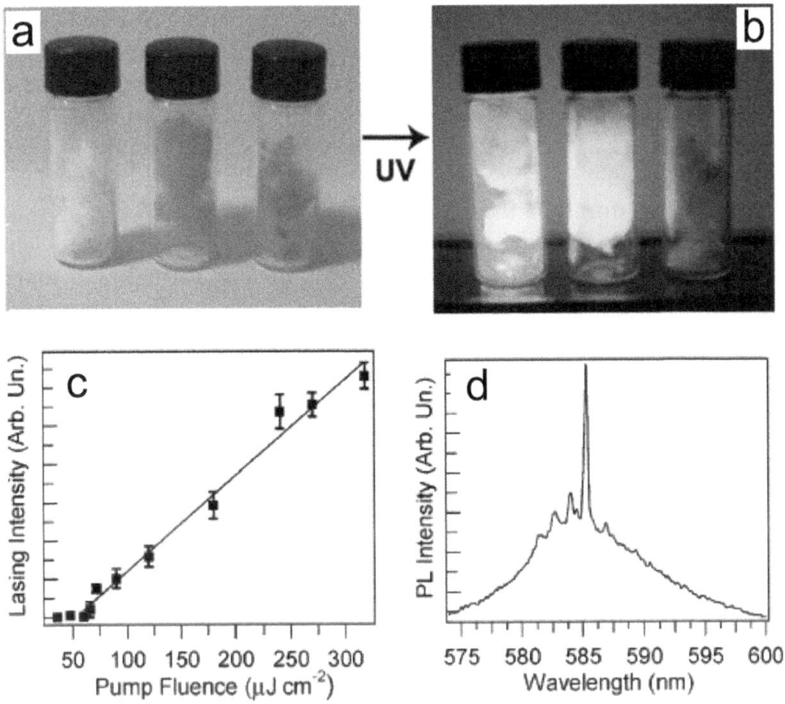


**Figure 5.17** (a, b) Photographs of electrospun, light-emitting nanofibers under white (a) and UV (b) light. The fibers are made by a PMMA matrix which embeds coumarin, rhodamine and Nile blue A Perchlorate (from left to right). (c) Laser emission intensity *versus* excitation fluence, for rhodamine-doped polymer nanofibers. The continuous line is a linear fit to the data collected by excitation fluences ranging from 60 to 350 μJ cm$^2$. (d) Exemplary single-fiber lasing spectrum. Excitation fluence = 150 μJ cm$^2$. Adapted with permission from Ref. 55, *Small*, 2009, **5**, 562–566. Doi: 10.1002/smll.200801165. Copyright © 2009, Wiley-VCH Verlag GmbH & Co. KGaA.

spacing observed not only in melt-processed, light-emitting semicrystalline PFO nanowires (Figure 5.16) fabricated by membrane-templating ($\Delta\lambda_c \cong 4.5$ nm, cavity length of 4.7 μm),[76,174] but also in lasing electrospun nanofibers (Figure 5.17) made of dye-doped PMMA ($\Delta\lambda_c \cong 1.3$ nm, cavity length of 87 μm).[55]

Another way to induce optical feedback in polymer nanofibers could consist in patterning the surface of the organic material by features whose periodicity roughly matches the emitted wavelength. This is the principle of so-called distributed-feedback microcavities, which are widely demonstrated in organic lasers,[36–39,185] but are still not well assessed in gaining polymer nanofibers. For this and other reasons, the motivations for developing specific methods to pattern light-emitting nanofibers are undoubtedly well grounded. Some possible approaches are described in the next section.

We now review a few numbers concerning the lasing performances of polymer nano- and microfibers, specifying for each case the used conditions of optical pulsed excitation, which may greatly influence the stimulated emission of gaining organic materials. Semicrystalline PFO nanowires exhibit lasing action from their tips, for pumping fluences above 1.7–3.0 mJ $cm^{-2}$ (excited by the third harmonic of a Neodymium-doped yttrium orthovanadate laser, $\lambda$ = 355 nm, pulse duration 0.7 ns, pulse repetition rate 1.25 kHz).[174] Non-cylindrical PS fibers doped by the red-emitting dye, 4-dicyanomethylene-2-methyl-6-*p*-dimethylaminostyryl-4H-pyran, and placed on a low-*n* material (Cytop©, $n$ = 1.34) to increase optical confinement, show a lasing threshold at 0.16–0.82 mJ $cm^{-2}$ (excited by the second harmonic of a Nd:YAG laser, $\lambda$ = 532 nm, pulse duration 0.5 ns, repetition rate 10 Hz).[177] PMMA electrospun nanofibers doped by rhodamine have lasing threshold of 60 μJ $cm^{-2}$ (again excited at $\lambda$ = 532 nm by a Nd:YAG microlaser, pulse duration 0.6 ns, repetition rate 100 Hz).[55]

## 5.6 Nanopatterned Nanofibers

Tourists travelling in the Middle East rapidly become acquainted with the beautiful and vivid color shadings observable when light is reflected by silk carpets, suddenly changing when looked at under different directions. This unique effect is due to a peculiar prismatic structure of *B. mori* silk fibers, which present a surface pattern making them able to reflect light at optical wavelengths according to a characteristic angular dependence. Nowadays, photonic nanostructures and periodic patterns with feature size of wavelength-scale have a lot of strategic applications in technologies including plasmonics,[11] enhancement of extraction efficiency of solid-state lightening devices such as OLEDs,[10–12] control of emission directionality,[186] selection of spectral modes in lasers by distributed optical feedback[185,186] and design of waveguides and reflectors by means of photonic crystals.[187] In all these applications, active polymer nanofibers can be potentially utilized as valuable and low-cost alternatives to standard inorganic components, thus opening interesting perspectives in the fields of patterned miniaturized light sources, optical amplifiers in the visible and near infrared, modulators and optical switches.[188] However, fully controlling the optical density of states and making active polymer nanofibers tailorable in terms of emission directionality and photon out-coupling needs not only gentle fabrication approaches for producing nanofibers retaining the light-emission properties of pristine materials, but also accurate patterning methods, by means of which one can texture the surface of the fibers by well-defined features.

How to control the surface pattern of polymer nanofibers? Fortunately, this can be done by a variety of both bottom-up and top-down methods, with good effectiveness. For example, the phase separation related to the rapid evaporation of solvents with low boiling point during electrospinning can be a smart approach to fabricate polymer nanofibers with controlled porosity and surface texturing (Section 4.2.1.3).[189] In addition, a few top-down technologies can be applied to polymer nanofibers. Room temperature NIL, introduced in Section 1.2.1, is an instructive example. Avoiding thermal cycles that could negatively impact both the shape of pre-formed nanofibers and their light emission efficiency, room temperature NIL, following either electrospinning or other techniques for fabricating polymer fibers, is indeed a powerful tool to pattern single nanostructures. The demonstrated resolution of the transferred features is up to the 100 nm scale,[62] although the cross-sectional variations, such as flattening eventually occurring in fibers due to the high pressures (up to 300 MPa) applied during the process, have not yet been investigated in sufficient detail. In general, the depth of imprinted features depends both on that of the master and on processing parameters, particularly on imprinting time, which is quite a lot longer compared to conventional, high-temperature NIL, according to the time-temperature super-position principle (Section 1.4.1). Furthermore, a proper choice of the imprinting parameters allows isolated features to be produced from individual nanofibers, without residual bottom polymer layer.[62] This is a quite general characteristic of soft molding and NIL methods,[190,191] and can be advantageous in view of application to nanophotonic devices where optical or electrical coupling between adjacent features should be as low as possible, in order to improve optical feedback across imprinted patterned cavities or to produce single electroluminescent point sources. By manual or automated positioning, the relative orientation of the features on the master with respect to the fiber axis can be controlled with various levels of precision. Two limit cases are especially interesting for nanophotonics. Imprinting with master gratings having 100 nm-scale, parallel and periodic features (as in one-dimensional photonic crystals)[187] can transfer the pattern onto nanofibers with the grating periodicity direction (i) parallel or (ii) perpendicular to the fiber axis, respectively.

In the first case, a nanopatterned cavity is produced along the axis, which consists in two media (polymer and air) alternating with the periodicity of the imprinted features, and exhibiting different refractive index. At each imprinted corrugation, a part of the energy carried by the electromagnetic wave is transmitted and another part is back-reflected, and the super-position of different waves counter-propagating and travelling along the fiber axis can lead to various intriguing photonic effects. Furthermore, polymers exhibit light absorption and eventually optical gain (Section 5.1.3), which extends the list of properties that are periodically modulated because of the imprinted pattern. A first consequence can be the enhancement of the emission out-coupling to free space from the side surface of the nanofiber. Such emission therefore becomes detectable at some angular direction (Figure 5.18a). Angle-resolved PL measurements on single fibers allow this enhancement to be observed, by moving a detection optical fiber perpendicularly to the direction of the patterned grooves. The physical law rationalizing this phenomenon is the conservation of the component of the photon wavevector along the longitudinal axis (in-plane) direction:[186]

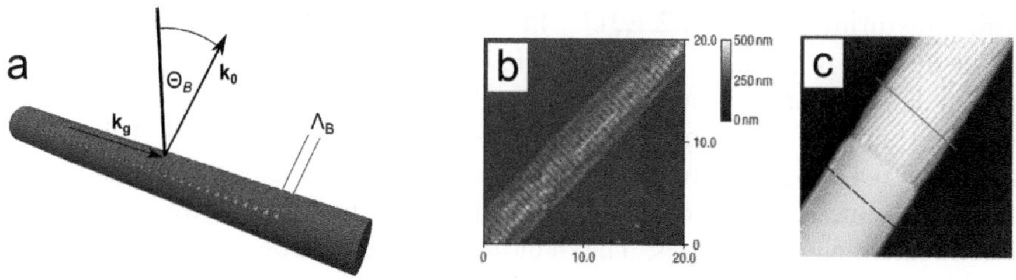


**Figure 5.18** (a) Scheme of Bragg scattering of light out from a patterned polymer nanofiber. (b, c) AFM micrographs of patterned polymer nanofibers with features perpendicular (b) or parallel (c) to the fiber length. (b) Scan size = 20 × 20 μm$^2$, vertical scale = 500 nm. (c) Scan size = 25 × 25 μm$^2$, vertical scale = 1 μm. (b) Adapted with permission from Ref. 62, *Nat. Nanotechnol.*, 2008, **3**, 614–619. Doi: 10.1038/nnano.2008.232. Copyright © 2008, Nature Publishing Group. (c) Adapted with permission from Ref. 193, *Nanotechnology*, 2010, **21**, 215304. Doi: 10.1088/0957-4484/21/21/215304. Copyright © 2010, IOP Publishing Ltd.

$$\mathbf{k}_0 \sin \Theta_B = \pm \mathbf{k_g} \pm m_B \mathbf{G}\,, \tag{5.22}$$

where $\mathbf{k}_0$ and $\mathbf{k_g}$ are the wavevectors of the free and of the in-fiber, guided electromagnetic waves, respectively, $\Theta_B$ is the photon out-coupling direction measured from the normal to the plane where the fiber lies, $m_B$ is an integer (0, $\pm 1$, $\pm 2$, ...) and **G** is the Bragg vector related to the imprinted, one-dimensional photonic crystal along the fiber axis. Due to the grating periodicity, in the previous equation one has to consider an integer number of involved Bragg vectors, analogously to the laws used for electrons in solid-state physics when dealing with crystalline semiconductors.[192] In terms of the photonic crystal produced on the surface of the light-emitting nanofiber, Equation (5.22) relates the observable angular-dependent emission to the crystal band-structure whose reciprocal lattice is described by the **G** vector. The involved wavevectors and angular coordinate are schematized in Figure 5.18a, and the wavevector magnitudes are:

$$|\mathbf{k}_0| = \frac{2\pi}{\lambda}\,, \tag{5.23a}$$

$$|\mathbf{k_g}| = \frac{2\pi n_{eff}}{\lambda}\,, \tag{5.23b}$$

$$|\mathbf{G}| = \frac{2\pi}{\Lambda_B}\,, \tag{5.23c}$$

where $\Lambda_B$ indicates the period of the on-fiber imprinted pattern, and $n_{eff}$ is the so-called effective refractive index, which takes into account the refractive index of air and of the polymer, the patterned geometry and the spatial distribution of the electric field associated to the guided light mode.

Also the polarization properties of emitted light depend on the periodic features patterned on the surface of polymer nanofibers. When a grating is present on nanofibers, one can distinguish between *s*-polarized and *p*-polarized light. These two states describe an electromagnetic wave, propagating in the free space (*i.e.* already escaped from the fiber body), whose electric field is parallel or perpendicular to the grooves of the grating, respectively. In other words, if the **G** vector is parallel (namely if the patterned, periodic features are perpendicular) to the longitudinal axis of the light-emitting fiber as shown in the micrograph in Figure 5.18b, *s*-polarized light corresponds to modes that are called transversal electric (TE) in the waveguide constituted by the polymer fiber deposited on a substrate, whereas *p*-polarized light corresponds to so-called transversal magnetic (TM) guided modes. In general, both *s*-polarized and *p*-polarized light is present in the fiber emission that is out-coupled to free space by Bragg scattering as described by Equation (5.22). From patterned MEH-PPV, *p*-polarized peaks are blue-shifted by 18–46 meV with respect to *s*-polarized emission maxima,[62] which indicates that $n_{eff}$ values are higher for Bragg-out-coupled TE modes. In particular, the difference in the refractive indexes of TE and TM modes is found to be of the order of a few percent, which is high enough to lead to well-measurable optical shifts.

Furthermore, experiments on poly[(9,9-dioctylfluorenyl-2,7-diyl)-*co*-(1,4-benzo-{2,1′-3}-thiadiazole)] light-emitting fibers show a decrease of the PL polarization ratio (measured with unpolarized excitation and selection of the *Z*-polarized emission), observable upon imprinting a grating with features perpendicular to the fiber length.[193] On the contrary, an enhancement of the polarization ratio of emitted light is found upon imprinting a grating with features parallel to the fiber length.[193] This demonstrates that the emission polarization properties of polymer nanofibers can be tailored by controlling the orientation and geometry of surface nanopatterns. In particular, the polarization enhancement can be a consequence of the mechanical response of conjugated polymer chains under NIL, which may lead to a further alignment of the polymer backbones (and ultimately of the optical transition dipoles) within the volume defined by the imprinted features.[194] This nanofluidic, viscoelastic effect can improve the degree of alignment along the fiber length, if features are parallel to this direction as displayed in Figure 5.18c. Another mechanism, inherently optical, is due to preferential Bragg scattering of *s*-polarized light (which here, being the **G** vector perpendicular to the longitudinal axis of the fiber, corresponds to a polarization direction along the fiber length).

## References

1. J. H. Burroughes, D. D. C. Bradley, A. R. Brown, R. N. Marks, K. Mackay, R. H. Friend, P. L. Burn and A. B. Holmes, *Nature*, 1990, **347**, 236–294.
2. R. Farchioni and G. Grosso (ed.), *Organic Electronic Materials*, Springer-Verlag, Berlin, Heidelberg, 2001.
3. W. P. Su and J. R. Schrieffer, *Proc. Natl Acad. USA*, 1980, **77**, 5626–5629.
4. A. J. Heeger, S. Kivelson, J. R. Schrieffer and W.-P. Su, *Rev. Mod. Phys.*, 1988, **60**, 781–850.
5. X.-Y. Zhu, Q. Yang and M. Muntwiler, *Acc. Chem. Res.*, 2009, **42**, 1779–1787.
6. M. T. Bernius, M. Inbasekaran, J. O'Brien and W. Wu, *Adv. Mater.*, 2000, **12**, 1737–1750.
7. S. Han, X. Feng, Z. H. Lu, D. Johnson and R. Wood, *Appl. Phys. Lett.*, 2003, **82**, 2715–2717.
8. H.-K. Kim, K.-S. Lee and J. H. Kwon, *Appl. Phys. Lett.*, 2006, **88**, 012103.
9. T.-H. Han, Y. Lee, M.-R. Choi, S.-H. Woo, S.-H. Bae, B. H. Hong, J.-H. Ahn and T.-W. Lee, *Nat. Photon.*, 2012, **6**, 105–110.
10. J. M. Lupton, B. J. Matterson, I. D. W. Samuel, M. J. Jory and W. L. Barnes, *Appl. Phys. Lett.*, 2000, **77**, 3340–3342.
11. P. A. Hobson, J. A. E. Wasey, I. Sage and W. L. Barnes, *IEEE J. Sel. Top Quantum Electron.*, 2002, **8**, 378–386.
12. J. M. Ziebarth, A. K. Saafir, S. Fan and M. D. McGehee, *Adv. Funct. Mater.*, 2004, **14**, 451–456.
13. Th. Förster, *Ann. Physik*, 1948, **437**, 55–75.
14. Th. Förster, *Discuss. Faraday Soc.*, 1959, **27**, 7–17.
15. R. M. Clegg, *Curr. Opin. Biotechnol.*, 1995, **6**, 103–110.
16. B. Schuler, E. A. Lipman and W. A. Eaton, *Nature*, 2002, **419**, 743–748.
17. L. M. Herz, C. Silva, R. H. Friend, R. T. Phillips, S. Setayesh, S. Becker, D. Marsitsky and K. Müllen, *Phys. Rev. B*, 2001, **64**, 195203.
18. J. Cabanillas-Gonzalez, A. M. Fox, J. Hill and D. D. C. Bradley, *Chem. Mater.*, 2004, **16**, 4705–4710.
19. A. A. R. Neves, A. Camposeo, R. Cingolani and D. Pisignano, *Adv. Funct. Mater.*, 2008, **18**, 751–757.
20. D. L. Dexter, *J. Chem. Phys.*, 1953, **21**, 836–850.
21. T.-Q. Nguyen, I. B. Martini, J. Liu and B. J. Schwartz, *J. Phys. Chem. B*, 2000, **104**, 237–255.
22. M. Yan, L. J. Rothberg, E. W. Kwock and T. M. Miller, *Phys. Rev. Lett.*, 1995, **75**, 1992–1995.
23. D. Hu, J. Yu, K. Wong, B. Bagchi, P. J. Rossky and P. F. Barbara, *Nature*, 2000, **405**, 1030–1033.
24. G. D. Scholes and G. Rumbles, *Nat. Mater.*, 2006, **5**, 683–696.

25. T.-Q. Nguyen, J. Wu, V. Doan, B. J. Schwartz and S. H. Tolbert, *Science*, 2000, **288**, 652–656.
26. I. B. Martini, I. M. Craig, W. C. Molenkamp, H. Miyata, S. H. Tolbert and B. J. Schwartz, *Nature Nanotech.*, 2007, **2**, 647–652.
27. E. Collini and G. D. Scholes, *Science*, 2009, **323**, 369–373.
28. I. Hwang and G. D. Scholes, *Chem. Mat.*, 2011, **23**, 610–620.
29. T.-Q. Nguyen, V. Doan and B. J. Schwartz, *J. Chem. Phys.*, 1999, **110**, 4068–4078.
30. P. Björk, D. Thomsson, O. Mirzov, J. Wigenius, O. Inganäs and I. G. Scheblykin, *Small*, 2009, **5**, 96–103.
31. H.-M. Liem, P. Etchegoin, K. S. Whitehead and D. D. C. Bradley, *Adv. Funct. Mater.*, 2003, **13**, 66–72.
32. J. Michl and E. W. Thulstrup, *Spectroscopy with Polarized Light*, VCH, Weinheim, 1987.
33. M. Richard-Lacroix and C. Pellerin, *Macromolecules*, 2012, **45**, 1946−1953.
34. S. Pagliara, M. S. Vitiello, A. Camposeo, A. Polini, R. Cingolani, G. Scamarcio and D. Pisignano, *J. Phys. Chem. C*, 2011, **115**, 20399–20405.
35. O. Svelto, *Principles of Lasers*, Plenum Press, New York, 1989.
36. N. Tessler, *Adv. Mater.*, 1999, **11**, 363–370.
37. G. Kranzelbinder and G. Leising, *Rep. Prog. Phys.*, 2000, **63**, 729–762.
38. M. D. McGehee and A. J. Heeger, *Adv. Mater.*, 2000, **12**, 1655–1668.
39. C. Grivas and M. Pollnau, *Laser Photonics Rev.*, 2012, **6**, 419–462.
40. R. Loudon, *The Quantum Theory of Light*, Oxford University Press, New York, 3rd edn, 2000.
41. M. Nisoli, S. Stagira, M. Zavelani-Rossi, S. De Silvestri, P. Mataloni and C. Zenz, *Phys. Rev. B*, 1999, **59**, 11328-11332.
42. S. V. Frolov, W. Gellerman, M. Ozaki, K. Yoshino and Z. V. Vardeny, *Phys. Rev. Lett.*, 1997, **78**, 729–732.
43. C. Zenz, W. Graupner, S. Tasch, G. Leising, K. Müllen and U. Scherf, *Appl. Phys. Lett.*, 1997, **71**, 2566–2568.
44. F. Hide, M. A. Díaz-García, B. J. Schwartz, N. R. Andersson, Q. Pei and A. J. Heeger, *Science*, 1996, **273**, 1833–1836.
45. F. Hide, B. J. Schwartz, M. A. Díaz-Garcia and A. J. Heeger, *Chem. Phys. Lett.*, 1996, **256**, 424–430.
46. U. Scherf, S. Riechel, U. Lemmer and R. F. Mahrt, *Curr. Opin. Solid State Mater. Sci.*, 2001, **5**, 143–154.
47. R. Xia, G. Heliotis and D. D. C. Bradley, *Appl. Phys. Lett.*, 2003, **82**, 3599–3601.
48. D. R. Reyes, D. Iossifidis, P.-A. Auroux and A. Manz, *Anal. Chem.*, 2002, **74**, 2623–2636.
49. P.-A. Auroux, D. Iossifidis, D. R. Reyes and A. Manz, *Anal. Chem.*, 2002, **74**, 2637–2652.
50. J. Wook Hong and S. R. Quake, *Nat. Biotechnol.*, 2003, **21**, 1179–1183.

51. D. Li, J. T. McCann and Y. Xia, *Small*, 2005, **1**, 83–86.
52. M. Yan, L. J. Rothberg, F. Papadimitrakopoulos, M. E. Galvin and T. M. Miller, *Phys. Rev. Lett.*, 1994, **73**, 744–747.
53. Y. Gao, A. M. Baca, B. Wang and P. R. Ogilby, *Macromolecules*, 1994, **27**, 7041–7048.
54. S. Madhugiri, A. Dalton, J. Gutierrez, J. P. Ferraris and K. J. Balkus, *J. Am. Chem. Soc.*, 2003, **125**, 14531–14538.
55. A. Camposeo, F. Di Benedetto, R. Stabile, A. A. R. Neves, R. Cingolani and D. Pisignano, *Small*, 2009, **5**, 562–566.
56. K. P. Dhakal, H. Lee, J. W. Lee, J. Joo, M. Guthold and J. Kim, *J. Appl. Phys.*, 2012, **111**, 123504.
57. J. M. Moran-Mirabal, J. D. Slinker, J. A. DeFranco, S. S. Verbridge, R. Ilic, S. Flores-Torres, H. Abruña, G. G. Malliaras and H. G. Craighead, *Nano Lett.*, 2007, **7**, 458–463.
58. A. Babel, D. Li, Y. Xia and S. A. Jenekhe, *Macromolecules*, 2005, **38**, 4705–4711.
59. A. Camposeo, F. Di Benedetto, R. Stabile, R. Cingolani and D. Pisignano, *Appl. Phys. Lett.*, 2007, **90**, 143115.
60. G. Kwak, S. Fukao, M. Fujiki, T. Sakaguchi and T. Masuda, *Chem. Mater.*, 2006, **18**, 5537–5542.
61. A. Bianco, C. Bertarelli, S. Frisk, J. F. Rabolt, M. C. Gallazzi and G. Zerbi, *Synth. Met.*, 2007, **157**, 276–281.
62. F. Di Benedetto, A. Camposeo, S. Pagliara, E. Mele, L. Persano, R. Stabile, R. Cingolani and D. Pisignano, *Nat. Nanotechnol.*, 2008, **3**, 614–619.
63. H. A. Liu, D. Zepeda, J. P. Ferraris and K. J. Balkus, *ACS Appl. Mater. Interfaces*, 2009, **1**, 1958–1965.
64. Y. Wang, J. S. Park, J. P. Leech, S. Miao and U. H. F. Bunz, *Macromolecules*, 2007, **40**, 1843–1850.
65. W. Zhong, F. Li, L. Chen, Y. Chen and Y. Wei, *J. Mater. Chem.*, 2012, **22**, 5523–5530.
66. D. Li, A. Babel, S. A. Jenekhe and Y. Xia, *Adv. Mater.*, 2004, **16**, 2062–2066.
67. T. Kim, J. H. Im, H. S. Choi, S. J. Yang, S. W. Kim and C. R. Park, *J. Mater. Chem.*, 2011, **21**, 14231–14239.
68. M. Campoy-Quiles, Y. Ishii, H. Sakai and H. Murata, *Appl. Phys. Lett.*, 2008, **92**, 213305.
69. D. M. Eisele, J. Knoester, S. Kirstein, J. P. Rabe and D. A. Vanden Bout, *Nat. Nanotechnol.*, 2009, **4**, 658–663.
70. G. A. O'Brien, A. J. Quinn, D. A. Tanner and G. Redmond, *Adv. Mater.*, 2006, **18**, 2379–2383.
71. C. De Marco, E. Mele, A. Camposeo, R. Stabile, R. Cingolani and D. Pisignano, *Adv. Mater.*, 2008, **20**, 4158–4162.
72. R. Zhou, W. Chen, X. Jiang, S. Wang and Q. Gong, *Appl. Phys. Lett.*, 2010, **96**, 133309.

73. F. Massuyeau, J. L. Duvail, H. Athalin, J. M. Lorcy, S. Lefrant, J. Wéry and E. Faulques, *Nanotechnology*, 2009, **20**, 155701.
74. G. Padmanaban and S. Ramakrishnan, *J. Am. Chem. Soc.*, 2000, **122**, 2244–2251.
75. D. Hu, J. Yu, G. Padmanaban, S. Ramakrishnan and P. F. Barbara, *Nano Lett.*, 2002, **2**, 1121–1124.
76. D. O'Carroll, I. Lieberwirth and G. Redmond, *Small*, 2007, **3**, 1178–1183.
77. C. L. Donley, J. Zaumseil, J. W. Andreasen, M. M. Nielsen, H. Sirringhaus, R. H. Friend and J.-S. Kim, *J. Amer. Chem. Soc.*, 2005, **127**, 12890–12899.
78. G. R. Hayes, I. D. W. Samuel and R. T. Phillips, *Phys. Rev. B*, 1997, **56**, 3838–3843.
79. Z. Cai, J. Lei, W. Liang, V. Menon and C. R. Martin, *Chem. Mater.*, 1991, **3**, 960–966.
80. W. Liang and C. R. Martin, *J. Am. Chem. Soc.*, 1990, **112**, 9666–9668.
81. S. Pagliara, A. Camposeo, A. Polini, R. Cingolani and D. Pisignano, *Lab Chip*, 2009, **9**, 2851–2856.
82. S. Moynihan, P. Lovera, D. O'Carroll, D. Iacopino and G. Redmond, *Adv. Mater.*, 2008, **20**, 2493–2502.
83. A. O'Riordan, D. Iacopino, P. Lovera, L. Floyd, K. Reynolds and G. Redmond, *Nanotechnology*, 2011, **22**, 105602.
84. W. C. Molenkamp, M. Watanabe, H. Miyata and S. H. Tolbert, *J. Am. Chem. Soc.*, 2004, **126**, 4476–4477.
85. C.-C. Kuo, C. T. Wang and W.-C. Chen, *Macromol. Mater. Eng.*, 2008, **293**, 999–1008.
86. C. R. L. P. N. Jeukens, P. Jonkheijm, F. J. P. Wijnem, J. C. Gielen, P. C. M. Christianen, A. P. H. J. Schenning, E. W. Meijer and J. C. Maan, *J. Am. Chem. Soc.*, 2005, **127**, 8280–8281.
87. S. Samitsu, Y. Takanishi and J. Yamamoto, *Macromolecules*, 2009, **42**, 4366–4368.
88. M. L. Chabinyc, D. T. Chiu, J. C. McDonald, A. D. Stroock, J. F. Christian, A. M. Karger and G. M. Whitesides, *Anal. Chem.*, 2001, **73**, 4491–4498.
89. S.-W. Ahn, K.-D. Lee, J.-S. Kim, S. H. Kim, J.-D. Park, S.-H. Lee and P.-W. Yoon, *Nanotechnology*, 2005, **16**, 1874–1877.
90. S. Moynihan, D. Iacopino, D. O'Carroll, H. Doyle, D. A. Tanner and G. Redmond, *Adv. Mater.*, 2007, **19**, 2474–2479.
91. K. Dawson, P. Lovera, D. Iacopino, A. O'Riordan and G. Redmond, *J. Mater. Chem.*, 2011, **21**, 15995–16000.
92. A. Camposeo, F. Di Benedetto, R. Cingolani and D. Pisignano, *Appl. Phys. Lett.*, 2009, **94**, 043109.
93. E. J. W. List, C. Creely, G. Leising, N. Schulte, A. D. Schlüter, U. Scherf, K. Müllen and W. Graupner, *Chem. Phys. Lett.*, 2000, **325**, 132–138.
94. T. Smith and J. Guild, *Trans. Opt. Soc.*, 1931, **33**, 73–134.

95. V. Vohra, A. Devaux, L.-Q. Dieu, G. Scavia, M. Catellani, G. Calzaferri and C. Botta, *Adv. Mater.*, 2009, **21**, 1146–1150.
96. V. Vohra, A. Devaux, L.-Q. Dieu, G. Scavia, M. Catellani, G. Calzaferri and C. Botta, *ACS Nano*, 2010, **4**, 1409–1416.
97. X. M. Sui, C. L. Shao and Y. C. Liu, *Appl. Phys. Lett.*, 2005, **87**, 113115.
98. H. Sun, H. Zhang, J. Zhang, H. Wei, J. Ju, M. Li and B. Yang, *J. Mater. Chem.*, 2009, **19**, 6740–6744.
99. C. Simbrunner, F. Quochi, G. Hernandez-Sosa, M. Oehzelt, R. Resel, G. Hesser, M. Arndt, M. Saba, A. Mura, G. Bongiovanni and H. Sitter, *ACS Nano*, 2010, **10**, 6244–6250.
100. www.rti.org/NLITe.
101. P. Anzenbacher, Jr. and M. A. Palacios, *Nat. Chem.*, 2009, **1**, 80–86.
102. P. Andrew and W. L. Barnes, *Science*, 2004, **306**, 1002–1005.
103. D. H. Park, M. S. Kim and J. Joo, *Chem. Soc. Rev.*, 2010, **39**, 2439–2452.
104. D. Hyuk Park, Y. Ki Hong, E. Hei Cho, M. Suk Kim, D.-C. Kim, J. Bang, J. Kim and J. Joo, *ACS Nano*, 2010, **4**, 5155–5162.
105. F. He, T. Gädt, I. Manners and M. A. Winnik, *J. Amer. Chem. Soc.*, 2011, **133**, 9095–9103.
106. A. Ashkin, *Phys. Rev. Lett.*, 1970, **24**, 156–159.
107. A. Ashkin, J. M. Dziedzic, J. E. Bjorkholm and S. Chu, *Opt. Lett.*, 1986, **11**, 288–290.
108. A. Ashkin, *Proc. Natl Acad. Sci. USA*, 1997, **94**, 4853–4860.
109. K. Dholakia, P. Reece and M. Gu, *Chem. Soc. Rev.*, 2008, **37**, 42–55.
110. J. R. Moffitt, Y. R. Chemla, S. B. Smith and C. Bustamante, *Ann. Rev. Biochem.*, 2008, **77**, 205–228.
111. A. Jonáš and P. Zemánek, *Electrophoresis*, 2009, **29**, 4813–4851.
112. O. M. Maragò, P. H. Jones, F. Bonaccorso, V. Scardaci, P. G. Gucciardi, A. G. Rohzina and A. C. Ferrari, *Nano Lett.*, 2008, **8**, 3211–3216.
113. F. Borghese, P. Denti, R. Saija, M. A. Iatì and O. M. Maragò, *Phys. Rev. Lett.*, 2008, **100**, 163903.
114. A. A. R. Neves, A. Camposeo, S. Pagliara, R. Saija, F. Borghese, P. Denti, M. A. Iatì, R. Cingolani, O. M. Maragò and D. Pisignano, *Opt. Exp.*, 2010, **18**, 822–830.
115. J. D. Jackson, *Classical Electrodynamics*, Wiley, New York, 3rd edn, 1998.
116. R. N. C. Pfeifer, T. A. Nieminen, N. R. Heckenberg and H. Rubinsztein-Dunlop, *Rev. Mod. Phys.*, 2007, **79**, 1197–1216.
117. F. Borghese, P. Denti, R. Saija and M. A. Iatì, *Opt. Exp.*, 2007, **15**, 11984–11998.
118. A. K. Chibisov and H. Görner, *Chem. Phys.*, 1998, **237**, 425–442.
119. J. T. C. Wojtyk, P. M. Kazmaier and E. Buncel, *Chem. Mater.*, 2001, **13**, 2547–2551.
120. C. Bohne, M. G. Fan, Z. J. Li, Y. C. Liang, J. Lusztyk and J. C. Scaiano, *J. Photochem. Photobiol. A*, 1992, **66**, 79–90.

121. A. Malatesta, C. Neri, M. L. Wis, L. Montanari and R. Millini, *J. Am. Chem. Soc.*, 1997, **119**, 3451–3455.
122. A. K. Chibisov and H. Görner, *J. Phys. Chem. A*, 1997, **101**, 4305–4312.
123. M. Levitus, M. Talhavini, R. M. Negri, T. D. Z. Atvars and P. F. Aramendía, *J. Phys. Chem. B*, 1997, **101**, 7680–7686.
124. F. Di Benedetto, E. Mele, A. Camposeo, A. Athanassiou, R. Cingolani and D. Pisignano, *Adv. Mater.*, 2008, **20**, 314–318.
125. D. Iacopino and G. Redmond, *Chem. Comm.*, 2011, **47**, 9170–9172.
126. K. Huang and M. Wan, *Chem. Mater.*, 2002, **14**, 3486–3492.
127. S. Il Cho and S. Bok Lee, *Acc. Chem. Res.*, 2008, **41**, 699–707.
128. S. I. Cho, K. J. Kwon, S.-J. Choi, P. Kim, S.-A. Park, J. Kim, S. J. Son, R. Xiao, S.-H. Kim and S. B. Lee, *Adv. Mater.*, 2005, **17**, 171–175.
129. S.-Y. Jang, V. Seshadri, M.-S. Khil, A. Kumar, M. Marquez, P. M. Mather and G. A. Sotzing, *Adv. Mater.*, 2005, **17**, 2177–2180.
130. H. Peng, X. Sun, F. Cai, X. Chen, Y. Zhu, G. Liao, D. Chen, Q. Li, Y. Lu, Y. Zhu and Q. Jia, *Nat. Nanotechnol.*, 2009, **4**, 738–741.
131. F. A. Boroumand, P. W. Fry and D. G. Lidzey, *Nano Lett.*, 2005, **5**, 67–71.
132. H. Yamamoto, J. Wilkinson, J. P. Long, K. Bussman, J. A. Christodoulides and Z. H. Kafafi, *Nano Lett.*, 2005, **5**, 2485–2488.
133. S. P. Price, J. Henzie and T. W. Odom, *Small*, 2007, **3**, 372–374.
134. M. Granström, M. Berggren and O. Inganäs, *Science*, 1995, **267**, 1479–1481.
135. D. O'Carroll, D. Iacopino, A. O'Riordan, P. Lovera, É. O'Connor, G. A. O'Brien and G. Redmond, *Adv. Mater.*, 2008, **20**, 42–48.
136. V. Vohra, U. Giovanella, R. Tubino, H. Murata and C. Botta, *ACS Nano*, 2011, **5**, 5572–5578.
137. H. Yang, C. R. Lightner and L. Dong, *ACS Nano*, 2012, **6**, 622–628.
138. H. Imahori, S. Kitaura, A. Kira, H. Hayashi, M. Nishi, K. Hirao, S. Isoda, M. Tsujimoto, M. Takano, Z. Zhe, Y. Miyato, K. Noda, K. Matsushige, K. Stranius, N. V. Tkachenko, H. Lemmetyinen, L. Qin, S. J. Hurst and C. A. Mirkin, *J. Phys. Chem. Lett.*, 2012, **3**, 478–481.
139. Y. Guo, Q. Tang, H. Liu, Y. Zhang, Y. Li, W. Hu, S. Wang and D. Zhu, *J. Amer. Chem. Soc.*, 2008, **130**, 9198–9199.
140. K. Kim, B. H. Kim, S.-H. Joo, J.-S. Park, J. Joo, J.-I. Jin, *Adv. Mater.*, 2005, **17**, 464–468.
141. M. C. Scharber, D. Mühlbacher, M. Koppe, P. Denk, C. Waldauf, A. J. Heeger and C. J. Brabec, *Adv. Mater.*, 2006, **18**, 789–794.
142. S. Gunes, H. Neugebauer and N. S. Sariciftci, *Chem. Rev.*, 2007, **107**, 1324–1338.
143. A. C. Mayer, S. R. Scully, B. E. Hardin, M. W. Rowell and M. D. McGehee, *Mater. Today*, 2007, **10**, 28–33.
144. R. Kroon, M. Lenes, J. C. Hummelen, P. W. M. Blom and B. De Boer, *Polym. Rev.*, 2008, **48**, 531–582.

145. J.-L. Brédas, J. E. Norton, J. Cornil and V. Coropceanu, *Acc. Chem. Res.*, 2009, **42**, 1691–1699.
146. G. Yu, J. Gao, J. C. Hummelen, F. Wudl and A. J. Heeger, *Science*, 1995, **270**, 1789–1791.
147. J. J. M. Halls, C. A. Walsh, N. C. Greenham, E. A. Marseglia, R. H. Friend, S. C. Moratti and A. B. Holmes, *Nature*, 1995, **376**, 498–500.
148. H.-S. Wang, L.-H. Lin, S.-Y. Chen, Y.-L. Wang and K.-H. Wei, *Nanotechnology*, 2009, **20**, 075201.
149. X. Yang, J. Loos, S. C. Veenstra, W. J. H. Verhees, M. M. Wienk, J. M. Kroon, M. A. J. Michels and R. A. J. Janssen, *Nano Lett.*, 2005, **5**, 579–583.
150. H. Xin, G. Ren, F. Sunjoo Kim and S. A. Jenekhe, *Chem. Mater.*, 2008, **20**, 6199–6207.
151. P.-T. Wu, H. Xin, F. Sunjoo Kim, G. Ren and S. A. Jenekhe, *Macromolecules*, 2009, **42**, 8817–8826.
152. H. Xin, O. G. Reid, G. Ren, F. Sunjoo Kim, D. S. Ginger and S. A. Jenekhe, *ACS Nano*, 2010, **4**, 1861–1872.
153. B.-G. Kim, M. S. Kim and J. Kim, *ACS Nano*, 2010, **4**, 2160–2166.
154. L. Li, G. Lu and X. Yang, *J. Mater. Chem.*, 2008, **18**, 1984–1990.
155. S. Berson, R. De Bettignies, S. Bailly and S. Guillerez, *Adv. Mater.*, 2007, **17**, 1377–1384.
156. H. Xin, F. Sunjoo Kim and S. A. Jenekhe, *J. Amer. Chem. Soc.*, 2008, **130**, 5424–5425.
157. S. Ren, L.-Y. Chang, S.-K. Lim, J. Zhao, M. Smith, N. Zhao, V. Bulović, M. Bawendi and S. Gradečak, *Nano Lett.*, 2011, **11**, 3998–4002.
158. S. Ramakrishna, R. Jose, P. S. Archana, A. S. Nair, R. Balamurugan, J. Venugopal and W. E. Teo, *J. Mater. Sci.*, 2010, **45**, 6283–6312.
159. S. Sundarrajan, R. Murugan, A. S. Nair and S. Ramakrishna, *Mater. Lett.*, 2010, **64**, 2369–2372.
160. J. Liu, M. A. G. Namboothiry and D. L. Carroll, *Appl. Phys. Lett.*, 2007, **90**, 063501.
161. B. O'Connor, K. P. Pipe and M. Shtein, *Appl. Phys. Lett.*, 2008, **92**, 193306.
162. Y. Li, W. Zhou, D. Xue, J. Liu, E. D. Peterson, W. Nie and D. L. Carroll, *Appl. Phys. Lett.*, 2009, **95**, 203503.
163. M. R. Lee, R. D. Eckert, K. Forberich, G. Dennler, C. J. Brabec and R. A. Gaudiana, *Science*, 2009, **324**, 232–235.
164. A. L. Briseno, T. W. Holcombe, A. I. Boukai, E. C. Garnett, S. W. Shelton, J. J. M. Fréchet and P. Yang, *Nano Lett.*, 2010, **10**, 334–340.
165. K. Kim, J. W. Shin, Y. B. Lee, M. Y. Cho, S. H. Lee, D. H. Park, D. K. Jang, C. J. Lee and J. Joo, *ACS Nano*, 2010, **4**, 4197–4205.
166. Y. Guo, Y. Zhang, H. Liu, S.-W. Lai, Y. Li, Y. Li, W. Hu, S. Wang, C.-M. Che and D. Zhu, *J. Phys. Chem. Lett.*, 2010, **1**, 327–330.

167. F. Balzer, V. G. Bordo, A. C. Simonsen and H.-G. Rubahn, *Phys. Rev. B*, 2003, **67**, 115408.
168. K. Takazawa, Y. Kitahama, Y. Kimura and G. Kido, *Nano Lett.*, 2005, **5**, 1293–1296.
169. N. Huby, J. L. Duvail, D. Duval, D. Pluchon and B. Bêche, *Appl. Phys. Lett.*, 2011, **99**, 113302.
170. C. Meng, Y. Xiao, P. Wang, L. Zhang, Y. Liu and L. Tong, *Adv. Mater.*, 2011, **23**, 3770–3774.
171. F. Gu, H. Yu, P. Wang, Z. Yang and L. Tong, *ACS Nano*, 2010, **4**, 5332–5338.
172. H. Liu, J. B. Edel, L. M. Bellan and H. G. Craighead, *Small*, 2006, **2**, 495–499.
173. J. C. Johnson, H. Yan, P. Yang and R. J. Saykally, *J. Phys. Chem. B*, 2003, **107**, 8816–8828.
174. D. O'Carroll, I. Lieberwirth and G. Redmond, *Nat. Nanotechnol.*, 2007, **2**, 180–184.
175. P. K. Tien, *Appl. Opt.*, 1971, **10**, 2395–2413.
176. A. J. Cox, A. J. DeWeerd and J. Linden, *Am. J. Phys.*, 2002, **70**, 620–625.
177. A. J. Das, C. Lafargue, M. Lebental, J. Zyss and K. S. Narayan, *Appl. Phys. Lett.*, 2011, **99**, 263303.
178. F. Quochi, F. Cordella, R. Orrù, J. E. Communal, P. Verzeroli, A. Mura, G. Bongiovanni, A. Andreev, H. Sitter and N. S. Sariciftci, *Appl. Phys. Lett.*, 2004, **84**, 4454–4456.
179. F. Quochi, A. Andreev, F. Cordella, R. Orrù, A. Mura, G. Bongiovanni, H. Hoppe, H. Sitter and N. S. Sariciftci, *J. Lumin.*, 2005, **112**, 321–324.
180. F. Quochi, F. Cordella, A. Mura, G. Bongiovanni, F. Balzer and H.-G. Rubahn, *J. Phys. Chem. B*, 2005, **109**, 21690–21693.
181. F. Quochi, F. Cordella, A. Mura, G. Bongiovanni, F. Balzer and H.-G. Rubahn, *Appl. Phys. Lett.*, 2006, **88**, 041106.
182. A. Andreev, F. Quochi, F. Cordella, A. Mura, G. Bongiovanni, H. Sitter, G. Hlawacek, C. Teichert and N. S. Sariciftci, *J. Appl. Phys.*, 2006, **99**, 034305.
183. F. Quochi, M. Saba, F. Cordella, A. Gocalinska, R. Corpino, M. Marceddu, A. Anedda, A. Andreev, H. Sitter, N. S. Sariciftci, A. Mura and G. Bongiovanni, *Adv. Mater.*, 2008, **20**, 3017–3021.
184. X. Duan, Y. Huang, R. Agarwal and C. M. Lieber, *Nature*, 2003, **421**, 241–245.
185. H. Kogelnik and C. V. Shank, *Appl. Phys. Lett.*, 1971, **18**, 152–154.
186. G. A. Turnbull, P. Andrew, M. J. Jory, W. L. Barnes and I. D. W. Samuel, *Phys. Rev. B*, 2001, **64**, 125122.
187. J. D. Joannopoulos, R. D. Meade and J. N. Winn, *Photonic Crystals – Molding the Flow of Light*, Princeton University Press, Princeton, 1995.
188. J. Clark and G. Lanzani, *Nat. Photon.*, 2010, **4**, 438–446.
189. M. Bognitzki, W. Czado, T. Frese, A. Schaper, M. Hellwig, M. Steinhart, A. Greiner and J. H. Wendorff, *Adv. Mater.*, 2001, **13**, 70–72.

190. K. Y. Suh, Y. S. Kim and H. H. Lee, *Adv. Mater.*, 2001, **13**, 1386–1389.
191. Z. Hu, B. Muls, L. Gence, D. A. Serban, J. Hofkens, S. Melinte, B. Nysten, S. Demoustier-Champagne and A. M. Jonas, *Nano Lett.*, 2007, **7**, 3639–3644.
192. N. W. Ashcroft and N. D. Mermin, *Solid State Physics*, Thomson Brooks/Cole, New York, 1976.
193. S. Pagliara, A. Camposeo, E. Mele, L. Persano, R. Cingolani and D. Pisignano, *Nanotechnology*, 2010, **21**, 215304.
194. Z. Zheng, K.-H. Yim, M. S. M. Saifullah, M. E. Welland, R. H. Friend, J.-S. Kim and W. T. S. Huck, *Nano Lett.*, 2007, **7**, 987–992.

CHAPTER 6

# *Electrical Properties and Their Applications*

## 6.1 Transport Phenomena in Organic Semiconductors

The first observations of metallic-like conductivity (*i.e.* with $\mu \cong 1$ cm$^2$ V s$^{-1}$) in doped polyacetylene date back to 1976.[1,2] Indeed, originally, conjugated polymers attracted a lot of attention especially because their conductivity can be increased by orders of magnitude by doping. However, undoped polymers have also been demonstrated to show significant conduction. Today, we can undoubtedly state that conjugated materials[1,3–5] have opened an entirely new field of science at the boundary of physics, chemistry, materials sciences and nanotechnology, and still offer exciting scientific and technological perspectives.

Given the band-gap energy of many conjugated materials (2–3 eV), these are naturally classified as semiconductors. Impurities (down to the order of ppm) are often left in conjugated polymers or in low-molar-mass organic semiconductor molecules during synthesis or purification.[6,7] Overall, many of these materials can be considered *intrinsic* semiconductors. For many years, scientists have been distinguishing between *p*-type (by far most numerous) and *n*-type organic semiconductors, on the basis of experimental evidences indicating the material ability of conducting only (or predominantly) positively or negatively charged charge-carriers, respectively.[8] For the sake of comparison, one could notice that this distinction is in striking contradiction with common findings in inorganic semiconductors, whose prevailing transport properties (and consequently *p*- or *n*-type classification) are not intrinsically related to the specific material, but are instead due to majority carriers mainly determined by extrinsic doping.[7] In the last decade, however,

RSC Nanoscience & Nanotechnology No. 29
Polymer Nanofibers: Building Blocks for Nanotechnology
By Dario Pisignano

Published by the Royal Society of Chemistry, www.rsc.org

an increasing amount of experiments and data have indicated the crucial importance of the device materials and architectural characteristics, such as the dielectric layer used in field-effect transistors (FETs),[9,10] in affecting the transport features of organic semiconductors. Recently, the distinction between *p*-type and *n*-type *compounds* has been replaced by that between *p*-type and *n*-type *devices*. Consequently, organic semiconductors could be considered as *ambipolar* materials in terms of their intrinsic properties.[11] Having both *n*-type[12] and *p*-type operation available is important, on one hand to realize complementary logics,[8,13,14] and on the other hand possibly to achieve the simultaneous transport of charge-carriers of opposite sign and lead them to radiative recombination, thus obtaining organic light-emitting transistors (OLETs).[15–18]

Though the properties of conduction of conjugated materials are still a matter of investigation, some features of transport phenomena in organic semiconductors can be easily, schematically drawn to provide an introductory description. First of all, it is clear that the conduction behavior in the solid state is largely related to the supramolecular assembly of organic molecules and not only to properties of individual molecules. Electrical conduction can be due to charge-carrier transport by band-like conduction (similarly to inorganic semiconductors), as in crystalline organics, or by hopping, as in amorphous systems. As a reference, one often considers a charge-carrier mobility, $\mu$, of $1\ \mathrm{cm^2\ V\ s^{-1}}$ as a boundary value between hopping (for $\mu < 1\ \mathrm{cm^2\ V\ s^{-1}}$) and band conduction (for $\mu > 1\ \mathrm{cm^2\ V\ s^{-1}}$). Charge trapping phenomena also play an important role in many organic semiconductors. Indeed, in most cases the disorder present in the material, which almost ubiquitously has a partially or completely amorphous supramolecular organization, the presence of chemical defects, and the effect of molecular distortions limiting the conjugation length, are factors that intrinsically limit the transport of charge-carriers. As a consequence, reaching mobilities above $1\ \mathrm{cm^2\ V\ s^{-1}}$ in organics is not an easy task. This is achievable only in low-molar-mass molecules in the crystalline state or in carefully optimized polymeric systems, and requires in any case significant efforts in molecular design, synthesis, control of processing conditions and device engineering.[19,20] Conductive polymers most often show mobilities of the order of $0.1\ \mathrm{cm^2\ V\ s^{-1}}$,[21–25] or lower (down to, say, $10^{-5}\ \mathrm{cm^2\ V\ s^{-1}}$). For the sake of comparison, amorphous silicon in thin-film transistors exhibits a mobility in the range of $0.5\text{–}1\ \mathrm{cm^2\ V^{-1}}$, whereas low-temperature polycrystalline silicon reaches $30\text{–}300\ \mathrm{cm^2\ V\ s^{-1}}$.[26]

We do not enter into the details of the many models proposed to explain the conduction properties in conjugated materials. These include, for instance, the variable range-hopping model (VRHM),[27–30] descriptions based on electronic tunneling effects and on the granular metallic model[31–34] and other variants, whose features are widely available in the literature.[35] We only mention a couple of examples. In the simplest case, as an effect of disorder, the mobility can follow an Arrhenius law, namely a thermally activated dependence on temperature:[36]

$$\mu = \mu^* e^{-\frac{\Delta E}{k_B T}}, \tag{6.1}$$

where $\Delta E$ indicates an activation energy that is around 100 meV, and $\mu^*$ is a pre-exponential factor describing the mobility at very high temperature ($k_B T >> \Delta E$). According to the VRHM, which is often used to describe the temperature dependence of conductivity ($\chi = Nq\mu$ where $q$ is the charge and $N$ is the concentration of the charge-carriers) in polymer nanofibers, one has instead the following, stretched-exponential expression:[30,35,37]

$$\chi = \chi^* e^{-\left(\frac{T_0}{T}\right)^n}, \tag{6.2}$$

or, according to another variant:[30,38]

$$\chi = \chi^* T^{-1/2} e^{-\left(\frac{T_0}{T}\right)^n}, \tag{6.3}$$

where $\chi^*$ is a pre-exponential constant that corresponds to the high-temperature limit of the conductivity, $T_0$ is a characteristic temperature (Mott temperature)[38] and the exponent $n$ depends on the dimensionality of the transport process ($n$ = 1/2, 1/3 or 1/4 for one-dimensional, two-dimensional and three-dimensional conduction, respectively).

### 6.1.1 Supramolecular Organization and Electrical Conduction in Polymers

The general idea behind the fact that an organic material can conduct electrons relies on the particular features of *conjugated* materials. In these molecules, a charge-carrier is highly delocalized along the conjugation length because of the overlap of adjacent carbon orbitals along molecular backbones.[1] Therefore, along the (*intra*-molecular) length of conjugation, the mobility is very high (>>1 $cm^2 V s^{-1}$).[39] Furthermore, under favorable conditions of supramolecular packing in the solid state, a certain degree of overlap of the electronic wavefunction of face-to-face, neighbor molecules, which is the π–π stacking, can also provide the system with a relevant (*inter*-molecular) mobility.[36,40,41] The π–π stacking can also contribute to the overall energy landscape, thus further affecting the structural assembly of molecules. This kind of packing is very important for polymer nanofiber materials and in general for conductive polymer nanostructures, whose molecules often assemble with π–π stacking.[42] The mobility of charge-carriers along the longitudinal axis of nanofibers often benefits greatly from this supramolecular organization as is described later.

An important representative of conductive polymers, which is also relevant in the framework of nanofibers (Section 6.2), is P3HT. This and other similar P3ATs are constituted by a rigid chain, which is rendered soluble by lateral functionalization of the backbone with alkyl groups $(CH_2)_n$ (molecular structure in Figure 6.1a). Following deposition by casting, spin-coating or production of nanofibers from solution, P3HT can take a crystalline form[43] with the lamellar structure schematized in Figure 6.1b. Different lamellae are stacked in one (100)-direction (*a*-axis of the unit cell of the P3HT microcrystalline structure), being separated by 1.6 nm and somehow electronically decoupled by the alkyl chains. On the contrary, charge transport effectively takes place in the other (010)-direction (*b*-axis of the unit cell), which is normal to the mutually parallel lamellar sheets defined by the aligned thiophene rings (π–π stacking). The inter-sheet distance along the *b*-axis is below 0.4 nm. Finally, the so-called *c*-axis corresponds to the orientation of the molecular backbones, which completes the orthorhombic unit cell of the crystal. In addition, different P3AT systems exhibit a beautiful richness of packing structures. The various aspects of the supramolecular organization of P3ATs and of other π-conjugated polymer systems, and how these impact on the charge transport properties, have recently been reviewed in Ref. 44. For instance, some studies on self-assembled nanowires of crystalline-crystalline diblock co-P3ATs highlight the possibility of specific supramolecular packing structures, with two different inter-layer spacing distances.[45] Among other compounds, the so-called *regioregular* P3HT is particularly intriguing since it has alkyl chains at fixed positions along the backbone. Such a higher degree of order is widely demonstrated to give the polymer higher values of mobility.[36] In fact, the degree of regioregularity is a main issue affecting the P3HT charge transport performances. For regioregular P3HT in FET devices, typical (maximum) hole mobility values are of the order of $10^{-2}$–$10^{-1}$ $cm^2\ V\ s^{-1}$ ($\sim$1 $cm^2\ V\ s^{-1}$) along the *b*-axis.[36,46] Other systems, proposed more recently, are based on a more symmetric monomeric unit than P3HT, thus enhancing the regioregularity and favoring the formation of highly ordered, crystalline domains in the material, as in poly(3,3‴-dialkylquater-thiophene)[47] and poly(2,5-bis(3-alkylthiophen-2-yl)thieno[3,2-*b*]thiophene).[24,48] Furthermore, solvents clearly matter. Using solvents with high boiling points (>200 °C) is demonstrated to improve the crystalline order because of the slower evaporation rate and of the longer time left to the polymer to assemble, and this impacts positively on the mobility of spin cast P3HT films.[49]

Methods for characterizing these supramolecular arrangements are various. The internal morphology of crystals can be determined by electron diffraction patterns obtained by TEM, as recalled briefly and very clearly in Ref. 43. Furthermore, how the microcrystalline structure orients on a solid substrate following film deposition (thus resulting in so-called molecular "edge-on" or "face-on" domains, or even in configurations in which the polymer backbones are not parallel but instead perpendicular[50] to the underlying substrate) is easily determined by means of X-ray diffraction (XRD).[36,49,51] In addition, the folding of the molecules and the lamellar structures on highly oriented

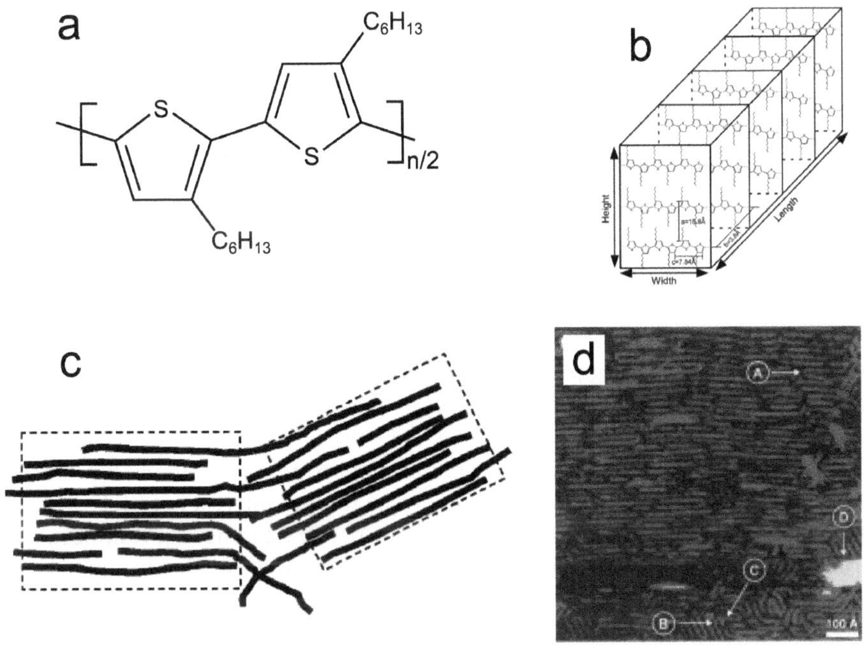


**Figure 6.1** Molecular structure (a) and typical supramolecular, crystalline organization of regioregular P3HT within nanofibers (b). (b) Adapted with permission from Ref. 46, *J. Phys. Chem. B*, 2004, **108**, 19169–19179. Doi: 10.1021/jp047023a. Copyright © 2004, American Chemical Society. (c) Example of schematics of two adjacent crystalline domains in a P3AT solid sample. The areas in and outside the dashed rectangles schematize the intra-domain volume and the domain boundaries, respectively. Bridging chains between the two domains can increase the overall conductivity providing further pathways for charge-carrier transport. Adapted with permission from Ref. 55, *Phys. Rev. B*, 2005, **71**, 165202. Doi: 10.1103/PhysRevB.71.165202. Copyright © 2005, The American Physical Society. (d) Experimental visualization of the lamellar and multidomain structure. STM micrograph of monolayers of P3ATs on highly oriented pyrolytic graphite (HOPG). Picture size: 1000 Å × 1000 Å. (A), (B) and (C) arrows indicate well-ordered areas, and chain folding of 120° and 60°, respectively. The area labeled with (D) indicates a multilayer of adsorbed molecules. Adapted with permission from Ref. 52, *Angew. Chem. Int. Ed.*, 2000, **39**, 2679–2684. Doi: 10.1002/1521-3773(20000804)39:15<2679::AID-ANIE2679>3.0.CO;2-2. Copyright © 2000, Wiley-VCH Verlag GmbH.

pyrolytic graphite (HOPG) can be directly visualized by means of STM.[52,53] Therefore, P3ATs are ideal systems to investigate the dependence of charge-carrier mobility and of its anisotropy on the molecular alignment and supramolecular packing.[25,54] Interesting studies exist as well that model the (hardly accessible) intra-crystalline domain and the inter-domain anisotropies expected in P3ATs.[51,55] For instance, taking into account two different mobilities values, $\mu_1$ and $\mu_2$, for the intra-domain and the inter-domain charge-transport, respectively (with $\mu_1$ expected to be significantly larger than $\mu_2$), the overall effective mobility, $\mu_{eff}$, could be figured out as:[51]

$$\frac{L_1 + L_2}{\mu_{eff}} = \frac{L_1}{\mu_1} + \frac{L_2}{\mu_2}. \tag{6.4}$$

In the previous expression, $L_1$ and $L_2$ respectively indicate the total lengths (or volume portions)[55] of the microcrystalline grains and of the domain boundaries (Figure 6.1c and d). It is clear that these concepts can be strategic for modeling the transport in nanofibers, whose structure can drive the orientation of the microcrystalline domains thus determining peculiar properties of enhanced directional mobility and transport anisotropy. In particular, the lamellar sheets can package in such a way that the *b*-axis coincides with the longitudinal axis of the fiber as schematized in Figure 6.1b.[46,56] More details on P3HT nanofibers are given in Section 6.2.

Another compound that has attracted an increasing interest is F8BT, which exhibits optical gain,[57,58] and a generally polycrystalline supramolecular structure, with μm-scale domains of uniform orientation.[59] This material shows a liquid crystalline phase that can be used to induce the alignment of polymeric chains in nano-confined volumes. Indeed, patterning F8BT by nanoimprint lithography (NIL) (Section 1.2.1) promotes the alignment of the polymer chains, which results in polarized light-emission and enhanced charge-carrier mobility compared to untextured polymer layers.[60] Polymer nanofibers allow the alignment and the resulting conduction properties of F8BT in nano-confined structures to be investigated. Especially appealing is the use of F8BT in *n*-type[11,61] and in ambipolar[18,60] devices. These can be realized as long as suitable dielectric materials are used to minimize electron trapping at the semiconductor/dielectric interface (Section 6.3).

## 6.2 Nanofibers by Conductive Polymers

Many of the nanofibers realized by light-emitting polymers and described in Section 5.2 are also conductive because of the conjugation properties of those materials. In addition, several other classes of conductive polymer nanofibers and nanowires can be produced by a variety of fabrication methods.[62,63] For their unique combination of size downscaling, chemical flexibility and at least partially controllable molecular ordering, polymer nanofibers are dramatically interesting in order to study fundamentals of charge transport at nanoscale.[35] In the following we give an idea of the accessible range of conductivities that one measures in polymer nanofibers, and then focus on a couple of examples that are particularly relevant. The basic properties of the charge-carrier transport in nanofibers certainly deserve further investigation, and extensive experiments on other conjugated polymer materials as well.

*General conductivity properties.* Since the 1990s, many polymer nanocylinders realized by template synthesis (Section 3.3) have been demonstrated to exhibit conductivities higher than their bulk counterpart,[64] and transport properties can be changed by controlling the diameter of the conductive nanostructures.[38] For example, the exponent $n$ of the VRHM expression (Equation (6.3)) is found to vary from the value typical of three-dimensional charge transport to that typical of lower dimensionality upon decreasing the fiber diameter. Studied systems include several conjugated polymers such as PPy, PAN, polythiophenes and others.[65–70] As mentioned in Section 3.3, measurements of the conduction properties for template synthesized nanostructures are carried out easily, and even without removing the external template membrane. In pioneering reports, the membrane filled with the conductive polymer fibers is simply positioned on a bed of Pt particles having a diameter of 0.8 μm[65] or of Ag particles having a diameter of 0.2 μm,[66] while the other side is contacted by a Pt wire sealed in glass. More generally, depositing metals on the two sides of the membrane can be sufficient to contact the two ends of the polymer nanocylinders across the sample.[71,72] In many cases, applying a pressure onto the membrane (up to ~50 MPa) greatly helps in achieving satisfactory electrical contacts with the polymer cylinders.[65,66,72] Two-points[73] and four-points[67,70] conductivity measurements are also possible, following the removal of the template membrane or in general the deposition of nanofibers on previously fabricated electrodes, or on insulating substrates where contacts are realized on top of the nanostructures.[70,74] Solution-processed, templated nanowires of the conjugated polymer, F8T2, have a measured resistivity down to $7 \times 10^3$ Ω cm (*i.e.* $\chi$ up to $1.4 \times 10^{-4}$ S cm$^{-1}$).[73] This resistivity is quite high, however this value is also affected by the contribution from the organic/metal junctions at the contact electrodes (Section 6.3). Other conjugated polymers allow much higher values of conductivity to be reached in their nanofiber form. For example, 50 nm wide lines of poly(3-octylthiophene), produced by direct EBL writing, reach $\chi$ of the order of 1 S cm$^{-1}$ upon $FeCl_3$ doping (with a value of resistance of the order of $10^{-6}$ Ω over a fiber length of 20 nm).[75] Nanofibrous films of helical PACT synthesized in a chiral nematic liquid crystal have conductivity as high as 1.5–1.8 $\times$ $10^3$ S cm$^{-1}$ following iodine doping.[76] Maximum reported values of conductivity for PPy nanostructures produced in this way largely exceed $2 \times 10^3$ S cm$^{-1}$, which is an order of magnitude more than reference films, and, more generally, the conductivity is found to increase upon decreasing the diameter of the fibers.[66] PEDOT nanowires with a diameter of 10 nm are reported to show a conductivity larger than in reference thin-films by more than 60 times.[77] Analogous findings are reported for PAN (with $\chi$ up to 50 S cm$^{-1}$)[67,68,72,78–80] and poly(3-methylthiophene) (with $\chi$ up to $1.5 \times 10^3$ S cm$^{-1}$).[65]

Different behaviors and sometimes controversial results have been reported for membrane-templated nanostructures of PEDOT, showing room-temperature conductivity from roughly 60 S cm$^{-1}$ to 550 S cm$^{-1}$, and a prevalently two-dimensional VRHM transport at very small diameters (35 nm).[70,81,82] This behavior can be rationalized by considering a heterogeneous internal structure in the wires, with an external, more conductive shell,[81] which is likely related to a higher conformational order of the polymer chains near the walls

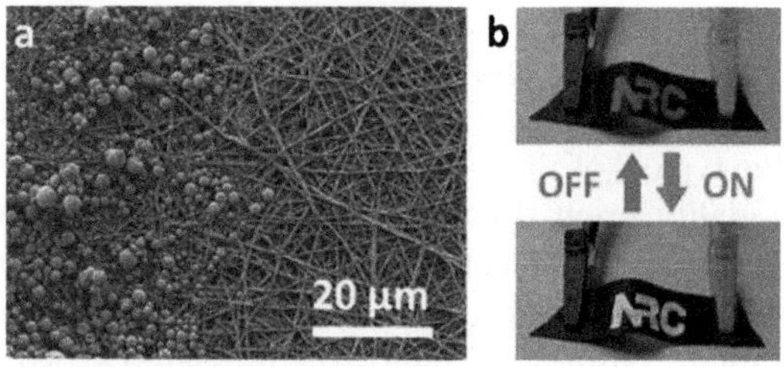


**Figure 6.2** (a) SEM micrograph of thermochromic microcapsules deposited on a mat of PEDOT-containing nanofibers, on its turn realized by electrospinning PVP and then polymerizing the conductive polymer according to the vapor-phase, pseudo-template method (Section 3.3.2). The thermochromic ink, here only present in the left part of the imaged area, is patterned onto the nanofiber mat by spraying through masks. (b) Electrochromic switching obtained by applying 0/120 mA to the electrospun mat. The thermochromic ink switches its color at 37 °C, a temperature that is reached within seconds due to heating induced by the applied current. Reproduced with permission from Ref. 86, *J. Mater. Chem.*, 2010, **20**, 8233–8235. Doi: 10.1039/c0jm02307h. Copyright © 2010, National Research Council Canada.

of membrane nanopores where the polymerization starts (Section 3.3). By performing four-probe measurements, the conductivity of these PEDOT nanowires is also found to increase upon decreasing the diameter of the fibers from 190 nm to 35 nm.[70]

Other combinations of nanofabrication and processing techniques are interesting as well, allowing conduction properties to be tailored. A conductivity up to about 1 S cm$^{-1}$ is measured in PS/PEDOT fibrous mats produced by electrospinning PS nanofibers containing ferric *p*-toluenesulfonate and then growing PEDOT from the monomer vapor phase, which improves the inter-fiber connectivity by melt-welding.[83] The conductivity of PEDOT-coated PAC nanofiber samples, obtained by electrospinning followed by vapor phase polymerization, is in the range of 1–10 S cm$^{-1}$,[84] whereas maximum values of $\chi$ measured for PEDOT nanofiber samples obtained by this class of methods reaches (60 $\pm$ 10) S cm$^{-1}$.[85] Such fibers also exhibit a significant Joule heating and can be consequently used as convenient substrates for electrically controlling the color of thermochromic particles deposited on the mats (Figure 6.2).[86]

*PAN conductive nanofibers.* PAN-based nanofibers* have conductivity values in the range from 10$^{-4}$ S cm$^{-1}$ to many tens of S cm$^{-1}$,[67,68,72,78–80,88–95] and can

* Some experiments with two-terminal devices and PAN nanofibers are quite simple to perform, and allow one to realize (for instance, by means of electrospinning) and electrically characterize the nanostructures in a few hours. For this reason, these experiments can be interesting for didactics and practical laboratories for undergraduate students in physics, materials sciences and engineering, as reported for the case of the undergraduate physics program at the University of Puerto Rico.[87]

be used to realize FETs[96] and Schottky diodes.[87,97] The fibers can be fabricated in many different ways. For instance, during the still early stages of development of modern electrospinning, researchers from the University of Pennsylvania and Drexel University have presented electrospun nanofibers made of sulfuric acid doped PAN and of its blends with a variety of other polymers including PEO.[98] Other reports are focused on similar fibers, doped by camphorsulfonic acid.[96] Once positioned between coplanar Au electrodes, fibers of 50% doped PAN in PEO exhibit conductivity of the order of 0.1 S $cm^{-1}$. Increasing the percentage of doped PAN to 72% leads the conductivity up to 33 S $cm^{-1}$, a value that is unexpectedly high given the insulating properties of the PEO component, which has suggested pioneering considerations on the possibility of internal alignment of conjugated molecules within the nanofibers.[98] Combining conductivity measurements with mobility values obtained by the FET characteristics allows one to estimate the linear concentration of charge-carriers in the PAN/PEO nanofibers to be of the order of $10^8$ holes per cm.[96] The conductivity increases upon increasing temperature (Section 6.1), and shows a significant dependence on a variety of chemicals, such as aliphatic alcohols,[99] to which nanofibers are exposed, which is promising for the realization of sensing devices.

Schottky diodes are fabricated even more straightforwardly because the two electrodes can be obtained directly by a doped silicon substrate. These devices can be realized in one single step by electrospinning a conductive PAN/PEO nanofiber directly onto a substrate consisting of *n*-doped Si, an electrically insulating silica layer and a pre-patterned Au electrode.[87,97] In this way, a Schottky junction is immediately formed between the nanofiber and the region of silicon exposed to air at the vertical edge of the substrate, with an energetic barrier ($\Delta E$) for charge-carriers. This is a first, simple but elegant example of nanofiber nanoelectronics. At the junction according to the thermoionic emission model the current density, $J_I$, at a bias, $V$, is given by Richardson's law:[87,97,100]

$$J_I = J_{I,0} e^{-\frac{qV}{f_I k_B T}}, \tag{6.5}$$

where $f_I$ is an ideality factor that can account for deviations from the standard model, and the pre-exponential factor, $J_{I,0}$, indicates the current density at saturation which is:

$$J_{I,0} = A^* T^2 e^{\frac{\Delta E}{k_B T}}. \tag{6.6}$$

In the previous equation, $A^*$ [A $m^{-2}$ $K^{-2}$] is the Richardson constant ($A^* = 4\pi q m^* k^2 / h_P^3$, where $m^*$ is the effective mass of the charge-carrier). PAN/PEO nanofiber-diodes exhibit good rectifying behavior, and in forward conditions they roughly behave as predicted by the thermoionic emission model, especially at

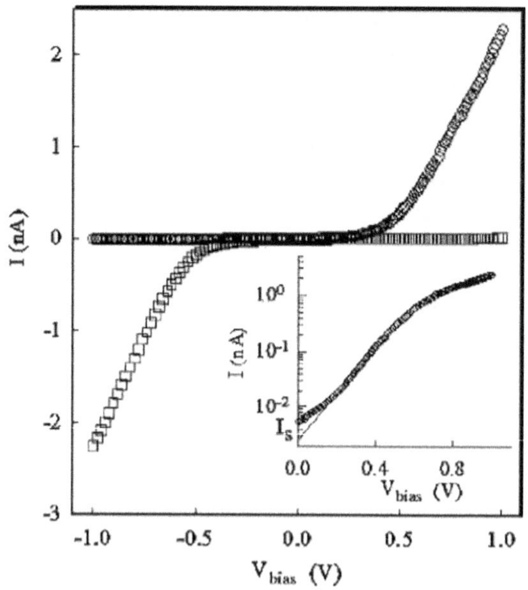


**Figure 6.3** Current-voltage characteristics at 300 K of a PAN/PEO nanofiber diode. The two curves (squares and circles, respectively) are obtained by inverting the polarization of the applied bias. Inset: Semilog plot of the forward bias current *vs.* forward bias voltage for the case when the positive terminal of the applied bias is connected to an Au electrode in contact with the nanofiber. The linear behavior at low voltage bias values is described by Equation (6.5). Reproduced with permission from Ref. 97, *Appl. Phys. Lett.*, 2006, **89**, 033505. Doi: 10.1063/1.2227758. Copyright © 2006, American Institute of Physics.

moderate bias (Figure 6.3). Moreover, the fibers present an improved molecular alignment compared to reference thin-films, as highlighted by EPR.[101]

Fibers of pure PAN can be electrospun as well, starting from solutions of sulfuric or formic acid and depositing onto a bath of water[98] or acetone.[102] As described in Section 3.1, nanofibers of PAN can also be produced by using $NaClO_2$[103] and $KH(IO_3)_2$ as oxidants,[104] or recombinant CiP as biocatalyst.[105] The resulting fibers have diameters between 50 and 100 nm, and lengths of a few μm. The doping state of biocatalyzed fibers can be controlled both during and after the production by using camphorsulfonic acid (to dope) and $NH_4OH$ (to undope), respectively. Dedoping can also be achieved by $NH_3$ gas, for which the nanofibers can constitute good chemical sensors.[106] In their doped state, single nanofibers carry currents of the order of 500 nA (~2 MΩ resistance) upon applying a bias of 1 V with Au electrodes,[105] which (for a fiber with diameter 100 nm and a length of 500 nm) would correspond to a conductivity below 1 S $cm^{-1}$. The synthesis using potassium biiodate allows instead higher conductivity values to be measured, up to about 100 S $cm^{-1}$.[104] PAN nanofibers produced by interfacial polymerization, which is another approach frequently used for this class of materials (Section 3.4),[79,80,91,107] show conductivity from $10^{-1}$ S cm$^{-1}$,[108] to 8 S cm$^{-1}$, depending, for instance, on the used APS/aniline molar ratio and molarity of the $HClO_4$ solution in which APS is dissolved.[109] Pressed pellets of nanofibers realized by using β-NSA as dopant, or by using twin-tailed anionic surfactants during interfacial polymerization, have conductivity up to 5.8 S cm$^{-1}$,[79] and in the range of 1–5 S cm$^{-1}$, respectively.[107] By β-NSA or camphorsulfonic acid, nanotubes of PAN have been obtained as well, with $\chi$ between 9.3 S cm$^{-1}$,[110] and 31–47 S cm$^{-1}$,[92,93] at room temperature. The temperature dependence of conductivity highlights a quasi-one-dimensional VHRM conduction (Equation (6.2)).[110] In addition, these nanostructures exhibit, below 60 K, a transition from small negative values to positive values of the magnetoresistance, MR, defined as:

$$\mathrm{MR} = \frac{\Delta R_X}{R_X(H=0)} = \frac{R_X(H) - R_X(H=0)}{R_X(H=0)}, \tag{6.7}$$

where $\Delta R_X$ indicates the variation of the resistance, $R_X$, measured in the fibers upon switching on a magnetic field of intensity, $H$.[93,110] These findings (and other analogous results in template synthesized PEDOT nanofibers below 10 K)[82] can be discussed in terms of wavefunction shrinkage in the presence of the magnetic field (for positive values of MR), and of quantum interference effects in the particular conduction regime within the nanostructures (for negative values of MR).[110]

By using D-10-camphorsulfonic acid as dopant, the conductivity of pellets of PAN nanofibers lowers to $3.4 \times 10^{-3}$–$3.5 \times 10^{-1}$ S cm$^{-1}$, again depending on the dopant/monomer relative concentration during the assembly process.[111] More generally, the conductivity of PAN nanofibers is reported to increase (from 170 to 640 S cm$^{-1}$) by increasing the number of COOH groups in used carboxylic acid dopants.[112] Pressed pellets of PAN nanofibers realized by the nanofiber seeding method described in Section 3.3.2 show conductivity values between 2 and 10 S cm$^{-1}$.[113] Other experiments are instead focused on improving the electrical properties of PAN nanofibers by doping with MWCNTs,[114] or with a highly conductive, inorganic component, such as ZnO or Ga-doped Zn nanoparticles.[115] Electrospinning (Section 2.3.3) has to be carefully optimized to achieve a uniform distribution of nanoparticles in these fibers, which then have a conductivity up to five times higher than in purely polymeric nanostructures. In addition, there is a threshold concentration (~20%) of the inorganic nanoparticles needed to observe an actual increase of conductivity, because the used dopants have *n*-type transport and consequently they have to compensate the opposite charges normally carried by PAN before leading the system to show improved electrical performances.[115] Nanoparticle doping opens also other interesting opportunities. For instance, embedding ferroelectric†, 50 nm particles of $BaTiO_3$, with

† The basics of piezo- and ferroelectric materials are introduced in Section 6.4.1.

relative concentrations of 0.1–5% in weight in electrospun PAN nanofibers, allows the carried current to be tuned effectively by means of an external polarizing electric field.[116] Without externally induced polarization, the nanocomposite fibers show currents up to 600 nA ($\sim$1.7 MΩ resistance) by applying a bias of 1 V with the usual Au electrodes. For fiber diameters in the range 200–700 nm and an inter-electrode distance of 10 μm, this would correspond to a conductivity in the range of 0.1–1.8 S cm$^{-1}$. With the application of the external electric field that orients the ferroelectric domains of the nanoparticles, the conduction properties of these nanocomposite fibers become more appealing, because this determines an increase of the two-terminal resistance (Figure 6.4a), which can be controlled in real time during the operation of the device.[116] The effect can be explained on the basis of both classical and quantum considerations. In the simplest picture (Figure 6.4b), the external field allows one to control the screening volume due to the ferroelectric domains, which asymmetrically attracts charge-carriers and, depending on the resulting orientation, can correspondingly reduce the cross-sectional area of the fiber available to conduction.[116] Envisaged applications include memory components, electrically controlled switches and various types of sensors.

*P3AT conductive nanofibers.* P3AT nanofibers are a very widely studied class of conjugated polymer nanostructures.[46,117] Upon fiber formation, the lamellar sheets of P3HT often package parallel to each other, thus forming a microcrystalline structure with relevant π–π stacking along the *b*-axis as introduced in Section 6.1.1. In other words, almost irrespective of the particular fabrication method (electrospinning, self-assembly through precipitation,[46,56,117] NIL,[118] *etc.*), the *b*-axis frequently coincides with the longitudinal axis of the fiber, which is at the base of the possibly enhanced mobility measured along this direction (Figure 6.1b). Given the inter-sheet distance, the supramolecular packing results in a linear density of the order of 2600 lamellar sheets per μm. Room-temperature conductivity values of P3AT nanofibers, measured by the four-probes method, are up to 89 S cm$^{-1}$.[27] The eventual effect of the length of the alkyl chain on the conductivity of P3AT nanofibers is another interesting matter of research. This feature can significantly affect the electrical performances of thin-films, with lower mobilities usually being associated to longer alkyl chains. However, recent studies indicate a substantial independence of the hole mobility from the alkyl chain length, which leads one to argue that the different conduction performances observed in many devices upon varying this length have not to be considered as intrinsic, molecular-related properties, but instead something due to the quality of the organic semiconductor/dielectric interface as produced in thin-film devices (see also the next Section).[117,119] Other works focus on amphiphilic, regioregular polythiophenes assemble in nanowires by Langmuir–Blodgett methods (Section 3.1.3), and then showing conductivities of 10$^{-5}$ S cm$^{-1}$ or less, which increases up to 40 S cm$^{-1}$ upon doping with vapor from a iodine crystal.[37] Among other applications, P3HT nanofibers can

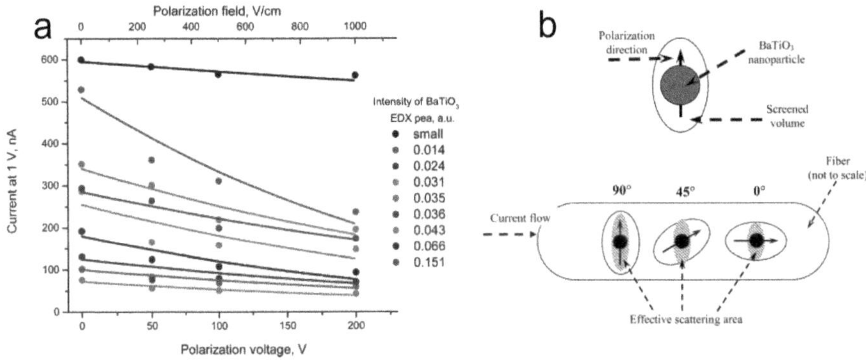


**Figure 6.4** (a) Two-terminal current flowing in $BaTiO_3$/PAN nanocomposite fibers *vs.* external voltage applied to orient the ferroelectric domains of the nanoparticles. Different curves indicate different amounts of nanoparticle loadings. The two-terminal voltage bias is kept fixed at 1 V in this plot. (b) Schematics of the orientation of the nanoparticle ferroelectric domains upon application of an external polarizing electric field, and of the resulting increase of the fiber resistance. A screened volume around the particle is generated because of the asymmetric attraction of free charge-carriers present in the nanocomposite, which obstructs the conductive path thus enhancing resistivity. Adapted with permission from Ref. 116, *Nano Lett.*, 2006, **6**, 896–900. Doi: 10.1021/nl052017r. Copyright © 2006, American Chemical Society.

be used to realize rectifying nanojunctions by crossing *n*-doped $SnO_2$ nanofibers.[120]

*PPy conductive nanofibers.* PPy nanofibers obtained by micelle-based approaches have good room-temperature conductivity ($\chi$ = 120–130 S cm$^{-1}$ for fibers, and $\chi$ = 19–30 S cm$^{-1}$ for nanotubes), *i.e.* slightly lower than the best values measured in membrane-templated PPy nanostructures, and moreover they show a quasi-three-dimensional conduction behavior according to Equation (6.2).[110,121,122] Pressed pellets of PPy nanofibers produced by nanofiber seeding (Section 3.3.2) have $\chi$ = 50 S cm$^{-1}$.[123] A power-law dependence of current on applied bias together with a threshold voltage for conduction are observed in PPy nanowires with diameters from 30 to 450 nm, realized by template synthesis (Section 3.3).[71] PPy nanowires with a diameter of 10 nm show a conductivity larger than in reference thin-films by almost two orders of magnitude,[77] and the conductivity of single PPy nanotubes increases from 0.13 S cm$^{-1}$ to 73 S cm$^{-1}$ upon decreasing the outer diameter from 560 nm to about 100 nm.[93] Conductive nanowires of PPy have also been realized by using copolymers with py-silane, and defined by EBL and lift-off for use in chemical sensors (Section 4.3.2).[124]

*PEDOT conductive nanofibers.* PEDOT nanotubes produced by reverse microemulsion polymerization exhibit $\chi$ values as high as 250 S cm$^{-1}$.[122] Nanofibers of PEDOT/PSS can be produced by the molecular combing method (Section 3.3.2).[125] These fibers are very thin, with diameters well below

10 nm, and exhibit conductivity in the range of 0.09–0.6 S cm$^{-1}$ and resistance around 1 GΩ. Pressed pellets of PEDOT nanofibers realized by nanofiber seeding (Section 3.3.2) have $\chi$ = 16 S cm$^{-1}$.[126]

*Other polymers.* A rich variety of different behaviors is possible for quasi-one-dimensional nanostructures based on conductive polymers. For instance, compared to P3ATs, nanobelts of BBL show striking differences, *i.e.* (i) the long axis of the nanostructures coinciding with the *c*-axis of the orthorhombic unit cell (instead of the *b*-axis), namely with the direction of the polymer chains (instead of the π–π stacking), and (ii) a favored electron transport.[127] The CDT-BTZ copolymer is effectively processed by the SVED method (Section 3.1.2),[128] and the resulting fibers have lateral dimensions between 80 and 600 nm and lengths up to 20 μm, exhibit a highly ordered supramolecular organization and are promptly usable in FETs.[128]

A detailed study reports on the dependence of the transport properties of conductive fibers made of helical PACT.[129] The lateral dimensions of these fibers are between 40 and 300 nm, and their length is up to 10 μm. The fibers, doped with iodine from vapor and connected to Pt electrodes, exhibit a threshold voltage ($V^*$) of the order of 1 V for conduction at low temperatures (roughly below 30 K). Above threshold (Figure 6.5a and b), the current increases with a scaling law:

$$I \sim [(V/V^*)1]^{\Xi}, \tag{6.8}$$

where Ξ is an exponent in the range 1.8–2.1 and weakly dependent on temperature. This behavior is attributed to the so-called Coulomb-blockade effects, which are more significant at low temperatures. The mechanism can be rationalized by schematizing the nanofibers as an array of spatially separated conducting regions that confine with nanoscale barriers (each of resistance >> $h_P/q^2$) disfavoring charge transport. Coulomb-blockade conduction can be seen as the combination of electrical charging of the single conducting regions (each of capacitance << $q^2/2k_BT$) by a single (or a few) charge-carriers, and of tunneling between adjacent conducting regions through the insulating barriers (Figure 6.5c). According to theory,[130] this results in the non-ohmic behavior expressed by Equation (6.8), with the Ξ exponent depending on the system dimensionality. At temperature above 30 K, the same nanofibers show instead significantly different transport properties, with a super-linear dependence of the current on the voltage bias ($I \sim V^{\beta}$ with $\beta$ = 2–5.7).[131] These are valuable results for discussing the possibility of electron–electron interactions of short-range (as in Lüttinger liquids)[132] or long-range (as in Wigner crystals)[133] in polymer nanofibers. In another framework, tunneling effects have been also invoked to explain the non-linear, stepwise *I-V* characteristics found in nanojunctions between EBL-defined metal electrodes and self-assembled nanowires of PPE with thioacetyl end groups and diameter in the range of 50–130 nm.[134]

The concept of modeling polymer nanofibers as somehow inhomogeneous sequences of conductive segments, separated by insulating barriers disfavoring conduction,[116,135] is based on general considerations on the largely amorphous nature of conjugated materials. From a theoretical viewpoint, this concept is also used in other studies treating conductive polymers as granular media and analyzing how stochastic nuclear motions impact on the transport properties by taking into account the electron-phonon interactions:[136] in this way, new states are predicted that assist the transport of charge-carriers along polymer fibers, leading to distinctive features concerning dissipative effects[136] and the temperature dependence of current and conductivity,[137] which may be peculiar of these nanomaterials.

Analogously to thin-films, those organic nanofibers (or nanowires) made of oligomeric conjugated materials, which show a large prevalence of highly ordered, crystalline state, may reach particularly high values of mobilities ($\mu \geq 1\ cm^2\ V\ s^{-1}$). For instance, this is the case of elongated crystalline aggregates of phenylene-thiophene (co-)oligomers, such as the *para*-hexaphenylene (p6P), 5,5′-di-4-biphenylyl-2,2′-bithiophene (PPTTPP) and 4-4′-di-2,2′-dithienyl-biphenyl (TTPPTT) molecules (Figure 6.6).[138]

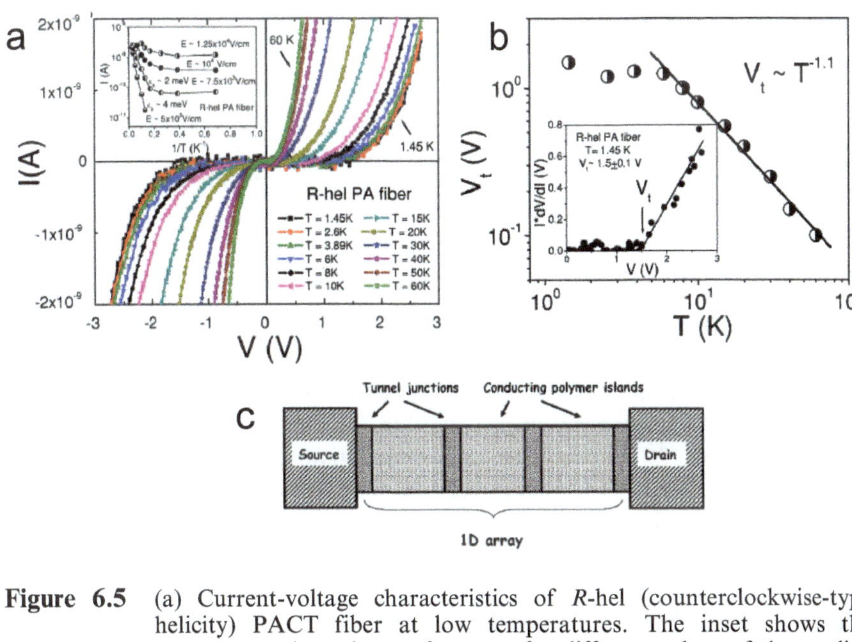


**Figure 6.5** (a) Current-voltage characteristics of *R*-hel (counterclockwise-type helicity) PACT fiber at low temperatures. The inset shows the temperature dependence of current for different values of the applied electric field. (b) Temperature dependence of the threshold voltage (in this figure, $V_t$). Inset: $IdV/dI$ *vs.* $V$ at $T$ = 1.45 K evidencing the threshold voltage. (c) Model scheme for a metal-insulator-metal tunneling junction structure along a conducting polymer nanofiber. Adapted with permission from Ref. 129, *Phys. Rev. B*, 2005, **72**, 153202. Doi: 10.1103/PhysRevB.72.153202. Copyright © 2005, The American Physical Society.

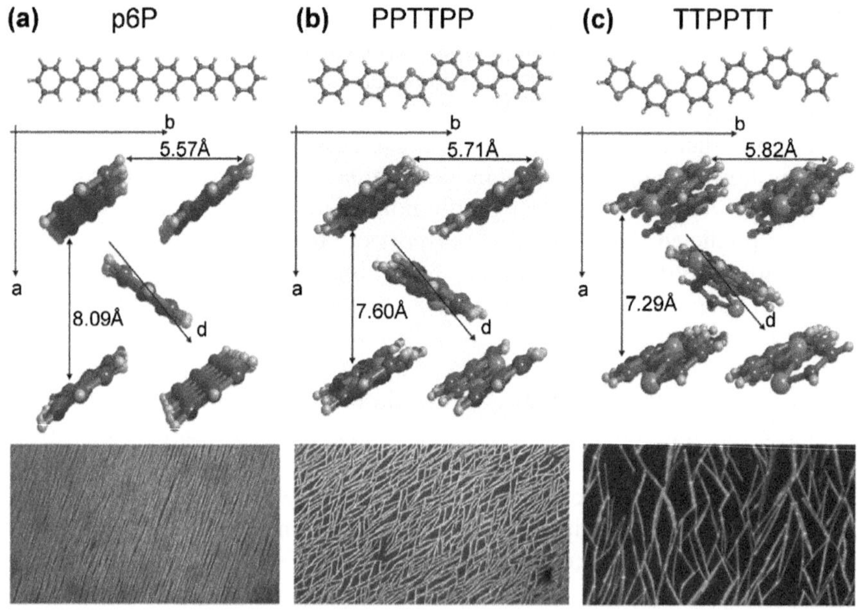


**Figure 6.6** Molecular structures and crystalline organization of (a) p6P, (b) PPTTPP and (c) TTPPTT. Arrows: crystal lattice directions. Bottom figures: fluorescence micrographs of the different types of nanofibers. Micrograph size: 112 × 63 μm$^2$ (p6P), 223 × 125 μm$^2$ (PPTTPP) nanofibers, 130 × 74 μm$^2$ (TTPPTT). Reproduced with permission from Ref. 138, *Org. Electron.*, 2009, **10**, 1228–1234. Doi: 10.1016/j.orgel.2009.06.015. Copyright © 2009, Elsevier B.V.

# 6.3 Applications in Nanoelectronics

The most studied electronic devices exploiting organic semiconductors as active materials are undoubtedly FETs,[13,26,139,140] which have important applications in flexible displays, backplanes and radio-frequency identification (RFID) tags.[141] FETs based on organic thin-films have been fabricated since the early 1980s.[142,143] The use of conductive polymer nanofibers in electrochemical devices for energy storage should also be mentioned in this framework. These devices and applications are described in Section 4.3.1.

The different device architectures of organic FETs are schematized in Figure 6.7, comparing film (Figure 6.7a–d) and nanofiber-based architectures (Figure 6.7e–h). Basically, the transport of charge-carriers occurs through a *channel* between two coplanar electrodes, named source (*S*) and drain (*D*), separated by a distance, *L*, that corresponds to the channel length, and each having width, $W_E$. A third electrode, which is the gate (*G*), is used to modulate the flowing current, just as in inorganic FETs.[100]

In most microelectronic devices, the coplanar electrodes (here *S* and *D*) are defined in metals by evaporation through shadow masks or by photolithography (in turn followed by etching or alternatively by metal evaporation and lift-off of the residual resist). $W_E$ and $L$ are generally in the range of 1–100 μm. In conventional *p*- or *n*-type thin-film organic FETs (Figure 6.7a–d),[8,12,26,61] the semiconductor compound is generally deposited by evaporation or, for solution-processable materials, by spin-coating or other wet methods, which produces a continuous organic active layer in contact with the source and drain electrodes. For this reason, in these devices the width ($W_E$) of the electrodes also corresponds to the width ($W$) of the transport channel in the semiconductor material. Instead, when source and drain are inter-connected by means of a polymer nanofiber (Figure 6.7e–h), whose radius, $r$, is much smaller than the width of the electrodes, the channel width coincides with the fiber diameter. In other words, using nanofibers as active media, FET devices are led from the micro-electronics into the nanoelectronics domain (though still retaining microscale channel lengths in most cases). In the following equations, the reader will always find the channel width, $W$, meaning $W = W_E$ and $W = 2r$ for film-based and nanofiber-based devices, respectively.

The position of the source and drain electrodes with respect to the semiconducting medium defines a first classification of the different possible FET geometries, allowing us to distinguish between *top-contact* (Figure 6.7a,b,e,f) and *bottom-contact* (Figure 6.7c,d,g,h) devices. In addition, the gate has to be electrically insulated from the other two electrodes in order to avoid undesired leakage currents, and this is accomplished by means of a dielectric layer ($SiO_2$, $Si_3N_4$, $Al_2O_3$, other oxides, various types of insulating plastic materials, *etc.*). The position of the gate and of the dielectrics with respect to the semiconducting medium defines a second classification of FETs, which can be realized in a top-gate (Figure 6.7a,c,e,g) or bottom-gate (Figure 6.7b,d,f,h) configuration.

FETs can work either in accumulation or in depletion mode,[100] namely the gate voltage can be used to either attract and concentrate charge-carriers at the semiconductor/dielectrics interface, thus determining an increase of the source-drain current ($I_{DS}$), or, on the contrary, to push charge carriers away and deplete the channel region, thus determining a decrease of $I_{DS}$‡. In fact, it is in a very thin, nm-scale layer (namely few molecular monolayers) near the interface with the dielectric film that the organic semiconductor transports

‡ Of course, as in OLEDs, the (source and drain) electrodes of an organic FET should have a set of energy levels as favorable as possible to inject charge-carriers into the semiconductor material. For instance, we already know from Section 5.1.1 that a high work function is preferred to inject holes in the HOMO level. Au, Ag,[144,145] Pd,[146] Pt[128,147] or ITO,[148] and also some conductive polymers that can be deposited from solution, such as PAN added with poly(2-acrylamido-2-methyl-1-propane-sulfonic acid)[149] or doped with SWCNTs,[150] and PEDOT:PSS,[151,152] are all available options in this case. Concerning the injection of electrons into the LUMO level, one would analogously expect that metals of lower work function, as Ca or Li, are to be preferred (though these are generally quite unstable to oxidation when exposed to air atmosphere). However, it should be noticed that *n*-type FET devices seem to be quite robust even when metals with reasonably high work function are used in electrodes.[26] Furthermore, it is well established that the work function can be engineered in several ways by functionalizing electrodes with self-assembled monolayers.[153–155] Overall, relatively high work function metals, and especially Au,[18,58] are largely used also for defining the source and the drain electrodes in *n*-type or in ambipolar FETs.[144,145,148]

charge-carriers in FETs§. Therefore, the conductive region is generally much thinner than (and it must not be confused with) the thickness of the organic material.[159] This results in the usually applied *charge sheet approximation*,[42] which requires that the carriers are transported and spread in the very thin layer (or sheet) very close to the organic/dielectric interface. A superficial density [C $m^{-2}$] of charges is generated, all affected by the same surface potential gradient (*i.e.* a voltage gradient that is only along the source-drain direction). The validity of this approximation is therefore critically related to the transversal thickness of the conductive sheet (in other words the thickness of the channel region), which has to be much smaller than the thickness ($d_D$) of the dielectric layer. In practically all the cases explored to date, the charge sheet approximation can be applicable also to FETs based on polymer nanofibers, for which the thickness of the organic material roughly corresponds to $2r$ (say, from 10 nm to hundreds of nm), $d_D$ is mostly in the range of many tens to a few hundreds of nm and the thickness of the conductive channel is still in the scale of nanometers near the interface with the dielectrics.

To infer the operational mode of FETs, one can have a look at the typical plots displaying the *output characteristics*, namely the dependence of $I_{DS}$ on the source-drain voltage bias ($V_{DS}$ = drain voltage minus source voltage) for different values of the source-gate voltage ($V_{GS}$ = gate voltage minus source voltage) as in Figure 6.8a. In most reports, and in the following, positive and negative $I_{DS}$ are associated with electron (*n*-type devices) or hole (*p*-type devices) transport, respectively. As a consequence, $V_{DS}$ values driving conduction are normally positive and negative, respectively, in the two cases. Concerning the field-effect, a condition of $V_{GS} < 0$, *i.e.* a gate negatively biased with respect to the source electrode, would determine an operation in accumulation mode for *p*-type devices (thus determining an increase of $|I_{DS}|$ upon increasing $|V_{GS}|$, $\partial|I_{DS}|/\partial|V_{GS}||_{V_{DS}} > 0$) and in depletion mode for *n*-type devices (thus determining a decrease of $|I_{DS}|$ upon increasing $|V_{GS}|$, $\partial|I_{DS}|/\partial|V_{GS}||_{V_{DS}} = \partial I_{DS}/\partial|V_{GS}||_{V_{DS}} < 0$). $V_{GS} > 0$ would correspond to the opposite situation (for instance, accumulation for *n*-type devices, with $\partial|I_{DS}|/\partial|V_{GS}||_{V_{DS}} = \partial I_{DS}/\partial V_{GS}|_{V_{DS}} > 0$). The different operational conditions and adopted signs for electric currents are summarized in Table 6.1.

§ For this reason, the dielectric/semiconductor interface[156,157] critically affects the device performances. Both the interface roughness,[158] and, even more importantly, the wettability from aqueous vapor negatively impact on the mobility measured in the organic semiconductor. In particular, hydrophilic dielectric surfaces easily attract aqueous vapor and other kinds of contaminations, which may act as trapping centers for charge-carriers. Commonly used methods to increase the hydrophobicity of the surface (*e.g.* $SiO_2$) on which the organics are deposited include the application of self-assembled monolayers of octadecyl trichlorosilane (OTS), HMDS, *etc.*[155] As an alternative, plastic materials that are able to guarantee lack of hydroxyls at the interface with the organic semiconductor, such as a cross-linked benzocyclobutene (BCB) derivative, are attracting increasing attention. The role of these dielectric layers is demonstrated to be especially critical for observing *n*-type conduction in organic FETs, and consequently to eventually find ambipolar behavior and light-emission.[11,61]

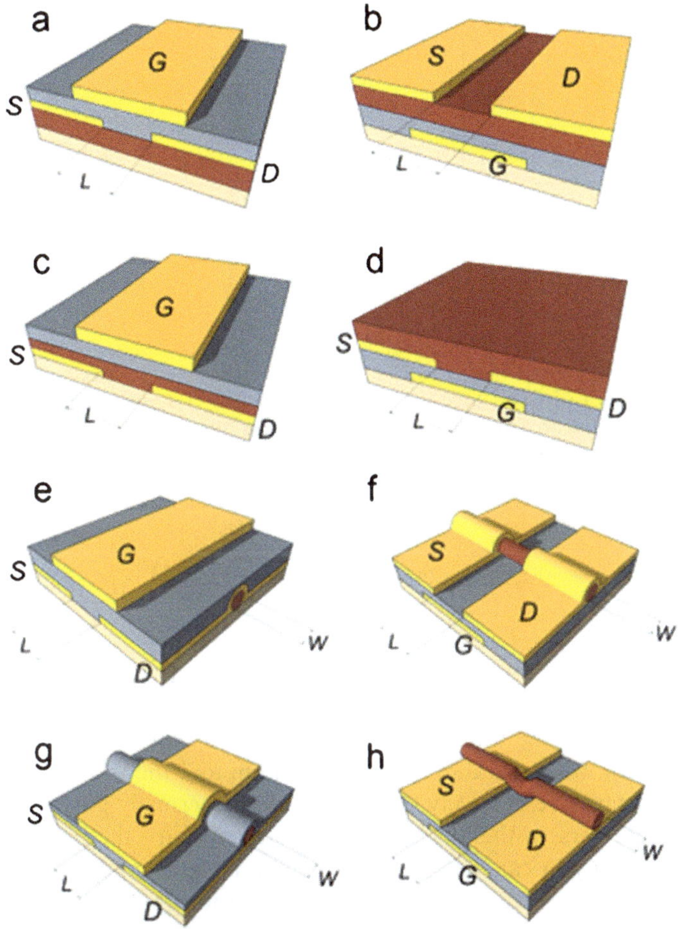


**Figure 6.7** Schemes of FET architectures based on polymer films (a–d) and on polymer nanofibers (e–h). Pink layer: substrate. Red: active material. Cyan: dielectric layer. Gold: electrodes. (a, e) Top-contact, top-gate. (b, f) Top-contact, bottom-gate. (c, g) Bottom-contact, top-gate. (d, h) Bottom-contact, bottom-gate. The (b, f) and (c, g) staggered configurations can be advantageous for some applications because here the surface of the coplanar electrodes through which charges are injected is larger than in other configurations.

In addition, from Figure 6.8a it is clear that different regimes can be observed in the $I_{DS}(V_{DS})$ curves. At low absolute values of the source-drain voltage ($|V_{DS}| << |V_{GS}|$), $|I_{DS}|$ increases almost linearly upon increasing $|V_{DS}|$. In the so-called, largely used gradual channel approximation ($L >> d_D$),[100,160] this behavior is described by the equation:

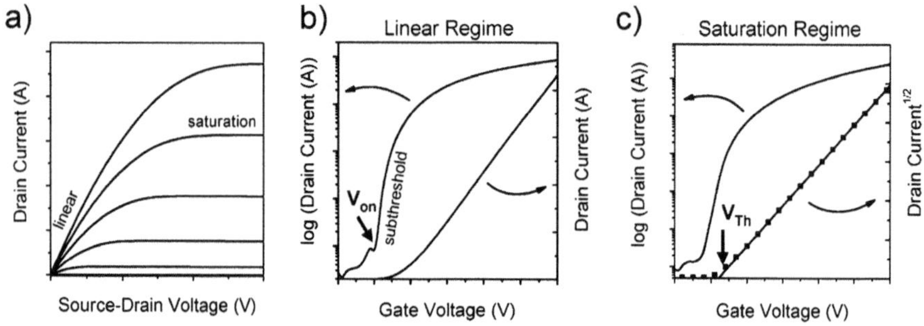


**Figure 6.8** (a) FET current-voltage characteristics. Drain current, $I_{DS}$ (A), *vs.* source-drain voltage bias, $V_{DS}$ (V). The different curves are collected by different values of the source-gate voltage, $V_{GS}$. (b, c) Transfer characteristics in the linear (b) and in the saturation (c) regime. Reproduced with permission from Ref. 61, *Chem. Rev.*, 2007, **107**, 1296–1323. Doi: 10.1021/cr0501543. Copyright © 2007, American Chemical Society.

$$I_{DS}^{(linear)} = \frac{W}{L} C_i \mu^{(linear)} \left( V_{GS} - V_{TH} - \frac{V_{DS}}{2} \right) V_{DS}. \tag{6.9a}$$

where $C_i$ [F m$^{-2}$, more often expressed in units of F cm$^{-2}$] is the gate dielectric capacitance per unit area ($\cong$ capacitance/$WL$), $\mu^{(linear)}$ is the linear-regime mobility and $V_{TH}$ is the threshold voltage. For thin-film devices, $C_i$ is simply given by $\varepsilon_0\varepsilon_r/d_D$, where $\varepsilon_0$ and $\varepsilon_r$ are the vacuum and the insulating material dielectric constants, respectively (for instance, for thermally grown $SiO_2$, $\varepsilon_r$ = 3.9). For nanofibers, in the ideal case of cylindrical shape the capacitance per unit length ($\cong C_iW$) can be estimated by the basic formula, $2\pi\varepsilon_0\varepsilon_r/\ln(2d_D/r)$.[74,96] Incidentally, since $|V_{DS}| << |V_{GS}|$, Equation (6.9a) can be approximated as:

**Table 6.1** Operation modes and usually adopted signs of currents and bias voltages for FET devices.

| *Operational mode* | $I_{DS}$ | $V_{DS}$ | $V_{GS}$ | $\left.\frac{\partial I_{DS}}{\partial V_{DS}}\right\|_{V_{GS}}$ | $\left.\frac{\partial \|I_{DS}\|}{\partial \|V_{DS}\|}\right\|_{V_{GS}}$ | $\left.\frac{\partial I_{DS}}{\partial V_{GS}}\right\|_{V_{DS}}$ | $\left.\frac{\partial \|I_{DS}\|}{\partial \|V_{GS}\|}\right\|_{V_{DS}}$ |
|---|---|---|---|---|---|---|---|
| *p*-type, accumulation | <0 | <0 | <0 | >0 | >0 | >0 | >0 |
| *p*-type, depletion | <0 | <0 | >0 | >0 | >0 | >0 | <0 |
| *n*-type, accumulation | >0 | >0 | >0 | >0 | >0 | >0 | >0 |
| *n*-type, depletion | >0 | >0 | <0 | >0 | >0 | >0 | <0 |

$$I_{DS}^{(linear)} = \frac{W}{L} C_i \mu^{(linear)} (V_{GS} - V_{TH}) V_{DS}. \tag{6.9b}$$

In this regime and under these approximations, the electric potential increases linearly in absolute value moving along the longitudinal coordinate, $x$, of the channel, passing from 0 (at $x = 0$, namely at the source that is usually grounded) to $|V_{DS}|$ (at $x = L$, namely at the drain electrode). Furthermore, the FET transconductance $g_m$ ($\Omega^{-1} = S$) is defined as the slope of the $I_{DS}$ dependence on $V_{GS}$ at a constant value of $V_{DS}$, *i.e.*:

$$g_m = \left. \frac{\partial I_{DS}^{(linear)}}{\partial V_{GS}} \right|_{V_{DS}} = \frac{WC_i}{L} \mu^{(linear)} V_{DS}. \tag{6.10}$$

At high $|V_{DS}|$ (more precisely for $|V_{DS}| > |V_{GS}| - |V_{TH}|$), the channel of the FET undergoes a pinch-off,[100] and the $|I_{DS}|$ saturates at constant values that are almost independent from the source-gate voltage, as described by the following equation:

$$I_{DS}^{(satur.)} = \frac{W}{2L} C_i \mu^{(satur.)} (V_{GS} - V_{TH})^2, \tag{6.11}$$

where $\mu^{(satur.)}$ is the mobility in the saturation regime**. The corresponding expression for the transconductance is:

$$g_m = \left. \frac{\partial I_{DS}^{(satur.)}}{\partial V_{GS}} \right|_{V_{DS}} = \frac{WC_i}{L} \mu^{(satur.)} (V_{GS} - V_{TH}) \tag{6.12}$$

In any case, one has to take into account that Equations (6.9) and (6.11) are just approximations, which can be used for a rough estimation of the charge-carrier mobilities. Indeed, while in these equations the mobility is usually taken as constant, in real devices it very often exhibits a dependence on the gate-source voltage bias.

Besides the output characteristics (Figure 6.8a), the *transfer characteristics* (Figure 6.8b and c) are widely used to display the FET performances in

** As is evident from Equations (6.9) and (6.11), a higher capacity per unit area means higher carried currents. For this reason, dielectric thicknesses as low as possible and dielectric constants as high as possible (as in so-called high-$k$ dielectrics,[161–163] where $k$ stands for $\varepsilon_r$) are highly desirable for realizing devices with high performances. Available options as high-$k$ dielectrics include metal oxides such as $Ta_2O_5$, $HfO_2$, $TiO_2$, or $Al_2O_3$, a few polymer dielectrics and a variety of nanocomposites made of polymeric matrices embedding nanoparticles.[163]

graphic form. These characteristics consist in the curves of $I_{DS}(V_{GS})$ at constant $V_{DS}$. The curves are generally displayed in a semilogarithmic plot. Furthermore, the square root of $I_{DS}$ is often plotted against $V_{GS}$ (in a linear scale), because a linear behavior of $I_{DS}^{1/2}(V_{GS})$ allows one immediately to visualize the saturation regime and to extract $\mu^{(satur.)}$ and $V_{TH}$ by fitting according to Equation (6.11) (Figure 6.8c). The different regimes of operation of an organic FET are summarized in the schemes displayed in Figure 6.9.

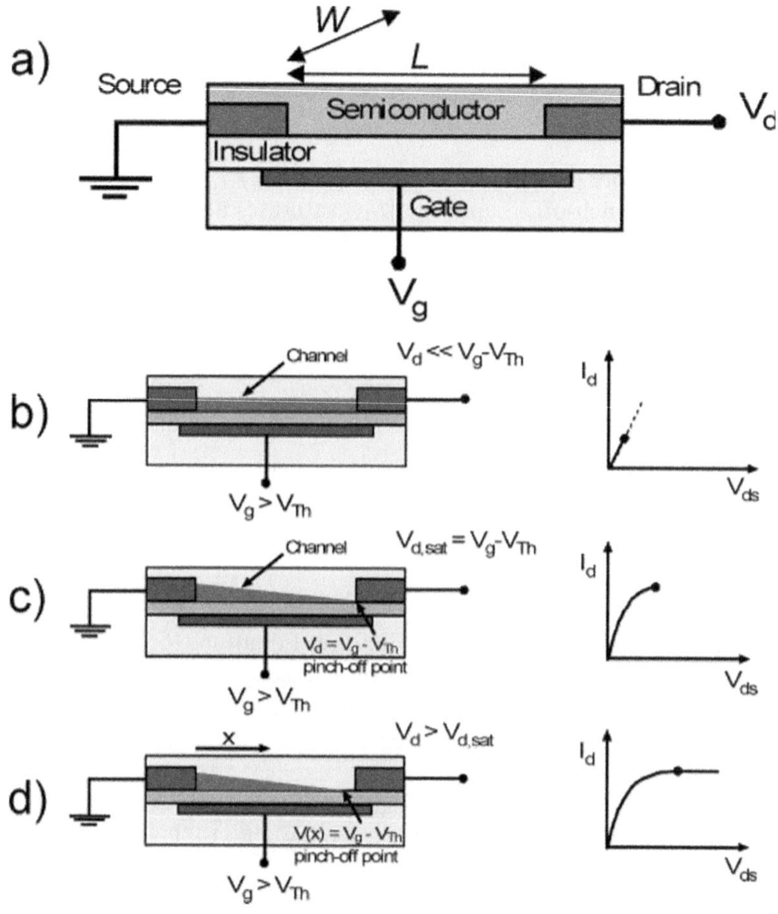


**Figure 6.9** Illustration of the FET structure (a) and regimes. In this figure the source electrode is grounded and symbols are as follows. Source-drain voltage bias: $V_d$ or equivalently $V_{ds}$. Source-gate voltage: $V_g$. Source-drain current: $I_d$. (b) Linear regime. (c) Start of saturation. (d) Saturation regime. Reproduced with permission from Ref. 61, *Chem. Rev.*, 2007, **107**, 1296–1323. Doi: 10.1021/cr0501543. Copyright © 2007, American Chemical Society.

Finally, any report on organic FETs normally includes the estimation of the $I_{on}/I_{off}$ ratio, where $I_{on}$ and $I_{off}$ indicate the $I_{DS}$ in the on state, for a particular value of $V_{GS}$, and in the off state, respectively. For thin-film FETs, under the simplifying hypotheses reported above, the order of magnitude of this ratio can be estimated as:[157]

$$\frac{I_{on}}{I_{off}} = C_i \mu^{(linear)} \frac{V_{GS}}{q \mu_0 N d_S}, \tag{6.13}$$

where $q$ is the electron charge, $d_S$ is the thickness of the semiconducting layer and $N$ and $\mu_0$ are the concentration of charge-carriers and the mobility in the off state, respectively. The $I_{on}/I_{off}$ ratio should be as high as possible (say, $10^6$–$10^8$).[8] However, in real devices it can be lowered due to unintentional doping and impurities, which may cause a significantly non-zero off current even at very low $|V_{DS}|$, exposure to ambient atmosphere, oxygen, humidity, *etc.* The latter mechanisms can also cause a gradual deterioration of the device performances during operation (particularly a temporal increase of $V_{TH}$), a degradation that is sometimes in part reversible as observed upon switching off the bias.[26,164–166] In this respect using nanostructured polymeric media and in particular polymer nanofibers can pose some extra problems for keeping the FET operation stable over time, because the very high surface-to-volume ratio can greatly favor the diffusion of oxygen into the organics. These effects will be further discussed in the next section. Proper device packaging methods can be considered to reduce air-induced degradation issues. Furthermore, one should keep in mind that in real organic FETs the contact resistance due to non-optimal conditions for charge injection is often a relevant issue. In particular, we recall that for short-channel transistors, extending the usual gradual channel approximation one can consider contact resistances in series at the source ($R_S$) and at the drain ($R_D$), respectively.[167] In this model, the linear and the saturation regimes are described by the following equations[††]:[167]

$$I_{DS}^{(linear)} = \frac{W}{L} C_i \mu^{(linear)} \left[ \left(V_{GS} - R_S I_{DS}^{(linear)}\right) - V_{TH} - \frac{V_{DS} - (R_S + R_D) I_{DS}^{(linear)}}{2} \right] \times \left[ V_{DS} - (R_S + R_D) I_{DS}^{(linear)} \right] \tag{6.14a}$$

†† In particular, neglecting the term $R_S I_{DS}$ in the linear regime, Equation (6.14a) can be rewritten as: $I_{DS}^{(linear)} = \frac{W}{L} C_i \mu^{(linear)} \left[ \frac{\left(V_{GS} - V_{TH} - \frac{V_{DS}}{2}\right) V_{DS}}{1 - \frac{r_C C_i \mu^{(linear)}}{L} \left(V_{GS} - V_{TH} - \frac{V_{DS}}{2}\right)} \right]$, where $r_C = W(R_S + R_D)$. This equation replaces Equation (6.9b) in a situation where the contact resistance makes the second term of the denominator in squared brackets significant.

$$I_{DS}^{(satur.)} = \frac{W}{2L} C_i \mu^{(satur.)} \left[ (V_{GS} - R_S I_{DS}^{(satur.)}) - V_{TH} \right]^2. \tag{6.14b}$$

In addition, in the linear regime and in the limit of small $V_{DS}$, the dependence of the total resistance, $R_{ON}$, of the device on the channel length is approximately expressed as:[150,168–170]

$$R_{ON} = \left. \frac{\partial V_{DS}}{\partial I_{DS}} \right|_{V_{DS} \cong 0}^{V_{GS}} = R_{CH} + R_S + R_D, \tag{6.15}$$

where the channel resistance, $R_{CH}$, is the only term depending on $L$, *i.e.* $R_{CH} = L/[WC_i\mu^{(linear)}(V_{GS} - V_{TH})]$. Under these assumptions, one can also define a critical gate voltage, $V_{G,c}$, which is the value of $V_{GS}$ at which the resistance of the transistor channel equals $R_S + R_D$.[167] For instance, for small $V_{DS}$ one has:

$$V_{G,c} = V_{TH} - \frac{L}{r_C C_i \mu^{(linear)}}. \tag{6.16}$$

In the exemplary case of devices working in accumulation mode, for large absolute values of the gate-source bias, $R_{CH}$ is decreased by channel opening, and the operation of the FET can be therefore dominated by the contact resistance. These considerations are particularly critical for nanoelectronics based on reduced channel lengths.[170] Indeed, small $L$ values, as in FETs with sub-100 nm gap between source and drain, determine correspondingly reduced $R_{CH}$, and such architectures are therefore especially demanding in terms of contact resistance.

### 6.3.1 Polymer Nanofiber Field-effect Transistors

Due to the ease of fabrication of organic micro- and nanoelectronic devices and to the background on nanoelectronics and conjugated materials of some of the main research groups involved in electrospinning research in the years 1995–2000 (Section 2.1), electrospun polymer nanofibers made of conductive polymers have been rapidly used in FETs. Nanofibers made of conductive polymers and realized with other techniques have been largely applied to FETs too. Nowadays, flexible electronics is rapidly maturing from laboratories to the market, offering new motivations for the development of organic FETs and in particular of nanofiber-based FETs. These components are promising candidates for next-generation nanoelectronics, for a number of reasons: (i) high potential for improved conduction performances, due to the in-fiber molecular orientation, which, as in the case of conjugated polymers aligning along the longitudinal axis of single nanofibers, may result in enhanced transport between source and drain; (ii) intrinsic potentiality of electrically active nanofibers for natural integration with textiles, functional fabrics, wearable electronics and electronic paper applications; (iii) wide compatibility with plastic substrates for bendable and stretchable electronics,[171] due to the fiber mechanical properties (Section 4.1) and to the availability of many fabrication methods that entirely work at room temperature for producing nanofibers; (iv) high potential for further miniaturization of the overall device, because the FET channel width is limited by the diameter of nanofibers, without need for complex and costly high-resolution lithographic tools. This can open new ways for increasing the device density. In addition, miniaturization opens possibilities for a faster operation compared to thin-film devices, due to the dimension dependence of the transition frequency:[100]

$$f_T = \frac{g_m}{2\pi C_i WL} \cong \frac{\mu}{2\pi L^2} V_{DS}; \tag{6.17}$$

(v) improved biocompatibility, hence higher potential for integration of polymer fiber-based nanoelectronics in devices implantable in the body; (vi) high susceptibility to external control of the charge-carrier transport by doping or other chemical methods, because of the large surface-to-volume ratio, which make nanofiber-based FETs ideal candidates for realizing chemical and bio-sensors. In the following we provide an excursus of reported performances for FETs based on polymer nanofibers. However, on the basis of the introduction above, the reader should be well aware that directly comparing the performances obtained through the different synthetic approaches for producing nanofibers is generally hard and often misleading, given the paramount importance of the individual device architectures in determining the measured conduction behavior.

#### 6.3.1.1 P-type Devices

*P*-type FETs with electrospun nanofibers of PAN/PEO blends have been reported since 2003.[96] Such devices have been described to show still moderate performances (hole mobility of $1.4 \times 10^{-4}$ cm$^2$ V s$^{-1}$, transconductance of 0.03 nS and $I_{on}/I_{off} = 2$). *P*-type FET devices with assembled PAN nanofibers have been demonstrated even earlier (in 2001), with reported charge-carrier mobilities of about $7.8 \times 10^{-2}$ cm$^2$ V s$^{-1}$ and a transconductance of 0.02 nS (at 233 K).[172] Early reports on *p*-type FETs by electrospun nanofibers made of regioregular P3HT have shown similar achievable performances, reporting values of the hole mobility ranging from $4 \times 10^{-4}$ cm$^2$ V s$^{-1}$[173] to $3 \times 10^{-2}$ cm$^2$ V s$^{-1}$,[174] and more encouraging values concerning $I_{on}/I_{off}$ ratios, ranging from 7[173] to $10^2$–$10^3$.[174,175] Nanofibrous films of regioregular P3HT, poly(3-pentylthiophene) and other P3ATs have hole mobility up to $10^{-2}$–$10^{-1}$ cm$^2$ V s$^{-1}$ and $I_{on}/I_{off}$ ratios up to $10^4$.[119,176,177] Nowadays, FETs based on P3HT electrospun nanofibers have reached outstanding performances, as demonstrated by U. Jeong and his collaborators at Yonsei University by using polyelectrolyte gate dielectrics, *i.e.* UV-cross-linkable PEGDA doped by an ionic liquid‡‡.[180] Deploying conjugated polymer nanofibers as active medium in top-gate FET devices is strategic to improve the device stability, because this allows one to have a large area available for the mutual contact between the cross-linkable polyelectrolyte dielectrics and an underlying methacrylated PET substrate. The main steps of the fabrication process are schematized in Figure 6.10. In this way, an irreversible sealing (possibly spatially selective by mask shadowing) is achieved upon fast (2 s) UV curing, through covalent bonding involving methacrylate groups in the substrate and acrylate groups in the gel.[180] It is evident that, on the contrary, this approach could not be successfully carried out for standard thin-film devices, because the organic semiconductor layer sandwiched between the substrate and the gate dielectrics would prevent the formation of irreversible covalent bonds and would finally lead to poor overall adhesion and device delamination even after UV curing. So, this method is an elegant demonstration of how active polymer nanofibers not only serve as functional medium when embedded in devices, but also favor micro- and nanofabrication processes throughout the successive steps of device fabrication. Of course, similar considerations may hold also for other devices for optoelectronics, sensing, *etc.*, described in other sections of this book. In particular, P3HT nanofiber FETs, realized in this way on flexible substrates, operate at voltage below 2 V, with a mobility as high as 2 $cm^2\ V\ s^{-1}$, and $I_{on}/I_{off}$ ratios of $10^5$.[180] Very high mobilities (>1 $cm^2\ V\ s^{-1}$) are also reported for P3HT features realized by electrohydrodynamic printing, using short nozzle-substrate distances and motorized stages to achieve high precision in writing deposition.[181]

Furthermore, as previously mentioned (Section 6.1.1), using solvents with high boiling point such as 1,2,4-trichlorobenzene has an improving effect on the crystalline order and consequently on the mobility of P3HT, as found in spin cast films.[49] Many of these films clearly show a surface topography rich in nanofibers when investigated by AFM.[25,54] Mobility values are also reported for P3HT FETs based on samples cast from toluene, dichloromethane, chloroform and THF solutions, ranging from $10^{-4}$ $cm^2\ V\ s^{-1}$ (for $CH_2Cl_2$) to $10^{-2}$ $cm^2\ V\ s^{-1}$ (for $CHCl_3$). These findings again highlight the importance of the used solvent in determining the nanofiber morphology, the supramolecular organization, internal to the nanostructures, and ultimately the conduction performances.[54] A different strategy, based, however, on the same idea of providing the polymer with ample time to self-organize in a more ordered structure, consists in assembling P3HT nanofibers through casting under conditions of high solvent vapor pressure (56.5 kPa).[182] Indeed, this approach results in a reliably increased molecular ordering because of the slower solvent evaporation rate, and leads to measured mobilities of $2 \times 10^{-2}$ $cm^2\ V\ s^{-1}$ and

‡‡ So-called ion-gel dielectrics[178,179] are rapidly gathering popularity in the organic electronics community because of their high capacitance per unit area (reaching tens of μF/$cm^2$ at frequencies of the order of 10 Hz, which is three orders of magnitude higher than the corresponding value in standard 100 nm-thick $SiO_2$ or PMMA layers), though they still show a few problems related to stability and delamination issues.

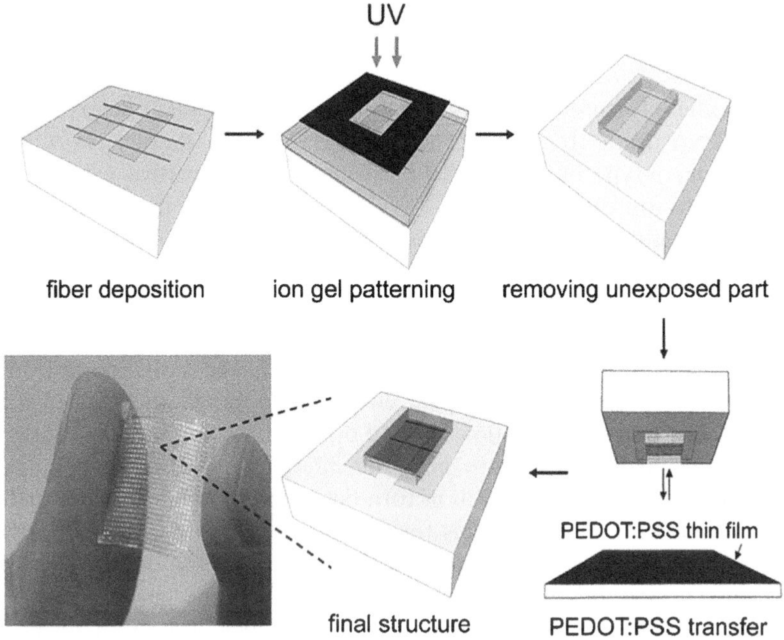


**Figure 6.10** Schematics of the process for fabricating an array of top-gate FETs using polyelectrolyte cross-linkable dielectric layers. The sequence of steps leading to the device realization is as follows. (i) Deposition of active polymer nanofibers across the gap (or the array of gaps) separating source and drain electrodes, (ii) deposition of the polyelectrolyte ion-gel layer and its patterning by UV-cross-linking through shadow masks. This step produces irreversible covalent bonding of the gate layer with the underneath polymer substrate, which is largely exposed and available to the reaction, due to the small size of deposited nanofibers. (iii) Removing unexposed regions and (iv) turning the resulting assembly upside down on PEDOT:PSS, which is transferred onto the dielectric layer thus serving as gate contact, complete the device structure. Reproduced with permission from Ref. 180, *Nano Lett.*, 2010, **10**, 347–351. Doi: 10.1021/nl903722z. Copyright © 2010, American Chemical Society.

$I_{on}/I_{off}$ ratio of $10^4$. Nanofibers of regioregular P3HT precipitated from solvents with strong temperature-dependent solubility seem to show even better performances in FETs, with $\mu = 2\text{–}6 \times 10^{-2}$ cm$^2$ V s$^{-1}$ and $I_{on}/I_{off}$ ratio up to $10^6$.[46,183] By using blends of polythiophene and PS and dichloromethane as selectively marginal solvent for P3HT, values of mobility of $4 \times 10^{-3}$ cm$^2$ V s$^{-1}$ are measured for the nanowires embedded in the insulating polymer matrix.[176] FETs with electrospun MEH-PPV/P3HT nanofibers have been reported as well, with measured mobilities in the range $10^{-4}$–$10^{-6}$ cm$^2$ V s$^{-1}$, depending on the blend composition.[184] In another work, the mobility in self-assembled P3HT nanofibers in FET devices shows an exponential increase upon increasing the width of the wire.[185] When measured, the temperature dependence of $\mu$ in P3AT nanofibers FETs is usually well described by Equation (6.1), with activation energies varying between 30 and 120 meV.[46,117,183]

Some other features of P3HT nanofiber-based FETs deserve to be mentioned. A quite frequent characteristic is that, though working in accumulation, *p*-type mode, these FETs may exhibit positive values of the threshold gate-source voltage, $V_{TH}$.[46,174,184] It has been pointed out that this finding indicates that the P3HT nanofiber is able to accumulate sufficient holes to form a conductive channel between source and drain even without applying negative $V_{GS}$. What is the source of such an intrinsic amount of available charge-carriers? Plausible explanations attribute their origin to intrinsic impurities due to synthetic residues in the conjugated polymers, or to an extrinsic contamination that has the effect of "doping" the nanofibers as mentioned in the previous section.[46,174] In other words, impurities of different origin can act as acceptor traps near the interface between the conjugated polymer and the dielectrics, which in turn determines a non-zero conductivity for holes even with $V_{GS} \geq 0$. This finally results in a worsening of the FET performances (particularly increasing the off-current). Importantly, possible causes of external impurities could be related to the specific fabrication process, since oxygen or aqueous vapor,[186,187] which heavily affect charge transport properties, can be incorporated in the nanofibers during the electrospinning process if this is performed in uncontrolled atmosphere. Indeed, how much the process of making nanofibers and in particular electrospinning can affect the intrinsic electrical properties of the nanostructures if carried out under not perfectly controlled synthetic conditions is something that will deserve further investigation in future. Nowadays, more sophisticated electrospinning experiments employing controlled process environments are beginning to be conceived and carried out (Section 2.2). This approach is promising to elucidate greatly the physico-chemical origin of extrinsic contaminations and to push the resulting electronic performances of electrospun nanofibers to their limit.

Metal nanoparticles doping is of course another way to improve the conduction performances. FETs based on Au-impregnated PAC/polythiophene core–shell nanofibers exhibit a *p*-behavior, with hole mobility as high as 2 cm$^2$ V s$^{-1}$ and $I_{on}/I_{off}$ ratio of $10^4$–$10^5$.[188] Recently presented *p*-type FETs based on CDT-BTZ have superior performances because of the improved supramolecular ordering.[128] These devices are realized with a standard $SiO_2$ dielectric layer with thickness 300 nm and an on-top HMDS functionalization, and Pt source and drain. Hole mobilities are as high as 3.3–5.5 cm$^2$ V s$^{-1}$, outperforming the value measured in thin-films by an order of magnitude, and the $I_{on}/I_{off}$ ratio is $10^6$. Such mobility value, which is to date one of the highest measured in polymers, suggests a mixed effect of reduced chemical defects in the nanofibers, enhanced crystallinity and at least a component of the charge transport carried along the length of conjugation over the polymer backbone.[128]

### 6.3.1.2 N-type Devices

*N*-type FETs using polymer nanofibers or analogous elongated nanostructures as active medium have been demonstrated as well. For instance, devices based on single nanobelts and on networks of filaments of the polymer BBL allow one to achieve values of electron mobilities of 3–7 $\times$ $10^{-3}$ $cm^2$ V $s^{-1}$, $I_{on}/I_{off}$ ratios of $10^4$ and, importantly, a very high stability after storing in ambient laboratory conditions for about 6 months.[127] We recall that other FETs have been demonstrated by nanofibers made of other classes of materials, achieved through various processing steps of native polymer nanofibers. For example, nanofibers of conductive metal oxides, such as CuO or ZnO, are obtained by calcination (at 500 °C) of electrospun samples of PVA and copper acetate or zinc acetate, and these fibers have been used in simple *p*-type and *n*-type FET devices, respectively.[189,190]

### 6.3.1.3 Photo-transistors and Other Devices

Polymer nanofiber transistors coupling electrical and optical functionalities are especially intriguing, since these hybrid devices can open the way toward low-cost, possibly flexible electro-optic switches and nanoscale OLETs. To this aim, light-emitting conjugated polymers (Chapter 5) are of course the material of choice for producing nanofibers. For instance, the performances of MEH-PPV nanofiber FETs are in line with or slightly better than those obtained in reference thin-film transistors. In nanofiber devices, the measured hole mobility is of the order of $10^{-3}$ $cm^2$ V $s^{-1}$, the transconductance is 0.26 nS and the on/off current ratios are up to 780.[191] But in addition to their relatively good figures of merit (not satisfactory, however, to compete with FETs using better-conducting conjugated polymers), these devices are interesting for the possibility of tuning the emission characteristics by means of the gate-source voltage bias. In fact, while $V_{GS}$ does not strongly influence the spectral shape of the PL from the MEH-PPV nanofibers, the application of a negative gate-source bias determines a remarkable decrease of the emission intensity. This tuning approach corresponds to the so-called *luministor mode* of operation of the FET device.[192] On one hand, it is clear that negative $V_{GS}$, inducing an accumulation of holes at the nanofiber/dielectric interface, can result in an enhanced, charge-carrier induced non-radiative decay of the singlet excitons responsible for the PL emission.[192] This mechanism can be rationalized by modeling the one-dimensional migration of excitons generated by optical excitation towards polarons/bipolarons (Section 5.1.1) annihilation centers, which can be induced along the molecular chains by electrochemical doping of the conducting polymers.[193] The diffusion of excitons towards quenching centers can take place even at low concentrations of polaron/bipolaron species, and mostly involves an inter-chain, three-dimensional mechanism[192] related to the above-mentioned accumulation of charges at the interface between the organic semiconductor and the dielectric layer. Finally, another possible explanation attributes the $V_{GS}$-induced quenching to the dissociation of the emitting species induced by the electric field, which reaches values as high as

$10^8$ V m$^{-1}$ through the FET dielectrics. This effect can also be observed in other device architectures, such as OLED metal layers sandwiching ladder-type poly(paraphenylene).[194] The exact attribution of the luministor behavior will deserve further investigation in future research. This will have to include the assessment of the radiative and non-radiative decay rates after photo-excitation in the active polymer nanofibers, which is possible by coupling absolute quantum yield measurements and time-resolved PL spectroscopy.

Photo-transistors are another class of devices that allow transport properties and light stimuli to be combined. Photo-transistors can be described by simply adding to the channel current, for instance to that expressed from Equation (6.9a) in the linear FET region, an extra, bulk component coming from photo-generated charge-carriers. Such bulk current can be approximated as $Nq\mu Wd_S V_{DS}/L$, where all the symbols have been defined previously. MEH-PPV nanofibers can be used in these devices, with demonstrated photo-sensitivity (difference between the current under illumination and the dark current, normalized to the dark current) up to 10, and photo-responsivity (amount of photo-generated current per incident light power) up to 100 mA W$^{-1}$ under an illumination intensity of 9.6 mW cm$^{-2}$.[195] In addition, the switching cycles of the photo-response under repeated on-off switching of the excitation light source (Figure 6.11) evidence that the relaxation time of the photo-current can be reduced down to about 40 s by increasing the fiber surface-to-volume ratio (*i.e.* decreasing the fiber diameter).[195] In fact, it is highly probable that thinner nanofibers may offer denser distribution of reacting sites (both at the surface and because of oxygen diffusion into the organic material) that determine a faster decay of photo-current after illumination ceases.[196–199] Hence, this is a further example of how the nanofiber geometry can be fruitfully utilized in devices. Moreover, it is very likely that better performances will be reachable by means of optimized photo-transistor architectures and nanofibers made of specifically designed photo-responsive blends or nanocomposites.

Another especially promising line in the field of organic FETs, stemmed from a 2003 report on polycrystalline thin-films of tetracene by researchers at the Darmstadt University of Technology,[15] is certainly the study of ambipolar transistors and of OLETs.[16–18,60] In these devices, the simultaneous conduction of charge-carriers of opposite signs allows the generation of light upon radiative recombination. The field of ambipolar FETs and OLETs is still basically unexplored with polymer nanofibers and may constitute an interesting field for research in the next years.

## 6.4 Piezoelectricity and Thermo-electricity in Polymer Nanofibers

### 6.4.1 Piezoelectric Properties

Piezoelectricity is the capability of a material, following an external stress that produces a physical deformation (Section 1.4.2), to (i) *polarize*, by means of

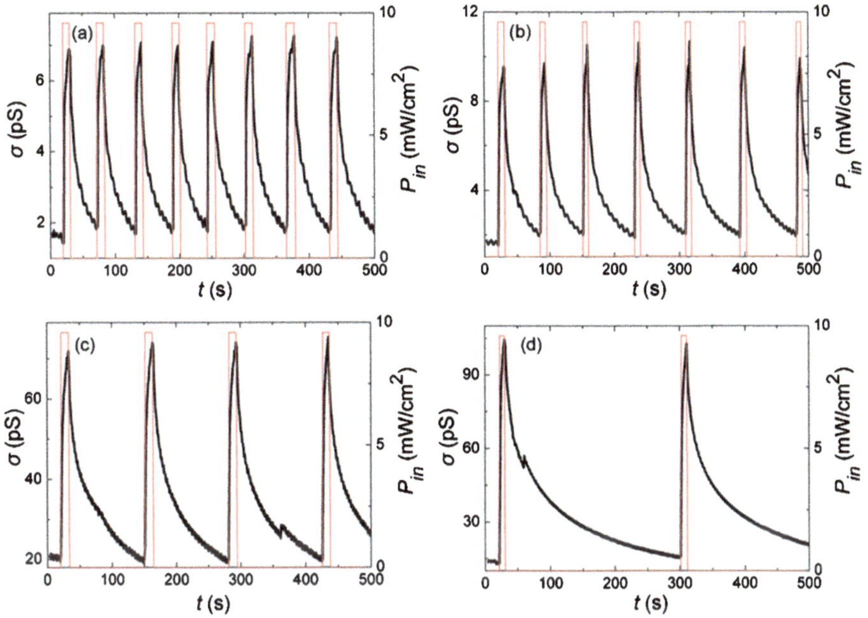


**Figure 6.11** Cycles of photo-conductivity (here, $\sigma$, black curves) for a photo-transistor based on an electrospun MEH-PPV nanofiber. $V_{DS}$ = −50 V, $V_{GS}$ = 0 V. $P_{in}$: incident light intensity (red curves highlighting the on-off illumination cycles). The transistor channel width is 0.5 μm (a), 5 μm (b), 40 μm (c) and 80 μm (d). Reproduced with permission from Ref. 195, *Appl. Phys. Lett.*, 2011, **98**, 023307. Doi: 10.1063/1.3534803. Copyright © 2011, American Chemical Society.

spatially separated electrical charges of opposite sign, which are generally accumulated at two opposite side surfaces of the material body, and (ii) when these two side surfaces are not short-circuited, to *generate a voltage bias* across the material body. Piezoelectricity is an important point of strength of a few polymers, allowing their electrical and mechanical properties to be coupled. This effect can be observed in materials whose crystalline state has no center of symmetry (so-called non-centrosymmetric). Furthermore, piezoelectricity is largely (though not exclusively) related to the presence of permanent electric dipoles, meaning that the material is polar.[200] These electric dipoles can orient randomly, thus resulting in the lack of a macroscopic polarization, or can be mutually aligned within certain spatial regions (Weiss domains), thus producing a neat macroscopic polarization vector, **P**. The alignment in turn can be spontaneous, or, for some materials (*ferroelectrics*), it can be forced by an externally applied, intense electric field, a procedure that is called *poling*. For favoring the configurational rearrangement of dipoles, the poling process can be carried out at high temperature.[200]

Here we briefly introduce the common formalism used to describe the piezoelectric effect and that also applies to polymer nanofibers. Following a mechanical stress ($\sigma$), the intensity and/or the direction of **P** may vary as a consequence of the stress-induced material strain, $\gamma$. We already know that for a Hookean (elastic) solid, which generally approximates well the behavior of crystalline materials, the simplest stress-strain relation is given by $\sigma(t) = G\gamma(t)$ (Equation (1.9b)) or, equivalently, $\gamma(t) = J\sigma(t)$, where $G$ and $J = G^{-1}$ are the elastic modulus and the corresponding compliance, respectively. In a matrix form (Voigt notation introduced in Section 1.4.2.1), one can write $\boldsymbol{\sigma} = \mathbf{G}\boldsymbol{\gamma}$ or $\boldsymbol{\gamma} = \mathbf{J}\boldsymbol{\sigma}$. At the same time, the relation between the electric displacement, **D**, and the electric field, **F**, is given by the well-known formula:[201]

$$\mathbf{D} = \varepsilon_0\mathbf{F} + \mathbf{P} = \varepsilon_0\varepsilon_r\mathbf{F}. \tag{6.18}$$

The previous equation can be put in a matrix form as well, introducing the so-called dielectric tensor ($\boldsymbol{\varepsilon}$), namely $\mathbf{D} = \boldsymbol{\varepsilon}\mathbf{F}$. Now, combining linearly the dependence of the electric displacement on **F** with a law cross-coupling **D** and the applied stress, $\boldsymbol{\sigma}$, one has:

$$\mathbf{D} = \boldsymbol{d}\boldsymbol{\sigma} + \boldsymbol{\varepsilon}^{\mathrm{T}}\mathbf{F}, \tag{6.19a}$$

where $\boldsymbol{d}$ is another (3 × 6) matrix that accounts for piezoelectricity and the superscript in $\boldsymbol{\varepsilon}^{\mathrm{T}}$ indicates that the dielectric tensor is considered at a zero, or constant value (T) of the stress (in other words, when this superscript is present the stress is seen as a "boundary condition"). For the sake of clarity, one may recall that by using the Voigt notation for the stress tensor, the previous equation reads as follows:

$$\begin{pmatrix} D_1 \\ D_2 \\ D_3 \end{pmatrix} = \begin{pmatrix} d_{11} & d_{12} & d_{13} & d_{14} & d_{15} & d_{16} \\ d_{21} & d_{22} & d_{23} & d_{24} & d_{25} & d_{26} \\ d_{31} & d_{32} & d_{33} & d_{34} & d_{35} & d_{36} \end{pmatrix} \begin{pmatrix} \sigma_1 \\ \sigma_2 \\ \sigma_3 \\ \sigma_4 \\ \sigma_5 \\ \sigma_6 \end{pmatrix} + \begin{pmatrix} \varepsilon_{11}^{\mathrm{T}} & \varepsilon_{12}^{\mathrm{T}} & \varepsilon_{13}^{\mathrm{T}} \\ \varepsilon_{21}^{\mathrm{T}} & \varepsilon_{22}^{\mathrm{T}} & \varepsilon_{23}^{\mathrm{T}} \\ \varepsilon_{31}^{\mathrm{T}} & \varepsilon_{32}^{\mathrm{T}} & \varepsilon_{33}^{\mathrm{T}} \end{pmatrix} \begin{pmatrix} F_1 \\ F_2 \\ F_3 \end{pmatrix} \tag{6.19b}$$

Here, the subscript 1, 2 and 3 of the components of **D** and **F** indicate three mutually orthogonal directions (as for the usual Cartesian system of axes, $x$, $y$ and $z$). By convention, the 3-direction is generally taken as the direction of pristine polarization in the material, as determined by the fabrication or by the poling processes. From the previous expression it is evident that the elements of the $\boldsymbol{d}$ matrix can be defined as:[200,202,203]

$$d_{ij} = \left(\frac{\partial D_i}{\partial \sigma_j}\right)^{\mathbf{F}}, \text{ for } i = 1, 2, 3 \text{ and } j = 1, 2, 3, 4, 5, 6 \tag{6.20}$$

where, as usual, the superscript (**F**) indicates that the partial derivative is calculated at a constant or zero value of the electric field. The $d_{ij}$ elements (measured in m V$^{-1}$ or equivalently C N$^{-1}$) defined by Equation (6.20) describe the so-called *direct* piezoelectric effect (electric displacement, hence ultimately voltage bias, caused in the material because of an external stress), which is the phenomenon we have been describing so far. The first subscript in the $d_{ij}$ coefficients indicates the direction of the electric displacement in the piezoelectric material, which corresponds to the direction along which a voltage bias is generated. The second subscript indicates the direction of application of stress (for $j$ = 1, 2 and 3) and also specifies if a shear-stress is involved (for $j$ = 4, 5 and 6). With respect to the pristine polarization direction of a ferroelectric body (which corresponds to $i$ = 3 as mentioned above), the $d_{31}$, $d_{32}$ and $d_{33}$ are the piezoelectric coefficients to be taken into consideration to describe the material behavior for non-shear-stresses, as schematized in Figure 6.12a–c. The coefficient, $d_{15}$, describing a voltage bias generated perpendicularly to the original poling direction, is also often used in many applications where a shear-stress is involved (Figure 6.12d).

One may notice that the $d_{ij}$ coefficients are also easily related to the surface charges accumulated on the opposite sides of the piezoelectric body over the $i$-direction. This is also arguable from dimensional considerations, since the $d_{ij}$ elements are measurable in units of C N$^{-1}$, which can be re-written as (C m$^{-2}$) × (m$^2$ N$^{-1}$). These clearly are the units of a surface charge density divided by an applied stress (the latter being identified by the $j$-subscript as usual). The charge

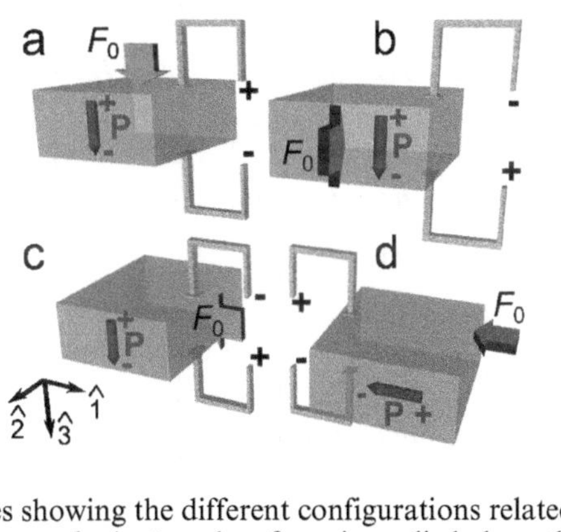


**Figure 6.12** Schemes showing the different configurations related to the piezoelectric coefficients, $d_{33}$ (a, *i.e.* when force is applied along the 3-direction, which is the direction of polarization), $d_{32}$ (b), $d_{31}$ (c). The shown coordinate axes are referred to (a c). (d) Scheme showing the generation of a voltage bias due to the application of a shear-stress (as described for instance by the $d_{15}$ coefficient).

accumulated can easily generate an electric current, $i$, flowing through an external circuit inter-connecting the two sides of the material body, which is one of the most used ways in which piezoelectric devices operate. For instance, connecting the side surfaces normal to the original poling direction, each of area $A_3$ and on which a charge is accumulated as a consequence of a $\sigma_3$ stress (*i.e.* of a force co-directional to the poling axis), the following current is generated:

$$\begin{aligned} I &= \frac{dq}{dt} = A_3 \frac{d}{dt} d_{33}\sigma_3 = A_3 d_{33} \frac{d}{dt}\sigma_3 = A_3 d_{33} \frac{d}{dt} E\gamma_3 \\ &= A_3 d_{33} E\dot{\gamma}_3 \end{aligned} \tag{6.21}$$

where we use Young's modulus, $E$, introduced in Section 1.4.2, and the strain rate along the 3-direction, $\dot{\gamma}_3$. The various relations among $I$ (or $q$) and $\sigma$ (or the applied force or the resulting strain, $\gamma$, or strain rate, $\dot{\gamma}$) expressed by the equalities in Equation (6.21) provide effective ways to determine experimentally the $d_{33}$ coefficient.

Of course, other effects are possible, such as the so-called *converse* piezoelectric effect, which describes a deformation in the material, *i.e.* a strain originated by the application of an electric field:

$$\boldsymbol{\gamma} = \mathbf{J}\boldsymbol{\sigma} + \mathbf{d}^{\mathbf{t}}\mathbf{F}, \tag{6.22a}$$

where the superscript, $\mathbf{t}$, indicates the transposed matrix, *i.e.*:

$$\begin{pmatrix} \gamma_1 \\ \gamma_2 \\ \gamma_3 \\ \gamma_4 \\ \gamma_5 \\ \gamma_6 \end{pmatrix} = \begin{pmatrix} J_{11} & J_{12} & J_{13} & J_{14} & J_{15} & J_{16} \\ J_{21} & J_{22} & J_{23} & J_{24} & J_{25} & J_{26} \\ J_{31} & J_{32} & J_{33} & J_{34} & J_{35} & J_{36} \\ J_{41} & J_{42} & J_{43} & J_{44} & J_{45} & J_{46} \\ J_{51} & J_{52} & J_{53} & J_{54} & J_{55} & J_{56} \\ J_{61} & J_{62} & J_{63} & J_{64} & J_{65} & J_{66} \end{pmatrix} \begin{pmatrix} \sigma_1 \\ \sigma_2 \\ \sigma_3 \\ \sigma_4 \\ \sigma_5 \\ \sigma_6 \end{pmatrix} + \begin{pmatrix} d_{11} & d_{21} & d_{31} \\ d_{12} & d_{22} & d_{32} \\ d_{13} & d_{23} & d_{33} \\ d_{14} & d_{24} & d_{34} \\ d_{15} & d_{25} & d_{35} \\ d_{16} & d_{26} & d_{36} \end{pmatrix} \begin{pmatrix} F_1 \\ F_2 \\ F_3 \end{pmatrix}. \tag{6.22b}$$

The analogy of Equation (6.20) for the converse effect is therefore:

$$d_{ij} = \left(\frac{\partial \gamma_j}{\partial F_i}\right)^{\mathrm{T}}, \text{ for } i=1, 2, 3 \text{ and } j=1, 2, 3, 4, 5, 6. \quad (6.23)$$

We conclude this introduction to piezoelectricity by mentioning that other constitutive equations are available for both the direct and the converse effect, providing several linear combinations of the four relevant physical quantities ($\gamma$, $\boldsymbol{\sigma}$, **D** and **F**). These other relations can be found elsewhere[200,202,203] and are not strictly needed for what follows.

We now specialize to the case of piezoelectric polymer nanofibers. Polymers exhibiting piezoelectric properties include nylon, poly(vinylchloride), PAC, natural biomaterials such as silk and various derivatives of poly(vinylidene fluoride) (PVDF). The PVDF, $(CH_2CF_2)_n$,[203–205] (Figure 6.13a) and its copolymers such as the commercially available poly(vinylidene fluoride-trifluoroethylene) [P(VDF–TrFE)][§§] (Figure 6.13b and c) have been widely applied for the realization of piezoelectric films,[207] and so far have been the materials of choice for fabricating organic piezoelectric nanofibers as well, mostly by electrospinning.[208–217] The PVDF crystallizes quite easily, and it presents several different polymorphic configurations depending on the chain organization. These include the $\alpha$-form (or $\alpha$-phase), which has randomly distributed electric dipoles (and therefore lacks macroscopic polarization), and the $\beta$-form which is instead non-centrosymmetric and ferroelectric and exhibits therefore remarkable piezoelectric properties. In the $\beta$-phase, the macromolecular chains assume a zigzag, all-*trans* configuration, which determines a ferroelectric orientation of the molecular dipoles related to the presence of the strongly electronegative F atoms. In this phase, typical values of the piezoelectric coefficients of PVDF are*** $d_{33}$ = –20–35 pC $N^{-1}$ and $d_{31}$ = 18–24 pC $N^{-1}$.[207,218,219] The $\beta$-, piezoelectric phase of PVDF is typically obtained (for instance, in film samples) by poling or by uniaxial stretching and drawing.[203] Nanofiber production technologies and in particular electrospinning are very interesting in this respect, aiming to induce the largest possible spatial anisotropy that may significantly favor piezoelectricity. This leads, for microfibers with diameter of 1–5 μm, to measured values of $d_{33}$ in the range of –40–80 pV $m^{-1}$, which is around a factor of two higher in absolute value than in PVDF films of analogous thickness.[210]

§§ The P(VDF–TrFE) copolymer is particularly useful since it exhibits a very stable piezoelectric phase at room temperature.[206]

*** From the background information presented above, it is clear that negative values of $d_{33}$ simply indicate that, *with respect to the original direction of poling* of the ferroelectric polymer, the application of a positive (negative) force originates an internal electric displacement pointing in the opposite (same) direction, according to Equation (6.20), whereas the application of a positive (negative) electric field determines a negative (positive) strain, namely a contraction (extension), according to Equation (6.23).

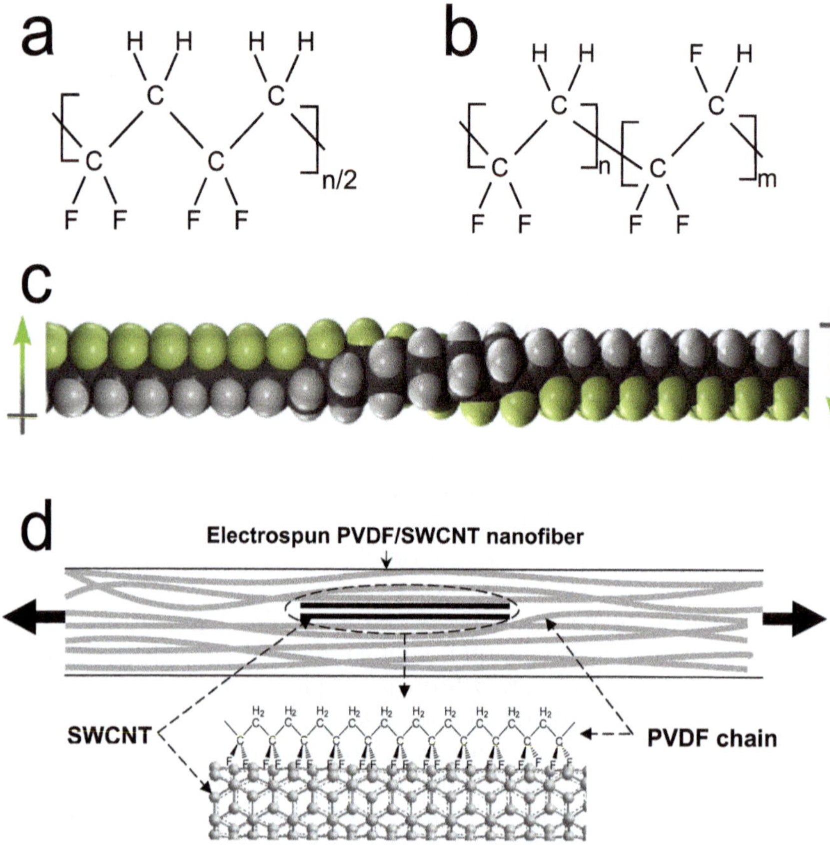


**Figure 6.13** Molecular structures of PVDF (a) and of P(VDF–TrFE) (b). (c) Schematics showing a dipole switching in the VDF part of the P(VDF–TrFE) copolymer. The polymer chain (horizontal direction in the figure) is generally oriented along the length of nanofibers. The fluorine and hydrogen atoms are larger (and in green), and smaller (and in gray), respectively. On the left of the figure, the larger fluorine atoms are on top and the associated, vertically oriented dipole moment points upwards. On the right of the figure there is an opposite situation. The dipole rotation can be induced by an external electric field. Adapted with permission from Ref. 206, *Mater. Today*, 2011, **14**, 592–599. Doi: 10.1016/S1369-7021(11)70300-5. Copyright © 2011, Elsevier Ltd. (d) Scheme of a composite nanofiber based on PVDF and carbon nanotubes, highlighting the preferential alignment of the polymer chains and of nanotubes along the fiber length. The nanotube–polymer interaction can favor the formation of extended β-phase at the interface. Adapted with permission from Ref. 216, *Langmuir*, 2008, **24**, 13621–13626. Doi: 10.1021/la8024183. Copyright © 2008, American Chemical Society.

The composition of PVDF samples in terms of its different crystallite phases can be investigated by means of XRD (the β-phase exhibits a peak in the XRD pattern at $2\theta = 20.5°$), and of Fourier-transform infrared (FTIR) spectroscopy (for instance, the β-phase shows strong vibration peaks at about 840 $cm^{-1}$ and 1280 $cm^{-1}$). To give an idea, in order to estimate the fraction, $\varphi_\beta$, of the β-phase from FTIR measurements, the following formula is often used, which comes from assuming Beer–Lambert absorption, and related absorption coefficients that are known for both the α- and the β-phases:[220–223]

$$\phi_\beta = \frac{A_{abs,\beta}}{1.26A_{abs,\alpha} + A_{abs,\beta}}. \tag{6.24}$$

In the previous equation, $A_{abs,\alpha}$ and $A_{abs,\beta}$ indicate the absorption intensity registered for the spectral peaks characteristic of the two phases (532 $cm^{-1}$ or 765 $cm^{-1}$ are often used as typical peak wavelengths for the α-phase).

How the molecular configuration of ferroelectric polymers and their polymorphic states relate to the nanofiber geometry, and how piezoelectric properties can be induced by nanofabrication technologies is certainly a very promising research topic. For example, the fraction of β-phase is found to increase upon decreasing the diameter of fibers electrospun from PVDF solutions in DMF, and upon increasing the voltage bias used during electrospinning.[222] These results agree well with the corresponding enhancement of the stretching undergone by the polymer molecules in the electrospun jet, which is more relevant for higher bias and thinner resulting fibers. The same authors also report on a strategy, consisting in the use of continuously dispersed nanoparticles of ferrite ($Ni_{0.5}Zn_{0.5}Fe_2O_4$), which further favors the arrangement of the polymeric chains in a fully extended configuration, thus enhancing the β-phase.[224] Another recent work, in analogy to results on films cast from blend solutions,[225] reports on the enhancement of the β-phase in nanofibers when PVDF is electrospun together with surface-oxidized SWCNTs, likely due to the concomitant effects of the alignment of nanotubes along the longitudinal fiber axis during electrospinning and of their interfacial interaction with the polymer chains (Figure 6.13d).[216] Adding the hydrated salt, nickel chloride hexahydrate ($NiCl_2 \cdot 6H_2O$) also leads to an enhancement of the formation of the β-phase, evaluated around 30%.[223] Overall, controlling at least to some extent the amount of the different phases of electrospun PVDF nanofibers through the process parameters (temperature, nozzle-to-collector distance and solution flow rate) appears viable.[226] Composite nanofibers can also be realized, with PVDF compounds and conjugated polymers. For instance, composite fibers made of P(VDF–TrFE) and PEDOT-PSS, that have a diameter as small as 15 nm, can be electrospun.[227] Differently from bare PVDF materials, which are electrically insulating, these fibers are conductive, and for this reason they could allow to concomitantly exploit ferroelectric properties together with charge transport. The ideality factor (Equation (6.5)) of junctions of these fibers with *n*-Si is 7.2, corresponding to an energetic barrier, $\Delta E$, of 0.58 eV.[227] In a different approach, the amino acid glycine is electrospun in PVA blends.[228] Glycine too presents several different crystalline polymorphic forms (α, β and γ) in bulk,[229–231] with piezoelectric properties generally associated with the γ-phase.[232,233] While piezoelectricity is hardly observed in the less stable β-phase of glycine, it is instead measured in β-glycine-PVA nanofibers (Figure 6.14) by means of piezoresponse force microscopy (PFM).[228] Finally, electrospun nanofibers of the liquid crystalline polymer, poly(γ-benzyl α-L-glutamate) (PBLG), have recently been demonstrated.[234] This material can be poled easily, and the produced fibers exhibit $d_{33}$ values in line with those of PVDF and, importantly, good stability at high temperatures (100 °C).

Once obtained, piezoelectric nanofibers may be used to realize many types of sensing devices and transducers. Strain[223] and force/pressure[235] sensors are obviously a possible targeted application of these nanomaterials. Data reported on the achievable sensitivity are of 42 mV generated per N of applied force, corresponding to about 0.8 μV per Pa of pressure.[235] Another direction consists in the development of polymer electrolytes based on PVDF, realized by immersions of nanofiber mats in $LiPF_6$ electrolyte solutions and motivated by interesting potential applications in Li-ion polymer batteries.[236]

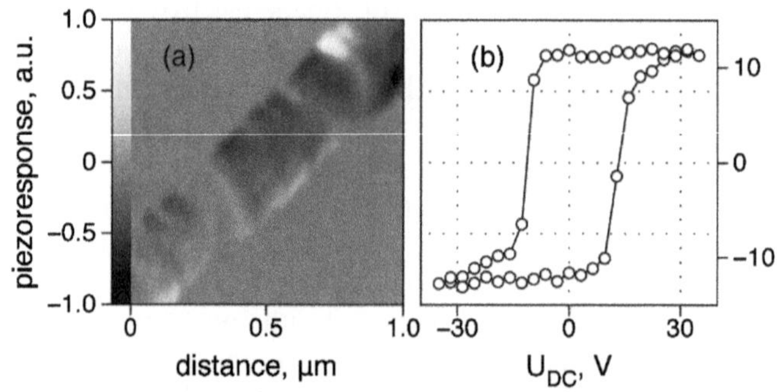


**Figure 6.14** (a) PFM micrograph of a glycine-PVA nanofiber. The fiber is deposited on a conductive substrate. The PFM technique is a powerful tool to detect the sample converse piezoelectric response (the local nanofiber deformation is measured upon applying an oscillating electric field through the conductive AFM tip). Here the micrograph brightness intensity is proportional to the amplitude times the cosine of the phase of the piezoresponse, allowing the pristine domain structure of the spun fiber to be estimated. (b) Measured hysteresis of the piezoresponse (here $d_{eff}$, measured in pm $V^{-1}$) of a single glycine-PVA nanofiber. In PFM, the hysteresis is measured by applying a sequence of direct voltage pulses on top of the application of the oscillating field. Coercive fields measured by the PFM hysteresis loops are typically of the order of $10^7$ V $m^{-1}$ for electrospun nanofibers (see also Ref. 217), consistently with the order of magnitude of poling electric fields of near-field electrospinning and post-deposition treatments used to impart or enhance dipole orientation and ferroelectric properties. Reproduced with permission from Ref. 228, *Cryst. Growth Des.*, 2011, **11**, 4288–4291. Doi: 10.1021/cg2009336. Copyright © 2011, American Chemical Society.

Conformal acoustic devices have been developed as well.[237] In addition, due to their shape and their structural properties, PVDF nanofibers are excellent candidates as active building blocks in nanogenerators (*i.e.* generators by nanomaterials) for mechanical energy scavenging from renewable sources. This class of devices, originally implemented by using inorganic nanowires,[238–243] is aimed at harvesting mechanical energy intrinsically present in the environment or associated with vibrations and movements of living organisms (due, for instance, to respiration, beat of the heart and other muscular movements), optimizing the efficiency of conversion to electrical energy especially at frequencies in the range from a few Hz to a few tens of Hz. Nanogenerators have been largely based on inorganic piezoelectric nanowires, mostly ZnO nanowires.[240] Since the nanowires grow vertically on GaN, sapphire or other substrates, to be connected to electrodes at both their ends they need to be embedded in complex devices, consisting of packaging polymers and partially suspended, sliding Pt-coated elements that provide Schottky contacts at the top terminations of the ZnO wires.[241,243] Due to output and mechanical stability, lifetime and fabrication complexity issues, researchers have looked for simpler, planar geometries.[238] A very important progress, with potentially enormous implications for the realization of functional fabrics for self-powering clothes, is given by integrating the nanogenerator concept with Kevlar microfibers, on which ZnO nanowires grow radially.[242] However, it is clear that a full development and widespread use of nanogenerators will be favored by using soft materials not just as substrates but also as piezoelectric active elements. In this respect, piezoelectric polymer nanofibers may have a bright future.

First of all, differently from most of the nanogenerators based on inorganics, the most suitable device architecture to use polymer nanofibers as piezoelectric medium is very simple, being based on an insulating, glass or elastomeric substrate, on which one deposits parallel electrodes. The nanofibers can simply lie on the substrate inter-connecting the metal pads as one would do between the source and drain electrodes of an organic FET. If flexible substrates are employed, an output voltage between the two electrodes is straightforwardly obtained as a consequence of *stretching* or *bending* the whole device, which is able to transmit the mechanical movement to the active nanomaterial. Many different plastic materials are available, which offer convenient options as substrates for building stretchable (such as PDMS and other silicone elastomers) or bendable (such as Kapton® by Dupont) flexible devices, respectively. As an alternative, one may also adopt sandwiched structures, in which the piezoelectric polymer layer is placed between two metals in a vertical stack (Figure 6.15).[211,213] This geometry may be an interesting approach for some applications, since it makes the device able to generate an output voltage as a result of a vertical *compression*.

The energy conversion efficiency of the nanogenerator devices can be estimated quite easily as the ratio between the generated electrical energy, $U_E$, and the mechanical energy, $U_M$, stored because of the induced deformation:[208,211,238]

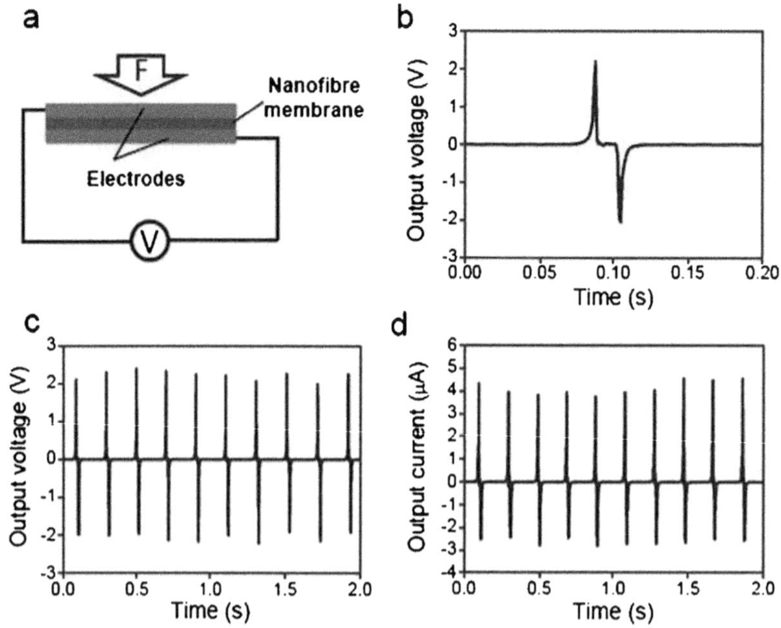


**Figure 6.15** (a) Scheme of a sandwich-architecture for electrical power generation from randomly oriented piezoelectric polymer nanofibers. (b) Typical voltage output due to a mechanical impact onto the device. (c) Voltage and (d) current outputs under repeated compressions at a frequency of 5 Hz. Reproduced with permission from Ref. 211, *J. Mater. Chem.*, 2011, **21**, 11088–11091. Doi: 10.1039/c1jm11445j.

$$\eta_E = \frac{U_E}{U_M}. \tag{6.25}$$

The electrical energy is simply given by:

$$U_E = \int VIdt, \tag{6.26}$$

where $V$ and $I$ are the output voltage and current, whereas $U_M$, for the exemplary case of a stretching deformation, is expressed by Equation (1.12). Demonstrated nanogenerators with PVDF nanofibers have $\eta_E$ up to 13.6% in compression modality[211] and up to 21.8% in stretching modality.[208] Furthermore, the frequencies of the deformations are also found to impact on the energy conversion efficiency performances, which improve upon increasing the frequency from 1 to 10 Hz.[211] Finally, recent developments of energy scavenging by PVDF fiber-based nanogenerators include the parallel integration in more complex devices capable of harvesting not only mechanical energy, but also energy from other sources available in the body, such as biochemical energy by means of enzymatic fuel cells.[209]

Both single PVDF nanofibers[208] and uniaxially aligned arrays[209] can be used to realize these nanogenerators. Single fibers can be directly positioned between pre-defined metal electrodes with high precision by near-field electrospinning (Section 2.3).[208,210] Instead, uniaxially aligned arrays of many fibers are straightforwardly electrospun onto parallel metal electrodes following the procedures described in Section 2.3.2, and then put in contact and transferred onto the final device substrate, which is generally slightly adhesive.[209] A crucial issue is understanding the direction of preferential alignment of the dipoles of the β-phase, which consequently allows the most suitable device geometry (planar or vertical) to be chosen and the nanogenerators to be run in the most suitable modality (bending, stretching, compression, *etc.*). When conventional electrospinning is applied, the alignment along the longitudinal fiber axis is usually achieved *a posteriori* by electric poling, applying an external bias between the planar device electrodes.[209] The order of magnitude of the intensity of the poling fields corresponds well to that of the electric fields of near-field electrospinning ($10^7$ V m$^{-1}$).[209] An electric field of the same order of magnitude is also used for poling membranes of P(VDF–TrFE) nanofibers (also in this case realized by standard electrospinning) sandwiched between vertically stacked electrodes.[213] In this geometry, poling is achieved with the 3-direction along the vertical stack, namely perpendicular to the longitudinal axis of the deposited nanofibers, however the process may be rendered a little ineffective by the leakage currents between electrodes due to the intrinsic porosity of the polymer nanofiber layers.[213] When *a posteriori* electric poling is carried out on the nanofibers, the assembled devices are usually immersed in silicone or paraffin oils to avoid electrical breakdown.[209,213] Interestingly, piezoelectricity can be obtained for PVDF nanofiber membranes realized by conventional electrospinning also *without a posteriori* poling.[211] This finding strongly suggests that electrospinning promotes a certain alignment of electric dipoles in the cross-sectional plane of the fibers, namely *perpendicular* to the fiber longitudinal axis. This can be elucidated by polarized FTIR spectroscopy,[212] in which higher absorption peaks are measured for light polarized along the relevant vibrational transition moments, thus allowing one to argue the resulting dipoles orientation (Figure 6.16). In fact, for P(VDF-TrFE) nanofibers realized by conventional electrospinning onto a rotating collector, a higher absorption is found for the so-called $A_1$ and $B_2$ bands (in which the vibrational transition moments are parallel to the $\vec{a}$ and $\vec{b}$ axes of the unit crystal lattice of the ferroelectric phase, Figure 6.16b and c) by using a beam polarization perpendicular to the fiber longitudinal axis.[212] On the contrary, a higher absorption is obtained from the so-called $B_1$ band (in which the transition moment is parallel to the $\vec{c}$ axis) by a beam polarization parallel to the fiber longitudinal axis.[212] This can attributed to preferential orientation of the chains of P(VDF-TrFE), which are mostly parallel to the collector rotation direction (namely parallel to the extruded jet), whereas the preferential orientations of the electro-active $CF_2$ dipoles are perpendicular to the macromolecular chains.

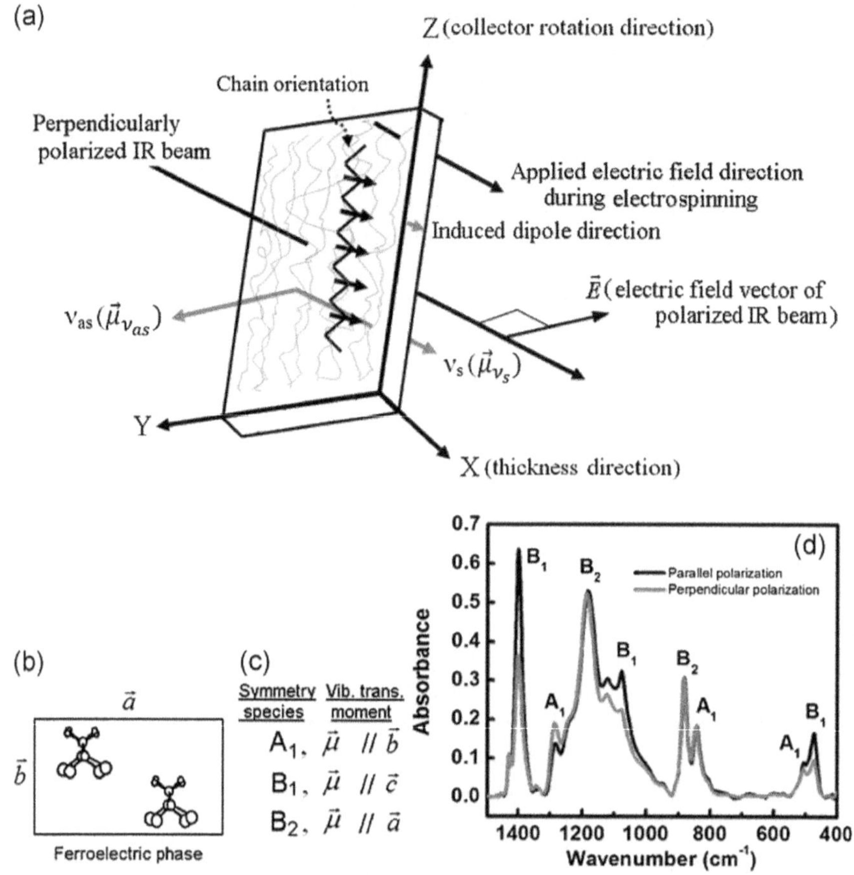


**Figure 6.16** (a) Scheme of partially aligned P(VDF-TrFE) nanofibers and of relevant directions for polarized FTIR analysis. $\vec{\mu}$ ($v_{as}$) indicates the vibrational transition moment associated with the $CH_2$ or $CF_2$ asymmetric stretching vibration mode, and $\vec{\mu}$ ($v_s$) indicates the moment associated with the $CH_2$ or $CF_2$ symmetric stretching vibration. (b) Unit crystal lattice of the ferroelectric phase of P(VDF-TrFE). (c) Relationship between the vibrational transition moment and the crystal unit axis for each vibrational symmetry species. (d) Polarized FTIR spectra of nanofibers. Reproduced with permission from Ref. 212, *Macromol. Rapid Commun.*, 2011, **32**, 831–837. Doi: 10.1002/marc.201100040. Copyright © 2011, WILEY-VCH Verlag GmbH & Co. KGaA.

Nowadays, the use of piezoelectric polymer nanofibers is spreading in even more exotic and inter-disciplinary applications. A possible direction for future research could be implementing nanofiber-based devices able to combine piezoresistivity with optoelectronic functionalities in nanoscale, non-volatile memories[244] or nano-OLEDs with integrated on-off switches as already demonstrated with thin-films.[245] Other intriguing applications rely on coupling piezoelectricity with biocompatibility to realize piezoelectric scaffolds for living cells.[246,247] Human skin fibroblasts can be cultured on PVDF-TrFE nanofibers. The gene expression for focal contact proteins (paxillin, vinculin, focal adhesion kinase, talin), measured by real-time reverse-transcription polymerase chain reaction (RT-PCR), is found at levels comparable with those of cells cultured on PS.[246] Also dorsal root ganglion (DRG) neurons adhere to PVDF-TrFE nanofibers, with neurite outgrowth aligned with the fibers underneath.[247] In addition, many studies demonstrate that electrical charges and applied bias affect the cellular behavior, growth and morphology. For these reasons, performing cell cultures on piezoelectric nanofibers may open new perspectives for regenerative medicine and tissue engineering assisted by electrical stimulation. The next chapter describes in more depth the application of polymer nanofibers to biological scaffolds.

### 6.4.2 Thermo-electric Properties

Thermo-electric properties of conductive polymer nanofibers can be interesting for energy generation. In this case, a temperature gradient across a material is exploited to generate electricity (Seebeck effect). To this aim, one of the most important performance parameters is the dimensionless figure of merit:

$$ZT = \frac{S_E^2 \chi T}{\kappa_T}, \tag{6.27}$$

which can be directly related to the maximum thermodynamic conversion efficiency achievable using a given material. In the previous formula, $S_E$ indicates the Seeback coefficient and $\kappa_T$ is the thermal conductivity. A good material for thermoelectric applications has to exhibit a $ZT$ parameter as high as possible, resulting from a smart combination of the values of the $S_E^2\chi$ product (so-called power factor, which is desired to be as large as possible) and $\kappa_T$ (desired to be low thus allowing the material to retain the difference between temperatures at its ends and avoiding wasting such a gradient in heat leaks). Of course, good electrical conductors are often also efficient in heat transport, and large Seebeck coefficients are often obtained in the low-doping/low-conduction regime, therefore maximizing $ZT$ is a matter of carefully looking for the best compromise for each class of materials. Besides the high flexibility and low cost, the advantage of conducting polymeric materials over inorganic conductors and semiconductors for thermoelectric applications is first of all the lower thermal conductivity.[248] The field is developing rapidly and polymer nanostructures, including nanofibers, are increasingly considered as thermoelectric materials. In general, nanostructures are appealing in this framework since the more effective boundary phonon scattering at surfaces can impact negatively on heat transfer processes while good properties are hopefully retained for what concerns electrical conduction. PAN nanotubes doped by β-NSA show $S_E$ values above 210 μV $K^{-1}$ at room temperature, with a good conductivity of $7.7 \times 10^{-3}$ S $cm^{-1}$ and $\kappa_T$ around $5 \times 10^{-2}$ cal/(K m s) [0.2 W/(K m)].[249] The resulting $ZT$ values approach $10^{-4}$. Electropolymerized PEDOT nanowires with thickness of 40–90 nm show $S_E$ values up to about 120 μV $K^{-1}$ at 310 K, outperforming thin-films, and also show very good conductivities (7–40 S $cm^{-1}$) allowing power factors of almost $10^{-5}$ W $m^{-1}$ $K^{-1}$ to be reached.[250] The collection of experimental evidences of thermoelectric power generation in polymer nanofibers is still at the beginning. Experiments need special care, particularly concerning measurements of thermal conductivity. Composite materials involving carbon nanofibers together with polymer nanofibers can be promising candidates for future studies. In particular, thermoelectric effects could be limited by (or at least should be validated considering) the high values of thermal conductivity [up to the order of 100 W/(K m)], which have been found in some species of polymer nanofibers due to the low linear density of defects.[251]

## References


1. C. K. Chiang, C. R. Fincher, Jr., Y. W. Park, A. J. Heeger, H. Shirakawa, E. J. Louis, S. G. Gau and A. G. MacDiarmid, *Phys. Rev. Lett.*, 1977, **39**, 295–348.
2. H. Shirakawa, E. J. Louis, A. G. MacDiarmid, C. K. Chiang and A. J. Heeger, *Chem. Comm.*, 1977, 578–580.
3. A. J. Heeger, S. Kivelson, J. R. Schrieffer and W.-P. Su, *Rev. Mod. Phys.*, 1988, **60**, 781–850.
4. A. J. Heeger, *J. Phys. Chem. B*, 2001, **105**, 8475–8491.
5. A. G. MacDiarmid, *Rev. Mod. Phys.*, 2001, **73**, 701–712.
6. M. S. A. Abdou, X. Lu, Z. W. Xie, F. Orfino, M. J. Deen and S. Holdcroft, *Chem. Mater.*, 1995, **7**, 631–641.
7. M. T. Bernius, M. Inbasekaran, J. O'Brien and W. Wu, *Adv. Mater.*, 2000, **12**, 1737–1750.
8. C. D. Dimitrakopoulos and P. R. L. Malenfant, *Adv. Mater.*, 2002, **14**, 99–117.
9. H. Klauk (ed.), *Organic Electronics, Materials, Manufacturing, and Applications*, Wiley-VCH, Weinheim, 2006.
10. Z. Bao and J. Locklin (ed.), *Organic Field-Effect Transistors*, CRC Press, New York, 2007.
11. L. L. Chua, J. Zaumseil, J. F. Chang, E. C. W. Ou, P. K. H. Ho, H. Sirringhaus and R. H. Friend, *Nature*, 2005, **434**, 194–199.
12. C. R. Newman, C. D. Frisbie, D. A. da Silva Filho, J.-L. Brédas, P. C. Ewbank and K. R. Mann, *Chem. Mater.*, 2004, **16**, 4436–4451.
13. G. Gelinck, P. Heremans, K. Nomoto and T. D. Anthopoulos, *Adv. Mater.*, 2010, **22**, 3778–3798.

14. A. L. Briseno, S. C. B. Mannsfeld, C. Reese, J. M. Hancock, Y. Xiong, S. A. Jenekhe, Z. Bao and Y. Xia, *Nano Lett.*, 2007, **7**, 2847–2853.
15. A. Hepp, H. Heil, W. Weise, M. Ahles, R. Schmechel and H. von Seggern, *Phys. Rev. Lett.*, 2003, **91**, 157406.
16. C. Rost, S. Karg, W. Riess, M. A. Loi, M. Murgia and M. Muccini, *Appl. Phys. Lett.*, 2004, **85**, 1613–1615.
17. M. Muccini, *Nat. Mater.*, 2006, **5**, 605–613.
18. J. Zaumseil, C. L. Donley, J.-S. Kim, R. H. Friend and H. Sirringhaus, *Adv. Mater.*, 2006, **18**, 2708–2712.
19. V. C. Sundar, J. Zaumseil, V. Podzorov, E. Menard, R. Willett, T. Someya, M. E. Gershenson, and J. A. Rogers, *Science*, 2004, **303**, 1644–1646.
20. O. D. Jurchescu, J. Baas and T. M. Palstra, *Appl. Phys. Lett.*, 2004, **84**, 3061–3063.
21. M. Zhang, H. N. Tsao, W. Pisula, C. Yang, A. K. Mishra and K. Müllen, *J. Am. Chem. Soc.*, 2007, **129**, 3472–3473.
22. Y. Li, Y. Wu, P. Liu, M. Birau, H. Pan and B. S. Ong, *Adv. Mater.*, 2006, **18**, 3029–3032.
23. I. Osaka, G. Sauvé, R. Zhang, T. Kowalewski and R. D. McCullough, *Adv. Mater.*, 2007, **19**, 4160–4165.
24. I. McCulloch, M. Heeney, C. Bailey, K. Genevicius, I. Macdonald, M. Shkunov, D. Sparrowe, S. Tierney, R. Wagner, W. Zhang, M. Chabinyc, R. J. Kline, M. D. McGehee and M. Toney, *Nat. Mater.*, 2006, **5**, 328–333.
25. J.-F. Chang, J. Clark, N. Zhao, H. Sirringhaus, D. W. Breiby, J. W. Andreasen, M. M. Nielsen, M. Giles, M. Heeney and I. McCulloch, *Phys. Rev. B*, 2006, **74**, 115318.
26. R. A. Street, *Adv. Mater.*, 2009, **21**, 2007–2022.
27. S. Samitsu, T. Shimomura, S. Heike, T. Hashizume and K. Ito, *Macromolecules*, 2008, **41**, 8000–8010.
28. J. Joo, S. M. Long, J. P. Pouget, E. J. Oh, A. G. MacDiarmid and A. J. Epstein, *Phys. Rev. B*, 1998, **57**, 9567–9580.
29. V. N. Prigodin, A. N. Samukhin and A. J. Epstein, *Synth. Met.*, 2004, **141**, 155–164.
30. N. F. Mott and E. A. Davis, *Electronic Processes in Non-Crystalline Materials*, Clarendon, Oxford, UK, 2nd edn, 1979.
31. F. Zuo, M. Angelopoulos, A. G. MacDiarmid and A. J. Epstein, *Phys. Rev. B*, 1987, **36**, 3475–3478.
32. P. Sheng, B. Abeles and Y. Arie, *Phys. Rev. Lett.*, 1973, **31**, 44–47.
33. Q. Li, L. Cruz and P. Phillips, *Phys. Rev. B*, 1993, **47**, 1840–1845.
34. R. Pelster, G. Nimtz and B. Wessling, *Phys. Rev. B*, 1994, **49**, 12718–12723.
35. A. N. Aleshin, *Adv. Mater.*, 2006, **18**, 17–27.
36. H. Sirringhaus, P. J. Brown, R. H. Friend, M. M. Nielsen, K. Bechgaard, B. M. W. Lanveld-Voss, A. J. H. Spiering, R. A. J. Jannsen,

E. W. Meijer, P. Herwig and D. M. de Leeuw, *Nature*, 1999, **401**, 685–688.

37. T. Bjørnholm, T. Hassenkam, D. R. Greve, R. D. McCullough, M. Jayaraman, S. M. Savoy, C. E. Jones and J. T. McDevitt, *Adv. Mater.*, 1999, **11**, 1218–1221.
38. J. P. Spatz, B. Lorenz, K. Weishaupt, H. D. Hochheimer, V. P. Menon, R. V. Parthasarathy, C. R. Martin, J. Bechtold and P.-H. Hor, *Phys. Rev. B*, 1994, **50**, 14888–14892.
39. P. Prins, F. C. Grozema, J. M. Schins, S. Patil, U. Scherf and L. D. A. Siebbeles, *Phys. Rev. Lett.*, 2006, **96**, 146601.
40. M. D. Curtis, J. Cao and J. W. Kampf, *J. Am. Chem. Soc.*, 2004, **126**, 4318–4328.
41. J. Cornil, D. Beljonne, J.-P. Calbert and J.-L. Brédas, *Adv. Mater.*, 2001, **13**, 1053–1067.
42. A. L. Briseno, S. C. B. Mannsfeld, S. A. Jenekhe, Z. Bao and Y. Xia, *Mater. Today*, 2008, **11**, 38–47.
43. J. Ah Lim, F. Liu, S. Ferdous, M. Muthukumar and A. L. Briseno, *Mater. Today*, 2010, **13**(5), 14–24.
44. P. M. Beaujuge and J. M. J. Fréchet, *J. Amer. Chem. Soc.*, 2011, **133**, 20009–20029.
45. P.-T. Wu, G. Ren, C. Li, R. Mezzenga and S. A. Jenekhe, *Macromolecules*, 2009, **42**, 2317–2320.
46. J. A. Merlo and C. D. Frisbie, *J. Phys. Chem. B.*, 2004, **108**, 19169–16179.
47. B. S. Ong, Y. Wu, P. Liu and S. Gardner, *Adv. Mater.*, 2005, **17**, 1141–1144.
48. B. H. Hamadani, C. A. Richter, D. J. Gundlach, R. J. Kline, I. McCulloch and M. Heeney, *J. Appl. Phys.*, 2007, **102**, 044503.
49. J.-F. Chang, B. Sun, D. W. Breiby, M. M. Nielsen, T. I. Sölling, M. Giles, I. McCullogh and H. Sirringhaus, *Chem. Mater.*, 2004, **16**, 4772–4776.
50. G. Lu, L. Li and X. Yang, *Adv. Mater.*, 2007, **19**, 3594–3598.
51. L. H. Jimison, M. F. Toney, I. McCulloch, M. Heeney and A. Salleo, *Adv. Mater.*, 2009, **21**, 1568–1572.
52. E. Mena-Osteritz, A. Meyer, B. M. W. Langeveld-Voss, R. A. J. Janssen, E. W. Meijer and P. Bäuerle, *Angew. Chem. Int. Ed.*, 2000, **39**, 2680–2684.
53. B. Grévin, P. Rannou, R. Payerne, A. Pron and J. P. Travers, *J. Chem. Phys.*, 2003, **118**, 7097–7102.
54. H. Yang, T. J. Shin, L. Yang, K. Cho, C. Y. Ryu and Z. Bao, *Adv. Funct. Mater.*, 2005, **15**, 671–676.
55. R. A. Street, J. E. Northrup and A. Salleo, *Phys. Rev. B*, 2005, **71**, 165202.
56. D. Hwan Kim, J. Tark Han, Y. Don Park, Y. Jang, J. H. Cho, M. Hwang and K. Cho, *Adv. Mater.*, 2006, **18**, 719–723.
57. M. A. Stevens, C. Silva, D. M. Russell and R. H. Friend, *Phys. Rev. B*, 2001, **63**, 165213.

58. M. C. Gwinner, S. Khodabakhsh, M. H. Song, H. Schweizer, H. Giessen and H. Sirringhaus, *Adv. Funct. Mater.*, 2009, **19**, 1360–1370.
59. B. Watts, T. Schuettfort and C. R. McNeill, *Adv. Funct. Mater.*, 2011, **21**, 1122–1131.
60. Z. Zheng, K.-H. Yim, M. S. M. Saifullah, M. E. Welland, R. H. Friend, J.-S. Kim and W. T. S. Huck, *Nano Lett.*, 2007, **7**, 987–992.
61. J. Zaumseil and H. Sirringhaus, *Chem. Rev.*, 2007, **107**, 1296–1323.
62. F. Sunjoo Kim, G. Ren and S. A. Jenekhe, *Chem. Mater.*, 2011, **23**, 682–732.
63. Y.-Z. Long, M.-M. Li, C. Gu, M. Wan, J.-L. Duvail, Z. Liu and Z. Fan, *Prog. Polym. Sci.*, 2011, **36**, 1415–1442.
64. C. R. Martin, *Chem. Mater.*, 1996, **8**, 1739–1746.
65. Z. Cai and C. R. Martin, *J. Am. Chem. Soc.*, 1989, **111**, 4138–4139.
66. Z. Cai, J. Lei, W. Liang, V. Menon and C. R. Martin, *Chem. Mater.*, 1991, **3**, 960–966.
67. C. R. Martin, R. Parthasarathy and V. Menon, *Synth. Met.*, 1993, **55**, 1165–1170.
68. R. V. Parthasarathy and C. R. Martin, *Chem. Mater.*, 1994, **6**, 1627–1632.
69. C. R. Martin, *Science*, 1994, **266**, 1961–1966.
70. J. L. Duvail, Y. Long, S. Cuenot, Z. Chen and C. Gu, *Appl. Phys. Lett.*, 2007, **90**, 102114.
71. A. Rahman and M. K. Sanyal, *Phys. Rev. B*, 2007, **76**, 045110.
72. C. R. Martin, *Acc. Chem. Res.*, 1995, **28**, 61–68.
73. G. A. O'Brien, A. J. Quinn, D. A. Tanner and G. Redmond, *Adv. Mater.*, 2006, **18**, 2379–2383.
74. R. Martel, T. Schmidt, H. R. Shea, T. Hertel and Ph. Avouris, *Appl. Phys. Lett.*, 1998, **73**, 2447–2449.
75. S. H. M. Persson, P. Dyreklev and O. Inganäs, *Adv. Mater.*, 1996, **8**, 405–408.
76. K. Akagi, G. Piao, S. Kaneko, K. Sakamaki, H. Shirakawa and M. Kyotani, *Science*, 1998, **282**, 1683–1686.
77. M. Granström and O. Inganäs, *Polymer*, 1995, **36**, 2867–2872.
78. M. Delvaux, J. Duchet, P.-Y. Stavaux, R. Legras and S. Demoustier-Champagne, *Synth. Met.*, 2000, **113**, 275–280.
79. Z. Wei, Z. Zhang and M. Wan, *Langmuir*, 2002, **18**, 917–921.
80. K. Huang and M. Wan, *Chem. Mater.*, 2002, **14**, 3486–3492.
81. J. L. Duvail, P. Rétho, V. Fernandez, G. Louarn, P. Molinié and O. Chauvet, *J. Phys. Chem. B*, 2004, **108**, 18552–18556.
82. Y. Z. Long, J. L. Duvail, Z. J. Chen, A. Z. Jin and C. Z. Gu, *Polym. Adv. Technol.*, 2009, **20**, 541–544.
83. S. Nair, E. Hsiao and S. H. Kim, *Chem. Mater.*, 2009, **21**, 115–121.
84. A. Laforgue and L. Robitaille, *Chem. Mater.*, 2010, **22**, 2474–2480.
85. A. Laforgue and L. Robitaille, *Macromolecules*, 2010, **43**, 4194–4200.
86. A. Laforgue, *J. Mater. Chem.*, 2010, **20**, 8233–8235.

87. N. J. Pinto, *Am. J. Phys.*, 2008, **76**, 1163–1167.
88. Y. Zhou, M. Freitag, J. Hone, C. Staii, A. T. Johnson, Jr., N. J. Pinto and A. G. MacDiarmid, *Appl. Phys. Lett.*, 2003, **83**, 3800–3802.
89. J. D. Norris, M. M. Shaker, F. K. Ko and A. G. MacDiarmid, *Synth. Met.*, 2000, **114**, 109–114.
90. R. C. Y. King and F. Roussel, *Synth. Met.*, 2005, **153**, 337–340.
91. Z. Zhang, Z. Wei and M. Wan, *Macromolecules*, 2002, **35**, 5937–5942.
92. Y. Long, Z. Chen, N. Wang, Y. Ma, Z. Zhang, L. Zhang and M. Wan, *Appl. Phys. Lett.*, 2003, **83**, 1863–1865.
93. Y. Long, L. Zhang, Z. Chen, K. Huang, Y. Yang, H. Xiao, M. Wan, A. Jin and C. Gu, *Phys. Rev. B*, 2005, **71**, 165412.
94. H. X. He, C. Z. Li and N. J. Tao, *Appl. Phys. Lett.*, 2001, **78**, 811–813.
95. M. Wan, Z. Wei, Z. Zhang, L. Zhang, K. Huang and Y. Yang, *Synth. Met.*, 2003, **135–136**, 175–176.
96. N. J. Pinto, A. T. Johnson, Jr., A. G. MacDiarmid, C. H. Mueller, N. Theofylaktos, D. C. Robinson and F. A. Miranda, *Appl. Phys. Lett.*, 2003, **83**, 4244–4246.
97. N. J. Pinto, R. González, A. T. Johnson, Jr. and A. G. MacDiarmid, *Appl. Phys. Lett.*, 2006, **89**, 033505.
98. A. G. MacDiarmid, W. E. Jones, Jr., I. D. Norris, J. Gao, A. T. Johnson, Jr., N. J. Pinto, J. Hone, B. Han, F. K. Ko, H. Okuzaki and M. Llaguno, *Synth. Met.*, 2001, **119**, 27–30.
99. N. J. Pinto, I. Ramosa, R. Rojas, P.-C. Wang and A. T. Johnson, Jr., *Sens. Actuators B: Chem.*, 2008, **129**, 621–627.
100. S. M. Sze, *Physics of Semiconductor Devices*, Wiley, New York, 2nd edn, 1981.
101. P. K. Kahol and N. J. Pinto, *Synth. Met.*, 2004, **140**, 269–272.
102. J. R. Cárdenas, M. G. O. de França, E. A. de Vasconcelos, W. M. de Azevedo and E. F. da Silva, Jr., *J. Phys. D: Appl. Phys.*, 2007, **40**, 1068–1071.
103. X. Li and X. Li, *Mater. Lett.*, 2007, **61**, 2011–2014.
104. A. Rahy and D. J. Yang, *Mater. Lett.*, 2008, **62**, 4311–4314.
105. B.-K. Kim, Y. Hwan Kim, K. Won, H. Chang, Y. Choi, K. Kong, B. Whan Rhyu, J.-J. Kim and J.-O Lee, *Nanotechnology*, 2005, **16**, 1177–1181.
106. H. Liu, J. Kameoka, D. A. Czaplewski and H. G. Craighead, *Nano Lett.*, 2004, **4**, 671–675.
107. X. Y. Zhang, R. Chan-Yu-King, A. Jose and S. K. Manohar, *Synth Met.*, 2004, **145**, 23–29.
108. J. Huang and R. B. Kaner, *J. Am. Chem. Soc.*, 2004, **126**, 851–855.
109. C. Su, G. Wang, F. Huang and X. Li, *J. Mater. Sci.*, 2008, **43**, 197–202.
110. Y. Long, Z. Chen, J. Shen, Z. Zhang, L. Zhang, K. Huang, M. Wan, A. Jin, C. Gu and J. L. Duvail, *Nanotechnology*, 2006, **17**, 5903–5911.
111. L. Zhang and M. Wan, *Nanotechnology*, 2002, **13**, 750–755.

112. L. Zhang, Y. Long, Z. Chen and M. Wan, *Adv. Funct. Mater.*, 2004, **14**, 693–698.
113. X. Zhang, W. J. Goux and S. K. Manohar, *J. Am. Chem. Soc.*, 2004, **126**, 4502–4503.
114. M. Kyoon Shin, Y. Jin Kim, S. I. Kim, S.-K. Kim, H. Lee, G. M. Spinks and S. Jeong Kim, *Sens. Actuators B: Chem.*, 2008, **134**, 122–126.
115. J. Hyoung Jun, K. Cho, J. Yun, K. S. Suh, T. Young Kim and S. Kim, *Org. Electron.*, 2008, **9**, 445–451.
116. M. Nikiforov, H. Liu, H. Craighead and D. Bonnell, *Nano Lett.*, 2006, **6**, 896–900.
117. S. Samitsu, T. Shimomura, S. Heike, T. Hashizume and K. Ito, *Macromolecules*, 2010, **43**, 7891–7894.
118. Z. Hu, B. Muls, L. Gence, D. A. Serban, J. Hofkens, S. Melinte, B. Nysten, S. Demoustier-Champagne and A. M. Jonas, *Nano Lett.*, 2007, **7**, 3639–3644.
119. W. D. Oosterbann, J.-C. Bolsée, A. Gadisa, V. Vrindts, S. Bertho, J. D'Haen, T. J. Cleij, L. Lutsen, C. R. McNeill, L. Thomsen, J. V. Manca and D. Vanderzande, *Adv. Funct. Mater.*, 2010, **20**, 792–802.
120. N. J. Pinto, K. V. Carrasquillo, C. M. Rodd and R. Agarwal, *Appl. Phys. Lett.*, 2009, **94**, 083504.
121. K. Huang, M. Wan, Y. Long, Z. Chen and Y. Wei, *Synth. Met.*, 2005, **155**, 495–500.
122. J. Jang and H. Yoon, *Chem. Commun.*, 2003, 720–721.
123. X. Zhang and S. K. Manohar, *J. Am. Chem. Soc.*, 2004, **126**, 12714–12715.
124. B. Dong, D. Zhong, L. Chi and H. Fuchs, *Adv. Mater.*, 2005, **17**, 2736–2741.
125. S. Samitsu, T. Shimomura, K. Ito, M. Fujimori, S. Heike and T. Hashizume, *Appl. Phys. Lett.*, 2005, **86**, 233103.
126. X. Zhang, A. G. MacDiarmid and S. K. Manohar, *Chem. Commun.*, 2005, 5328–5330.
127. A. L. Briseno, S. C. B. Mannsfeld, P. J. Shamberger, F. S. Ohuchi, Z. Bao, S. A. Jenekhe and Y. Xia, *Chem. Mater.*, 2007, **20**, 4712–4719.
128. S. Wang, M. Kappl, I. Liebewirth, M. Müller, K. Kirchhoff, W. Pisula and K. Müllen, *Adv. Mater.*, 2012, **24**, 417–420.
129. A. N. Aleshin, H. J. Lee, S. H. Jhang, H. S. Kim, K. Akagi and Y. W. Park, *Phys. Rev. B*, 2005, **72**, 153202.
130. A. A. Middleton and N. S. Wingreen, *Phys. Rev. Lett.*, 1993, **71**, 3198–3201.
131. A. N. Aleshin, H. J. Lee, Y. W. Park and K. Akagi, *Phys. Rev. Lett.*, 2004, **93**, 196601.
132. J. Voit, *Rep. Prog. Phys.*, 1995, **58**, 977–1116.
133. H. J. Schulz, *Phys. Rev. Lett.*, 1993, **71**, 1864–1867.
134. W. P. Hu, H. Nakashima, K. Furukawa, Y. Kashimura, K. Ajito and K. Torimitsu, *Appl. Phys. Lett.*, 2004, **85**, 115–118.

135. N. A. Zimbovskaya, A. T. Johnson, Jr. and N. J. Pinto, *Phys. Rev. B*, 2005, **72**, 024213.
136. N. A. Zimbovskaya, *J. Phys. Chem.*, 2007, **126**, 184901.
137. N. A. Zimbovskaya, *J. Phys. Chem.*, 2009, **129**, 114705.
138. J. Kjelstrup-Hansen, J. E. Norton, D. A. da Silva Filho, J.-L. Brédas and H.-G. Rubahn, *Org. Electron.*, 2009, **10**, 1228–1234.
139. G. Horowitz, *J. Mater. Res.*, 2004, **19**, 1946–1962.
140. C. Wang, H. Dong, W. Hu, Y. Liu and D. Zhu, *Chem. Rev.*, 2012, **112**, 2208–2267.
141. S. Steudel, K. Myny, V. Arkhipov, C. Deibel, S. De Vusser, J. Genoe and P. Heremans, *Nat. Mater.*, 2005, **4**, 597–600.
142. F. Ebisawa, T. Kurokawa and S. Nara, *J. Appl. Phys.*, 1983, **54**, 3255–3259.
143. K. Kudo, M. Yamashina and T. Moriizumi, *Jpn J. Appl. Phys.*, 1984, **23**, 130.
144. J. Tate, J. A. Rogers, C. D. W. Jones, B. Vyas, D. W. Murphy, W. J. Li, Z. A. Bao, R. E. Slusher, A. Dodabalapur and H. E. Katz, *Langmuir*, 2000, **16**, 6054–6060.
145. X. Y. Cai, M. W. Burand, C. R. Newman, D. A. da Silva, T. M. Pappenfus, M. M. Bader, J. L. Brédas, K. R. Mann and C. D. Frisbie, *J. Phys. Chem. B*, 2006, **110**, 14590–14597.
146. H. Klauk, D. J. Gundlach, J. A. Nichols and T. N. Jackson, *IEEE Trans. Electron. Dev.*, 1999, **46**, 1258–1263.
147. L. A. Majewski, R. Schroeder and M. Grell, *Appl. Phys. Lett.*, 2004, **85**, 3620–3622.
148. H. Kajii, K. Koiwai, Y. Hirose and Y. Ohmori, *Org. Electron.*, 2010, **11**, 509–513.
149. K. S. Lee, T. J. Smith, K. C. Dickey, J. E. Yoo, K. J. Stevenson and Y. L. Loo, *Adv. Funct. Mater.*, 2006, **16**, 2409–2414.
150. M. Lefenfeld, G. Blanchet and J. A. Rogers, *Adv. Mater.*, 2003, **15**, 1188–1191.
151. H. Sirringhaus, T. Kawase, R. H. Friend, T. Shimoda, M. Inbasekaran, W. Wu and E. P. Woo, *Science*, 2000, **290**, 2123–2126.
152. C. W. Sele, T. von Werne, R. H. Friend and H. Sirringhaus, *Adv. Mater.*, 2005, **17**, 997–1001.
153. B. de Boer, A. Hadipour, M. M. Mandoc, T. van Woudenbergh and P. W. M. Blom, *Adv. Mater.*, 2005, **17**, 621–625.
154. B. H. Hamadani, D. A. Corley, J. W. Ciszek, J. M. Tour and D. Natelson, *Nano Lett.*, 2006, **6**, 1303–1306.
155. Y. Wen, Y. Liu, Y. Guo, G. Yu and W. Hu, *Chem. Rev.*, 2011, **111**, 3358–3406.
156. J. Veres, S. Ogier, G. Lloyd and D. de Leeuw, *Chem. Mater.*, 2004, **16**, 4543–4555.
157. A. Facchetti, M. H. Yoon and T. J. Marks, *Adv. Mater.*, 2005, **17**, 1705–1725.

158. M. L. Chabinyc, R. Lujan, F. E. Endicott, M. F. Toney, I. McCulloch and M. Heeney, *Appl. Phys. Lett.*, 2007, **90**, 233508.
159. C. Tanase, P. W. M. Blom, D. M. de Leeuw and E. J. Meijer, *Phys. Status Solidi (a)*, 2004, **201**, 1236–1245.
160. G. Horowitz, *Adv. Mater.*, 1998, **10**, 923–927.
161. C. D. Dimitrakopoulos, I. Kymissis, S. Purushothaman, D. A. Neumayer, P. R. Duncombe and R. B. Laibowitz, *Adv. Mater.*, 1999, **11**, 1372–1375.
162. C. D. Dimitrakopoulos, S. Purushothaman, I. Kymissis, A. Callegari and J. M. Shaw, *Science*, 1999, **283**, 822–824.
163. R. Ponce Ortiz, A. Facchetti and T. J. Marks, *Chem. Rev.*, 2010, **110**, 205–239.
164. R. A. Street, *Phys. Rev. B*, 2008, **77**, 165311.
165. D. Knipp, A. Benor, V. Wagner and T. Muck, *J. Appl. Phys.*, 2007, **101**, 044504.
166. R. A. Street, M. C. Chabinyc and F. Endicott, *Phys. Rev. B*, 2007, **76**, 045208.
167. A. Hoppe, T. Balster, T. Muck and V. Wagner, *Phys. Stat. Sol. (a)*, 2008, **205**, 612–625.
168. S. Luan and G. W. Neudeck, *J. Appl. Phys.*, 1992, **72**, 766–772.
169. Q. J. Cai, M. B. Chan-Park, Q. Zhou, Z. S. Lu, C. M. Li and B. S. Ong, *Org. Electron.*, 2008, **9**, 936–943.
170. D. Tu, S. Pagliara, A. Camposeo, G. Potente, E. Mele, R. Cingolani and D. Pisignano, *Adv. Funct. Mater.*, 2011, **21**, 1140–1145.
171. J. A. Rogers, T. Someya and Y. Huang, *Science*, 2010, **327**, 1603–1607.
172. J. G. Park, G. T. Kim, J. H. Park, H. Y. Yu, G. McIntosh, V. Kristic, S. H. Jhang, B. Kim, S. H. Lee, S. W. Lee, M. Burghard, S. Roth and Y. W. Park, *Thin Solid Films*, 2001, **393**, 161–167.
173. R. Gonzáles and N. J. Pinto, *Synth. Met.*, 2005, **151**, 275–278.
174. H. Liu, C. H. Reccius and H. G. Craighead, *Appl. Phys. Lett.*, 2005, **87**, 253106.
175. S. W. Lee, G. D. Moon and U. Jeong, *J. Mater. Chem.*, 2009, **19**, 743–748.
176. L. Qiu, W. H. Lee, X. Wang, J. S. Kim, J. A. Lim, D. Kwak, S. Lee and K. Cho, *Adv. Mater.*, 2009, **21**, 1349–1353.
177. P.-T. Wu, H. Xin, F. Sunjoo Kim, G. Ren and S. A. Jenekhe, *Macromolecules*, 2009, **42**, 8817–8826.
178. M. J. Panzer and C. D. Frisbie, *J. Amer. Chem. Soc.*, 2007, **129**, 6599–6607.
179. J. H. Cho, J. Lee, Y. Xia, B. S. Kim, Y. He, M. J. Renn, T. P. Lodge and C. D. Frisbie, *Nat. Mater.*, 2008, **7**, 900–906.
180. S. W. Lee, H. J. Lee, J. H. Choi, W. G. Koh, J. M. Myoung, J. H. Hur, J. J. Park, J. H. Cho and U. Jeong, *Nano Lett.*, 2010, **10**, 347–351.
181. S.-Y. Min, T. S. Kim, J. H. Cho, T.-W. Lee, presented at the *2nd International Conference of Electrospinning*, Jeju, Korea 2012.

182. D. Hwan Kim, Y. Don Park, Y. Jang, S. Kim and K. Cho, *J. Phys. Chem. B*, 2006, **110**, 15763–15768.
183. J. A. Merlo and C. D. Frisbie, *J. Polym. Sci. Part B: Polym. Phys.*, 2003, **41**, 2674–2680.
184. A. Babel, D. Li, Y. Xia and S. A. Jenekhe, *Macromolecules*, 2005, **38**, 4705–4711.
185. R. Zhang, B. Li, M. C. Iovu, M. Jerreries-El, G. Sauvé, J. Cooper, S. Jia, S. Tristram-Nagle, D. M. Smilgies, D. N. Lambeth, R. D. McCullough and T. Kowalewski, *J. Am. Chem. Soc.*, 2006, **128**, 3480–3481.
186. M. S. A. Abdou, F. P. Orfino, Y. Son and S. Holdcroft, *J. Am. Chem. Soc.*, 1997, **119**, 4518–4524.
187. S. Hoshino, M. Yoshida, S. Uemura, T. Kodzasa, N. Takada, T. Kamata and K. Yase, *J. Appl. Phys.*, 2004, **95**, 5088–5093.
188. L. Wang, J. Hu, H. Zhang and T. Zhang, *Chem. Commun.*, 2011, **47**, 6837–6839.
189. H. Wu, D. Lin and W. Pan, *Appl. Phys. Lett.*, 2006, **89**, 133125.
190. H. Wu, D. Lin, R. Zhang and W. Pan, *J. Am. Ceram. Soc.*, 2008, **91**, 656–659.
191. D. Tu, S. Pagliara, A. Camposeo, L. Persano, R. Cingolani and D. Pisignano, *Nanoscale*, 2010, **2**, 2217–2222.
192. P. Dyreklev, O. Inganäs, J. Paloheimo and H. Stubb, *J. Appl. Phys.*, 1992, **71**, 2816–2820.
193. K. Kaneto, S. Hayahi and K. Yoshino, *J. Phys. Soc. Jpn*, 1988, **57**, 1119–1126.
194. S. Tasch, G. Kranzelbinder, G. Leising and U. Scherf, *Phys. Rev. B*, 1997, **55**, 5079–5083.
195. D. Tu, S. Pagliara, R. Cingolani and D. Pisignano, *Appl. Phys. Lett.*, 2011, **98**, 023307.
196. C. H. Lee, G. Yu and A. J. Heeger, *Phys. Rev. B*, 1993, **47**, 15543–15553.
197. E. Frankevich, A. Zakhidov, K. Yoshino, Y. Maruyama and K. Yakushi, *Phys. Rev. B*, 1996, **53**, 4498–4508.
198. B. Dulieu, J. Wéry, S. Lefrant and J. Bullot, *Phys. Rev. B*, 1998, **57**, 9118–9127.
199. A. Camposeo, F. Di Benedetto, R. Stabile, R. Cingolani and D. Pisignano, *Appl. Phys. Lett.*, 2007, **90**, 143115.
200. D. Dragan, *Rep. Prog. Phys.*, 1998, **61**, 1267–1324.
201. J. D. Jackson, *Classical Electrodynamics*, Wiley, New York, 3rd edn, 1998.
202. T. Furukawa, *IEEE Trans. Electr. Insul.*, 1989, **24**, 375–394.
203. V. Kochervinski?, *Crystallogr. Rep.*, 2003, **48**, 649–675.
204. G. Eberle, H. Schmidt and W. Eisenmenger, *IEEE Trans. Dielectr. Electr. Insul.*, 1996, **3**, 624–646.
205. V. Kochervinski?, *Polym. Sci. Ser. C*, 2008, **50**, 93–121.
206. K. Asadi, M. Li, P. W. M. Blom, M. Kemerink and D. M. de Leeuw, *Mater. Today*, 2011, **14**, 592–599.

207. J. Sirohi and I. Chopra, *J. Intell. Mater. Syst. Struct.*, 2000, **11**, 246–257.
208. C. Chang, V. H. Tran, J. Wang, Y.-K. Fuh and L. Lin, *Nano Lett.*, 2010, **10**, 726–731.
209. B. J. Hansen, Y. Liu, R. Yang and Z. L. Wang, *ACS Nano*, 2010, **4**, 3547–3652.
210. J. Pu, X. Yan, Y. Jiang, C. Chang and L. Lin, *Sens. Actuators A: Phys.*, 2010, **164**, 131–136.
211. J. Fang, X. Wang and T. Lin, *J. Mater. Chem.*, 2011, **21**, 11088–11091.
212. D. Mandal, S. Yoon and K. J. Kim, *Macromol. Rapid Commun.*, 2011, **32**, 831–837.
213. F. He, M. Sarkar, S. Lau, J. Fan and L. H. Chan, *Polym. Test.*, 2011, **30**, 436–441.
214. W. A. Yee, M. Kotaki, Y. Liu and X. Lu, *Polymer*, 2007, **48**, 512–521.
215. Z. Zhao, J. Li, X. Yuan, X. Li, Y. Zhang and J. Sheng, *J. Appl. Polym. Sci.*, 2005, **97**, 466–474.
216. S. Huang, W. A. Yee, W.C. Tjiu, Y. Liu, M. Kotaki, Y. C. F. Boey, J. Ma, T. Liu and X. Lu, *Langmuir*, 2008, **24**, 13621–13626.
217. A. Baji, Y.-W. Mai, Q. Lib and Y. Liu, *Nanoscale*, 2011, **3**, 3068–3071.
218. P. Ueberschlag, *Sens. Rev.*, 2001, **21**, 118–125.
219. D. Vatansever, R. L. Hadimani, T. Shah and E. Siores, *Smart Mater. Struct.*, 2011, **20**, 055019.
220. R. Gregorio, Jr. and M. Cestari, *Polym. Sci.: Part B: Polym. Phys.*, 1994, **32**, 859–870.
221. A. Salimi and A. A. Yousefi, *Polym. Test.*, 2003, **22**, 699–704.
222. J. S. Andrew and D. R. Clarke, *Langmuir*, 2008, **24**, 670–672.
223. D. Dhakras, V. Borkar, S. Ogale and J. Jog, *Nanoscale*, 2012, **4**, 752–756.
224. J. S. Andrew and D. R. Clarke, *Langmuir*, 2008, **24**, 8435–8438.
225. N. Levi, R. Czerw, S. Xing, P. Iyer and D. L. Carroll, *Nano Lett.*, 2004, **4**, 1267–1271.
226. J. Zheng, A. He, J. Li and C. C. Han, *Macromol. Rapid. Comm.*, 2007, **28**, 2159–2162.
227. O. Martínez, A. G. Bravo and N. J. Pinto, *Macromolecules*, 2009, **42**, 7924–7929.
228. D. Isakov, E. de Matos Gomes, I. Bdikin, B. Almeida, M. Belsley, M. Costa, V. Rodrigues and A. Heredia, *Cryst. Growth Des.*, 2011, **11**, 4288–4291.
229. J. W. Chew, S. N. Black, P. S. Chow, R. B. H. Tan and K. J. Carpenter, *Cryst. Eng. Comm.*, 2007, **9**, 128–130.
230. S. K. Poornachary, P. S. Chow and R. B. H. Tan, *Cryst. Growth Des.*, 2008, **8**, 179–185.
231. R. Dowling, R. J. Davey, R. A. Curtis, G. Han, S. K. Poornachary, P. S. Chow and R. B. H. Tan, *Chem. Commun.*, 2010, **46**, 5924–5926.
232. Y. Iitaka, *Acta Crystallogr.*, 1961, **14**, 1–10.
233. R. A. Kumar, R. E. Vizhi, N. Vijayan and D. R. Babu, *Physica B*, 2011, **406**, 2594–2600.

234. D. Farrar, K. Ren, D. Cheng, S. Kim, W. Moon, W. L. Wilson, J. E. West and S.-M. Yu, *Adv. Mater.*, 2011, **23**, 3954–3958.
235. Y. R. Wang, J. M. Zheng, G. Y. Ren, P. H. Zhang and C. Xu, *Smart Mater. Struct.*, 2011, **20**, 045009.
236. S. Won Choi, J. Rae Kim, Y. Rack Ahn, S. Mu Jo and E. J. Cairns, *Chem. Mater.*, 2007, **19**, 104–115.
237. N. Chocat, G. Lestoquoy, Z. Wang, D. M. Rodgers, J. D. Joannopoulos and Y. Fink, *Adv. Mater.*, 2012, **24**, 5327–5332.
238. R. S. Yang, Y. Qin, L. Dai and Z. L. Wang, *Nat. Nanotechnol.*, 2009, **4**, 34–39.
239. Z. L. Wang, *Adv. Funct. Mater.*, 2008, **18**, 3553–3567.
240. Z. L. Wang and J. H. Song, *Science*, 2006, **312**, 242–245.
241. X. Wang, J. Song, J. Liu and Z. L. Wang, *Science*, 2007, **316**, 102–105.
242. Y. Qin, X. Wang and Z. L. Wang, *Nature*, 2008, **451**, 809–813.
243. S. Xu, Y. Qin, C. Xu, Y. Wei, R. Yang and Z. L. Wang, *Nat. Nanotechnol.*, 2010, **5**, 366–373.
244. K. Asadi, D. M. de Leeuw, D. de Boer and P. W. M. Blom, *Nat. Mater*, 2008, **7**, 547–550.
245. K. Asadi, P. W. M. Blom and D. M. de Leeuw, *Adv. Mater*, 2011, **23**, 865–868.
246. N. Weber, Y.-S. Lee, S. Shanmugasundaram, M. Jaffe and T. L. Arinzeh, *Acta Biomater.*, 2010, **6**, 3550–3556.
247. Y.-S. Lee, G. Collins and T. L. Arinzeh, *Acta Biomater.*, 2011, **7**, 3877–3886.
248. N. Dubey and M. Leclerc, *J. Polym. Sci., Part B: Polym. Phys.*, 2011, **49**, 467–475.
249. Y. Sun, Z. Wei, W. Xu and D. Zhu, *Synth. Met.*, 2010, **160**, 2371–2376.
250. D. K. Taggart, Y. Yang, S.-C. Kung, T. M. McIntire and R. M. Penner, *Nano Lett.*, 2011, **11**, 125–131.
251. S. Shen, A. Henry, J. Tong, R. Zheng and G. Chen, *Nat. Nanotechnol.*, 2010, **5**, 251–255.

CHAPTER 7

# Biological Functional Properties and Their Applications

## 7.1 Control of Cell Behavior by Nanostructures: Motivations and General Aspects

Among other applications, the largest amount of literature and technological solutions reported for polymer nanofibers is in the fields of biology and biotechnology. There are several motivations at the basis of this success. On one hand, some reasons reside in the general points of strength of polymer nanostructures, such as the availability of production methods offering good throughput (particularly electrospinning),[1] the low cost of fabrication and especially the chemical (and biochemical) and structural flexibility provided by the astonishingly wide variety of natural and synthetic, organic and nanocomposite nanofibers. On the other hand, some advantages of polymer nanofibers are peculiar to biotechnological applications handling living cells such as tissue engineering (Figure 7.1).[1–3] In particular, the resemblance of polymer nanofiber architectures with the fibrous protein components of ECM environments[4] of living organisms makes these nanostructures especially promising for regenerative medicine. Presenting tissue engineering, R. Langer and J. P. Vacanti displayed what was then still an emerging and revolutionary scientific concept by sketching the typical biodegradable polymer scaffold as a schematic assembly of fibers.[2] An example of how much hierarchically organized, protein-based and nanocomposite elongated structures are important in the human body is given by the bone. Here, the fundamental building block, namely the collagen macromolecule, assembles

RSC Nanoscience & Nanotechnology No. 29
Polymer Nanofibers: Building Blocks for Nanotechnology
By Dario Pisignano

Published by the Royal Society of Chemistry, www.rsc.org

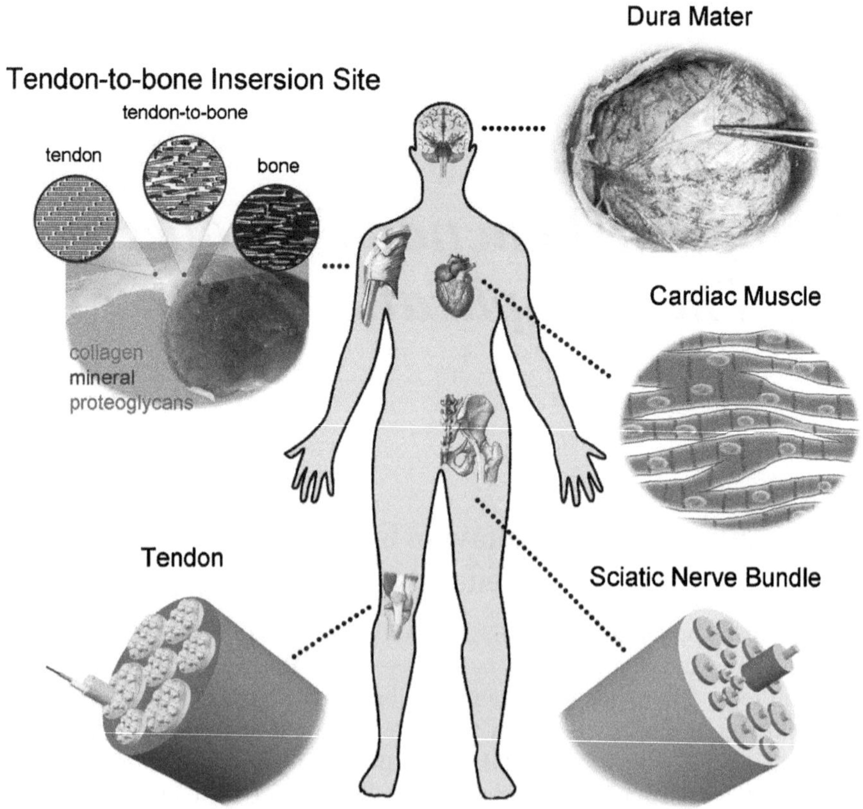


**Figure 7.1** Schematics showing typical examples of polymer nanofiber-assisted tissue engineering targets in the human body. The highlighted tissues and organs would particularly benefit from approaches able to produce spatially or compositionally anisotropic nanofiber scaffolds. Reproduced with permission from Ref. 3, *Adv. Healthcare Mater.*, 2012, **1**, 10–25. Doi: 10.1002/adhm.201100021. Copyright © 2012, Wiley-VCH Verlag GmbH & Co. KGaA.

in fibrils, which are intercalated with hydroxyapatite [$Ca_{10}(PO_4)_6(OH)_2$] crystals, and these fibers in turn organize in osteons, which ultimately form the bone.

More generally, the ECM surrounding living cells in tissues and organs comprises many fibrils, which have a diameter in the range from a few tens of nm (say, 50 nm) to a few hundreds of nm (say, 500 nm), and which support the cells both physically and biochemically.[4,5] These fibrils are made of several proteins such as collagen, glycoproteins including laminin, fibronectin, elastin, in a complex hydrated microenvironment also populated by soluble macromolecules such as growth factors.[6] Providing cells with structural support is only one of the many functions of the ECM. The physical and chemical signals generated by such an environment and by neighbor cells are captured by surface receptors and affect the cellular response and behavior through a variety of intra-cellular signaling pathways. Gene expression and regulation is also influenced in this way, as well as the ultimate differentiation, migration, apoptosis or replication of cells. On the other hand, cells continuously build and re-shape the ECM, thus playing a direct role in the protein fibrillogenesis,[7] which can be induced by cytoskeletally mediated mechanical stresses and movement of relevant linking domains on the cell surface.[8–10]

Laminin is a large ($MW$ = 900 kDa), cruciform shaped, heterotrimeric glycoprotein, exhibiting a peculiar molecular structure with four arms of different lengths in the range of roughly 30–80 nm.[11,12] The arms contain central and terminal globular domains alternated by rod-like regions, a molecular structure suited for bridging the protein with both cell receptors and other ECM proteins. Laminin self-assembles in networks of polygonal structures, with a supramolecular polymerization process involving the terminal globular domains.[13,14] The process is observed also *in vitro*,[13] and it can be mediated by biocompatible lithographic procedures.[15]

Another well-known and relevant example of supramolecular fibrils in the ECM is given by the fibrillogenesis of fibronectin. Fibronectin[7,16] is a flexible glycoprotein ($MW >$ 450 kDa), produced by several kinds of cells, such as hepatocytes and fibroblasts. It plays a role in crucial biological phenomena, including early development,[17] carcinoma phenomenology,[18] osteogenesis,[19] *etc.* Fibronectin is produced as a disulfide-bond dimer, *i.e.* with two, nearly identical sub-units, and each of these units comprises three types of domains ($FN_I$, $FN_{II}$ and $FN_{III}$). Analogously to laminin, these domains serve as binding sites to cell transmembrane receptors and to other ECM proteins, and as linkers at the base of self-assembly processes undergone by the same fibronectin molecules. Among other involved cell receptors, we mention for instance the $\alpha_5\beta_1$ integrins that target the arginine-glycine-aspartic acid tripeptide (RGD) sequence in the $FN_{III}$ domain. We will describe in the following a few methods that exploit RGD as tool to enhance the cellular growth, migration and differentiation on nanofiber scaffolds (Section 7.7). While in blood fibronectin takes a compact conformation, in the ECM this protein organizes in well-defined fibrils,[8,9,16] and fibrillogenesis through cell-mediated unfolding and translocation of the $\alpha_5\beta_1$ integrins (schematized in Figure 7.2) is proved to be involved in other important processes, such as neovessel formation and tissue neovascularization.[20] Even *in vitro*, the fibrillogenesis of fibronectin can be promoted by surface–protein interactions, and suitably functionalized surfaces can be used to control the resulting protein conformation, and eventually to induce the protein to expose its RGD domains, which are normally cryptic in solution, thus making the active sites properly accessible for the adhesion of cultured cells.[21,22]

We have already mentioned above collagen, which is a biopolymer and another well-known protein of the ECM, where it is naturally present in form

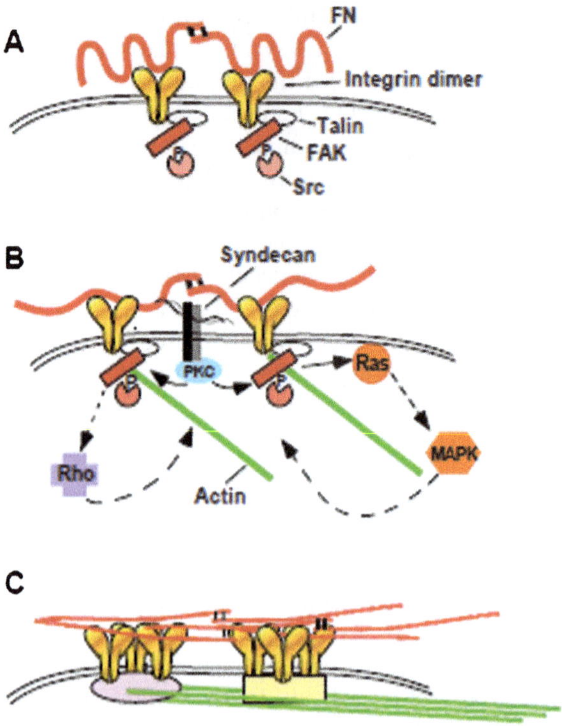


**Figure 7.2** Natural formation of protein fibers by cell-mediated processes: Schematics of a cell-mediated fibrillogenesis model for fibronectin. Top panel: Binding of compactly folded protein to integrins induces receptor clustering and co-localization of talin (white ovals) and focal adhesion kinase (abbreviated as FAK in the figure and indicated by red rectangles). FAK autophosphorylation (P) recruits Src (pink circles). Middle panel: Clustered integrins with co-localized syndecan (gray and black bars) organize the actin cytoskeleton (green lines) and activate a variety of signaling molecules, such as Ras/MAP kinase (orange), Rho GTPase (violet) and protein kinase C (PKC) (blue). The resulting signals reinforce organization of actin and focal complexes, whereas contractile forces aid in converting inactive fibronectin into the extended form. Bottom panel: Concentration of active fibronectin dimers at integrin clusters promotes the interactions between neighbor protein molecules and ultimately fibril assembly. Indeed, the formation of fibrils is favored by the movement of $\alpha_5\beta_1$ integrins and of other proteins along stress fibers towards the cell center, which may redistribute intra-cellular components into paxillin-rich focal adhesions (pink oval) and tensin-rich fibrillar adhesions (yellow rectangle). Reproduced with permission from Ref. 9, *J. Cell Sci.*, 2003, **116**, 3269–3276. DOI: 10.1242/jcs.00670. 

of fibers. It is an important component especially in connective tissues, bones, skin, tendons and ligaments, *etc.* As for other ECM proteins, it is well established that the fibrillar architecture formed by collagen molecules plays an important role for many cellular functions.[4,23–25]

### 7.1.1 Nanostructures–cell Interactions, Advantages and Issues with Polymer Nanofibers

Mimicking some of the topological, structural and biochemical functionalities of ECM fibrils, many classes of nanostructures, and particularly nanostructures elongated along a preferential direction such as polymer nanofibers, are able to induce specific behaviors in living cells. Biomaterials with nanostructured surfaces interfacing with cells can regulate the phenotype, and control the spatial distribution of focal adhesions. For instance, these mechanisms have been studied by means of periodic nanotopographies, realized by top-down, highly accurate nanofabrication methods such as EBL and NIL (Section 1.2). These nanostructures allow the orientation and the maturation of focal adhesions of neurons to be finely modulated, and the overall cell polarity through the direction of neurite outgrowth.[26] The underlying mechanisms, based on the surface nanotopography–cell interactions, have effects with surfaces made of nanofibers as well. As mentioned above, in nanofiber-based materials (for instance in electrospun non-woven mats) the diameter and entanglement assembly of fibers often resemble the natural, fibrous supramolecular architectures present in the ECM. Other advantages come from the possibility of tailoring the biochemical cues by which the nanostructures interact with cells.

Of course, for the utilization of nanofibers in biomedical applications there are also some issues. In general, the intrinsic porosity of assemblies of nanofibers (Section 4.2.1.1) is significantly lower than in other microstructured systems used as scaffolds for tissue engineering, and much smaller than the diameter of a cell.[6] Therefore one may need suitable methods to realize interconnected channels or controlled voids promoting cellular positioning and liquid and gas flows into and out of meshes.[27] Fortunately, there are many techniques allowing the porosity of mats of nanofibers to be varied in a somehow controlled way (Section 4.2.1). For example, the capability of engineering the final scaffold porosity by a careful choice of polymers, solvents and fabrication parameters is often mentioned as an advantage of electrospinning compared to other nanotechnologies possibly applicable to tissue reconstruction and regenerative medicine.[28] Electrospinning on coagulation baths and freeze-drying are effective in this respect.[28] In addition, the infiltration of cells may be facilitated by salt leaching methods, which allow scaffolds to be produced where intentional and engineered delaminations occur.[29] After three weeks of culture, one observes a significant infiltration of cells, with a resulting cellular coverage up to 70% in the delaminations.[29] To increase the final porosity of aligned fiber scaffolds and consequently favor cellular infiltration, it is also possible to co-electrospin two polymer species of different solubility in a given solvent.[30] This allows one to use one of the two components as sacrificial layer, to be then removed to reduce the resulting spatial density of the fibers.[30] Another possibility is given by pre-defined systems (with feature size in the μm-range and voids up to hundreds of μm), onto which dense nanofiber assemblies can be deposited. This allows the final resulting porosity to be carefully controlled, with excellent flexibility concerning the biochemical functionality of the integrated micro- and nanostructures (which can be made by different polymers or by a single material). In this framework, three-dimensional microfiber deposition and electrospinning have been combined to realize scaffolds, made by block-copolymers of polyethyleneoxide-terephtalate and polybutylene-terephtelate, which have been seeded by primary bovine articular chondrocytes.[31] After about one month, a rounded morphology of chondrocytes, together with an enhanced infiltration of cells and a higher ECM production are observable in these systems compared to conventional scaffolds realized by three-dimensional fiber deposition.[31] Cells can even be seeded in a web of nanofibers by *in situ* approaches, allowing a surrounding scaffold architecture to be built, or included in electrospun solutions (Section 7.7.5). In these cases, the possible cellular damage due to the fabrication processes has to be assessed and minimized whenever possible. It is worthwhile to mention that, independently of the specific engineering approach, controlling the matrix porosity, especially if concomitantly to the local chemical composition and biochemical function of the scaffold, can be finally aimed to achieve addressable *gradients* (both of porosity and of biochemical cues). This would result in obtaining so-called *functionally graded* scaffolds mimicking much more realistically ECM structures and complex natural tissues.[32–35]

Another issue is related to the toxicity of solvents used in the various steps of the nanofiber fabrication process. This can pose serious problems after seeding living cells. To tackle this drawback, continuous efforts are being made in order to improve the processability of polymers allowing nanofibers to be produced by fully aqueous solutions, and to find alternative, biologically and environmentally benign solvents to replace toxic solutions during nanofabrication. As an alternative, solvent-free approaches exist that allow nanofibers to be fabricated in a safe, eco-friendly way. For instance, melt electrospinning (Section 2.3) is performable directly onto *in vitro* cultured fibroblasts, to realize layered bio-constructs where cells gradually adhere to the on-top spun nanofibers.[36]

Finally, some post-fabrication treatments of the deposited meshes can be hardly compatible with living cells and biologically functional compounds, especially when these processes involve (chemical, thermal, UV, *etc.*) cross-linking to enhance the mechanical strength of the scaffold and its stability in aqueous environments. These procedures, besides complicating the overall fabrication process, may have in principle a detrimental effect on the level of biocompatibility and on the functionality of embedded molecular dopants, which should be carefully taken into account and assessed during experiments. Residual cross-linkers, not fully reacted upon curing, may increase toxicity towards living cells and organisms as well.

In general, there are at least two complementary approaches that one can easily figure out in order to produce biologically active nanofibers. On one side, one can use some biological molecules (hyaluronic acid, collagens, and many others) that can be processed directly. On the other side, one can use synthetic, biocompatible, possibly biodegradable polymers, which in most cases will need additional steps of surface functionalization after fiber production. Such methods provide the desired biological functionality and the capacity properly to support the growth of cells. These two approaches will be described in the next two sections.

## 7.2 Nanofibers by Biomolecules

Given the wide variety of viscoelastic behaviors or phase-transition pathways exhibited by biological molecules both in living organisms and under *in vitro* conditions, it is not surprising that such extraordinarily rich phenomenology can be exploited directly to shape many of these compounds in the form of supramolecular nanofibers, alone or in blends with other polymers. The advantage of having biological molecules directly processable by electrospinning or other techniques is clear, since many of the so-obtained biomaterials possess an intrinsic biofunctionality, and the resulting nanostructured surfaces can then be used as scaffolds without additional surface functionalization treatments. The criticalities of this approach consist in (i) the very careful optimization of the processing conditions and solution concentrations, which is needed given the often poor characteristics of biomolecules in terms of plastic behavior and molecular entanglements, (ii) the very frequent need for additional cross-linking to make the produced materials insoluble in water and (iii) the possible denaturation of so-processed biomolecules. Other issues for implementing large-scale production are related to the high cost of most biomolecular materials, and to the reproducibility of the developed fabrication recipes due to the intrinsic variability among different batches of biomolecules (which is of course much higher than in synthetic compounds). Finally, a specific criticality for implant applications is related to the possible immunogenicity and transmission of pathogens from materials of direct biological origin. However, many different biological molecules have been studied so far as materials to realize nanofibers, particularly by electrospinning, and the extent to which they are susceptible to each of these issues varies a great deal from one compound to another.

Hyaluronic acid is a linear polysaccharide that is naturally present in the ECM of mammalian connective tissues. It is composed of alternating disaccharide units of α-1,4-D-glucuronic acid and β-1,3-*N*-acetyl-D-glucosamine, and it shows a very peculiar rheological behavior including remarkable viscoelasticity and shear-dependence, which have been related to a lubricating function in the body,[37] in addition to high biocompatibility. It is also noticed[38] that the weak acid in the molecule making this polymer a polyelectrolyte leads to a high degree of inter-molecular interaction, and to the high viscosity shown by hyaluronic acid solutions even at relatively low concentrations. For all these reasons, hyaluronic acid and its derivatives have been extensively applied as biomedical materials.[39–42] Furthermore, this compound can be electrospun to produce nanofibers, although due to the high viscosity it requires carefully optimized solution parameters and set-up.[38,43,44] This has led to blending of thiolated derivatives with PEO or even to specifically modified versions of classical electrospinning, involving reactive depositions[44] or high-temperature electro-blowing[38] (Section 2.3). In particular, reactive electrospinning can be applied to mix hyaluronic acid derivatives with different molecular weight and with functional terminations inducing supramolecular cross-linking, which is demonstrated to improve remarkably the material processability.[44] Upon cross-linking by poly(ethylene glycol)-diacrylate (PEGDA) mediated conjugate addition and consequent functionalization by adsorption of fibronectin, the group of M. H. Rafailovich has studied the attachment and spreading of NIH 3T3 fibroblasts, observing the formation of a dendritic morphology.[43,44]

Another well-known polysaccharide is chitin, a structural biopolymer. Chitin is present in mollusks, crustaceans, insects and fungi, and more generally it is the second most abundant organic material in nature next to cellulose. Its *N*-deacetylated derivative, chitosan, is quite interesting in the framework of tissue engineering, since it is biocompatible and biodegradable, besides showing antibacterial activity. Chitosan in various grades is cross-linkable by glutaraldehyde, both as vapors[45] and as supplement directly added in the electrospun solution.[46] In general, electrospinning pure chitosan nanofibers presents some intrinsic difficulties because of the relatively poor solubility of this molecule, and because of the presence of hydrogen bonds leading to a three-dimensional, networked structure in the solution. Some experiments have been focused on electrospinning chitosan at carefully calibrated concentrations with PEO, eventually using suitable surfactant such as Triton X-100™.[47] However, pure chitosan too can be eventually electrospun, by using sufficiently concentrated solutions.[48–52] For instance, an approach consists in using chitosan of sufficiently high *MW* (at least 106,000 Da), and a concentration of at least 7% in aqueous 90% acetic acid solutions.[49] In particular, the relative concentration of acid acetic in water is found to be critical to produce chitosan nanofibers. While pure acetic acid is hardly usable to electrospin chitosan directly, a quite high relative concentration (≥30%) of acid acetic in water is found to favor the electrospinning process, reducing the overall solution surface tension and increasing the charge density carried by the electrified jet (Section 2.1.2) without concomitantly varying significantly the solution viscosity.[49] Of course, acid solutions can in principle reduce the overall process biocompatibility, which motivates research to produce water-soluble chitosan derivatives. To this aim, carboxymethyl chitosan with varying molecular weights (40-405 kDa) and degrees of substitution have been synthesized by alkalization of chitosan and subsequent carboxymethylation with monochloroacetic acid.[53] Such a procedure leads to water-soluble chitosan species, which can be electrospun in fully aqueous solutions though by the addition of highly processable polymers such as PEO, PVA, *etc.* Furthermore, the produced fibers can be cross-linked, thus becoming water insoluble, by heat-induced esterification (at 140 °C for 30 min), which completes the fabrication process by an entirely solvent-free route.[53] Another chitosan derivative, obtained by grafting L-lactide oligomers *via* ring opening polymerization, is instead electrospinnable by ethyl acetate and by 2-butanone.[54] An overview of other experiments focused on the realization of chitosan nanofibers can be found in Ref. 55. Importantly, chitosan nanofibers exhibit good biocompatibility, supporting the growth of human osteoblasts and chondrocytes.[47]

Fibrinogen is a soluble, globular glycoprotein, which is converted into fibrin by thrombin, thus playing an important role in blood coagulation and wound healing. Fibrinogen is a hexamer defined by six chains organized in two identical halves, and a material of interest for tissue engineering. The production of electrospun fibrinogen fibers with diameter down to 80 nm is reported by mixing 1,1,1,3,3,3-hexafluoro-2-propanol (HFIP, which alone is not able properly to dissolve the molecule) and minimal essential medium (Earle's salts).[56]

Collagen too allows electrospun fibers to be directly fabricated.[57,58] To this aim, type I collagen is commonly dissolved in HFIP.[27,57–59] Very similar processes are demonstrated for gelatin, which is a denaturated derivative of collagen, by using HFIP or 2,2,2-trifluoroethanol.[57,60] Formic acid can also be employed.[61] Layer-by-layer and two-stream electrospinning deposition (Section 2.3) have been investigated in depth for producing multilayered or mixed meshes of nanofibers of collagen with other fibers made of styrenate-gelatin (spun from HFIP solutions as well) and polyurethanes.[27] Some evidence, obtained by a variety of methods including circular dichroism, second-harmonic microscopy and differential scanning calorimetry, however, reports that electrospinning collagen by HFIP and other fluoro-alcohols can result in a heavy denaturation of the protein towards gelatin.[62] This finding, together with the toxicity of HFIP, motivates looking for alternative, environmentally friendly solvents to electrospin collagen. Some good results have been obtained by an ethanol/PBS solution.[63] Gelatin too can be electrospun with eco-friendly solvents, such as a mixture of ethyl acetate and acetic acid in water which allows fibers to be produced with diameter down to 47 nm depending on the gelatin concentration.[64] Phase separation methods (Section 3.6) can be valuable alternatives for producing scaffolds based on gelatin.[65,66] According to reported recipes, gelatin is dissolved in ethanol/water or methanol/water, then cooled at –76 °C for phase separation, and the resulting gel is immersed into cold ethanol and finally into 1,4-dioxane for solvent exchange.[65] The removal from the solvent, freezing and drying complete the fabrication of the scaffold. Independently from the specific fabrication approach, following the realization of nanofibers the cross-linking of collagen and of its derivatives can be carried out through glutaraldehyde, 1-ethyl-3-(3-dimethylaminopropyl) carbodiimide hydrochloride (EDC) and several other methods.[57–60,63,67] Elastin, another protein that is an important constituent of connective tissue where it exhibits native properties of compliance and elasticity, is also processable by electrospinning and then cross-linkable through analogous procedures.[57,68,69]

Fibroin is one of the proteins constituting silk, which is produced by several species of insects including silkworms (*Bombyx mori*) and spiders (Figure 1.3). A strategic advantage of silk for application in biotechnology, besides its well-known mechanical properties, is the intrinsic thermal stability (up to >200 °C), which makes it possible to use sterilization processes by autoclave. In particular, artificially produced nanofibers made of spider silks, resisting up to 280 °C under nitrogen, seem to be even more thermally stable than filaments made by silkworms.[70] Silk fibroin nanofibers can be realized by electrospinning from formic acid solutions, made insoluble by a treatment with a methanol aqueous solution, and then used for culturing several types of cells such as keratinocytes and fibroblasts.[71] Electrospinning can be favored by blending fibroin with water-soluble polymers such as PEO, whose fraction can be eventually removed from the produced fibers at a later stage.[72–74] Following such cues, other cells have been successfully cultured on silk-based electrospun nanofibers, such as human endothelial cells and smooth muscle cells.[72] When doped with bone morphogenetic protein (BMP)-2 growth factors* or with hydroxyapatite nanoparticles, silk fibroin-based fibers are able strongly to support osteoinduction in marrow-derived human mesenchymal stem cells.[74] Fibers can be electrospun from silk HFIP solutions as well.[70,75] These solutions can be prepared in several ways starting from the biological silk material, dissolving spider silk in HFIP in a few minutes, dissolving degummed silk from *B. mori*, which may take up to months at room temperature[70] or starting from degummed silk yarns dissolved in a ternary solvent system (calcium chloride, ethanol and water).[75] With their cross-sectional dimension down to a few nm corresponding to $\sim 1$–$2 \times 10^2$ molecules,[70] silk electrospun nanofibers are among the thinner nanostructures fabricated with biological polymers. Recipes have also been reported for electrospinning stable regenerated silk fibroin aqueous solutions at high concentrations (28%–37% by weight), even without blending with water-soluble synthetic polymers.[76] Ref. 77 gives a complete overview on electrospun nanofibers made of silk proteins.

Recently, even pure fibers of low-*MW* ECM proteins such as laminin have been obtained by electrospinning.[78] In this process, laminin purified from Engelbreth–Holm–Swarm tumor is dissolved in HFIP, leading to a uniform fiber morphology without the presence of beads and with retained bioactivity. Interestingly, these mats are stable in water even without preliminary

* BMP-2 is a protein that has an important role in inducing the formation of bone and cartilage. Today it is largely used in biomedical sciences, and also at clinical level in orthopedics and dentistry.

cross-linking.[78] Finally, DNA can be directly electrospun from water solutions to form nanofibers having diameter down to few tens of nm.[79] Other examples, involving plasmid DNA embedded in polymer nanofiber scaffolds, are given in Section 7.7.6.

## 7.3 Biofunctionalization of Polymer Nanofibers

Biofunctionalization techniques are fruitfully applied to polymer nanofibers for making them more biocompatible and more suitable for cell adhesion and growth, or for providing them with other specific biological functionalities. The advantage of this approach is clearly the possibility of fully exploiting synthetic polymers for the production of nanofibers to be then biofunctionalized. Synthetic polymers generally have (i) much more favorable processability, *i.e.* much better stretching and plastic behavior while realizing nanofibers, and (ii) more suitable structural and mechanical properties compared to pristine biomolecules, which allows mechanically viable scaffolds to be fabricated (Figure 7.3).[80,81] In general, fibers made of synthetic polymers may be rendered bio-functional (or their bio-functionality may be enhanced)

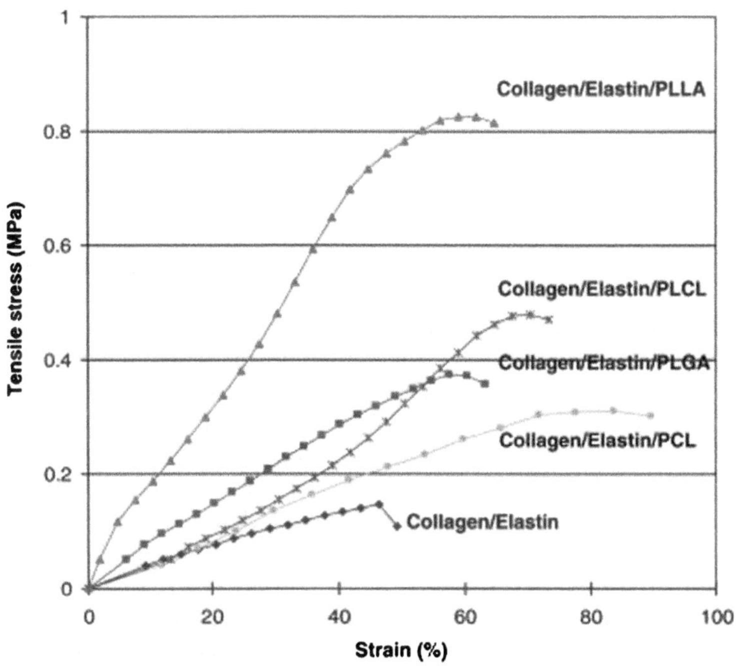


**Figure 7.3** Stress-strain curves from tensile testing experiments performed on electrospun nanofiber scaffolds made by blends of collagen/elastin and various synthetic polymers. Reproduced with permission from Ref. 80, *J. Biomed. Mater. Res. A*, 2007, **83**, 999–1008. Doi: 10.1002/jbm.a.31287. Copyright © 2007, Wiley Periodicals, Inc.

by means of *in situ* methods or of post-synthesis surface treatments based on physisorption or on covalent bonding. Of course, the most immediate choice for blending a biofunctional dopant within the fibers, or for surface functionalization, is that of ECM proteins. Again, the cases of collagen and gelatin (here as functionalizing surface modifiers) are likely the best studied. For these molecules, blend nanofibers are reported with a large number of synthetic polymers, and the cellular response on scaffolds made of such blends is generally found to be significantly better than on purely synthetic samples used as controls.[82–84] For instance, collagen is used for blending in nanocomposite nanofibers of PVA also embedding hydroxyapatite nanoparticles for bone tissue engineering,[85] or in nanofibers of poly(L-lactide-*co*-ε-caprolactone).[86] PCL/gelatin electrospun fibers with a relative concentration of 70:30 have good properties as scaffolds for the growth of neural stem cells (NSCs), promoting and directing the neurite outgrowth.[83] Poly(L-lactide acid)/gelatin nanofibers are produced by different blending ratios using HFIP as solvent.[87] Another molecule largely used for blending is elastin.[67,80]

Surface functionalization of polymer nanostructures by proteins and enzymes[88] is generally a complex task with sometimes serious issues of reproducibility and uniformity, due to the intrinsic difficulty of linking biomolecules on organic surfaces that are amorphous at nano- and microscales (Section 1.4.1). Readers interested in the fundamentals of molecular adsorption processes can refer to the excellent book by Myers.[89] In the framework of materials science and biotechnology, a widely used approach consists simply in physical absorption (*physi-sorption*). Here, the surface-molecule interaction is not based on covalent bonding but is instead mediated by relatively weak, mostly van der Waals inter-molecular forces. The method is especially easy from an operational viewpoint.[90,91] Physisorption leads to protein adhesion in a generally aspecific way, it can be reversible and hence it can be subject to subsequent protein detachment. On the other hand, this method can be useful for many applications, and it is often considered as a quite mild functionalization route that preserves the native configuration of pristine biomolecules.[92] For example, fibronectin has been used for surface functionalization with several classes of polymer fibers, by incubation in 0.1–1 mg $mL^{-1}$ PBS or serum-free Dulbecco's modified eagle's medium solutions over times of the order of one to a few hours.[44,93] Electrostatic interactions can be used too, such as in the formation of polysaccharide-protein multilayers, which allow polymer nanofibers to be functionalized through layer-by-layer self-assembly, following surface activation with poly(ethylene imine).[94]

*Chemi-sorption* is a complementary approach that somehow involves a chemical reaction and hence the formation of ionic or covalent bonds between the functionalizing molecule and linking groups at the surface. In this way, the surface coating is largely irreversible, and protein binding can be rendered highly specific by molecular design. Among less specific variants, a well-known method makes use of glutaraldehyde, which can be directly added in the protein solution where the target scaffold is incubated, thus adsorbing onto polymer nanofibers. The covalent cross-linking properties of glutaraldehyde[95] with amine-containing molecules may then allow further proteins present in solution to link with previously adsorbed biomolecules. Immobilization on nanofibers through glutaraldehyde cross-linking is largely applied to many ECM proteins including fibronectin[96] and collagen.[97] Other recipes are based on EDC, which is a zero-length linking molecule, leaving in principle no residue on newly grafted structures and binding proteins covalently without being detrimental to biomolecular functionality.[98,99] EDC mediates the formation of amide linkages between carboxylates and amines,[95,98–100] and can be utilized through several different experimental routes. The polymer scaffold can be simply incubated in a solution containing both EDC and proteins, or firstly treated with EDC to generate highly reactive O-acylisourea intermediates. The latter will in turn react with available primary amines on protein molecules during a second incubation step. EDC-based linking is used, for instance, for surface treatments of electrospun poly(lactic acid) nanofibers with cationized gelatin, which improves the material compatibility with chondrocytes.[101] Recombinant human epidermal growth factor can be immobilized on amine-terminated PCL/PEO block copolymers nanofibers as well.[102] It has to be mentioned that not only synthetic polymers, but also biomolecule nanofibers such as electrospun collagen or gelatin, can be enhanced or specialized in their biological functionality by such surface treatments. In particular, processing these fibers with perlecan domain I favors the subsequent binding of basic fibroblast growth factor, which in turn links to the heparan sulfate chains present on the fragment of perlecan.[103] Of course, many other absorption routes and procedures are possible in order to link proteins to the surface of polymers. An extensive list of possible protocols is given in Ref. 95. It should be pointed out that in general the largest amount of proteins adsorbed on nanofibers does not determine necessarily the best performance in terms of cells attachment and proliferation *in vitro*, as recently demonstrated for PMMA electrospun scaffolds seeded by Human Kidney 2 (HK-2) cells (Figure 7.4a).[97] Instead, it is likely that the biofunctional properties of protein-treated polymer nanofibers are mostly related to the specific type of linking used to attach the biomolecules onto the surface of the nanostructures (Figure 7.4b), and that they are also affected by the adsorption of serum biomolecules on the scaffold.

Let us see some final examples of bio-functionalization. Surface functionalization with laminin can be carried out through a preliminarily adsorbed coating of poly-L-lysine[104] or poly-L-ornithine,[105] according to very well-established recipes of cell culturing. This is typically performed by incubating nanofibers in poly-L-lysine or poly-L-ornithine solutions in PBS (for time intervals ranging from, say, tens of minutes to overnight), and vigorously washing samples and again incubating in laminin solutions. Analogously, laminin is also used to functionalize nanofibers of PLA, then used for culturing neuronal cells with improved neurite outgrowth performances, *via* intermediate heparin binding.[106] This process is carried out through multiple covalent

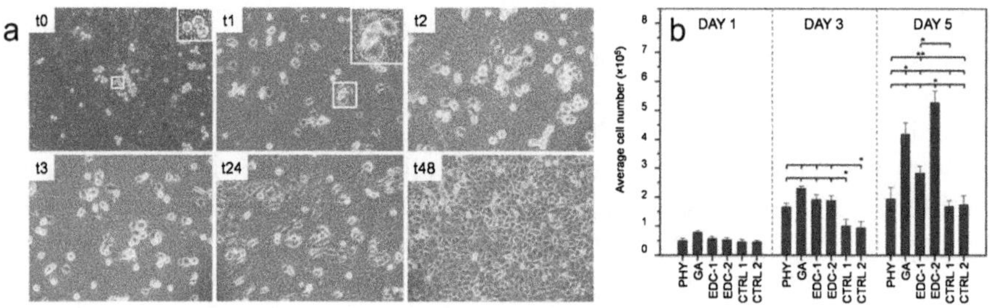


**Figure 7.4** (a) Optical micrographs of HK-2 cells growing on PMMA nanofibers at different culture times: just after seeding (t0) and after 1 h (t1), 2 h (t2), 3 h (t3), 24 h (t24), 48 h (t48). (b) Number of HK-2 cells proliferated on the collagen-functionalized scaffold. Collagen is attached onto the scaffolds by various processes, *i.e.* physisorption (PHY), glutaraldehyde cross-linking (GA) and two different EDC-based methods carried out by one single step (EDC-1), or two sequential steps of surface linking (EDC-2). Data on as-produced (CTRL 1) and oxygen plasma-treated (CTRL 2) nanofibers are also shown for comparison. The initial seeding is of $4 \times 10^4$ cells $mL^{-1}$. The vertical bars show standard deviations. *: $p < 0.01$; **: $p < 0.05$. Adapted with permission from Ref. 97, *Soft Matter*, 2010, **6**, 1668–1674. Doi: 10.1039/b921932c.

bonding steps, in which NaOH-treated PLA is firstly coated by di-amino-poly(ethylene glycol) *via* EDC-based methods. Then heparin is covalently linked to di-amino-poly(ethylene glycol), and so-treated nanofiber samples are finally incubated in laminin solutions in PBS.[106] A substantially identical recipe can be applied for functionalizing nanofibers by basic fibroblast growth factor.[106] Surface functionalization by collagen can be achieved *via* plasma treatment of poly(L-lactic acid)-*co*-poly(ε-caprolactone) nanofibers.[107]

In addition to the above-mentioned physical or chemical absorption methods, the fabrication routes available for polymer nanostructures leave a lot of other space for building protein surface coatings on nanofibers, wrapping a polymer core already during fiber drawing. Coaxial electrospinning (Section 2.3.1) is the most used and maybe more intuitive technique in this respect. This allows one to straightforwardly realize, in a single run, biodegradable fibrous scaffolds based on fibers with a core made of a synthetic polymer and an external shell of other electrospinnable materials, possibly of biological origin such as collagen[108] or gelatin.[109]

## 7.4 Drug Delivery

Three-dimensional assemblies of nanofibers can be effectively used as biomaterial medium to deliver functional molecules to a surrounding environment where living cells are present, both *in vitro* and *in vivo*. In general, the molecules of interest can be directly embedded as dopants within the fabricated nanofibers (for instance by directly adding them in the solution for electrospinning), reversibly linked (for instance by physisorption) to the surface of polymer nanostructures or incorporated as temporary cores of core-shell fiber architectures. Hollow nanostructures are also particularly suited as drug-delivery systems, and have attracted a lot of interest in the polymer and materials science communities in the last years.[110]

Ideally, the drug delivery should be *intelligent*, namely it should target with high accuracy a desired organ, tissue or point in the body, and the release of the drug should be triggered by a precise chemical, biochemical, thermic, electrical, magnetic or optical stimulus, thus being fully adjustable depending on the body's needs. For instance, the release of bioactive molecules can be triggered by using electrical stimulation of drug-loaded, hollow nanotubes made of conductive polymers.[111] In all these cases, the possible damage to the molecule biofunctionality, due to solution processing or surface immobilization, should be carefully estimated, hopefully ruled out[112] and ultimately prevented whenever possible by properly engineering the process. Released molecules can include heparin to be used as anticoagulant,[86,113] other proteins,[114–118] antibiotics such as cefoxitin sodium,[119–121] tetracycline hydrochloride,[122,123] rifampin,[124] ornidazole[125] and gentamycin sulfate,[126] triethyl benzyl ammonium chloride,[124] sodium dodecyl sulfate,[124] itraconazole,[127] ketanserin,[127] dexamethasone,[111,128] paracetamol,[129] anti-cancer agents such as paclitaxel,[124,130] antioxidant molecules such as resveratrol,[126] plasmid DNA,[131,132] small-interfering RNA and transfection reagent complexes[133] or growth factors.[134] The sustained release of human β-nerve growth factor stabilized in bovine serum albumin and embedded in electrospun fibers of a copolymer of ε-caprolactone and ethyl ethylene phosphate occurs over three months.[135] Similarly, the release of recombinant human BMP-2 embedded within an electrospun poly(D,L-lactide-*co*-glycolide)/hydroxyapatite system occurs over periods of up to two months.[112] Interesting perspectives also come from process engineering, and from implementing novel colloidal chemistry with conventional methods for nanofiber production. In particular, electrospinning can be smartly developed in order to dope the inner part of core-shell nanofibers with multiple colloidal species, such as poly(*N*-isopropylacrylamide) hydrogel or PMMA particles, in turn loaded with different active agents to be ultimately released.[136] The colloids can be designed with different polymer species to enable an independent control of the release of the single drugs, which opens the way for the realization of nanostructured devices for the intelligent, programmable delivery of multiple agents.[136] Two-phase, emulsion electrospinning (Section 2.3) and particle-loading are other approaches allowing drug-delivery systems encapsulating chemical agents within separated liquid or solid domains embedded in the polymer fibers to be produced.[116,117,137]

In general, the first goal of such systems is of attenuating the initial burst, which is typically observed at the beginning of the release kinetics, possibly because of molecules at or near the surface of delivering fibers. Reducing burst release, the system would support a *sustained* delivery of agents into the surrounding medium. By properly optimized scaffolds, chemicals can be released over time intervals as long as several months. Encapsulating strategies and core-shell nanofibers may help in this respect.[138,139] The second important objective is to achieve a *full* release of the encapsulated drug at sufficiently long times, thus making 100% of the dopant molecules, initially loaded in the scaffold, available and functional in solution at the end. This is not easy to obtain in general. Steric effects are always possible in polymer nanostructures and may determine undesired, unspecific interactions of the loaded drug with the scaffold matrix molecules, thus preventing a complete delivery of the dopants. For all these reasons the release kinetics deserves to be investigated in depth, aiming to suppress burst release and ultimately reach a condition as close as possible to full drug delivery. In most cases, the release temporal profile has to be clearly described by diffusive laws, whose rationalization has both a technology-related and a fundamental interest given the complex geometries of the releasing fibers assembly and the multiphysics involved. Some attempts have been made, for example, for modeling the release behavior by expressions of the form:[135]

$$m(t)/m_{\infty} = kt^{n_D} \, , \tag{7.1}$$

where $m$ indicates the amount released at time $t$, $m_{\infty}$ is the mass released at very long times, $k$ is a polymer- and drug-dependent constant and $n_D$ is a diffusional exponent.[140] A Fickian kinetics would result in a roughly square-root dependence of the released mass on time, as is sometimes observed in experiments.[113,127] In particular, a one-dimensional Fickian diffusion of drugs from a monodispersion of non-swelling cylindrical objects would lead to $n$ = 0.45,[140] whereas $n$ = 0.34 is found for the release of human β-nerve growth factor from poly(ethylene phosphate)-PCL copolymer fibers, and $n$ = 0.31 is found for the release of paclitaxel from PLAGA nanofibers.[130] Some mechanisms have been proposed for explaining the deviation from a Fickian diffusion, which involve the presence of a distribution of fiber sizes in the releasing mesh, the possible formation of protein aggregates and the existence of polymer–drug interactions in the scaffold, which as mentioned above can provide additional barriers slowing down the release kinetics.[130,135] In other systems, the degradation of polymer nanofibers drives the release instead of agent diffusion, which may lead to a nearly zero-order kinetics.[124] Also in this case, however, the morphological features of the scaffold, particularly the fiber diameter, are crucial to determine the resulting release kinetics.[129]

Another detailed and recent description of the diffusion processes for molecules released from polymer nanofibers is based on the desorption-limited theory of admixture release.[118,141] In this case, the release is mainly driven by desorption, or dissolution, from nanopores on the surface of the polymer nanostructures. To describe this mechanism, one firstly defines a so-called *nanoporosity factor*, $N_P$, as the ratio of the initial amount of drug admixture on the nanofiber surface, $m_{surf}(0)$, and the total initial amount of drug loaded in the nanofibers, $m_T(0)$ [*i.e.* the drug molecules on the surface plus those in the bulk, $m_{bulk}(0)$]:

$$N_P = m_{surf}(0)/m_T(0) = m_{surf}(0)/\left[m_{surf}(0) + m_{bulk}(0)\right]. \tag{7.2}$$

$N_P$ therefore weights the amount of initial molecules carried at the surface and is directly related to the surface to volume ratio of the scaffold. This analysis provides the following expression for the temporal behavior of the released amount from a non-swelling sample:[118]

$$m(t) = N_P m_\infty \left[1 - \exp\left(-\frac{\pi^2 t}{8\tau_R}\right)\right], \tag{7.3}$$

where $\tau_R$ is a characteristic time that depends on the dynamics of drug molecules while diffusing in the pores at the surface of nanofibers. Longer pores, or less effective diffusional dynamics in the pores (for instance because of lower temperature, higher desorption activation energy or smaller radius of the pores), would determine of course a slower release, because the dopants would be able to reach free solution space more difficultly. This can be summarized by the following equation:

$$\tau_R = l/D_{eff}, \tag{7.4}$$

where $l$ is the pore length and $D_{eff}$ is the effective diffusion coefficient for the dopant in the pores. $\tau_R$ is in the range of 10–20 h for prototype proteins such as bovine serum albumin delivered by PCL-based nanofibers, whereas $D_{eff}$ is reported to be of the order of magnitude of $10^{-13}$ cm$^2$ s$^{-1}$.[118] In addition, as is clear from Equation (7.3), the nanoporosity factor is the parameter that mostly affects the amount of drug delivered at the end, at least for non-swelling polymer systems (directly through the pre-exponential term $N_P m_\infty$), whereas scaffolds degrading over time also release dopants initially embedded in the bulk of nanofibers, which should be described by a different release mechanism. As physically expectable, enhancing nanoporosity, for instance by the inclusion of porogen systems (possibly other proteins)[118] in polymer nanofibers, turns out to be an effective way to optimize the amount of delivered drug.

Finally, other drug-delivery systems not addressing nanomedicine but instead agricultural applications deserve to be mentioned. Here, the idea is exploiting polymer nanofibers as carriers to diffuse drugs or molecules in the air, for plant protection. The Greiner group has performed an extensive work in this regard, focusing on the diffusion of pheromones inducing sex confusion of parasitic insects.[142,143]

### 7.4.1 Cancer Treatment

Anti-tumoral strategies based on polymer micro- and nanoparticles, including nanofibers, are gathering increasing interest for a number of reasons. In fact, these nanoparticles and filaments have some properties, as described throughout this book, that make them very promising candidates for intelligent drug delivery. They can be accumulated in the target tissue or organ, eventually by an external control such as in the case of nanofibers loaded by magnetic nanoparticles. Doping by metallic nanoparticles can be even strategic for locally destroying cells by hyperthermia, namely by strongly localized heating. Swelling polymers can be used to deliver drugs in a controlled way by means of internal or external stimuli.

Several anti-cancer drugs can be incorporated in polymer nanofibers for subsequent sustained release into the surrounding microenvironment. To date, paclitaxel is the agent most widely investigated in this framework.[124,130] Doxorubicin has also been studied as a model for encapsulation within biodegradable nanofibers of poly(ethylene glycol)-PLA copolymers.[144]

### 7.4.2 Anti-bacterial Molecules and Other Applications

Antibiotic compounds that have been embedded in polymer nanofibers have been listed in Section 7.4. Among other molecules, tetracycline hydrochloride[122,123] has been used in nanofibers made of poly(lactic acid), poly(ethylene-*co*-vinyl acetate), poly(ester urethane) urea and poly(lactide-*co*-glycolide), and its release kinetics and the resulting antibacterial properties of nanofibers have been studied both *in vitro* and *in vivo*. The antibiotic effects of cefoxitin sodium released from electrospun scaffolds have been tested with good success in the inhibition of the growth of *Staphylococcus aureus* (>90%),[121] and in preventing contaminations and post-surgery-induced abdominal adhesions in a rat model.[119,120] Similar promising effects have been observed with ornidazole released from nanofibrous membranes of PCL.[125]

The antibacterial properties of silver have been known since antiquity. These properties strongly motivate the production of various Ag-based or Ag-embedding nanostructures, which is possible through many different routes due to the huge development of nanoparticle research. In particular, embedding $AgNO_3$ in polymer nanofibers[145,146] allows the release kinetics of $Ag^+$ ions to be studied, which is found to depend on thermal treatments or UV irradiation undergone by samples.[146] These fibers constitute a valuable material disfavoring the proliferation of bacteria on wounds.[147]

## 7.5 Wound Dressing

Sheets and membranes made of nanofibers have been used in experiments for accelerating wound healing. Wound healing is a complex process involving a variety of cells (neutrophils, epithelial cells, fibroblasts, *etc.*) and a sequence of consecutive stages, which include hemostasis and inflammation leading to the formation of a fibrin network, migration and proliferation of epithelial cells in the damaged region and maturation of the newly formed tissue.[148] Mats made by polymer nanostructures have been applied on injured skin to promote a more rapid evolution of this cascade, acting as antibacterial barrier at microscale, or for locally delivering healing agents, antibiotics *etc.* Polymer nanofibers can be very useful in this respect, because their assemblies, mats, membranes, *etc.* exhibit a suitable porosity protecting wounds against dust or other contaminating agents, but at the same time they show good water permeability thus being "breathable" as described in Sections 4.3.1 and 4.3.3. In addition, once colonized by living cells these mats can serve as actual implantable scaffolds working like skin substitutes, a principle that raises wound healing applications to tissue engineering (Section 7.7).

For instance, collagen nanofibrous mats positioned on wounds in experiments on rats are found to promote the healing at its early stages, with a fast formation of capillaries and growth of fibroblasts observed in the first week.[58] PHBV nanofibers co-cultured with human hair follicular dermal sheath cells and epithelial outer root sheath cells favor wound closure and re-epithelization in a mouse model.[149] Also silk fibroin nanofibers[71] functionalized by collagen or other ECM proteins have been studied in wound healing experiments, and found to favor the adhesion and spreading of human oral keratinocytes.[150] Recombinant human epidermal growth factor linked on the surface of PEO/PCL copolymer electrospun nanofibers has been investigated *in vivo* in a mouse model, and found to improve wound healing of diabetic ulcers.[102] In addition to these examples, the reader may refer to reviews that are available in the literature about the preparation and characterization of wound healing membranes composed of polymer nanofibers.[148]

Another interesting, more sophisticated application consists in temporary abdominal closure following laparotomy and intra-abdominal packing for surgical treatment of abdominal traumas.[123] In laparotomy, the abdominal wall is incised for diagnostic or therapeutic purposes, and the patient outcome may be influenced by a number of critical complications such as infections, postsurgical adhesions, fascial necrosis, abnormal intra-abdominal pressure and so on. Dressing materials, hopefully biodegradable, are therefore needed to provide temporary barriers for reducing external contaminations or enabling highly controlled drug-release. Therefore, this is a typical example of biomedical application where the multifunctionality of polymer nanofibers can be very useful, since open abdominal field naturally (and critically) benefits from dressing material elasticity (because of the continuous swelling and contractions of exposed tissues), biodegradability, high strength, high antimicrobial activity and suturability. *In vivo* experiments with electrospun, composite fibrous sheets made of PLAGA and poly(ester urethane) urea loaded with tetracycline hydrochloride have been carried out in an infected abdominal wall rat model.[123] In other studies, electrospun fibers of PLAGA, PEO/poly(D,L-lactide) and PCL loaded with antibiotics are found to be effective in reducing post-surgery induced abdominal adhesions in rat models.[120,125]

## 7.6 Other Biomedical Devices

Some aspects of polymer nanofibers used as chemical sensors are presented in Section 4.3.2 and a related mathematical model[151] is summarized in Equation (4.23) for the temporal variation of the measured conductivity following exposure to a molecular agent diffusing into the fiber. These concepts are general and also applicable to the case of biological sensors.

Electrospinning has also been explored as a technology for the production of highly sophisticated bio-components such as cardiac valves. Experiments carried out at the Italian Istituto Superiore di Sanità and at the Tor Vergata University in Rome have been focused on the implementation of electrospun PCL bioresorbable trileaflet heart valve prosthesis.[152] Indeed, templated, second-generation valves can be fabricated by electrospinning onto an Al trileaflet heart valve-shaped target, which slowly rotates (0.3–0.4 rpm) to favor a homogeneous and continuous coverage by fibers.[152] *In vitro* tests, performed under external solicitation by means of a pulse duplicator mimicking pulmonary conditions, evidence that the three parts of the valve open synchronously (Figure 7.5), showing a substantially correct coapation in the diastolic phase.[152]

A parallel study from researchers at Eindhoven University of Technology has compared electrospun PCL scaffolds with knitted scaffolds for tissue engineering of the aortic valve, again testing the components under physiological flow conditions.[153] Both the systems allow a good cellular proliferation upon seeding human myofibroblasts. However, knitted architectures are found to be generally more structurally stable, whereas electrospun matrices exhibit a higher efficiency in cell seeding, likely because of their spatially denser topology.[153] More generally, the authors point out that, due to the largely non-uniform electric field intensity and directions around the complex metal collectors that have to be used to fabricate these systems by spinning polymers on templates, the resulting deposition of nanofibers and consequently scaffold thickness may be significantly different from one point to another. This can lead to inhomogeneous stress distributions and ultimately valve failure. Electrospinning biomedical components onto metallic templates will therefore need a more in-depth investigation in the future, possibly coupling process engineering through

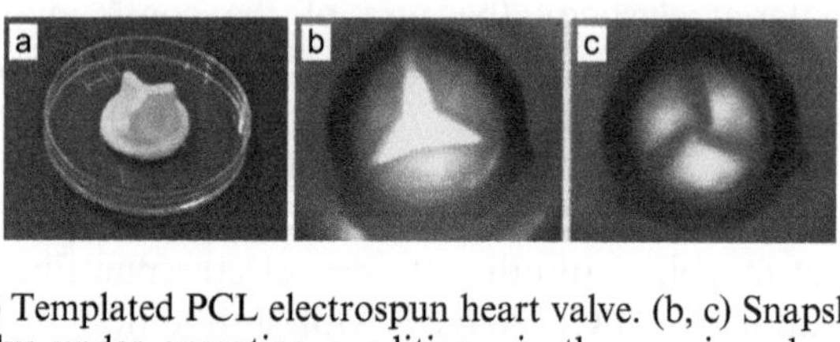


**Figure 7.5** (a) Templated PCL electrospun heart valve. (b, c) Snapshots showing the valve under operating conditions, in the opening phase at the systolic peak (b) and in the diastolic phase (c). Adapted with permission from Ref. 152, *Ann. Ist. Super. Sanità*, 2007, **44**, 178–186. Copyright © 2007, Istituto Superiore di Sanità.

locally controlled manipulation of involved electric fields by real-time monitoring and dielectric screening, together with a careful mechanical analysis of the resulting scaffolds.

## 7.7 Tissue Engineering: Nanofibers Scaffolds and Their Applications

The research on nanofiber-based scaffolds,[3,154–157] aimed to develop new supports for implants and regenerative medicine, is today vital like no other in the scenario of biologically functional nanostructures, and this is attracting a new and enthusiastic generation of young nano-scientists, materials engineers and biotechnologists. The ultimate objective of regenerative medicine is of repairing or replacing lost or damaged tissues, initiating and promoting the natural regeneration process, and polymer nanostructures may be very advantageous in this respect.[158] The aim of this section is to introduce schematically the well-defined sequence of logical steps and experiments of tissue engineering research, with a focus on polymer nanofiber-based scaffolds. Many excellent reviews and texts are available on tissue engineering and regenerative medicine. For instance, the reader may find complete overviews on the related books edited by Atala, Lanza, Thomson and Nerem,[159] and by Lanza, Langer and Vacanti.[160]

In principle, scaffolds should be able to produce or induce the entire and complex set of biochemical signals at all the different length-scales (macro-, micro- and nanoscale) that are relevant in regulating the cooperative behavior of cells at tissue level down to molecular level, directing the cell colonization, attachment, proliferation and differentiation mechanisms towards specific phenotypes. To this aim, the typical tissue engineering experiment on a nanostructured scaffold follows a characteristic logical flow, which is highlighted in hundreds of papers published on polymer nanofibers:

(i) *Scaffold fabrication*. This step includes (i-a) all the needed production steps for obtaining nanofibers, (i-b) the eventual optimization of the processing parameters aimed to obtain a specific distribution or average value of size, porosity, *etc.* (as determined by characterization experiments carried out at the following step, (ii), and through a proper feedback on the nanofabrication platform, and possibly iterative improvement stages), (i-c) biofunctionalization procedures (Section 7.3), whenever this is needed or desired to enhance the subsequent biological response of cells, and the cross-linking treatments that can be required to render the scaffold insoluble in aqueous solutions and (i-d) the sterilization and eventually packaging procedures needed before cell seeding, or ultimately for transport and commercialization of the realized scaffolds; several different sterilization procedures are available including the use of UV light, $\gamma$ rays, plasmas, autoclaves, *etc.*, and the choice of the most suitable method will critically depend on the photo-

and thermal stability of both the polymer matrix and the employed biofunctionalization agents;

(ii) *Scaffold material characterization.* Physico-chemical methods for inferring information on the properties of the obtained biomaterial include:

- (ii-a) morphological investigation (mainly by SEM, but also optical imaging, AFM, TEM, *etc.* can be used in order to extract some complementary morphological, local nanomechanical or internal compositional information). TEM allows the internal structure of nanofibers doped by inorganic nanocrystals to be studied, which are frequently used in scaffolds for bone tissue engineering, whereas AFM is a powerful tool for probing surface topographies especially at single-nanofiber scale, and local mechanical properties *via* force-distance indentation measurements;
- (ii-b) wettability measurements to determine the achieved degree of hydrophilicity, and porosity measurements;
- (ii-c) mechanical measurements, such as tensile tests; stress-strain testing is particularly relevant and sometimes critical since for most applications the scaffold should mimic as closely as possible the mechanical behavior of the target tissues, and different tissues in the body exhibit enormously different stiffness and strength;
- (ii-d) a variety of possible approaches to find the surface or internal chemical composition of the nanofibers. These include optical absorption, FTIR, immunofluorescence or confocal methods to assess the effectiveness and uniformity of protein functionalization through fluorescence staining with suitable antigens, elemental mapping through energy-dispersive spectroscopies, *etc.* The protein content on the nanofibrous scaffolds can also be evaluated by bicinchoninic acid assays;

(iii) *Cell culturing.* Cells for research purposes can be obtained from commercially available lines, or harvested from a donor or patient. Most cell cultures are performed over intervals ranging from two to three days up to a few weeks. Differentiation experiments generally imply the addition of suitable supplementing factors in the growth medium according to well-defined delivery recipes;

(iv) *Determination of the integrated performance of the scaffold-cell system.* The resulting cell morphology, growth, viability, possible migration phenomena, differentiation, *etc.* are then investigated in order to assess the biofunctional performance of the produced biomaterial. At the bare minimum, the cell activity and proliferation can be analyzed afterwards by tetrazolium salt colorimetric assays, which allow one to roughly evaluate the number of viable, metabolically active cells through the conversion of tetrazolium salts, 3-(4,5-dimethylthiazol-2-yl)-2,5-diphenyltetrazolium bromide (MTT) or 3-(4,5-dimethylthiazol-2-yl)-5-(3-carboxymethoxyphenyl)-2-(4-sulfophenyl)-2H-tetrazolium (MTS) to colored formazan with a strong absorption that can be easily quantified by a spectrophotometer. Water-soluble tetrazolium salts (WSTs), used for viability assays according to the same working principle, can be advantageous since highly water-soluble formazans are directly produced by metabolically active cells, thus allowing a more user-friendly and often more reliable spectrophotometric reading;

Live/dead assays are also available, and alkaline phosphatase (ALP) activity can be measured by commercial kits. An initial assessment of cell morphology can be obtained by optical micrographs, whereas time-lapse microscopy allows cell migration phenomena that lead to scaffold colonization to be investigated. SEM of the resulting constructs provides further information on how cells spread on nanofibers underneath. The focal adhesions resulting from the cell-scaffold interactions can be localized and quantified through fluorescence staining of vinculin and F-actin, by means of immunofluorescent or confocal microscopy. Gene expression can be detected using RT-PCR analysis for specific genes, which allows the obtained cell phenotype to be studied. Other physical measurements (morphological, mechanical, *etc.*) can be repeated on the scaffold-cell system to study the evolution of the bio-hybrid sample following cell culturing;

(v) *In vivo experiments*. This is of course the most advanced, sophisticated and crucial part of tissue engineering. The scaffold is implanted with the final objective of properly supporting the growth, organization and activity of cells up to the ultimate regeneration of a damaged tissue, without inducing rejection, inflammatory response or immune activation in the organism. Ideally, the scaffold should have the right degradation rate, being biodegraded in the body over a time interval long enough to elicit a proper recruitment of populating cells and the local regeneration of tissues, but not so long to result in undesired barriers against physiological tissue development and transport of nutrients. The scaffold should therefore degrade over the same timescale of the formation of the new functional tissue. Several different animal models (rat, hamster, mouse, rabbit, goat, swine and so on) can be used. The smallest animals are cheaper and more easily available, whereas bigger animals such as goat and pig can provide models better resembling human characteristics. Post-implantation, the evolution of the scaffold is monitored up to a few months (most often a few weeks in nanofiber-focused research). Immunohistochemical staining with the targeted antibodies is used for the histological analysis of the scaffolds infiltrated by cells or of the newly formed tissues that are removed at the end of the experiment. SEM, mechanical and other analyses are also possible at this stage on the removed constructs.

In conclusion, there are tens of complementary possibilities in terms of physico-chemical and biological characterization methods that can be applied in order to assess the properties of the scaffold before, during and after cell culturing *in vitro*, or before and after *in vivo* implantation. The final choice for each experimental campaign depends on the type of polymer, on the biofunctionalization characteristics, on the investigated cell lineages, on the targeted tissue and of course on the amount of resources available in each laboratory. In any case, however, an inter-disciplinary expertise, merging polymer science and biology is essential. In the following, some relevant examples of nanofiber-based tissue engineering are reviewed, for several target applications.

### 7.7.1 Interplay between Structural Properties and Biological Functionality of Nanofibers

As mentioned above, the interaction between a nano- or micro-structured surface and living cells is the result of a complex interplay between topography, biochemical interactions and cellular phenomenology, in which also the natural cues, such as ECM and adhesion molecules, produced *in situ* by the cultured cells take an important role. In the related scientific literature, the "*cue*" term means the very rich variety of possible interactions and stimulation provided by the surrounding microenvironment, and particularly by the scaffold material, to living cells. The study of these mechanisms is gathering increasing interest in the tissue nanoengineering field.

In particular, though the size of the scaffold structures certainly plays a pivotal role, the details of how this size affects biofunctional mechanisms is still a matter of investigation. For several polymers, the diameter of fabricated fibers can be quite precisely controlled from tens of μm to tens of nm, therefore these materials can provide a strategic bench platform for experiments on topography–cell interactions. For instance, human umbilical vein endothelial cells (HUVEC) and outgrowth endothelial cells have been investigated in this respect, and found to exhibit an interesting and varied phenomenology.[161] The expression levels of platelet/endothelial cell adhesion molecule-1 (PECAM-1) and vascular endothelial (VE)-cadherin, quantified by real-time PCR, do not seem to show relevant differences when cells are cultured on silk fibroin fibers with micro- or nanometric sizes, whereas transmembrane receptors such as integrin-$\beta_1$ are significantly upregulated in cells on nanometric fibers compared to those grown on microfibers.[161] In addition, such enhanced integrin expression can be involved in the slight polarization and formation of new focal adhesions observed on nanofibers.[161] Other studies focus instead on mixed nano- and microfibrous scaffolds, whose hybrid geometries could combine effective cuing mechanisms with favorable mechanical properties and porosity.[162,163] Endothelial cells on these hybrid scaffolds are found to be sensitive to pro-inflammatory stimuli and, under pro-angiogenic conditions, to migrate and organize into well-defined, circular architectures that resemble capillary-like structures.[162]

Highly aligned arrays of polymer nanofibers are very advantageous for cell guidance, directing, *etc.*, which is crucial for many applications in neuronal, vascular, cartilage and bone tissue engineering as is better described in the following. Many studies report enhanced cellular maturation, differentiation and functionality on aligned nanofibers.[156] Importantly, as is seen in Section 2.3.2, methods are available, such as FFT analysis,[164,165] that allow one to determine quantitatively the degree of mutual alignment of nanofibers in the scaffold, which can then be more precisely correlated to the outcome of biological experiments.

### 7.7.2 Neuronal Tissue Engineering

The possibility of directing the growth and of differentiating selectively neural progenitor cells on micropatterned polymer substrates is a very well-established result.[166] Nowadays, polymer nanofibers are widely investigated as new, low-cost, biochemically flexible and abundant nanostructures for realizing patterned surfaces aiming at aligning neuronal cells, enhancing and controlling their growth and maturation and eventually offering innovative scaffold solutions for improving the regenerative capability of nerves.[167] Given the tremendous social importance of healthcare addressing neurodegenerative disorders and peripheral nerve injuries, the enormous challenges of repairing processes and the generally limited self-regeneration capacity of the human adult nervous system, it is easy to foresee that this class of applications will stimulate a special, widely inter-disciplinary and continuously increasing interest in researchers in the forthcoming years.

Several elegant experiments have demonstrated the potential of electrospun nanofibers for supporting peripheral nerve guidance and for improving nerve regeneration. A fairly complete list of existing reports can be found in Ref. 168. For example, the incorporation of aligned protein-polymer composite fibers, made of human glial cell-derived neurotrophic factor embedded in an electrospun copolymer of caprolactone and ethyl ethylene phosphate, has been tested in a nerve conduit as an approach for enhancing the regeneration of the sciatic nerve, over a defect length as large as 15 mm in a rat model, after three months post-implantation.[169] Chitosan fiber scaffolds have been studied in similar test-cases, and found to support a partial motor function recovery after two months post-implantation (for a sciatic nerve gap of 10 mm) when the fibers are functionalized by proper peptide sequences.[51,52] Also electrospun tubular structures made of PLAGA/PCL and collagen/PCL have been implanted and found to favor the recovery of 8–10 mm nerve gaps in a rat sciatic nerve, without inducing significant inflammatory response.[170,171] Much evidence has been collected that shows that scaffolds embedding aligned nanofibers offer important advantages and enable better recovery performances than randomly oriented nanofiber mats. For example, a 17 mm sciatic nerve gap (again in a rat model) clearly exhibits a functional recovery after four months post-implantation of a scaffold made of aligned nanofibers of the random copolymer poly(acrylonitrile-*co*-methylacrylate).[172] The recovery performances are clearly enhanced compared to control experiments with random nanofibers (Figure 7.6). Indeed, electrophysiological and behavioral (by grid walking tests) analyses highlight that aligned fibers, differently from randomly oriented scaffolds, are able to favor both sensory and motor nerve regeneration.[172] Electrospun collagen scaffolds in spiral tubular structures, tested for the treatment of acute spinal cord injury in a rat model, show

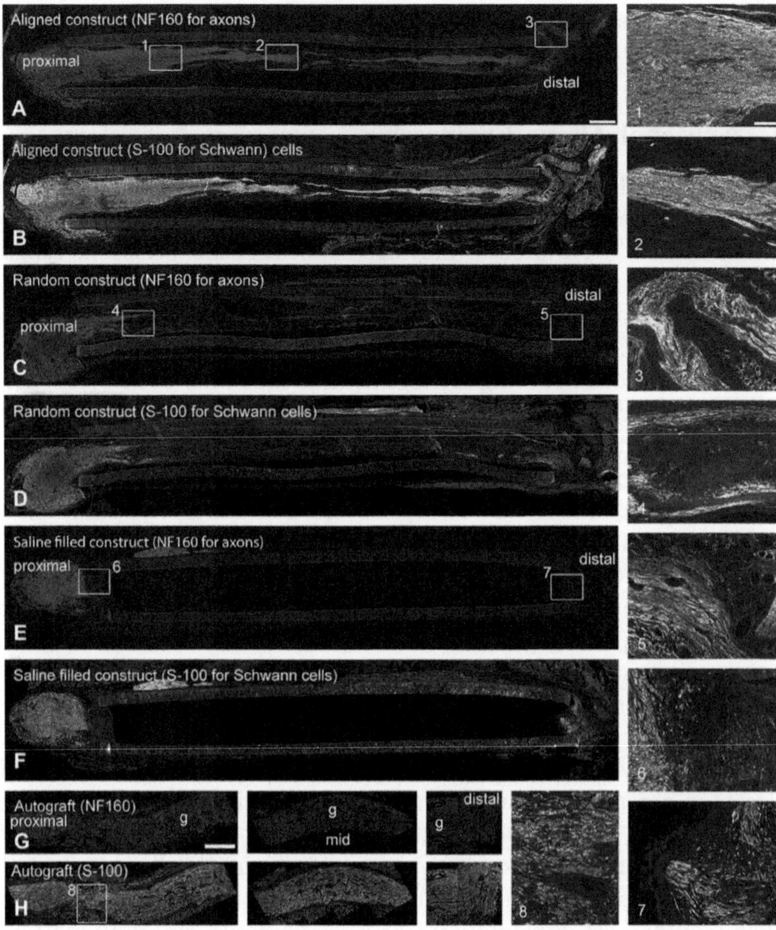


**Figure 7.6** Immunohistochemical analysis of nerve regeneration through implants of poly(acrylonitrile-*co*-methylacrylate) electrospun fibers, *in vivo* across a 17 mm nerve gap. (a), (b) and (1–3): Nerve regeneration through aligned fiber constructs. (c), (d) and (4, 5): Nerve regeneration through randomly oriented fibers. (e), (f) and (6, 7): Nerve regeneration through saline filled polymeric construct. (g), (h), (8): Nerve regeneration through autograft. (a, b): Double immunostained nerve regeneration (a, NF160) and Schwann cells infiltration from both proximal and distal nerve stump (b, S-100). Scale bar = 1 mm. Numbered images are NF160/S-100 overlapped magnifications of the corresponding boxes in (a). Scale bar = 200 μm. (c, d): Double immunostained nerve regeneration (c) and Schwann cells (d). Numbered images are from the highlighted boxes in (c). (e) and (f) double immunostained nerve (e) and Schwann cells (f). Numbered images are NF160/S-100 overlapped magnifications from boxes in (e). (g, h): Double immunostained nerve regeneration (g) and Schwann cells (h). Numbered images are from the highlighted boxes in (h). Scale bar = 800 μm. Reproduced with permission from Ref. 172, *Biomaterials*, 2008, **29**, 3117–3127. Doi: 10.1016/j.biomaterials.2008.03.042. Copyright © 2008, Elsevier Ltd.

excellent cellular colonization, no accumulation of astrocytes[173] in the implant† and better structural resistance after one month post-implantation when one uses aligned nanofibers.[174] Laminin-, fibronectin- and basic fibroblast growth factor-functionalized, aligned nanofibers of PLA and poly-acrylonitrile methylacrylate have been studied as supports for culturing rat dorsal root ganglions (DRGs), and were found to enhance significantly the neurite growth compared to purely polymeric or randomly oriented fibers or to reference films.[106,175] In other studies, neurites from DRG explants have been investigated on electrospun PLA[165] and PCL[104] nanofibers, and on core-shell nanofibers realized by *in situ* polymerization of pyrrole on electrospun PLA and PCL samples,[157] under various conditions of nanofiber alignment and even applying electrical stimulation.[157] Neurites have been found to be clearly guided and most extended on highly aligned scaffold. Nanofiber-coating of surfaces can be used to decrease the impedance of neural microelectrodes, as demonstrated by using conjugated polymer nanotubes.[111] Finally, the growth of nerve stem cells has also been investigated on PCL/gelatin electrospun fibers with a relative concentration of 70:30, and these scaffolds too are found to promote and orient the neurite outgrowth.[83]

Overall, scaffolds made of aligned polymer nanofibers, for instance directly produced by electrospinning on rotating collectors (Section 2.3.2), can be used as effective conduits for nerve re-growth (for example, within the lumen of the so-obtained nanostructured capillary). These systems are very promising for the surgical treatment of small nerve transection lesions of the peripheral nervous system.[170] We will give a few other important details on how aligned nanofiber constructs seem to affect the cell growth and differentiation when dealing with stem cells cultured on nanofiber scaffolds in Section 7.7.6. To investigate the possible biological mechanisms at the base of these results, Schwann cells have been proposed as valuable probes because of their important role in nerve regeneration and of the close contact that they can establish with nanofibers surfaces.[84,165,167,175–178] Schwann cells are able to favor the recovery of damage in peripheral nerves by means of neurotrophin secretions and by producing ECM proteins, however their migration and alignment towards injured regions, which is an important prerequisite for infiltrating axons, often exhibit slow rates that limit a fast and effective regeneration. *In vitro*, strongly elongating along the preferential axis of alignment of the micro- or nanostructures, Schwann cells show a significantly faster and enhanced neurite outgrowth compared to unguided cultures. The cellular orientation is also assisted by nanofibers with parallel line surface topographies.[179] These can be effective methods to mimic the formation of bands of Büngner‡ (prior to the eventual implantation) and ultimately to

† When astrocytes are cultured on polymer nanofibers, the cell proliferation is suppressed compared to cultures on reference films, and apoptosis and gene-silencing efficiency are correspondingly enhanced.[173]

‡ Bands of Büngner are formed following the proliferation of Schwann cells and have an important role in nerve regeneration.

improve nerve regeneration. Aligned electrospun PCL fibers are demonstrated to induce such strong elongation in the cytoskeleton and nuclei of human Schwann cells in 7 days (Figure 7.7a–f).[177] On PCL/collagen aligned fibers, Schwann cells exhibit an interesting migration behavior, tending to move along the preferential direction of alignment of the fibers.[84] Other interesting results come from the analysis of the cell phenotype by means of gene and protein expression, that allows information about the degree of maturation of the cultured cells to be inferred. Neurotrophin and neurotrophic receptors are found to be down-regulated in such aligned cells compared to cells grown on PCL films, whereas a myelin-associated glycoprotein (early myelination marker) and the NCAM-1 (indicating immature Schwann cells) are correspondingly up-regulated and down-regulated, respectively.[177] In particular, while these genes show analogous, or even higher, expression also in cells seeded on random fiber mats, the slight up-regulation of the myelin-specific gene, P0 (myelin protein zero), is found only in elongated Schwann cells grown on aligned fiber scaffolds (Figure 7.7g), which suggests that such oriented microstructure can actually favor the cell maturation.[177] In this respect other new perspectives have recently been opened by studying mouse embryonic stem cells cultured on PCL nanofibers, whose neurite outgrowth directing and differentiation capability have enormous potential for several nerve regeneration therapies.[180]

Other specific features of organic nanofibers can be exploited as well, to design tools to interface with neuronal cells. For example, the controlled motion of short segments of electrospun polymer fibers, embedding superparamagnetic nanoparticles, can be used to inter-connect neurons with high accuracy in positioning the individual nanostructures.[181] For their excellent electrical conductivity and their potential interfacing with neurons, carbon nanofibers are promising as neural biomaterial. For instance, the capability of these nanostructures to decrease the adhesion of astrocytes has been investigated.[182] Finally, other interesting findings deserve to be mentioned in this framework, which have been collected by a complementary approach based on the interaction of neuronal cells with self-assembled, three-dimensional nanofibrous architectures of peptides.[183–188] In particular, arginine–alanine–aspartate (RAD)16 self-assembled scaffolds have been widely studied with several classes of cells, and found to promote significantly the neurite outgrowth and ultimately the formation of synapses.[183] RAD16 scaffolds can be further specialized by the insertion of other specific bio-functional groups, such as those from RGD sequences from fibronectin or collagen.[185] For *in vivo* treatments, an advantage of three-dimensional peptide fiber architectures is that they can be easily injected from solutions into the injured region. Scaffolds made of self-assembled peptides have been tested as supplement medium for recovery from brain damage in a hamster model,[188] and found to favor the growth of axons in a few weeks. Functional recovery has been investigated by studying the visually elicited orienting behavior, after a transection of the optic tract and subsequent application of a solution of the

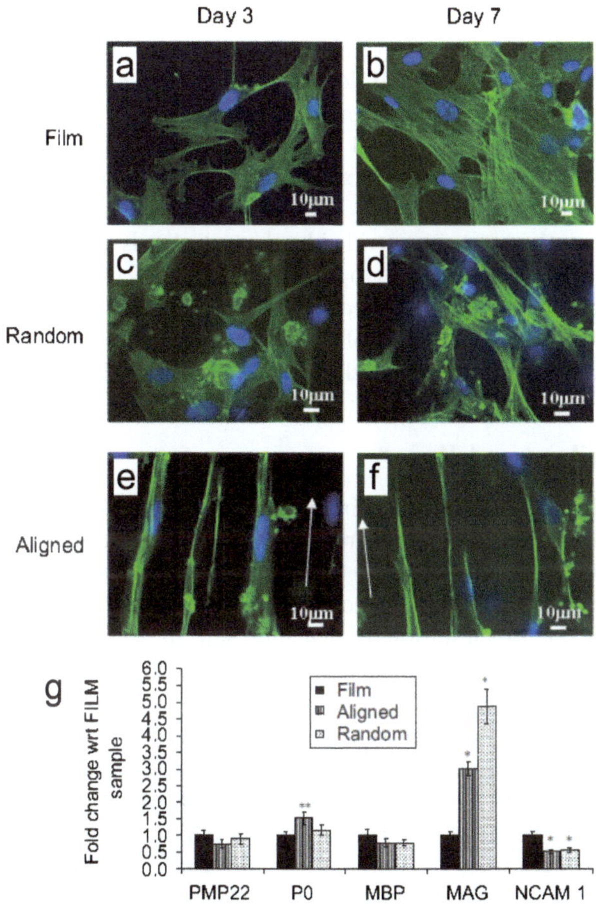


**Figure 7.7** Confocal fluorescent micrographs of human Schwann cells grown on PCL scaffolds for 3 days (a, c and e) and 7 days (b, d and f), for films (top panels), randomly oriented (middle panels) and aligned (bottom panels) electrospun fibers, respectively. The arrows in (e) and (f) evidence the direction of fiber alignment. Green and blue colors indicate the actin cytoskeleton and 4′,6-diamidino-2-phenylindole (DAPI), respectively. (g) Expression of myelin-specific genes and of immature Schwann cell marker, NCAM 1, determined by real-time PCR. Results are normalized with respect to film sample and suggest the enhancement in maturation of human Schwann cells cultured on electrospun fibers. Statistical parameters: * $p < 0.05$, ** $p < 0.01$, $t$-test between electrospun fibers and films, $n = 3$. Adapted with permission from Ref. 177, *Biomaterials*, 2008, **29**, 653–661. Doi: 10.1016/j.biomaterials.2007.10.025. Copyright © 2007, Elsevier Ltd.

self-assembling peptide nanofiber scaffold.[188] Neural progenitor cells and Schwann cells, cultured in scaffolds made of self-assembled peptide nanofibers thus producing transplantable bio-constructs, are demonstrated to favor recovery of a transected segment of injured spinal cord, in a rat model.[189] Significant migration of host cells, vascularization and axon growth are observed in these implanted scaffolds, which results in effectively bridging the injured tract.[189] Filamentous nanostructures using injectable self-assembling peptides have been designed as well to target myocardium,[190] and to mimic the activity of vascular endothelial growth factor in favoring the growth of new blood vessels.[191]

In conclusion, the scenario provided by these studies is really encouraging for the medium- to long-term development of effective nanoproducts for the recovery of nerve injuries. At the present stage of research, there is certainly the need for a much larger number of *in vivo* experiments, in order to validate recovery performances in more depth and on a higher number of animal models.

### 7.7.3 Cartilage and Bone

Cartilage and bone tissue engineering is especially suitable to be implemented and developed by means of polymer nanofibers used as scaffold material.[192,193] Indeed, remodeling phenomena in adult bone tissues are driven by the early formation of fibrous collagen structures entrapping and spatially organizing cells, and by the precipitation of calcium salt in these assemblies. These processes are well suited to be effectively mimicked by bio-artificial constructs made by cells colonizing artificially realized nanofiber materials.

On one hand, promising findings come from material development. A few innovative materials such as the degradable poly(urethane urea) elastomer, Artelon®, have recently been proposed for tissue engineering applications and for treating osteoarthritis, and they have been demonstrated to produce nanofibers when electrospun from DMF solutions.[194] Carbon nanofibers have also been investigated concerning the viability and functionality of cultured obsteoblasts.[195]

On the other hand, the proliferation and differentiation of osteoprogenitor cells on electrospun fibers appear very promising. On polymers such as PLA and PEO, and their diblock copolymers and blends, these mechanisms have been studied since 2006.[93] MC3T3-E1, in absence of osteogenic factors, shows a slower growth on fibers compared to spin-coated surfaces, whereas the density of cells after one or two weeks is comparable or slightly higher for fiber surfaces when an osteogenic factor (2mM β-glycerophosphate and 0.13 mM L-ascorbate-2-phosphate) is used. Comparable results are also obtained on nanofibers and on spin-coated substrates concerning the ALP activity after two weeks, in the presence of osteogenic factor.[93] Human bone marrow stromal cells attach and proliferate on electrospun silk fibers electrospun with the aid of a PEO added component (silk/PEO blends with 80%/20% relative composition in weight), and especially on scaffolds from which the PEO fraction is preliminarily extracted in water.[73] Mesenchymal stem cells have been extensively studied on electrospun PCL fibers, again in the presence of osteogenic supplements, and these scaffolds have been tested also *in vivo*,[196,197] and observed to evolve toward bone tissues in four weeks post-implantation. Mouse embryonic stem cells cultured on PLA nanofibers show enhanced interaction with the scaffold (again in comparison with cultures on flat films) and osteogenic differentiation in the presence of osteoinductive supplements (1 mM dexamethasone, 50 μg $mL^{-1}$ ascorbic acid and 10 mM β-glycerol phosphate).[198] Dexamethasone has also been incorporated in electrospun PCL.[128] How the influence on cells of osteogenic factors supplemented in culture media is modulated by the surface nanotopography is still matter of investigation. A recent study would indicate that biomimetic, bioactive composite scaffolds made of PCL nanofibers loaded with a hydroxyapatite and tricalcium phosphate mineral phase (Figure 7.8) are able to induce

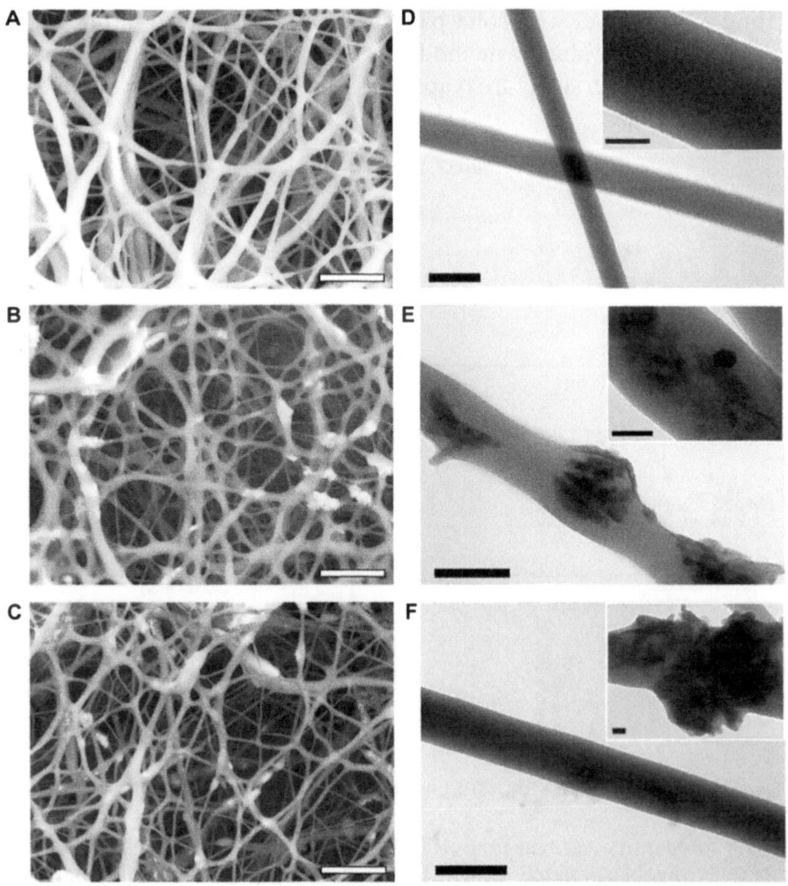


**Figure 7.8** Electron microscopy characterization of electrospun nanofibrous scaffolds embedding inorganic mineral particles. SEM (A–C) and TEM (D–F) micrographs for pure PCL (A, D), PCL-hydroxyapatite (B, E) and PCL/ tricalcium phosphate (C, F). Bar = 2 μm (A–C), 200 nm (D–F) or 50 nm (insets D–F). Reproduced with permission from Ref. 199, *PLoS ONE*, 2011, **6**, e26211. Doi: 10.1371/journal.pone.0026211. Copyright © Polini *et al.*

osteogenic differentiation of human mesenchymal stem cells even in the absence of supplemental, exogenous osteogenic growth factors.[199] In fact, a mature osteogenic differentiation is observed in basal growth medium conditions, with an increased ALP activity and mRNA expression of typical osteoblast-related genes (Figure 7.9) such as core binding factor alpha 1, collagen type 1 and bone sialoprotein (BSP).[199] Results from another work carried out on human bone marrow stromal cells on scaffolds with three-dimensional texturing strengthen the conclusion that nanofibrous polymer architectures have unique capacities in driving osteogenic differentiation in the absence of osteogenic supplements.[200]

As mentioned above, doping nanofibers by an inorganic phase, such as nano- or microparticles of hydroxyapatite and tricalcium phosphate, or fabricating architectures with solid particles densely deposited in the voids of nanofibrous mats are valuable methods for producing scaffolds for bone tissue engineering.[66] In general, hydroxyapatite is more stable and exhibits much

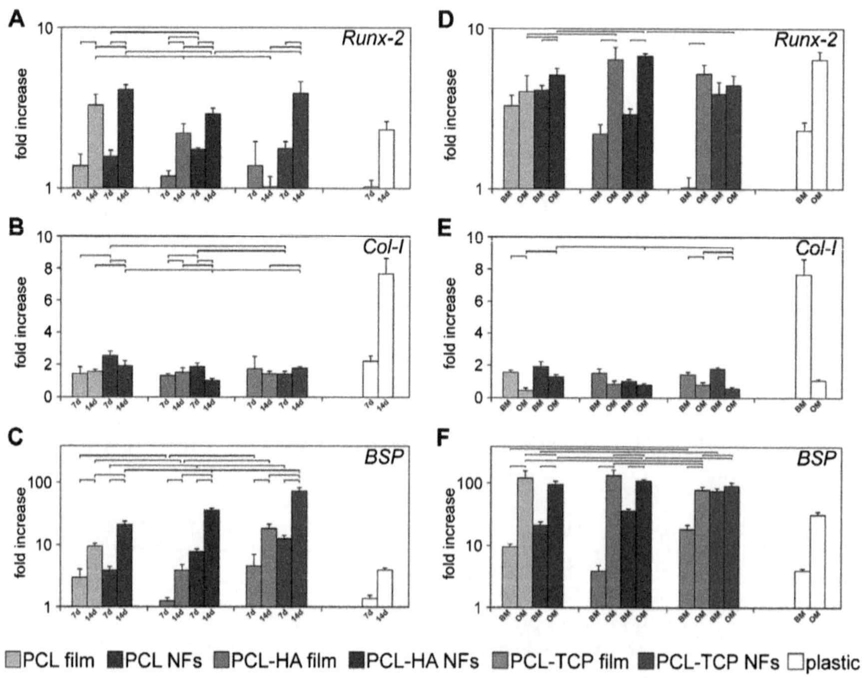


**Figure 7.9** Quantitative real-time RT-PCR gene expression analysis of some osteoblast-related genes. (A–C) Gene expression levels for human mesenchymal stem cells cultured in basal medium (BM) after 7 and 14 days of culture. (D–F) The effect of culture medium on gene expression levels is highlighted comparing results obtained at day 14 for cultures either in basal medium or in osteogenic medium. Expression levels of each gene are here normalized to averaged levels previously measured at day 1. Reproduced with permission from Ref. 199, *PLoS ONE*, 2011, **6**, e26211. DOI: 10.1371/journal.pone.0026211. Copyright © Polini *et al.*

lower dissolution rates, whereas β-TCP is more soluble and its degradation products ($Ca^{2+}$ and $PO_4^{3-}$) can be released into the surrounding environment, thus contributing to the overall induced bioactivity.[201,202] Electrospun scaffolds of PCL/$CaCO_3$ have also been tested concerning osteoblast attachment and proliferation.[203] Demineralized bone powder is another option, whose osteoinduction capability has been studied with scaffolds based on electrospun PLA fibers.[204] In fact, it is possible that simply embedding solid particles in the bodies of nanofibers leads to, at least partially, the osteoconductive properties of hydroxyapatites to be hindered. In addition, electrospinning blends of polymers and nanoparticles is often complicated by issues related to needle clogging, agglomeration and overall poor plastic behavior of the spun solutions (Section 2.3.3). To overcome these problems, various strategies can be pursued that are often based on smartly combining the fabrication of nanofibers with complementary methods, thus addressing issues related to making dispersions uniform and inorganic nanoparticles actually available to interact with cells. For example, nanofibers of poly(L-lactic acid)-*co*-poly(3-caprolactone), gelatin and hydroxyapatite nanoparticles can be produced by combining electrospinning with electrospraying, the latter technique being more suitable to deposit suspensions of solid particles.[205] The resulting rough surface morphology is well suited for promoting the growth and mineralization of osteoblasts, as tested with human fetal osteoblast (HFOB) cells, and the nanocomposite scaffold architectures also show high pore size (up to ~6 μm) and porosity up to 90%.[205] Another approach combines electrospinning with a twin-screw extruder (Section 2.3.3) with fully inter-meshing and co-rotating screws,[206] allowing functionally graded scaffold composed of PCL micro- and nanofibers and continuously varying embedded concentrations of β-tricalcium phosphate (Figure 7.10)[32] or of insulin and β-glycerophosphate to be realized.[207] When tested with MC3T3-E1 cells, after four weeks the scaffolds lead to mechanical properties, ECM formation and mineralization resembling typical bone-cartilage interfaces.[32] Furthermore, nanocomposite hydroxyapatite/chitosan nanofibers can be produced by combining in various ways co-precipitation approaches with electrospinning.[33,208] These methods are developed in two sequential steps. For instance, the preparation of the nanocomposite by a co-precipitation synthesis approach[209] can be followed by electrospinning of an aqueous DMSO solution of the resulting compound, upon PEO addition.[208] These scaffolds have been tested *in vitro* by HFOB cells as well, over two weeks, showing better bone formation performances compared to purely chitosan fibers.[208] Apatite microparticles can be deposited on the surface of nanofibrous scaffolds by various incubation processes in simulated body fluids.[210] Even gradients of the mineral phase can be obtained in this way by immersing different regions of samples for different time intervals, which can be done very easily by gradually adding more solution in a beaker where the scaffold is tilted and partially immersed.[34] Electrodeposition is another valuable alternative to treat nanofibrous scaffolds and to deposit calcium phosphate, which is then

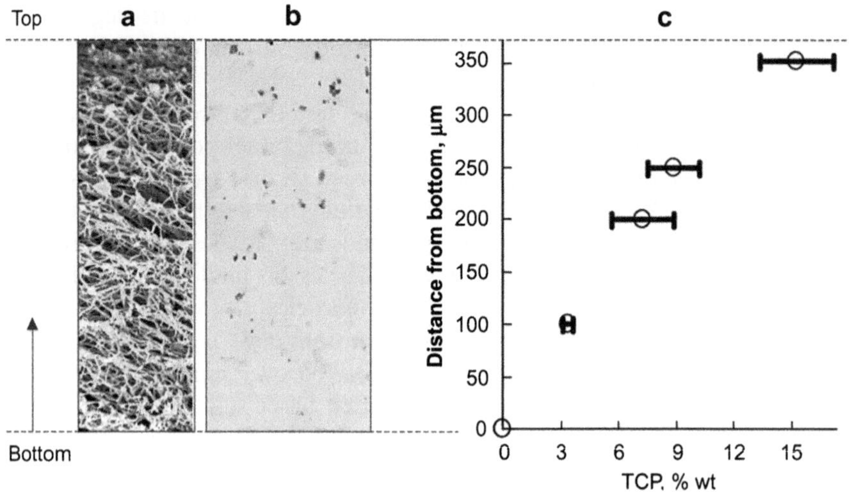


**Figure 7.10** (a) Cross-sectional SEM micrograph of the functionally graded scaffold, (b) von Kossa staining allowing the corresponding distribution of β-tricalcium phosphate to be estimated (black spots). The staining qualitatively agrees with β-tricalcium phosphate quantification by thermo-gravimetric analysis (c). Reproduced with permission from Ref. 32, *Biomaterials*, 2008, **29**, 4065–4073. Doi: 10.1016/j.biomaterials.2008.06.022. Copyright © 2008, Elsevier Ltd.

promptly available at the surface of fibers.[211] The electrodeposition coating can be performed by simple, two-electrode electrochemical systems under potentiostatic conditions, with a working electrode preliminarily covered with fibers, another metal counter-electrode and a mixture of calcium and phosphate ions with a hydroxyapatite-like Ca/P ratio of 1.67.[211]

In general, intriguing findings come from visualizing focal adhesions and from studying how these relate to the underlying topography of polymer nanofibers. While the focal adhesion contacts (imaged as punctate spots by immunofluorescent staining of vinculin) have been observed as largely spread throughout the cell periphery when osteoprogenitor cells are grown on spin-coated films (Figure 7.11a), when a fiber scaffold made of poly(ethylene-*b*-lactide) copolymers is used the focal adhesions often become denser and organize in clusters modulated by the fibers underneath (Figure 7.11b).[93]

Hybrid systems have also been proposed, using self-assembling peptide amphiphile nanofibers combined with a collagen sponge reinforced by polymeric fibers,[212] or with a porous and otherwise biologically inert titanium foam.[213] While biocompatible, porous metallic implants are interesting for orthopedics because of their mechanical properties, peptide amphiphile nanofibers may render the system highly bioactive in view of cell encapsulation. These architectures have been tested *in vivo* in a rat model, where bone formation is found around and inside the scaffolds after four weeks post-implantation, together with vascularization and absence of

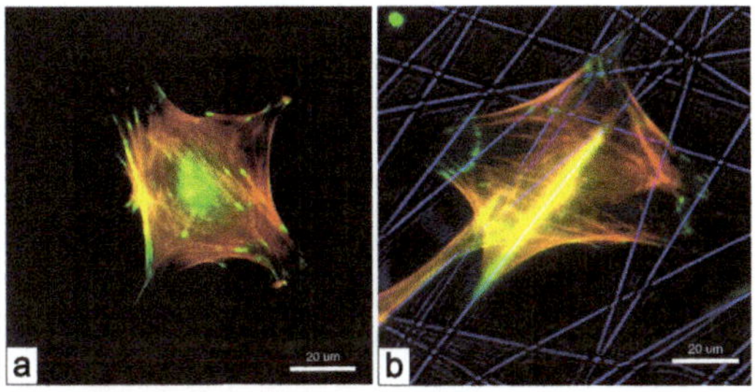


**Figure 7.11** (a) Immunofluorescent micrograph of a MC3T3-E1 osteoprogenitor cell, grown on a spin-coated film of a poly(ethylene-*b*-lactide) copolymer functionalized by fibronectin. Green spots: vinculin highlighting focal adhesion contacts. Red: actin. (b) The cell image is superimposed to a phase contrast image of poly(ethylene-*b*-lactide) fibers (blue), highlighting the clustering of focal adhesions in correspondence of underlying nanostructures. Adapted with permission from Ref. 93, *Biomaterials*, 2006, **27**, 596–606. Doi: 10.1016/j.biomaterials.2005.05.084. Copyright © 2005, Elsevier Ltd.

cytotoxic response, which suggests such a hybrid route as particularly promising.[213]

Besides bones, cartilage tissue engineering has been studied using human mesenchymal stem cells on three-dimensional nanofiber scaffolds.[214] These cells have been found to evolve towards multiple differentiated lineages on the scaffolds.[215,216] Indeed, on PCL nanofibers, mesenchymal stem cells evolve towards adipogenic, chondrogenic or osteogenic lineages upon culturing with specific differentiation media (Figure 7.12).[215] Such biocompatibility and inductive capacity is quite general and observed in a number of nanofiber species. For instance, analogous results are found for electrospun poly(D,L-lactide-*co*-glycolide).[216] In particular, aligned PCL fibers have been fabricated to realize engineered constructs for treating the fibrocartilaginous menisci of the knee, and found to induce an augmented stiffness (*i.e.* with a modulus >7 MPa) in tissues generated by seeding mesenchymal stem cells compared to scaffolds made of randomly oriented fibers (for which the modulus is about 1 MPa).[217] *In vivo* tests of PCL nanofibers seem to give promising indications of articular cartilage repair in a swine model, especially when the scaffolds are seeded with human mesenchymal stem cells.[218] Ectopic cartilaginous formations are also observed in rabbits after four weeks of subcutaneous implantation of chondrocyte-seeded poly(lactic acid) nanofibers, surface-modified with cationized gelatin.[101]

### 7.7.4 Dermal Tissues

Regenerative approaches targeting dermal tissue closely relate to cutaneous wound healing strategies (Section 7.5), and have special relevance for the treatment of chronic problems, such as those related to ulcers of various origin

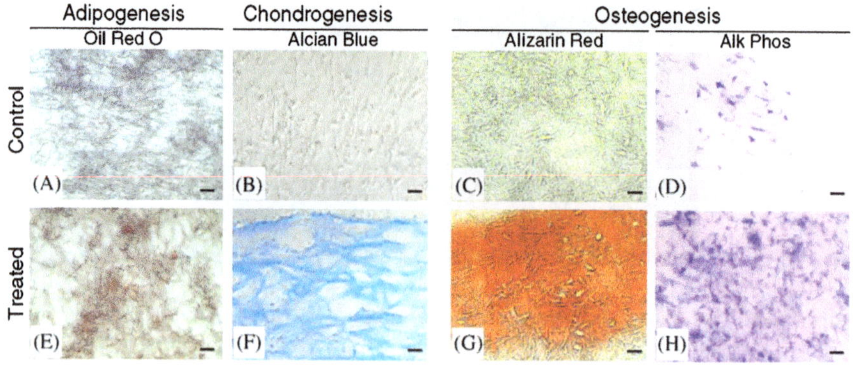


**Figure 7.12** Histological analysis of constructs made of mesenchymal stem cells and PCL nanofibers maintained in adipogenic, chondrogenic or osteogenic medium. Staining is carried out with oil red O (A, E), alcian blue (B, F) or alizarin red (C, G), or histochemically for ALP (D, H). For example, in adipogenic cultures (E) one can appreciate oil red O-positive lipid droplets, compared to the lack of staining in the control culture (A). In chondrogenic cultures (F), intense alcian blue staining highlights cells surrounded by a sulfated proteoglycan-rich ECM, compared to control cultures (B). Analogously, in osteogenic cultures mineralization (G) and ALP activity (H) are higher than in control cultures (C, D). Scale bar = 20 μm (B, F), 40 μm (A, C, E, G) or 80 μm (D, H). Reproduced with permission from Ref. 215, *Biomaterials*, 2005, **26**, 5158–5166. Doi: 10.1016/j.biomaterials.2005.01.002. Copyright © 2005, Elsevier Ltd.

including diabetic diseases, or of severe burns. Human skin equivalents exist, which can be exploited to mimic the natural tissue and for testing dermal scaffolds *in vitro*.[219] Experiments with scaffolds made of polymer nanofibers are numerous and generally promising for the development of innovative and effective skin-mimicking systems. For example, functionalizing the surface of PLA nanofibers with basic fibroblast growth factor is shown to favor slightly the migration of normal human dermal fibroblasts across fiber-based, nanostructured gaps.[106] Electrospun PS has been studied with cultures and co-cultures of fibroblasts, keratinocytes and endothelial cells.[220] In particular, for dermal tissue engineering what is especially desirable are layered scaffold architectures colonized by multiple classes of cells and with good overall elasticity. These requirements greatly highlight the importance of material fabrication methods producing or engineering multilayer and graded architectures of polymer nanofibers. Electrospinning PLAGA and Dextran allows fibers to be produced showing favorable interaction with dermal fibroblasts, which would ultimately lead to the formation of a multilayered bio-construct resembling the structure of derma.[221] PCL and collagen nanofibers are also studied in this respect.[222,223] Alternating seeding of human dermal fibroblasts and electrospinning steps allows cell-integrated, bilayered systems to be generated.[223] Electrospinning gelatin is another option, for which inter-fiber distances between 5 μm and 10 μm are found to be optimal in inducing the best cellular infiltration and organization,[224] a conclusion that is likely generalizable to many other tissue engineering applications relying on polymer nanofiber scaffolds. This sort of stratification involving a layer of basal keratinocytes at the dermal–epidermal junction is also highlighted *in vivo* as promoted by electrospun collagen implanted in mice.[225]

Another comprehensive, *in vivo* study in a rat model confirms the generally good performances of electrospun scaffolds: an effective cellular colonization is observed without significant inflammatory response and the scaffolds are found to be stable and functional for many months post-implantation.[226] In particular, PLAGA random multiblock copolymers fibers with lactide/glycolide mole fractions of 85:15 and 75:25 show a 50% loss of mass after four and three months, respectively (a good duration performance for the scaffolds to be gradually replaced by neo-formed dermal tissue). Supporting the growth of keratinocytes, fibroblasts and endothelial cells with physiological ECM generation, these scaffolds are particularly promising.[226]

Several plasma treatments (with oxygen and methane) have been tested as well, particularly on electrospun silk fibroin fibers, and the resulting biofunctional properties studied with human epidermal keratinocytes and fibroblasts.[75] Finally, self-assembling peptide nanofiber scaffolds are also utilizable for dermal tissue engineering. Stably covering the wound surface, these architectures can be designed to release epidermal growth factor, which enhances the rate of wound coverage by five times compared to untreated controls.[219]

## 7.7.5 Applications to Other Tissues and Hybrid Architectures

Pioneered in the 1970s,[227] the application of electrospun tubular architectures to vascular tissue engineering stimulates increasing research efforts, with the main goal of achieving compliance matching with native arteries. Tubular constructs have been produced in a large variety of electrospinning experiments, taking advantage of cylindrical collectors that allow hollow capillaries to be directly formed during the fiber deposition and the caliber of the produced vessels to be carefully controlled (Section 2.3.2).[27,68,228] The mechanical properties of these scaffolds can also be favorable, and solid–liquid nanostructured interfaces with anti-thrombogenic properties can be potentially designed, which is another, important motivation for developing new nanofiber-based vascular grafts. The most demanding requirement for these vessels, from a structural viewpoint, is of supporting sufficiently high internal pressures, compatible to physiological blood flows, while keeping some flexibility and compliance. Another important characteristic is the suture retention strength, which needs to be carefully assessed while testing the scaffolds. As usual, performances of natural tissues are hard to emulate. The burst pressure of saphenous vein reaches the astonishing value of 1800 mm Hg.[77] Electrospun silk capillaries reach an average burst pressure of 811 mm Hg with maximum of 919 mm Hg,[72] and thermally treated electrospun PCL vessels can support about 1300 mm Hg.[229] Other PCL and PCL/collagen fibers are reported to reach values as high as 4000 mm Hg[230] and 4900 mm Hg,[231] respectively. Specific protocols have been developed to increase the burst pressure. In particular, treating PCL nanofibers in a Pluronic F127 solution, at temperatures above $T_g$ and slightly below the melting temperature, allows the fibers to be bound in a microscopically interconnected architecture, thus significantly improving mechanical properties while maintaining the original structure and shape (Figure 7.13).[229]

The behavior of several cellular species, such as human aortic endothelial cells (HAECs) and human coronary artery smooth muscle cells (HCASMCs), has been investigated on electrospun, *B. mori* silk fibroin scaffolds.[232] HCASMCs are found to elongate and align even on random nanofiber scaffolds in five days of culture, and cord-like structures are formed from HAECs in four days, respectively, leading to the formation of capillaries in a week. Evaluated markers in this study include CD146, VE-cadherin, PECAM-1 and vWF for HAECs, and SM-MHC2 and SM-actin for HCASMCs, and the formation of ECM features too is demonstrated in the cell cultures, which is promising in view of applications.[232] Rat mesenchymal stem cells obtained from bone marrow, modified with vasculoprotective gene endothelial nitric oxide synthase or marker gene green fluorescent protein through gene transfection, have been cultured on vascular scaffolds made of electrospun poly(propylene carbonate).[233] The cells colonize the biomaterial and properly produce nitric oxide and endothelial nitric oxide synthase protein.[233] In the near future, combining gene therapy methods and engineered scaffolds based on polymer nanostructures will likely represent an excellent example of merging of the most advanced technologies from complementary scientific disciplines, and will likely open new and exciting perspectives.

The investigation of electrospun vascular prostheses in *in vivo* experiments dates back to the 1990s. Clinical tests on Vascugraft® polyurethane arterial

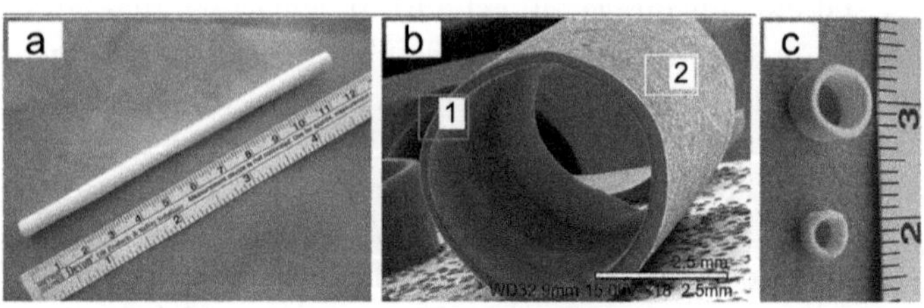


**Figure 7.13** Photograph (a) and SEM micrograph (b) of an electrospun PCL nanofiber vessel, and (c) corresponding appearance of the scaffold after thermal fiber bonding in water (bottom sample) and Pluronic F127 solution (top sample). In (b) the numbers 1 and 2 indicate the vessel cross-section and outer surface, respectively. The thermal treatment is carried out at 55 °C for 30 min. Adapted with permission from Ref. 229, *Biomaterials*, 2008, **29**, 1422–1430. DOI: 10.1016/j.biomaterials.2007.11.024. Copyright © 2007, Elsevier Ltd.

prostheses (B. Braun Melsungen AG), used for below-knee implantation in patients and surgically removed after periods ranging from three weeks to about one year due to thrombosis, did not evidence better performances compared to other biomaterials such as expanded PTFE.[234] The related development program was terminated.[234] Later reports highlight instead more convincing *in vivo* performances for nanofiber-based vascular implants, however they are still largely limited to animal models. *In vivo* implanted vascular graft scaffolds made of electrospun blends of type I collagen, elastin and PLAGA copolymers show good biocompatibility in a mouse model and satisfactory mechanical properties, increasing their diameter by 12–14% under a supported pressure of 120 mm Hg, which is a compliance behavior comparable with the physiological mechanics of bovine arteries (Figure 7.14).[67] Antithrombogenic properties are suggested by bone marrow mesenchymal stem cells in poly(L-lactic acid) aligned nanofiber grafts.[235] After the cells grow to confluence, the scaffold can be rolled around a mandrel, which is subsequently removed, thus producing a hollow capillary whose walls embed the culture. This structure is then implantable, and its post-implantation evolution can be analyzed. In a rat model, after 60 days post-implantation researchers have found a CD31$^+$ and myosin heavy chain positive staining in the formed tissue, collagen and elastin deposition, physiological cellular organization, together with antithrombogenic features (*i.e.* very limited platelet aggregation in the lumen of the tubular structures) exhibited by mesenchymal stem cells on the nanofibers.[235] This is a relevant finding since thrombosis is a main cause of failure of small-diameter vascular grafts.

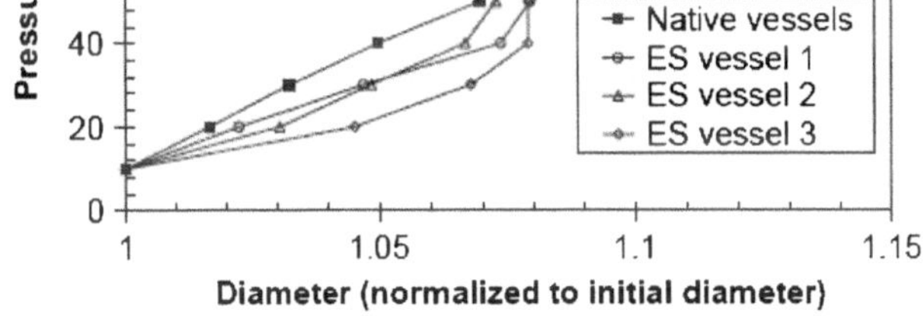


**Figure 7.14** Compliance curve (*i.e.* relative increment of the vessel diameter *vs.* internally applied pressure) for native (*i.e.* from bovine iliac) vessels and for three electrospun vessels, cross-linked by glutaraldehyde after production. A photograph of the collagen/elastin/PLAGA electrospun scaffold before cross-linking is displayed in the inset. Adapted with permission from Ref. 67, *Biomaterials*, 2006, **27**, 1088–1094. Doi: 10.1016/j.biomaterials.2005.07.048. Copyright © 2005, Elsevier Ltd.

*In vivo* studies (or prolonged *in vitro* experiments under physiological testing conditions including solicitations by pulsatile liquid flows)[236] aimed at assessing the stability of implanted electrospun scaffolds are going to become more and more important in the forthcoming years. This work will open the way to possible clinical tests, through the evaluation of the reliability and mechanical stability of scaffolds under actual operation conditions. Early results are quite promising. For example, electrospun PCL/collagen vascular constructs are found to maintain patency in a rabbit aortoiliac bypass model, and to retain their structural and mechanical integrity one month post-implantation.[237] Furthermore, the mechanical stimulation, provided by pulsatile perfusion, enhances the proliferation of smooth muscle cells and endothelial cells cultured on PLAGA fibers electrospun on a jellyfish collagen porous matrix.[238] Concomitantly, the cells align in a direction radial to the distending direction, similarly to physiological vessels *in vivo*, whereas such alignment is not observed in cultures under static conditions.[238] Smooth muscle cells have been grown on electrospun PS, in which Argon plasma treatments significantly enhance the cell attachment.[239] Skeletal muscle cells have also been successfully cultured on electrospun scaffolds of aligned PCL/collagen fibers.[240] Different myogenic cells (*i.e.* C2C12, L6 and primary human satellite cells) have been investigated together with aligned electrospun nanofibers made of a degradable block polyesterurethane.[241] Other cellular systems studied on aligned nanofibers include human ligament fibroblasts,[242] and primary cardiomyocytes that are able to develop mature contractile features and electrical activity,[243] thus suggesting that even delicate, mechanical and electro-mechanical bio-functionalities can be achieved in these constructs. In fact, evaluating how different cellular species grow and organize when co-cultured on nanofibers is another important step for mimicking physiological tissue operation. For example, primary cardiac ventricular cells cultured on electrospun aligned polyurethane fibers show not only alignment but also numerical percentage stability of their cardiomyocyte sub-population over two weeks.[244] This can be related to a more rapid cellular phenotype maturation as evidenced by a reduced expression of atrial natriuretic peptide.[244]

Concerning dental tissues, the odontogenic differentiation of human dental pulp stem cells is demonstrated on nanofibrous PLA scaffolds, both *in vitro* and (following *in vitro* culture for two weeks in presence of inductive supplements) *in vivo* by dentin sialoprotein immunoistochemical staining performed two months post-implantation.[245,246]

Self-assembled peptide nanofibers have been studied as scaffolds for culturing porcine hepatocytes and found to maintain differentiated functions, which is promising due to the usual rapid decrease of function of these cells *in vitro* and their interest for xenografts and adjuvant therapies of liver failure.[247]

Urologic tissue engineering by polymer nanofiber scaffolds is still in its infancy, however some interesting evidences have already been reported. Experiments have been focused on biomaterials with potential application to the reconstruction of the urinary tract. Human bladder smooth muscle cells, seeded on electrospun fibrinogen mats and assessed in their viability and proliferation by WST-assays and histology, are found to be able to remodel the scaffold after two weeks, through collagen production.[248] The process can also be modulated by means of the aprotin concentration in the culturing medium.

Finally, we mention here another interesting, somehow exotic class of microstructures, which are realized by embedding living cells or microorganisms, directly in electrospun solutions (a process sometimes called *cell electrospinning*).[249] This approach generates hybrid fibers in which cells are incorporated in the polymer, with possible application in tissue engineering and in various therapeutic strategies.[250] These fibers present a typical structure with beads corresponding to incorporated cells, distributed along the longitudinal axis of the nanofibers (Figure 7.15). Some of the polymers used for producing the hybrid fibers can be rendered insoluble in water,[251] thus allowing stable biofilms and membranes embedding bacteria to be realized. Examples of embedded microorganisms include rod-shaped M13 viruses in PVP fibers,[252] *E. coli* and *Staphylococcus albus* in PVA,[253] *M. luteus* and *E. coli* in PEO,[254] *Pseudomonas fluorescens*, *Zymomonas mobilis* and *E. coli* in PEO-polypropylene oxide-PEO triblock polymer (Pluronic F127 dimethacrylate),[251] *Bifidobacterium animalis* in PVA,[255] *Pseudomonas ADP* and *Pseudomonas putida*,[256] *Corynebacterium glutamicum*,[257] *etc.*, and experiments are even reported about the incorporation of mammalian cells (human astrocytoma 1321N1 cells,[258] 3T3-L1 mouse fibroblasts[259]) within electrospun fibers. In general, the biological activity of living microorganisms seems to be satisfactory and maintained after electrospinning, which is a non-trivial result given the very intense electric field and strong mechanical solicitations undergone by electrified jets. Mammalian cells seem to be much more delicate in this respect.[259] In any case, once the viability of encapsulated cells is assessed, a second issue is understanding how long cells survive in such an

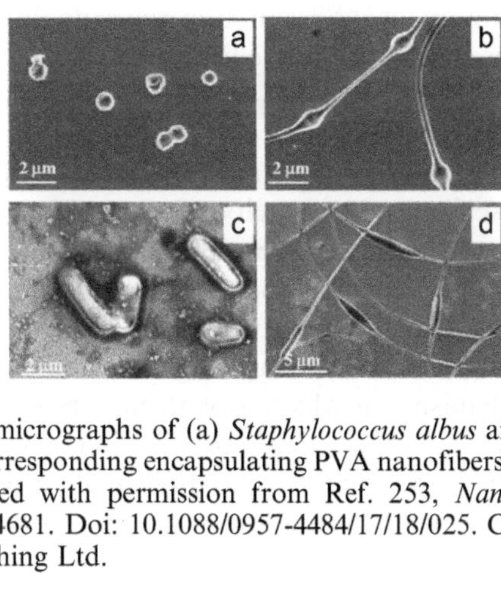


**Figure 7.15** SEM micrographs of (a) *Staphylococcus albus* and (c) *E. coli* cells, and the corresponding encapsulating PVA nanofibers (b and d, respectively). Adapted with permission from Ref. 253, *Nanotechnology*, 2006, **17**, 4675–4681. Doi: 10.1088/0957-4484/17/18/025. Copyright © 2006, IOP Publishing Ltd.

embedded state. The normal lifetime of bacteria within the polymer is usually limited by several factors, the dense packing of macromolecules in the spun fibers being of course very different from a physiological ECM. This problem can be overcome by a double encapsulation strategy, consisting in firstly placing bacteria within PVA-hydrogel capsules, and subsequently embedding the so-loaded particles in polyvinyl butyral, PLA and PCL fibers electrospun from organic solvents, such as ethanol, methanol and dichloromethane.[260] In this way, living cells are stored in a more friendly microenvironment, and survive for more than one month at 4 °C.[260] In electrospun fibers based on PEO-polypropylene oxide-PEO triblock polymers and PEO, bacteria are found to be viable for more than one week at 4 °C, and for more than two months if stored at −70 °C.[251] *B. animalis* Bb12 cells in PVA fibers are viable for 40 days at room temperature and for more than four months following storage at −20 °C,[255] and so on. For improving the life conditions of electrospun cells in the fibers and rendering the overall system bio-functional as long as possible, one can also use core-shell fibers, in which the inner part embedding cells is made of an aqueous medium and the outer part is made of a polymer phase making the overall microstructure insoluble in water. This approach is being developed especially by the Zussman group at Technion.[256,257] Immobilizing bacteria in the inner part of core-shell, electrospun polymeric microtubes can even be a valuable method for producing hybrid active biomaterials for water decontamination treatments, as demonstrated by using *C. glutamicum* for the biodegradation of phenol-contaminated wastewater.[257] Combining biologically active systems with polymer nanofiber architectures will certainly lead to elegant solutions for next-generation filtration technologies.

## 7.7.6 Effects on Differentiation of Stem Cells

Stem cells are very important in the framework of regenerative medicine, because of their capability of differentiating towards different lineages. For this reason, enormous efforts in biomedical research have been oriented to looking for suitable sources of stem cells from the human body. Besides embryonic stem cells, several niches of stem cells have been localized in adult tissues and organs. These cells, initially in an immature (*undifferentiated*) state, show a remarkable self-renewal capacity, and work as precursors able to differentiate into tissue-specific cell types. These features render them the most promising candidates for successful application in tissue and organ repair. The differentiation paths of stem cells are regulated by a variety of biochemical or topological interactions with the surrounding environment, through complex and interplaying mechanisms in which growth factors play a crucial role. In this scenario, polymer nanofiber scaffolds may serve to support both the growth and proliferation and the differentiation of stem cells towards desired directions. Indeed, among other molecules possibly incorporated and then released in a controlled fashion by nanofibers architectures, growth factors deserve special attention because these molecules can be employed for recruiting and differentiating various classes of stem cells. For example, BMP-2 growth factors embedded within silk fibroin fibers are demonstrated to strongly support osteoinduction in marrow-derived human mesenchymal stem cells, determining a higher expression of BMP-2, BSP-II and collagen I together with a four-fold higher deposition of calcium compared to undoped scaffolds.[74] A second, more sophisticated and modern approach is aimed to develop gene therapy methods that involve the transfection of cells. The drug-delivery capability of nanofiber-based systems (Section 7.4) is strategic in this respect, since cell transfection can be performed through plasmid DNA released by the same scaffold material. Of course, to this aim plasmid DNA should still be functional after nanofiber fabrication. Early studies have been carried out by electrospun PLAGA and poly(D,L-lactide)–PEO copolymer fibers, releasing bioactive plasmid DNA over 20 days with a maximum release after about 2 h and leading to transfection of MC3T3-E1 cells.[131] Pioneering experiments targeting stem cells have been performed with PLAGA/hydroxylapatite composite fibers incorporating DNA-loaded chitosan nanoparticles.[132] The subsequent, gradual release of the particles (and hence of BMP-2 plasmid DNA) over two months and transfection of cultured human marrow stem cells have been demonstrated.

Other reports have described the effects of pristine or bio-functionalized polymer nanofibers on the differentiation of stem cells. A quite general feature highlighted in the literature is that scaffolds composed of highly aligned nanofibers,[156] which typically induce significant morphological elongation in the cultured cells, can consequently determine an enhanced differentiation of cells and ultimately improved biological functionalities. For instance, aligned fibers lead to a significantly enhanced myotube formation in C2C12 myoblasts compared to random fibers.[261,262] On nylon electrospun fibers C2C12 are found to express α-actinin, adult myosin heavy chain and nicotinic acetylcholine receptors as muscle specific marker proteins.[262] Though possibly related to an enhanced response of the elongated, coherently organized cells to differentiating cues and supplements provided in the culture medium or grafted or self-produced at the supporting surface, the mechanisms at the base of these results certainly need to be investigated in more depth for being completely rationalized. In the following, a few relevant examples are mentioned, concerning the behavior of stem cells on nanofiber scaffolds. A review entirely dedicated to stem-cell engineering by electrospun scaffolds is available in Ref. 263.

Mesenchymal stem cells from the bone marrow of neonatal rats have been studied on electrospun PCL scaffolds.[196] The cells have been cultured and differentiated in the presence of osteogenic supplements for up to four weeks, in a specifically designed rotational, oxygen-permeable bioreactor. *In vivo*, a strong resemblance is found between how these systems rapidly evolve and the natural bone tissue.[197] Other studies have been focused on human mesenchymal stem cells, which can be effectively cultured on nanofiber scaffolds made of PCL or of its copolymers. Here cells exhibit chondrogenic gene expression and even capability to differentiate toward multiple lineages (chondrogenic, osteogenic and adipogenic),[214,215,264] or they can be differentiated to show neuronal morphologies and to express neurofilament and nestin protein.[264] An up-regulated expression of neural markers is also found culturing human mesenchymal stem cells on PCL/gelatin electrospun nanofibers loaded with retinoic acid.[265] The culture of human, bone marrow derived mesenchymal stem cells on electrospun type I collagen nanofibers evidences enhanced viability compared to conventional PS, together with high expression levels of osteogenic markers.[266] A comprehensive study of the response of human bone marrow stromal cells on scaffolds with different three-dimensional textures highlights that each type of scaffold structure is able to induce a unique gene expression signature, and that the structure predominates over the polymer chemical composition in affecting the cell shape and fate (Figure 7.16), with nanofibrous architectures uniquely driving osteogenic differentiation even in the absence of osteogenic supplements.[200] *In vivo* experiments have also been carried out by PCL nanofibers, human mesenchymal stem cells and porcine chondrocytes, in a swine model.[218]

Neural stem/progenitor cells exhibit remarkable properties of self-renewal and of differentiation into neural lineages upon proper biochemical stimulation[267–270] or even as a consequence of the interaction with specific micro or nanoscale surface topographies.[166] For example, the proliferation, morphology, and differentiation of rat, hippocampus-derived adult NSCs placed on electrospun fibers of polyethersulfone (PES) functionalized by laminin show a rich phenomenology and a particularly interesting dependence on the fiber diameter.[105] Differentiation is obtained with 1 μM retinoic acid and 1% fetal bovine serum, and an increased proliferation and cell spreading are in fact observed on fibers that have a smaller diameter. In addition, the observed differentiation behavior is quite complex, with a 40% increase in oligodendrocyte differentiation on fibers with diameter of about 280 nm, and a 20% increase in neuronal differentiation on fibers with diameter of about 750 nm (compared to standard PS culture surfaces).[105] PCL[271] and PCL/gelatin electrospun fibers[83] have been used too as scaffolds for the growth and directed differentiation of NSCs. In particular, these scaffolds are able to promote and direct the neurite outgrowth.[83] Taken together, these and other findings confirm how complex are the mechanisms regulating the interactions between stem cells and topographical and biochemical cues, and undoubtedly stimulate further research concerning NSCs and their differentiation paths on polymer nanofiber materials.

Another opportunity in the wide field of regenerative medicine is provided by embryonic stem cells, which are pluripotent cells, although their use is debated because of their possible tumorogenity, accessibility and reliability issues, and of their ethical implications. In a study, mouse embryonic stem cells

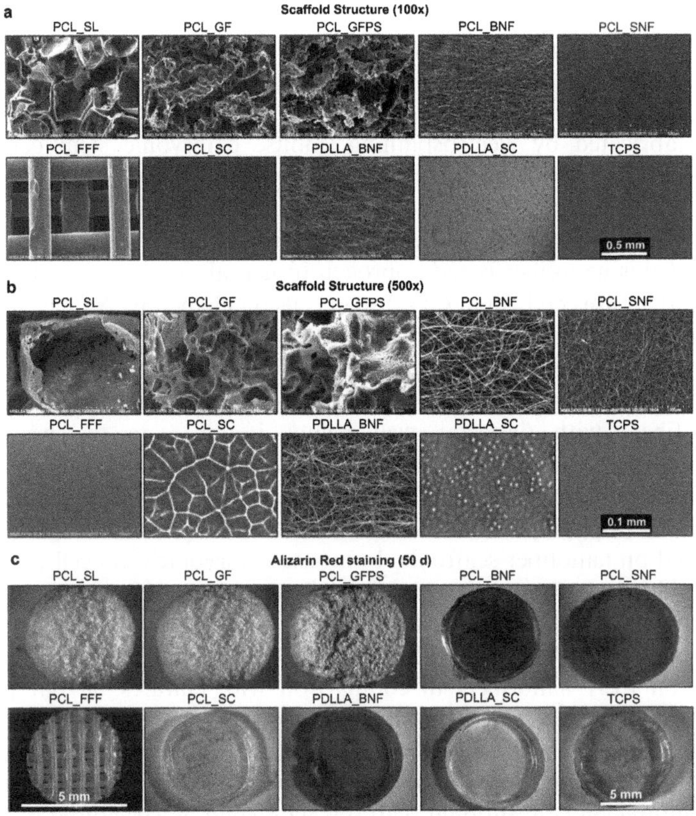


**Figure 7.16** (a, b) Comparison of several microscopic structures of polymer scaffolds. Scaffolds are realized by PCL or on poly(D,L-lactic acid) (PDLLA) through different fabrication methods, *i.e.* salt-leaching (PCL_SL), gas foaming (PCL_GF), thermally induced phase-separation combined with gas foaming (PCL_GFPS), electrospinning with relatively more concentrated solutions [30% PCL by mass in 9:1 volume ratio chloroform:methanol, injection rate = 0.5 mL $h^{-1}$, or 15% poly(D,L-lactic acid) by mass in HFIP, injection rate = 1.5 mL $h^{-1}$, and $V$ = 15 kV] leading to fibers of diameter around 900 nm (PCL_BNF and PDLLA_BNF), electrospinning with relatively less concentrated solutions (10% by mass in 3:1 volume ratio chloroform:methanol, injection rate = 3 mL $h^{-1}$, $V$ = 16.5 kV) leading to fibers of diameter around 300 nm (PCL_SNF), precision extrusion deposition (PCL_FFF), spin coating (PCL_SC and PDLLA_SC). Tissue culture PS (TPCS) is also used as reference. The scaffold morphology is imaged by SEM, or by phase-contrast microscopy for spin cast and control samples. Scale bar in (a) applies to all images in (a) and scale bar in (b) applies to all images in (b). (c) Stereomicrographs of Ca staining through alizarin red for the osteogenesis of human bone marrow stromal cells on the scaffolds, in the absence of osteogenic supplements after 50 days of culture, clearly evidencing the superior performances of fibrous scaffolds. The scale bar in the TCPS photograph applies to all panels except PCL_FFF. Reproduced with permission from Ref. 200, *Biomaterials*, 2011, **32**, 9188–9196. DOI: 10.1016/j.biomaterials.2011.08.054.

have been induced to become neural progenitors by adding retinoic acid to embryoid body cultures on PCL electrospun nanofibers for four days.[180] Several specific neural lineages, namely neurons, oligodendrocytes and astrocytes, have been obtained upon culturing. On commercial nanofibrillar surfaces fabricated by electrospinning, mouse embryonic stem cells show enhanced proliferation and self-renewal and activation of specific signaling pathways.[272] The combination of enhanced differentiation and neurite outgrowth directing capacity has to be underlined as a remarkable point of strength of this nanofiber-based approach to neural tissue engineering.

Neonatal mouse cerebellum C17.2 stem cells on aligned PLA fibers show the best viability and proliferation on nanostructures that have a diameter of 500 nm,[273] where they polarize and elongate effectively,[274] whereas a different trend and different morphologies are observed on randomly oriented fibers. Overall, fibers with diameter well below 1 μm seem to show better differentiating performances than microfibers,[274] which is another effect of the fascinating interplay between differentiation mechanisms and the several different cues (topographic, biochemical, ECM-related, *etc.*) acting when cells are cultured on nanofiber scaffolds. In mouse embryonic stem cells, enhanced cell-scaffold interactions (evidenced by cell spreading and extending protrusions) and a greater degree of differentiation in the osteogenic sense (by using suitable osteoinductive supplements in the culture media) have also been observed on PLA nanofibers in comparison with cultures on flat polymer supports.[198] Human embryonic stem cells on PLA nanofibrous scaffolds realized by phase-separation express collagen type 1, runt related transcription factor 2 (Runx2) and osteocalcin (OCN) mRNA at higher levels than in control cultures on conventional scaffolds with solid, unstructured walls after one week of culture, and a clearly enhanced mineralization is shown after three weeks (Figure 7.17).[275] Another study has demonstrated that nanofibrous scaffolds made of PCL/collagen and PCL/gelatin are also able to support

**Figure 7.17** Calcium staining, after three weeks of culture of human embryonic stem cells under osteogenic differentiation conditions on a nanofiber scaffolds (Nano), on flat films (Solid) and on gelatin-coated tissue culture plastics (Control). Adapted with permission from Ref. 275, *Biomaterials*, 2010, **31**, 5526–5535. Doi: 10.1016/j.biomaterials.2010.03.065. Copyright © 2010, Elsevier Ltd.

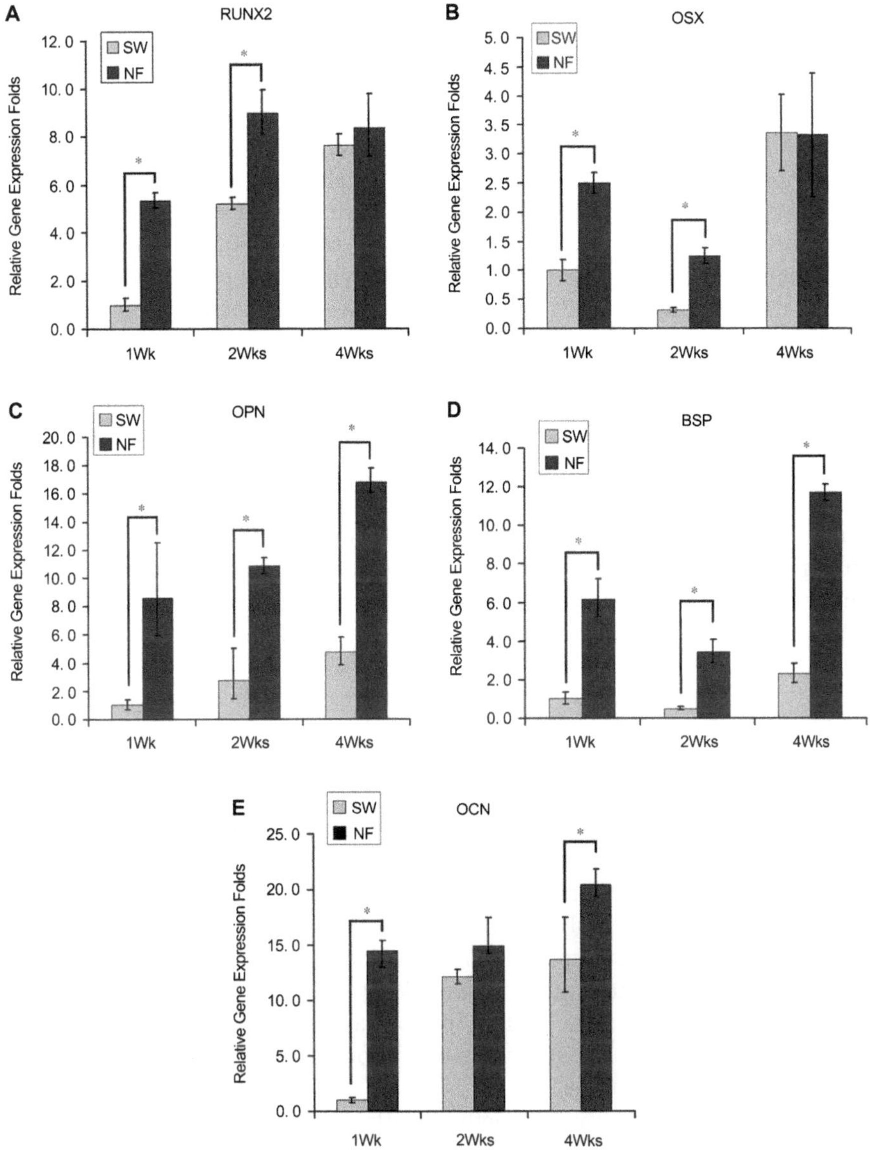


**Figure 7.18** Expression of genes associated with osteoblastic differentiation, determined by real time PCR, by human AFSCs on nanofibrous PLA scaffolds (NF) and on unstructured scaffolds (SW, *i.e.* solid-walled) at one, two and four weeks. rhBMP-7 is used for osteogenic induction. * $p < 0.05$. Reproduced with permission from Ref. 277, *Biomaterials*, 2010, **31**, 1133–1139. Doi: 10.1016/j.biomaterials.2009.10.030. Copyright © 2009, Elsevier Ltd.

effectively the expansion of human embryonic stem cells in an undifferentiated condition, retaining features of stemness.[276]

Amniotic fluid-derived stem cells (AFSCs) are very appealing since they exhibit clear multipotentiality and rapid proliferation with fewer accessibility and ethical issues than embryonic stem cells. To date, the interaction of AFSCs with nanostructured scaffolds has been very rarely explored. Human AFSCs derived from back-up amniocentesis cultures have been effectively induced for osteoblastic differentiation by recombinant human BMP-7 (rhBMP-7) responding more strongly than human mesenchymal stem cells, and culturing on PLA nanofibrous scaffolds have led to enhanced ALP activity, Ca content, von Kossa staining and expression of osteogenic genes [RUNX2, osterix (OSX), BSP, osteopontin (OPN) and OCN] compared to unstructured scaffolds (Figure 7.18).[277] A better mineralization has been observed following subcutaneous implantation of the AFSC-scaffold system (after rhBMP-7 induction for one week *in vitro*) in nude mice, and examining the bio-constructs after six weeks.[277] With respect to dental tissues, analogous results are obtained for the odontogenic differentiation and mineralization of human dental pulp stem cells.[246] Electrospun scaffolds of aminated PAA-grafted PES nanofibers have been tested as supports for expansion cultures of human umbilical cord blood hematopoietic stem/progenitor cells.[278] This is another interesting class of multipotent cells, whose use can be limited by the low number of available cells thereby benefiting greatly from the availability of suitable expansion supports. Functionalizing the surface of PAA-grafted PES nanofibers with different amine groups (1,2-ethanediamine, 1,4-butanediamine, 1,6-hexanediamine) allows to be shown that even the different spacer groups of the linked layers have an influence on the stem cell proliferation, with ethylene and butylene spacers leading to higher expansion efficiency (up to about 800-fold).[278] Human umbilical cord blood-derived hematopoietic stem/progenitor cells also show an enhanced adhesion and proliferation on PAA-grafted, PES electrospun fibers functionalized by ethylene diamine through a 2-step carbodiimide cross-linking method.[279] These aminated nanofibrous scaffolds show better performances in supporting $CD34^+$, $CD45^+$ cell growth over 10 days, compared to both positive control substrates and reference aminated films.

Finally, nanofibers realized by methods different from electrospinning have been used as well, in combination with neural progenitor and stem cells. The selective differentiation of neural progenitor cells has been studied *in vitro* in a three-dimensional network of high laminin-epitope density ($\cong 7 \times 10^{14}$ cm$^{-2}$) nanofibers, realized by self-assembling of peptide amphiphiles.[280] A rapid and selective differentiation into neurons is observed. In fact, peptide amphiphile biomaterials are made very interesting by their molecular design and by the possibility of direct *in vivo* injection. Then they can re-assemble and form nanofibers in a highly programmable manner. For example, some peptide nanofiber scaffolds have been designed specifically for osteoblasts, utilizing several biologically active motifs such as osteogenic growth peptide, OPN cell adhesion and RGD domains.[281] The resulting self-assembled nanofibers can further enhance MC3T3-E1 cell proliferation and differentiation compared to RAD16 scaffolds.[281]

## References

1. A. Szentivanyi, T. Chakradeo, H. Zernetsch and B. Glasmacher, *Adv. Drug Deliv. Rev.*, 2011, **63**, 349–409.
2. R. Langer and J. P. Vacanti, *Science*, 1993, **260**, 920–926.
3. W. Liu, S. Thomopoulos and Y. Xia, *Adv. Healthcare Mater.*, 2012, **1**, 10–25.
4. E. D. Hay, *J. Cell Biol.*, 1981, **91**, S205–S223.
5. V. Beachley and X. Wen, *Prog. Polym. Sci.*, 2010, **35**, 868–892.
6. M. P. Lutolf and J. A. Hubbell, *Nat. Biotechnol.*, 2005, **23**, 47–55.
7. Y. Mao and J. E. Schwarzbauer, *Matrix Biol.*, 2005, **24**, 389–399.
8. R. Pankov, E. Cukierman, B. Z. Katz, K. Matsumoto, D. C. Lin, S. Lin, C. Hahn and K. M. Yamada, *J. Cell. Biol.*, 2000, **148**, 1075–1090.
9. I. F. Wierzbicka-Patynowski and J. E. Schwarzbauer, *J. Cell Sci.*, 2003, **116**, 3269–3276.
10. G. Baneyx, L. Baugh and V. Vogel, *Proc. Natl Acad. Sci. USA*, 2002, **99**, 5139–5143.
11. J. Engel, E. Odermatt, A. Engel, J. A. Madri, H. Furthmayr, H. Rohde and R. Timpl, *J. Mol. Biol.*, 1981, **150**, 97–120.
12. C. H. Chen, D. O. Clegg and H. G. Hansma, *Biochemistry*, 1998, **37**, 8262–8267.
13. P. D. Yurchenco, E. F. Tsilibary, A. Charonis and H. Furthmayr, *J. Biol. Chem.*, 1985, **260**, 7636–7644.
14. P. D. Yurchenco and Y.-S. Cheng, *J. Biol. Chem.*, 1993, **268**, 17286–17299.
15. N. Sgarbi, D. Pisignano, F. Di Benedetto, G. Gigli, R. Cingolani and R. Rinaldi, *Biomaterials*, 2004, **25**, 1349–1353.
16. R. O. Hynes, *Fibronectins*, Springer-Verlag, New York, 1990.
17. E. L. George, E. N. Georges-Labouesse, R. S. Patel-King, H. Rayburn and R. O. Hynes, *Development*, 1993, **119**, 1079–1091.
18. S. Han, F. R. Khuri and J. Roman, *Cancer Res.*, 2006, **66**, 315–323.
19. R. K. Globus, S. B. Doty, J. C. Lull, E. Holmuhamedov, M. J. Humphries and C. H. Damsky, *J. Cell Sci.*, 1998, **111**, 1385–1393.
20. X. Zhou, R. Grant Rowe, N. Hiraoka, J. P. George, D. Wirtz, D. F. Mosher, I. Virtanen, M. A. Chernousov and S. J. Weiss, *Genes & Dev.*, 2008, **22**, 1231–1243.
21. T. Miller and D. Boettiger, *Langmuir*, 2003, **19**, 1723–1729.
22. M. Bergkvist, J. Carlsson and S. Oscarsson, *J. Biomed. Mater. Res. A*, 2003, **64A**, 349–356.
23. T. Elsdale and J. Bard, *J. Cell Biol.*, 1972, **54**, 626–637.

24. D. R. Eyre, *Science*, 1980, **207**, 1315–1322.
25. F. Grinnell and M. H. Bennett, *Methods Enzymol.*, 1982, **82**, 535–544.
26. A. Ferrari, M. Cecchini, A. Dhawan, S. Micera, I. Tonazzini, R. Stabile, D. Pisignano, F. Beltram, *Nano Lett.*, 2011, **11**, 505–511.
27. S. Kidoaki, I. K. Kwon and T. Matsuda, *Biomaterials*, 2005, **26**, 37–46.
28. S. Zhong, Y. Zhang and C. T. Lim, *Tissue Eng. B*, 2012, **18**, 77–87.
29. J. Nam, Y. Huang, S. Agarwal and J. Lannutti, *Tissue Eng.*, 2007, **13**, 2249–2257.
30. B. M. Baker, A. O. Gee, R. B. Metter, A. S. Nathan, R. A. Marklein, J. A. Burdick, R. L. Mauck, *Biomaterials*, 2008, **29**, 2348–2358.
31. L. Moroni, R. Schotel, D. Hamann, J. R. de Wijn and C. A. van Blitterswijk, *Adv. Funct. Mater.*, 2008, **18**, 53–60.
32. C. Erisken, D. M. Kalyon and H. Wang, *Biomaterials*, 2008, **29**, 4065–4073.
33. H. W. Kim, J. H. Song and H. E. Kim, *Adv. Funct. Mater.*, 2005, **15**, 1988–1994.
34. X. Li, J. Xie, J. Lipner, X. Yuan, S. Thomopoulos and Y. Xia, *Nano Lett.*, 2009, **9**, 2763–2768.
35. A. G. Mikos, S. W. Herring, P. Ochareon, J. Elisseeff, H. H. Lu, R. Kandel, F. J. Schoen, M. Toner, D. Mooney, A. Atala, M. E. Van Dike, D. Kaplan and G. Vunjak-Novakovic, *Tissue Eng.*, 2006, **12**, 3307–3339.
36. P. D. Dalton, K. Klinkhammer, J. Salber, D. Klee and M. Möller, *Biomacromolecules*, 2006, **7**, 686–690.
37. T. C. Laurent, U. G. Laurent and J. E. Fraser, *Ann. Rheum. Dis.*, 1995, **54**, 429–432.
38. I. C. Um, D. F. Fang, B. Hsiao, A. Okamoto and B. Chu, *Biomacromolecules*, 2004, **5**, 1428–1436.
39. L. Lapčik, Jr., L. Lapčik, S. D. Smedt, J. Demeester and P. Chabreček, *Chem. Rev.*, 1998, **98**, 2663–2684.
40. D. Campoccia, P. Doherty, M. Radice, P. Brun, G. Abatangelo and D. F. Williams, *Biomaterials*, 1998, **19**, 2101–2127.
41. K. P. Vercruysee, D. M. Marecek, J. F. Marecek and G. D. Preswich, *Bioconjug. Chem.*, 1997, **8**, 686–694.
42. D. Girotto, S. Urbani, P. Bruna, D. Renier, R. Barbucci and G. Abatangelo, *Biomaterials*, 2003, **24**, 3265–3275.
43. Y. Ji, K. Ghosh, X. Z. Shu, B. Li, J. C. Sokolov, G. D. Prestwich, R. A. F. Clark and M. H. Rafailovich, *Biomaterials*, 2006, **27**, 3782–3792.
44. J. Ji, K. Ghosh, B. Li, J. C. Sokolov, R. A. F. Clark and M. H. Rafailovich, *Macromol. Biosci.*, 2006, **6**, 811–817.
45. J. D. Schiffman and C. L. Schauer, *Biomacromolecules*, 2007, **8**, 594–601.
46. J. D. Schiffman and C. L. Schauer, *Biomacromolecules*, 2007, **8**, 2665–2667.
47. N. Bhattarai, D. Edmondson, O. Veiseh, F. A. Matsen and M. Zhang, *Biomaterials*, 2005, **26**, 6176–6184.

48. K. Ohkawa, D. Cha, H. Kim, A. Nishida and H. Yamamoto, *Macromol. Rapid Commun.*, 2004, **25**, 1600–1605.
49. X. Geng, O. H. Kwon and J. Jang, *Biomaterials*, 2005, **26**, 5427–5432.
50. K. Ohkawa, K. Minato, G. Kumagai, S. Hayashi and H. Yamamoto, *Biomacromolecules*, 2006, **7**, 3291–3294.
51. W. Wang, S. Itoh, A. Matsuda, S. Ichinose, K. Shinomiya, Y. Hata and J. Tanaka, *J. Biomed. Mater. Res. A*, 2008, **84A**, 557–566.
52. W. Wang, S. Itoh, A. Matsuda, T. Aizawa, M. Demura, S. Ichinose, K. Shinomiya and J. Tanaka, *J. Biomed. Mater. Res. A*, 2008, **85A**, 919–928.
53. J. Du and Y.-L. Hsieh, *Nanotechnology*, 2008, **19**, 125707.
54. M. Skotak, A. P. Leonov, G. Larsen, S. Noriega and A. Subramanian, *Biomacromolecules*, 2008, **9**, 1902–1908.
55. C. K. S. Pillai, W. Paul and C. P. Sharma, *Prog. Polym. Sci.*, 2009, **34**, 641–678.
56. G. Wnek, M. Carr, D. G. Simpson and G. L. Bowlin, *Nano Lett.*, 2003, **3**, 213–216.
57. M. Li, M. J. Mondrinos, M. R. Gandhi, F. K. Ko, A. S. Weiss and P. I. Lelkes, *Biomaterials*, 2005, **26**, 5999–6008.
58. K. S. Rho, L. Jeong, G. Lee, B. M. Seo, Y. J. Park, S. D. Hong, S. Roh, J. J. Cho, W. H. Park and B.-M. Min, *Biomaterials*, 2006, **27**, 1452–1461.
59. J. A. Matthews, G. E. Wnek, D. G. Simpson and G. L. Bowlin, *Biomacromolecules*, 2002, **3**, 232–238.
60. Y. Z. Zhang, J. Venugopal, Z. M. Huang, C. T. Lim and S. Ramakrishna, *Polymer*, 2006, **47**, 2911–2917.
61. C. Seok Ki, D. Hyun Baek, K. Don Gang, K. Hoon Lee, I. Chul Um and Y. Hwan Park, *Polymer*, 2005, **46**, 5094–5102.
62. D. I. Zeugolis, S. T. Khew, E. S. Yew, A. K. Ekaputra, Y. W. Tong, L.-Y. Yung, D. W. Hutmacher, C. Sheppard and M. Raghunath, *Biomaterials*, 2008, **29**, 2293–2305.
63. B. Dong, O. Arnoult, M. E. Smith and G. E. Wnek, *Macromol. Rapid Commun.*, 2009, **30**, 539–542.
64. J. H. Song, H. E. Kim and H. W. Kim, *J. Mater. Sci.: Mater. Med.*, 2008, **19**, 95–102.
65. X. Liu and P. X. Ma, *Biomaterials*, 2009, **30**, 4094–4103.
66. I. O. Smith, X. H. Liu, L. A. Smith and P. X. Ma, *WIREs Nanomed. Nanobiotechnol.*, 2009, **1**, 226–236.
67. J. Stitzel, J. Liu, S. J. Lee, M. Komura, J. Berry, S. Soker, G. Lim, M. Van Dyke, R. Czerw, J. J. Yoo and A. Atala, *Biomaterials*, 2006, **27**, 1088–1094.
68. E. D. Boland, J. A. Matthews, K. J. Pawlowski, D. G. Simpson, G. E. Wnek and G. L. Bowlin, *Front. Biosci.*, 2004, **9**, 1422–1432.
69. L. Buttafoco, N. G. Kolkman, P. Engbers-Buijtenhuijs, A. A. Poot, P. J. Dijkstra, I. Vermes and J. Feijen, *Biomaterials*, 2006, **27**, 724–734.

70. S. Zarkoob, R. K. Eby, D. H. Reneker, S. D. Hudson, D. Ertley and W. W. Adams, *Polymer*, 2004, **45**, 3973–3977.
71. B.-M. Min, G. Lee, S. H. Kim, Y. S. Nam, T. S. Lee and W. H. Park, *Biomaterials*, 2004, **5**, 1289–1297.
72. L. Soffer, X. Wang, X. Zhang, J. Kluge, L. Dorfmann, D. L. Kaplan and G. Leisk, *J. Biomater. Sci. Polymer Ed.*, 2008, **19**, 653–664.
73. H. J. Jin, J. Chen, V. Karageorgiou, G. H. Altman and D. L. Kaplan, *Biomaterials*, 2004, **25**, 1039–1047.
74. C. Li, C. Vepari, H. J. Jin, H. J. Kim and D. L. Kaplan, *Biomaterials*, 2006, **27**, 3115–3124.
75. L. Jeong, I.-S. Yeo, H. N. Kim, Y. I. Yoon, D. H. Jang, S. Y. Jung, B.-M. Min and W. H. Park, *Intern. J. Biol. Macromol.*, 2009, **44**, 222–228.
76. C. Chen, C. Chuanbao, M. Xilan, T. Yin and Z. Hesun, *Polymer*, 2006, **47**, 6322–6327.
77. X. Zhang, M. R. Reagan and D. L. Kaplan, *Adv. Drug Deliv. Rev.*, 2009, **61**, 988–1006.
78. R. A. Neal, S. G. McClugage III, M. C. Link, L. S. Sefcik, R. C. Ogle and E. A. Botchwey, *Tissue Eng. Part C Methods*, 2009, **15**, 11–21.
79. X. Fang and D. H. Reneker, *J. Macromol. Sci., Part B: Phys.*, 1997, **36**, 169–173.
80. S. J. Lee, J. J. Yoo, G. J. Lim, A. Atala and J. Stitzel, *J. Biomed. Mater. Res. A*, 2007, **83**, 999–1008.
81. T. K. Dash and V. B. Konkimalla, *J. Control. Release*, 2012, **158**, 15–33.
82. H. W. Kim, H. S. Yu and H. H. Lee, *J. Biomed. Mater. Res. A*, 2008, **87**, 25–32.
83. L. Ghasemi-Mobarakeh, M. P. Prabhakaran, M. Morshed, M. H. Nasr-Esfahani and S. Ramakrishna, *Biomaterials*, 2008, **29**, 4532–4539.
84. E. Schnell, K. Klinkhammer, S. Balzer, G. Brook, D. Klee, P. Dalton and J. Mey, *Biomaterials*, 2007, **28**, 3012–3025.
85. W. Song, D. C. Markel, S. Wang, T. Shi, G. Mao and W. Ren, *Nanotechnology*, 2012, **23**, 115101.
86. I. K. Kwon and T. Matsuda, *Biomacromolecules*, 2005, **6**, 2096–2105.
87. S. Yan, L. Xiaoqiang, L. Shuiping, W. Hongsheng and H. Chuanglong, *J. Appl. Polym. Sci.*, 2010, **117**, 542–547.
88. Z.-G. Wang, L.-S. Wan, Z.-M. Liu, X.-J. Huang and Z.-K. Xu, *J. Mol. Catal. B: Enzym.*, 2009, **56**, 189–195.
89. D. Myers, *Surfaces, Interfaces and Colloids*, Wiley-VCH Publishers, New York, 1999.
90. J. van den Dolder, G. N. Bancroft, V. I. Sikavitsas, P. H. M. Spauwen, A. G. Mikos and J. A. Jansen, *Tissue Eng.*, 2003, **9**, 505–515.
91. Z. Ma, Z. Mao and C. Gao, *Colloids Surf., B*, 2007, **60**, 137–157.
92. R. A. Williams and H. W. Blanch, *Biosens. Bioelectron.*, 1994, **9**, 159–167.
93. A. S. Badami, M. R. Kreke, M. S. Thompson, J. S. Riffle and A. S. Goldstein, *Biomaterials*, 2006, **27**, 596–606.

94. L. He, Y. Shi, Q. Han, Q. Zuo, S. Ramakrishna, W. Xue and L. Zhou, *J. Mater. Chem.*, 2012, **22**, 13187–13196.
95. G. T. Hermanson, *Bioconjugate Techniques*, Academic Press, New York, 2nd edn, 2008.
96. Y. Zhu, M. F. Leong, W. F. Ong, M. B. Chan-Park and K. S. Chian, *Biomaterials*, 2007, **28**, 861–868.
97. A. Polini, S. Pagliara, R. Stabile, G. S. Netti, L. Roca, C. Prattichizzo, L. Gesualdo, R. Cingolani and D. Pisignano, *Soft Matter*, 2010, **6**, 1668–1674.
98. W. S. Li, Y. Guo, H. Wang, D. J. Shi, C. F. Liang, Z. Ye, F. Qing and J. Gong, *J. Mater. Sci.: Mater. Med.*, 2008, **19**, 847–854.
99. M. Gümüsderelioglu and H. Türkoglu, *Biomaterials*, 2002, **23**, 3927–3935.
100. Z. W. Ma, W. He, T. Yong and S. Ramakrishna, *Tissue Eng.*, 2005, **11**, 1149–1158.
101. J.-P. Chen and C.-H. Su, *Acta Biomater.*, 2011, **7**, 234–243.
102. J. S. Choi, K. W. Leong and H. S. Yoo, *Biomaterials*, 2008, **29**, 587–596.
103. C. L. Casper, W. Yang, M. C. Farach-Carson and J. F. Rabolt, *Biomacromolecules*, 2007, **8**, 1116–1123.
104. J. Xie, M. R. MacEwan, X. Li, S. E. Sakiyama-Elbert and Y. Xia, *ACS Nano*, 2009, **3**, 1151–1159.
105. G. T. Christopherson, H. Song and H.-Q. Mao, *Biomaterials*, 2009, **30**, 556–564.
106. S. Patel, K. Kurpinski, R. Quigley, H. Gao, B. S. Hsiao, M. M. Poo and S. Li, *Nano Lett.*, 2007, **7**, 2122–2128.
107. W. He, Z. W. Ma, T. Yong, W. E. Teo and S. Ramakrishna, *Biomaterials*, 2005, **26**, 7606–7615.
108. Y. Z. Zhang, J. Venugopal, Z.-M. Huang, C. T. Lim and S. Ramakrishna, *Biomacromolecules*, 2005, **6**, 2583–2589.
109. P. Zhao, H. Jiang, H. Pan, K. Zhu and W. Chen, *J. Biomed. Mater. Res. A*, 2007, **83**, 372–382.
110. G.-D. Fu, G. L. Li, K. G. Neoh and E. T. Kang, *Prog. Polym. Sci.*, 2011, **36**, 127–167.
111. M. R. Abidian, D.-H. Kim and D. C. Martin, *Adv. Mater.*, 2006, **18**, 405–409.
112. A. H. Nie, B. W. Soh, Y. C. Fu and C. H. Wang, *Biotechnol. Bioeng.*, 2008, **99**, 223–234.
113. E. Luong-Van, L. Grøndahl, K. N. Chua, K. W. Leong, V. Nurcombe and S. M. Cool, *Biomaterials*, 2006, **27**, 2042–2050.
114. Y. Z. Zhang, X. Wang, Y. Feng, J. Li, C. T. Lim and S. Ramakrishna, *Biomacromolecules*, 2006, **7**, 1049–1057.
115. S. Maretschek, A. Greiner and T. Kissel, *J. Control. Release*, 2008, **127**, 180–187.
116. E. Sanders, R. Kloefkorn, G. Bowlin, D. Simpson and G. Wnek, *Macromolecules*, 2003, **36**, 3803–3805.

117. H. Qi, P. Hu, J. Xu and A. Wang, *Biomacromolecules*, 2006, **7**, 2327–2330.
118. M. Gandhi, R. Srikar, A. L. Yarin, C. M. Megaridis and R. A. Gemeinhart, *Mol. Pharmaceut.*, 2009, **6**, 641–647.
119. X. Zong, K. Kim, D. Fang, S. Ran, B. S. Hsiao and B. Chu, *Polymer*, 2002, **43**, 4403–4412.
120. X. H. Zong, S. Li, B. Garlick, K. Kim, D. F. Fang, J. Chiu, T. Zimmerman, C. Brathwaite, B. S. Hsiao and B. Chu, *Ann. Surg.*, 2004, **240**, 910–915.
121. K. Kim, Y. K. Luu, C. Chang, D. Fang, B. S. Hsiao, B. Chu and M. Hadjiargyrou, *J. Control. Release*, 2004, **98**, 47–56.
122. E.-R. Kenawy, G. L. Bowlin, K. Mansfield, J. Layman, D. G. Simpson, E. H. Sanders and G. E. Wnek, *J. Control. Release*, 2002, **81**, 57–64.
123. Y. Hong, K. Fujimoto, R. Hashizume, G. Ryotaro, S. Jianjun, J. John, K. Tobota and W. R. Wagner, *Biomacromolecules*, 2008, **9**, 1200–1207.
124. J. Zeng, X. Xu, X. Chen, Q. Liang, X. Bian, L. Yang and X. Jing, *J. Control. Release*, 2003, **92**, 227–231.
125. N. Bölgen, I. Vargel, P. Korkusuz, Y. Z. Menceloğlu and E. Pişkin, *J. Biomed. Mater. Res. Part B*, 2007, **81B**, 530–543.
126. Z. M. Huang, C. L. He, A. Z. Yang, Y. Z. Zhang, X. J. Han, J. L. Yin and Q. S. Wu, *J. Biomed. Mater. Res., Part A*, 2006, **77A**, 169–179.
127. G. Verreck, I. Chun, J. Rosenblatt, J. Peeters, A. V. Dijck, J. Mensch, M. Noppe and M. E. Brewster, *J. Control. Release*, 2003, **92**, 349–360.
128. A. Martins, A. R. C. Duarte, S. Faria, A. P. Marques, R. L. Reis and N. M. Neves, *Biomaterials*, 2010, **31**, 5875–5885.
129. W. G. Cui, X. H. Li, X. L. Zhu, G. Yu, S. B. Zhou and J. Weng, *Biomacromolecules*, 2006, **7**, 1623–1629.
130. J. W. Xie and C. H. Wang, *Pharm. Res.*, 2006, **23**, 1817–1826.
131. Y. K. Luu, K. Kim, B. S. Hsiao, B. Chu and M. Hadjiargyrou, *J. Control. Release*, 2003, **89**, 341–353.
132. H. Nie and C. H. Wang, *J. Control. Release*, 2007, **120**, 111–121.
133. P.-o. Rujitanaroj, Y.-C. Wang, J. Wang and S. Y. Chew, *Biomaterials*, 2011, **32**, 5915–5923.
134. J. Hu and P. X. Ma, *Pharm. Res.*, 2011, **28**, 1273–1281.
135. S. Y. Chew, J. Wen, E. K. F. Yim and L. W. Leong, *Biomacromolecules*, 2005, **6**, 2017–2024.
136. E. Jo, S. Lee, K. T. Kim, Y. S. Won, H.-S. Kim, E. C. Cho and U. Jeong, *Adv. Mater.*, 2009, **21**, 968–972.
137. B. Dong, M. E. Smith and G. E. Wnek, *Small*, 2009, **5**, 1508–1512.
138. S. K. Tiwari, R. Tzezana, E. Zussman and S. S. Venkatraman, *Int. J. Pharm.*, 2010, **392**, 209–217.
139. J. Zheng, A. Aigner, F. Czubayko, T. Kissel, J. H. Wendorff and A. Greiner, *Biomacromolecules*, 2005, **6**, 1484–1488.
140. P. L. Ritger and N. A. Peppas, *J. Control. Release*, 1987, **5**, 23–36.

141. R. Srikar, A. L. Yarin, C. M. Megaridis, A. V. Bazilevsky and E. Kelley, *Langmuir*, 2008, **24**, 965–974.
142. A. Greiner, J. H. Wendorff, M. Becker, R. Dersch and H. E. Hummel, International Patent Publication No. WO2005/115143, 2005.
143. C. Hellmann, A. Greiner and J. H. Wendorff, *Polym. Adv. Technol.*, 2011, **22**, 407–413.
144. X. Xu, L. Yang, X. Xu, X. Wang, X. Chen, Q. Liang, J. Zeng and X. Jing, *J. Control. Release*, 2005, **108**, 33–42.
145. W. K. Son, J. H. Youk, T. S. Lee and W. H. Park, *Macromol. Rapid Commun.*, 2004, **25**, 1632–1637.
146. K. H. Hong, J. L. Park, I. H. Sul, J. H. Youk and T. J. Kang, *J. Polym. Sci. Part B: Polym. Phys.*, 2006, **44**, 2468–2474.
147. P.-o. Rujitanaroj, N. Pimpha and P. Supaphol, *Polymer*, 2008, **49**, 4723–4732.
148. P. Zahedi, I. Rezaeian, S.-O. Ranaei-Siadat, S.-H. Jafari and P. Supaphol, *Polym. Adv. Technol.*, 2010, **21**, 77–95.
149. I. Han, K. J. Shim, J. Y. Kim, S. U. Im, Y. K. Sung, M. Kim, I.-K. Kang and J. C. Kim, *Artif. Organs*, 2007, **31**, 801–808.
150. B.-M. Min, L. Jeong, Y. S. Nam, J.-M. Kim, J. Y. Kim and W. H. Park, *Int. J. Biol. Macromol.*, 2004, **34**, 223–230.
151. H. Liu, J. Kameoka, D. A. Czaplewski and H. G. Craighead, *Nano Lett.*, 2004, **4**, 671–675.
152. C. Del Gaudio, A. Bianco and M. Grigioni, *Ann. Ist. Super. Sanità*, 2007, **44**, 178–186.
153. M. I. Van Lieshout, C. M. Vaz, M. C. M. Rutten, G. W. M. Peters and F. P. T. Baaijens, *J. Biomater. Sci. Polymer Edn*, 2006, **17**, 77–89.
154. Z. Ma, M. Kotaki, R. Inai and S. Ramakrishna, *Tissue Eng.*, 2005, **11**, 101–109.
155. S. Agarwal, J. H. Wendorff and A. Greiner, *Adv. Mater.*, 2009, **21**, 3343–3351.
156. R. Murugan and S. Ramakrishna, *Tissue Eng.*, 2007, **13**, 1845–1866.
157. J. Xie, M. R. MacEwan, S. M. Willerth, X. Li, D. W. Moran, S. E. Sakiyama-Elbert and Y. Xia, *Adv. Funct. Mater.*, 2009, **19**, 2312–2318.
158. T. Dvir, B. P. Timko, D. S. Kohane and R. Langer, *Nat. Nanotech.*, 2011, **6**, 13–22.
159. A. Atala, R. Lanza, J. Thomson and R. Nerem (ed.), *Principles of Regenerative Medicine*, Elsevier/Academic Press, San Diego, 2008.
160. R. Lanza, R. Langer and J. Vacanti (ed.), *Principles of Tissue Engineering*, Elsevier/Academic Press, San Diego, 3rd edn, 2007.
161. B. Bondara, S. Fuchsa, A. Mottab, C. Migliaresib and C. J. Kirkpatricka, *Biomaterials*, 2008, **29**, 561–572.
162. M. I. Santos, K. Tuzlakoglu, S. Fuchs, M. E. Gomes, K. Peters, R. E. Unger, E. Piskin, R. L. Reis and C. J. Kirkpatrick, *Biomaterials*, 2008, **29**, 4306–4313

163. A. Thorvaldsson, H. Stenhamre, P. Gatenholm and P. Walkenström, *Biomacromolecules*, 2008, **9**, 1044–1049.
164. C. Ayres, G. L. Bowlin, S. C. Henderson, L. Taylor, J. Shultz, J. Alexander, T. A. Telemeco and D. G. Simpson, *Biomaterials*, 2006, **27**, 5524–5534.
165. J. M. Corey, D. Y. Lin, K. B. Mycek, Q. Chen, S. Samuel, E. L. Feldman and D. C. Martin, *J. Biomed. Mater. Res., Part A*, 2007, **83A**, 636–645.
166. J. B. Recknor, D. S. Sakaguchi and S. K. Mallapragada, *Biomaterials*, 2006, **27**, 4098–4108.
167. X. Gu, F. Ding, Y. Yang and J. Liu, *Prog. Neurobiol.*, 2011, **93**, 204–230.
168. H. Cao, T. Liu and S. Y. Chew, *Adv. Drug Deliv. Rev.*, 2009, **61**, 1055–1064.
169. S. Y. Chew, R. Mi, A. Hoke and K. W. Leong, *Adv. Funct. Mater.*, 2007, **17**, 1288–1296.
170. S. Panseri, C. Cunha, J. Lowery, U. D. Carro, F. Taraballi, S. Amadio, A. Vescovi and F. Gelain, *BMC Biotechnol.*, 2008, **8**, 39.
171. W. Yu, W. Zhao, C. Zhu, X. Zhang, D. Ye, W. Zhang, Y. Zhou, X. Jiang and Z. Zhang, *BMC Biotechnol.*, 2011, **12**, 68.
172. Y.-T. Kim, V. K. Haftel, S. Kumar and R. V. Bellamkonda, *Biomaterials*, 2008, **29**, 3117–3127.
173. H. Cao, G. Marcy, E. L. K. Goh, F. Wang, J. Wang and S. Y. Chew, *Macromol. Biosci.*, 2012, **12**, 666–674.
174. T. Liu, J. D. Houle, J. Xu, B. P. Chan and S. Y. Chew, *Tissue Eng. Part A*, 2012, **18**, 1057–1066.
175. V. J. Mukhatyar, M. Salmerón-Sánchez, S. Rudra, S. Mukhopadaya, T. H. Barker, A. J. García and R. V. Bellamkonda, *Biomaterials*, 2011, **32**, 3958–3968.
176. P. Sangsanoh, S. Waleetorncheepsawat, O. Suwantong, P. Wutticharoenmongkol, O. Weeranantanapan, B. Chuenjitbuntaworn, P. Cheepsunthorn, P. Pavasant and P. Supaphol, *Biomacromolecules*, 2007, **8**, 1587–1594.
177. S. Y. Chew, R. Mi, A. Hok and K. W. Leong, *Biomaterials*, 2008, **29**, 653–661.
178. A. Subramanian, U. M. Krishnan and S. Sethuraman, *Ann. Biomed. Eng.*, 2012, **40**, 2098–2110.
179. C. Huang, Y. Tang, X. Liu, A. Sutti, Q. Ke, X. Mo, X. Wang, Y. Morsic and T. Lin, *Soft Matter*, 2011, **7**, 10812–10817.
180. J. Xie, S. M. Willerth, X. Li, M. R. Macewan, A. Rader, S. E. Sakiyama-Elbert and Y. Xia, *Biomaterials*, 2009, **30**, 354–362.
181. O. Kriha, M. Becker, M. Lehmann, D. Kriha, J. Krieglstein, M. Yosef, S. Schlecht, R. B. Wehrspohn, J. H. Wendorff and A. Greiner, *Adv. Mater.*, 2007, **19**, 2483–2485.
182. J. L. McKenzie, M. C. Waid, R. Shi and T. J. Webster, *Biomaterials*, 2004, **25**, 1309–1317.

183. T. C. Holmes, S. De Lacalle, X. Su, G. Liu, A. Rich and S. Zhang, *Proc. Natl Acad. Sci. USA*, 2000, **97**, 6728–6733.
184. C. E. Semino, J. Kasahara, Y. Hayashi and S. Zhang, *Tissue Eng.*, 2004, **10**, 643–655.
185. F. Gelain, D. Bottai, A. Vescovi and S. Zhang, *PLoS ONE*, 2006, **1**, e119.
186. A. Firth, A. Aggeli, J. L. Burke, X. Yang and J. Kirkham, *Nanomedicine*, 2006, **1**, 189–199.
187. E. Kokkoli, A. Mardilovich, A. Wedekind, E. L. Rexeisen, A. Garg and J. A. Craig, *Soft Matter*, 2006, **2**, 1015–1024.
188. R. G. Ellis-Behnke, Y. X. Liang, S. W. You, D. K. C. Tay, S. Zhang, K. F. So and G. E. Schneider, *Proc. Natl Acad. Sci. USA*, 2006, **103**, 5054–5059.
189. J. Guo, H. Su, Y. Zeng, Y.-X. Liang, W. M. Wong, R. G. Ellis-Behnke, K.-F. So and W. Wu, *Nanomedicine*, 2007, **3**, 311–321.
190. M. E. Davis, J. P. M. Motion, D. A. Narmoneva, T. Takahashi, D. Hakuno, R. D. Kamm, S. Zhang and R. T. Lee, *Circulation*, 2005, **111**, 442–450.
191. M. J. Webber, J. Tongers, C. J. Newcomb, K.-T. Marquardt, J. Bauersachs, D. W. Losordo and S. I. Stupp, *Proc. Natl Acad. Sci. USA*, 2011, **108**, 13438–13443.
192. J. M. Holzwarth and P. X. Ma, *Biomaterials*, 2011, **32**, 9622–9629.
193. J. M. Holzwarth and P. X. Ma, *J. Mater. Chem.*, 2011, **21**, 10243–10251.
194. E. Borg, A. Frenot, P. Walkenström, K. Gisselfält, C. Gretzer and P. Gatenholm, *J. Appl. Polym. Sci.*, 2008, **108**, 491–497.
195. K. L. Elias, R. L. Price and T. J. Webster, *Biomaterials*, 2002, **23**, 3279–3287.
196. H. Yoshimoto, Y. M. Shin, H. Terai and J. P. Vacanti, *Biomaterials*, 2003, **24**, 2077–2082.
197. M. Shin, H. Yoshimoto and J. P. Vacanti, *Tissue Eng.*, 2004, **10**, 33–41.
198. L. A. Smith, X. Liu, J. Hu, P. Wang and P. X. Ma, *Tissue Eng. Part A*, 2009, **15**, 1855–1864.
199. A. Polini, D. Pisignano, M. Parodi, R. Quarto and S. Scaglione, *PLoS ONE*, 2011, **6**, e26211.
200. G. Kumar, C. K. Tison, K. Chatterjee, P. Scott Pine, J. H. McDaniel, M. L. Salit, M. F. Young and C. G. Simon Jr., *Biomaterials*, 2011, **32**, 9188–9196.
201. E. Kon, A. Muraglia, A. Corsi, P. Bianco, M. Marcacci, I. Martin, A. Boyde, I. Ruspantini, P. Chistolini, M. Rocca, R. Giardino, R. Cancedda and R. Quarto, *J. Biomed. Mater. Res.*, 2000, **49**, 328–337.
202. R. Z. LeGeros, *Chem. Rev.*, 2008, **108**, 4742–4753.
203. K. Fujihara, M. Kotaki and S. Ramakrishna, *Biomaterials*, 2005, **26**, 4139–4147.
204. E. K. Ko, S. I. Jeong, N. G. Rim, Y. M. Lee, H. Shin and B.-K. Lee, *Tissue Eng. Part A*, 2008, **14**, 2105–2119.

205. D. Gupta, J. Venugopal, S. Mitra, V. R. Giri Dev and S. Ramakrishna, *Biomaterials*, 2009, **30**, 2085–2094.
206. C. Erisken, D. M. Kalyon and H. Wang, *Nanotechnology*, 2008, **19**, 165302.
207. C. Erisken, D. M. Kalyon, H. Wang, C. Örnek-Ballanco and J. Xu, *Tissue Eng. Part A*, 2011, **17**, 1239–1252.
208. Y. Zhang, J. R. Venugopal, A. E. Turki, S. Ramakrishna, B. Su and C. T. Lim, *Biomaterials*, 2008, **29**, 4314–4322.
209. I. Yamaguchi, K. Tokuchi, H. Fukuzaki, Y. Koyama, K. Takakuda, H. Monma and J. Tanaka, *J. Biomed. Mater. Res.*, 2001, **55**, 20–27.
210. X. Liu, L. A. Smith, J. Hu and P. X. Ma, *Biomaterials*, 2009, **30**, 2252–2258.
211. C. He, G. Xiao, X. Jin, C. Sun and P. X. Ma, *Adv. Funct. Mater.*, 2010, **20**, 3568–3576.
212. H. Hosseinkhani, M. Hosseinkhani, F. Tian, H. Kobayashi and Y. Tabata, *Tissue Eng.*, 2007, **13**, 11–19.
213. T. D. Sargeant, M. O. Guler, S. M. Oppenheimer, A. Mata, R. L. Satcher, D. C. Dunand and S. I. Stupp, *Biomaterials*, 2008, **29**, 161–171.
214. W.-J. Li, R. Tuli, C. Okafor, A. Derfoul, K. G. Danielson and D. J. Hall, *Biomaterials*, 2005, **26**, 599–609.
215. W.-J. Li, R. Tuli, X. Huang, P. Laquerriere and R. S. Tuan, *Biomaterials*, 2005, **26**, 5158–5166.
216. X. Xin, M. Hussain and J. J. Mao, *Biomaterials*, 2007, **28**, 316–325.
217. B. M. Baker and R. L. Mauck, *Biomaterials*, 2007, **28**, 1967–1977.
218. W.-J. Li, H. Chiang, T. F. Kuo, H. S. Lee, C. C. Jiang and R. S. Tuan, *J. Tissue Eng. Regen. Med.*, 2009, **3**, 1–10.
219. A. Schneider, J. A. Garlick and C. Egles, *PLoS ONE*, 2008, **3**, e1410.
220. T. Sun, S. Mai, D. Norton, J. W. Haycock, A. J. Ryan and S. MacNeil, *Tissue Eng.*, 2005, **11**, 1023–1033.
221. H. Pan, H. Jiang and W. Chen, *Biomaterials*, 2006, **27**, 3209–3220.
222. J. Venugopal and S. Ramakrishna, *Tissue Eng.*, 2005, **11**, 847–854.
223. X. Yang, J. D. Shah and H. Wang, *Tissue Eng. Part A*, 2009, **15**, 945–956.
224. H. M. Powell and S. T. Boyce, *J. Biomed. Mater. Res. A*, 2008, **84A**, 1078–1086.
225. H. M. Powell, D. M. Supp and S. T. Boyce, *Biomaterials*, 2008, **29**, 834–843.
226. K. A. Blackwood, R. McKean, I. Canton, C. O. Freeman, K. L. Franklin, D. Cole, I. Brook, P. Farthing, S. Rimmer, J. W. Haycock, A. J. Ryan and S. MacNeil, Biomaterials, 2008, **29**, 3091–3104.
227. D. Annis, A. Bornat, R. O. Edwards, A. Higham, B. Loveday and J. Wilson, *Trans. Am. Soc. Artif. Intern. Organs*, 1978, **24**, 209–214.
228. S. A. Sell and G. L. Bowlin, *J. Mater. Chem.*, 2008, **18**, 260–263.

229. S. Jin Lee, S. Heang Oh, J. Liu, S. Soker, A. Atala and J. J. Yoo, *Biomaterials*, 2008, **29**, 1422–1430.
230. S. Drilling, J. Gaumer and J. Lannutti, *J. Biomed. Mater. Res., Part A*, 2009, **88**, 923–934.
231. S. Jin Lee, J. Liu, S. Heang Oh, S. Soker, A. Atala and J. J. Yoo, *Biomaterials*, 2008, **29**, 2891–2898.
232. X. Zhang, C. B. Baughman and D. L. Kaplan, *Biomaterials*, 2008, **29**, 2217–2227.
233. J. Zhang, H. Qi, P. Hu, S. Guo, J. Li, Y. Che, Y. Yu and D. Kong, *Artif. Organs*, 2006, **30**, 898–905.
234. Z. Zhang, Y. Marois, R. G. Guidon, P. Bull, M. Marois, T. How, G. Laroche and M. W. King, *Biomaterials*, 1997, **18**, 113–124.
235. C. K. Hashi, Y. Zhu, G. Y. Yang, W. L. Young, B. S. Hsiao, K. Wang, B. Chu and S. Li, *Proc. Natl Acad. Sci. USA*, 2007, **104**, 11915–11920.
236. K. Inoguchi, I. K. Kwon, E. Inoue, K. Takamizawa, Y. Maehara and T. Matsuda, *Biomaterials*, 2006, **27**, 1470–1478.
237. B. W. Tillman, S. K. Yazdani, S. J. Lee, R. L. Geary, A. Atala and J. J. Yoo, *Biomaterials*, 2009, **30**, 583–588.
238. S. I. Jeong, S. Y. Kim, S. K. Cho, M. S. Chong, K. S. Kim, H. Kim, S. B. Lee and Y. M. Lee, *Biomaterials*, 2007, **28**, 1115–1122.
239. S. C. Baker, N. Atkin, P. A. Gunning, N. Granville, K. Wilson, D. Wilson and J. Southgate, *Biomaterials*, 2006, **27**, 3136–3146.
240. J. S. Choi, S. J. Lee, G. J. Christ, A. Atala and J. J. Yoo, *Biomaterials*, 2008, **29**, 2899–2906.
241. S. A. Riboldi, M. Sampaolesi, P. Neuenschwander, G. Cossu and S. Mantero, *Biomaterials*, 2005, **26**, 4606–4615.
242. C. H. Lee, H. J. Shin, I. H. Cho, Y.-M. Kang, I. A. Kim, K.-D. Park and J.-W. Shin, *Biomaterials*, 2005, **26**, 1261–1270.
243. X. Zong, H. Bien, C.-Y. Chung, L. Yin, D. Fang, B. S. Hsiao, B. Chu and E. Entcheva, *Biomaterials*, 2005, **26**, 5330–5338.
244. D. N. Rockwood, R. E. Akins Jr, I. C. Parrag, K. A. Woodhouse and J. F. Rabolt, *Biomaterials*, 2008, **29**, 4783–4791.
245. J. Wang, X. Liu, X. Jin, H. Ma, J. Hu, L. Ni and P. X. Ma, *Acta Biomater.*, 2010, **6**, 3856–3863.
246. J. Wang, H. Ma, X. Jin, J. Hu, X. Liu, L. Ni and P. X. Ma, *Biomaterials*, 2011, **32**, 7822–7830.
247. N. Navarro-Alvarez, A. Soto-Gutierrez, J. D. Rivas-Carrillo, Y. Chen, T. Yamamoto, T. Yuasa, H. Misawa, J. Takei, N. Tanaka and N. Kobayashi, *Cell Transplant.*, 2006, **15**, 921–927.
248. M. C. McManus, E. D. Boland, S. Sell, W. Bowen, H. Koo, D. G. Simpson and G. L. Bowlin, *Biomed. Mater.*, 2007, **2**, 257–262.
249. E. Zussmann, *Polym. Adv. Technol.*, 2011, **22**, 366–371.
250. S. Arumuganathar and S. N. Jayasinghe, *Biomacromolecules*, 2008, **9**, 759–766.

251. Y. Liu, M. H. Rafailovich, R. Malal, D. Cohn and D. Chidambaram, *Proc. Natl Acad. Sci. USA*, 2009, **106**, 14201–14206.
252. S.-W. Lee and A. M. Belcher, *Nano Lett.*, 2004, **4**, 387–390.
253. W. Salalha, J. Kuhn, Y. Dror and E. Zussman, *Nanotechnology*, 2006, **17**, 4675–4681.
254. M. Gensheimer, M. Becker, A. Brandis-Heep, J. H. Wendorff, R. K. Thauer and A. Greiner, *Adv. Mater.*, 2007, **19**, 2480–2482.
255. A. López-Rubio, E. Sanchez, Y. Sanz and J. M. Lagaron, *Biomacromolecules*, 2009, **10**, 2823–2829.
256. S. Klein, J. Kuhn, R. Avrahami, S. Tarre, M. Beliavski, M. Green and E. Zussman, *Biomacromolecules*, 2009, **10**, 1751–1756.
257. A. Nardi, R. Avrahami, E. Zussman, J. S. Rokem and C. L. Greenblatt, *J. Environm. Protect.*, 2012, **3**, 164–168.
258. A. Townsend-Nicholson and S. N. Jayasinghe, *Biomacromolecules*, 2006, **7**, 3364–3369.
259. M. F. Canbolat, C. Tang, S. H. Bernacki, B. Pourdeyhimi and S. Khan, *Macromol. Biosci.*, 2011, **11**, 1346–1356.
260. M. Gensheimer, A. Brandis-Heep, S. Agarwal, R. K. Thauer and A. Greiner, *Macromol. Biosci.*, 2011, **11**, 333–337.
261. N. F. Huang, S. Patel, R. G. Thakar, J. Wu, B. S. Hsiao, B. Chu, R. J. Lee and S. Li, *Nano. Lett.*, 2006, **6**, 537–542.
262. A. Huber, A. Pickett and K. M. Shakesheff, *Eur. Cells Mater.*, 2007, **14**, 56–63.
263. S. H. Lim and H.-Q. Mao, *Adv. Drug Deliv. Rev.*, 2009, **61**, 1084–1096.
264. M. P. Prabhakaran, J. R. Venugopal and S. Ramakrishna, *Biomaterials*, 2009, **30**, 4996–5003.
265. X. Jiang, H. Q. Cao, L. Y. Shi, S. Y. Ng, L. W. Stanton and S. Y. Chew, *Acta Biomater.*, 2012, **8**, 1290–1302.
266. Y. R. Shih, C. N. Chen, S. W. Tsai, Y. J. Wang and O. K. Lee, *Stem Cells*, 2006, **24**, 2391–2397.
267. F. H. Gage, P. W. Coates, T. D. Palmer, H. G. Kuhn, L. J. Fisher, J. O. Suhonen, D. A. Peterson, S. T. Suhr and J. Ray, *Proc. Natl Acad. Sci. USA*, 1995, **92**, 11879–11883.
268. K. Nakashima, M. Yanagisawa, H. Arakawa, N. Kimura, T. Hisatsune, M. Kawabata, K. Miyazono and T. Taga, *Science*, 1999, **284**, 479–482.
269. G. L. Ming and H. J. Song, *Ann. Rev. Neurosci.*, 2005, **28**, 223–250.
270. X. Zhang, J. Cai, K. M. Klueber, Z. Guo, C. Lu, W. I. Winstead, M. Qiu and F. J. Roisen, *Stem Cells*, 2006, **24**, 434–442.
271. D. R. Nisbet, L. M. Y. Yu, T. Zahir, J. S. Forsythe and M. S. Shoichet, *J. Biomater. Sci. Polym. Ed.*, 2008, **19**, 623–634.
272. E. K. A. Nur, I. Ahmed, J. Kamal, M. Schindler and S. Meiners, *Stem Cells*, 2006, **24**, 426–433.
273. L. He, S. Liao, D. Quan, K. Ma, C. Chan and S. Ramakrishna, J. Lu, *Acta Biomater.*, 2010, **6**, 2960–2969.

274. F. Yang, R. Murugan, S. Wang and S. Ramakrishna, *Biomaterials*, 2005, **26**, 2603–2610.
275. L. A. Smith, X. Liu, J. Hu and P. X. Ma, *Biomaterials*, 2010, **31**, 5526–5535.
276. K. Gauthaman, J. R. Venugopal, F. C. Yee, G. S. L. Peh, S. Ramakrishna and A. Bongso, *J. Cell. Mol. Med.*, 2009, **13**, 3475–3484.
277. H. Sun, K. Feng, J. Hu, S. Soker, A. Atala and P. X. Ma, *Biomaterials*, 2010, **31**, 1133–1139.
278. K. N. Chua, C. Chai, P. C. Lee, S. Ramakrishna, K. W. Leong and H. Q. Mao, *Exp. Hematol.*, 2007, **35**, 771–781.
279. K.-N. Chua, C. Chai, P.-C. Lee, Y.-N. Tang, S. Ramakrishna, K. W. Leong and H.-Q. Mao, *Biomaterials*, 2006, **27**, 6043–6051.
280. G. A. Silva, C. Czeisler, K. L. Niece, E. Beniash, D. A. Harrington, J. A. Kessler and S. I. Stupp, *Science*, 2004, **303**, 1352–1355.
281. A. Horii, X. Wang, F. Gelain and S. Zhang, *PLoS ONE*, 2007, **2**, e190.

# List of Acronyms

| Acronym | Full name | Introduced |
|---|---|---|
| 6-$NO_2$-BIPS | 1′,3′-dihydro-1′,3′,3′-trimethyl-6-nitrospiro [2H-1-benzopyran-2,2′-(2H)-indole] | 4.2.2 |
| AFM | Atomic force microscopy | 1.1 |
| AFSCs | Amniotic fluid-derived stem cells | 7.7.6 |
| ALP | Alkaline phosphatase | 7.7 |
| APS | Ammonium persulfate | 3.1.4 |
| ASE | Amplified Spontaneous Emission | 5.1.2 |
| BBL | Poly(benzobisimidazobenzophenanthroline) | 3.1.1 |
| BCB | Benzocyclobutene | 6.3 |
| BET | Brunauer–Emmett–Teller | 1.2 |
| BM | Basal medium | 7.7.3 |
| BMP | Bone morphogenetic protein | 7.2 |
| BSP | Bone sialoprotein | 7.7.3 |
| CDT-BTZ | Cyclopentadithiophene-benzothiadiazole | 3.1.2 |
| CFL | Continuous flow lithography | 1.2.1 |
| CIE | Commission Internationale de l'Eclairage | 5.2.3 |
| CiP | *Coprinus cinereus* peroxidase | 3.1.4 |
| CTAB | Cetyltrimethylammonium bromide | 3.3.2 |
| CVD | Chemical vapor deposition | 3.3.1 |
| DAPI | 4′,6-diamidino-2-phenylindole | 7.7 |
| DMF | Dimethylformamide | 2.1 |
| DMSO | Dimethyl sulfoxide | 5.1.2 |
| DPN | Dip-pen nanolithography | 3.5 |
| DRG | Dorsal root ganglion | 6.4.1 |
| DSC | Differential scanning calorimetry | 1.4.1 |
| EBL | Electron-beam lithography | 1.2.1 |
| ECM | Extracellular matrix | 2.3.1 |
| EDC | 1-ethyl-3-(3-dimethylaminopropyl) carbodiimide hydrochloride | 7.2 |
| EPR | Electron paramagnetic resonance | 2.1.5 |
| F8BT | Poly(9,9-di-*n*-octylfluorene-alt-benzothiadiazole) | 5.2.1 |
| F8T2 | Poly[(9,9-dioctylfluorenyl-2,7-diyl)-*co*-(bithiophene)] | 3.3.1 |

RSC Nanoscience & Nanotechnology No. 29
Polymer Nanofibers: Building Blocks for Nanotechnology
By Dario Pisignano

Published by the Royal Society of Chemistry, www.rsc.org

(*Continued*)

| *Acronym* | *Full name* | *Introduced* |
|---|---|---|
| FET | Field-effect transistor | 4.3.2 |
| FeTos | Ferric tosylate | 3.3.2 |
| FFT | Fast Fourier transform | 2.3.2 |
| FIB | Focused ion-beam | 3.2 |
| FRET | Fluorescence Resonant Energy Transfer | 5.1.2 |
| FTIR | Fourier-transform infrared | 6.4.1 |
| HAECs | Human aortic endothelial cells | 7.7.5 |
| HCASMCs | Human coronary artery smooth muscle cells | 7.7.5 |
| HEPA | High-efficiency particulate air | 4.3.1 |
| HFIP | 1,1,1,3,3,3-hexafluoro-2-propanol | 7.2 |
| HFOB | Human fetal osteoblast | 7.7.3 |
| HK-2 | Human Kidney 2 | 7.3 |
| HMDS | Hexamethyldisilazane | 3.1.2 |
| HOMO | Highest occupied molecular orbital | 5.1.1 |
| HOPG | Highly oriented pyrolytic graphite | 6.1.1 |
| HUVEC | Human umbilical vein endothelial cells | 7.7.1 |
| ITO | Indium-tin oxide | 3.3.2 |
| LUMO | Lowest unoccupied molecular orbital | 5.1.1 |
| MEH-PPV | Poly[2-methoxy-5-(2′-ethyl-hexyloxy)-1,4 phenylenevinylene] | 2.3.1 |
| MTS | 3-(4,5-dimethylthiazol-2-yl)-5-(3-carboxymethoxyphenyl)-2-(4-sulfophenyl)-2H-tetrazolium | 7.7 |
| MTT | 3-(4,5-dimethylthiazol-2-yl)-2,5-diphenyltetrazolium bromide | 7.7 |
| MWCNT | Multi-walled carbon nanotube | 2.3.3 |
| NIL | Nanoimprint lithography | 1.2.1 |
| Nd:YAG | Neodymium-yttrium-aluminum-garnet | 5.2.4 |
| NSA | Naphthalene sulfonic acid | 3.4 |
| NSCs | Neural stem cells | 7.3 |
| OCN | Osteocalcin | 7.7.6 |
| OLED | Organic light-emitting diode | 5.1 |
| OLET | Organic light-emitting transistor | 6.1 |
| OPN | Osteopontin | 7.7.6 |
| OSX | Osterix | 7.7.6 |
| OTS | Octadecyl trichlorosilane | 6.3 |
| P2VP | Poly(2-vinylpyridine) | 3.3.2 |
| P3AT | Poly(3-alkylthiophene) | 1.4 |
| P3HT | Poly-3-hexylthiophene | 3.1.1 |
| p6P | *Para*-hexaphenylene | 6.2 |
| PAA | Polyacrylic acid | 2.3 |
| PAC | Polyacrylonitrile | 2.3.1 |
| PACT | Polyacetylene | 3.3.2 |
| PAN | Polyaniline | 2.1.5 |
| PBI | Poly(benzimidazole) | 4.3.3 |
| PBLG | Poly(γ-benzyl α-L-glutamate) | 6.4.1 |
| PBS | Phosphate-buffered saline | 2.3 |
| PCBM | Phenyl-$C_{61}$-butyric acid methyl ester | 5.3.2 |
| PCL | Polycaprolactone | 2.3 |
| PDMS | Polydimethylsiloxane | 1.2.1 |

(*Continued*)

| *Acronym* | *Full name* | *Introduced* |
|---|---|---|
| PECAM-1 | Platelet/endothelial cell adhesion molecule-1 | 7.7.1 |
| PEDOT | Poly(3,4-ethylenedioxythiophene) | 3.3.1 |
| PEGDA | Poly(ethylene glycol)-diacrylate | 7.2 |
| PEO | Polyethylene oxide | 2.1.3 |
| PES | Polyethersulfone | 7.7.6 |
| PET | Poly(ethylene terephthalate) | 5.1.2 |
| PFM | Piezoresponse force microscopy | 6.4.1 |
| PFO | Poly(9,9-dioctylfluorene) | 3.3.1 |
| PFPE | Perfluoropolyether | 1.2.1 |
| PHBV | Poly(3-hydroxybutyrate-*co*-3-hydroxyvalerate) | 2.3.2 |
| PIV | Particle imaging velocimetry | 2.1.2 |
| PL | Photoluminescence | 4.3.2 |
| PLA | Poly(L-lactide) | 2.3.1 |
| PLAGA | Poly(lactide-*co*-glycolide) | 3.3.2 |
| PMMA | Poly(methyl methacrylate) | 2.5.4 |
| PPE | Poly(phenilene ethylene) | 3.1.2 |
| PPTTPP | 5,5′-di-4-biphenylyl-2,2′-bithiophene | 6.2 |
| PPV | Poly(*p*-phenylenevinylene) | 3.3.1 |
| PPy | Polypyrrole | 3.1.4 |
| PS | Polystyrene | 2.5.2 |
| PSS | Poly(styrenesulfonate) | 3.3.2 |
| PTFE | Polytetrafluoroethylene | 3.3.1 |
| PVA | Polyvinyl alcohol | 2.3.1 |
| PVAC | Poly(vinyl acetate) | 4.3.2 |
| PVDF | Poly(vinylidene fluoride) | 2.3 |
| P(VDF–TrFE) | Poly(vinylidene fluoride-trifluoroethylene) | 6.4.1 |
| PVK | Poly(*N*-vinylcarbazole) | 3.3.1 |
| PVP | Poly(vinyl pyrrolidone) | 2.3.1 |
| RAD | Arginine–alanine–aspartate | 7.7.2 |
| REM | Replica molding | 1.2.1 |
| RFID | Radio-frequency identification | 6.3 |
| RGD | Arginine-glycine-aspartic acid tripeptide | 7.1 |
| RIE | Reactive ion etching | 3.3.2 |
| RT-PCR | Reverse-transcription polymerase chain reaction | 6.4.1 |
| RUNX2 | Runt related transcription factor 2 | 7.7.6 |
| SEM | Scanning electron microscopy | 1.3 |
| STM | Scanning tunneling microscopy | 1.1 |
| SVED | Solvent vapor enhanced drop casting | 3.1.2 |
| SWCNT | Single-walled carbon nanotube | 2.3.3 |
| TE | Transversal electric | 5.6 |
| TEM | Transmission electron microscopy | 1.3 |
| THF | Tetrahydrofuran | 3.1.2 |
| TM | Transversal magnetic | 5.6 |
| TTPPTT | 4–4′-di-2,2′-dithienyl-biphenyl | 6.2 |
| TUFT | Tubes by fiber templates | 3.3.2 |
| VE | Vascular endothelial | 7.7.1 |
| VFT | Vogel–Fulcher–Tamman | 1.4.1 |
| VRHM | Variable range-hopping model | 6.1 |
| WST | Water-soluble tetrazolium salt | 7.7 |
| XRD | X-ray diffraction | 2.1.5 |

# List of Symbols

| *Symbol* | *Definition* | *Unit* | *First used* |
|---|---|---|---|
| $A$ | Surface area | $m^2$ | 1.4.2 |
| $A_{abs}$ | Absorption intensity | a.u. | 5.1.2 |
| $A_{em}$ | Emission intensity | a.u. | 5.1.2 |
| $\mathbf{B}$ | Magnetic field | T | 5.2.4 |
| $c$ | Concentration (general) | $m^{-3}$, mol $m^{-3}$, kg $m^{-3}$, wt%, *etc.* | 1.4.1 |
| $c_D$, $c_A$ | Concentration (donor, acceptor) | $m^{-3}$, mol $m^{-3}$, kg $m^{-3}$, wt%, *etc.* | 5.2.3 |
| $C_g$ | Specific capacitance | F $kg^{-1}$ | 4.3.1 |
| $C_i$ | Capacitance per unit area | F $m^{-2}$ | 6.3 |
| $C_p$ | Heat capacity | J $K^{-1}$ | 1.4.1 |
| $c_Q$ | Quencher concentration | mol $m^{-3}$ | 4.3.2 |
| $c_s$ | Solvent mass fraction | wt% | 2.5.1 |
| $\mathbf{D}$ | Electric displacement | C $m^{-2}$ | 6.4.1 |
| $D$ | Diffusion coefficient | $m^2$ $s^{-1}$ | 1.4.1 |
| $\boldsymbol{d}$ | Piezoelectricity matrix | C $N^{-1}$ | 6.4.1 |
| De | Deborah number | | 2.1.5 |
| $d_D$, $d_S$ | Thickness | m | 6.3 |
| $d_{FT}$ | Förster radius | m | 5.1.2 |
| $D_S$ | Strength parameter | | 1.4.1 |
| $E$ | Young's modulus | Pa | 1.2.1 |
| $F$, $\mathbf{F}$ | Electric field | V $m^{-1}$ | 2.1.1 |
| $f$ | Electrospinning jet cross-sectional area | $m^2$ | 2.5.2 |
| $F_0$, $F_N$, $\mathbf{F_N}$ | Force | N | 1.4.2 |
| $f_I$ | Ideality factor for Schottky junctions | | 6.2 |
| $f_T$ | Transition frequency | Hz | 6.3.1 |
| $F^*$ | Generalized applied force | | 1.4.1 |
| $g$ | Gravity acceleration | m $s^{-2}$ | 2.5.3 |
| $G$, $G_S$ | Modulus (elastic or shear) | Pa | 1.4.2 |

RSC Nanoscience & Nanotechnology No. 29
Polymer Nanofibers: Building Blocks for Nanotechnology
By Dario Pisignano

Published by the Royal Society of Chemistry, www.rsc.org

(*Continued*)

| *Symbol* | *Definition* | *Unit* | *First used* |
|---|---|---|---|
| $\tilde{G}$ | Optical gain coefficient | $m^{-1}$ | 5.1.3 |
| $g_F$ | Form factor | | 2.5.4 |
| $g_m$ | Transconductance | S | 6.3 |
| $G'$ | Modulus (storage) | Pa | 1.4.2 |
| $G''$ | Modulus (loss) | Pa | 1.4.2 |
| $H$, $\mathbf{H}$ | Magnetic field intensity | $A\ m^{-1}$ | 6.2 |
| $h$ | Electrospinning nozzle-collector distance | m | 2.1.2 |
| $H_0$ | Enthalpy | J | 1.4.3 |
| $h_P$ ($\hbar_P = h_P/2\pi$) | Planck constant | $J\ s^{-1}$ | 1.1 |
| $I$ | Electric current | A | 2.1.2 |
| $I_{DS}$ | Source-drain current | A | 6.3 |
| $I_{on}/I_{off}$ | On current/off current ratio in FETs | | 6.3 |
| $I_{Z\|Z}$, $I_{Z\|X}$, $I_{X\|Z}$, $I_{X\|X}$ | Polarized absorption/emission intensity | a.u. | 5.1.2 |
| $J$ | Shear creep compliance | $Pa^{-1}$ | 1.4.2 |
| $\mathbf{J}$ | Compliance matrix | $Pa^{-1}$ | 1.4.2 |
| $J_I$ | Current density | $A\ m^{-2}$ | 6.2 |
| $k$, $\mathbf{k}$ | Wavevector, wavenumber | $m^{-1}$ | 2.5.3 |
| $k_B$ | Boltzmann's constant | $J\ K^{-1}$ | 1.2 |
| $K_{FT}$ | FRET energy transfer rate | $s^{-1}$ | 5.1.2 |
| Kn | Knudsen number | | 4.3.1 |
| $K_P$ | Permeability | $m^2$ | 4.3.1 |
| $L$ | Channel length in field-effect transistors | m | 6.3 |
| $l$ | Length (of single fiber, wire, pores, *etc.*) | m | 4.2.1 |
| $\mathbf{M}$ | Torque | N m | 5.2.4 |
| $m$ | Mass | kg | 2.5 |
| $m_0$ | Mass (reduced) | kg | 1.1 |
| $m_F$ | Fragility index | | 1.4.1 |
| MR | Magnetoresistance | | 6.2 |
| $MW$ | Molecular weight | $kg\ mol^{-1}$ (or a.m.u.) | 1.4.2 |
| $m^*$ | Mass (effective) | kg | 6.2 |
| $N$ | Charge carrier density | $m^{-3}$ | 6.1 |
| $n$, $n_{in}$, $n_{out}$ | Refractive index | | 5.1.2 |
| $N_{AV}$ | Avogadro's number | | 1.2 |
| $N_D$ | Dean number | | 2.1.3 |
| $n_D$ | Diffusional exponent | | 7.4 |
| $n_{Do}$, $n_{Ac}$ | Populations (donor, acceptor) | | 5.2.3 |
| $n_{eff}$ | Effective refractive index | | 5.6 |
| $N_p$ | Total number of polymer molecules (Flory–Huggins model) | | 1.4.3 |
| $N_r$ | Number of repeat units in the polymer chain (Flory–Huggins model) | | 1.4.3 |

(*Continued*)

| Symbol | Definition | Unit | First used |
|---|---|---|---|
| $N_s$ | Total number of solvent molecules (Flory–Huggins model) | | 1.4.3 |
| $\mathbf{P}$ | Polarization | $C\ m^{-2}$ | 6.4.1 |
| $P$ | Porosity | % | 4.1 |
| $p$ | Pressure | Pa, atm, bar | 1.2 |
| $Q$ | Flow rate | $L\ s^{-1}$, $m^3\ s^{-1}$ | 2.1.2 |
| $q$ | Electric charge | C | 1.1 |
| $R$ | Hydrodynamic radius | m | 2.1.1 |
| $r$, $r_F$ | Radius (fiber, jet) | m | 1.2 |
| $R_B$ | Exciton Bohr Radius | m | 1.1 |
| $r_B$ | Electrospinning whipping radius | m | 2.5.3 |
| $R_{col}$ | Radius (cylindrical collector) | m | 2.3.2 |
| Re | Reynolds number | | 2.1.2 |
| $R_F$ | Flow resistance | $m^{-1}$ | 4.3.1 |
| $R_T$ | Taylor cone apex radius | m | 2.1.2 |
| $R_X$ | Resistance | Ω | 6.2 |
| $r_Z$ | Polarization anisotropy | | 5.1.2 |
| $S$ | Orientational order parameter | | 2.3 |
| $S_0$ | Entropy | $J\ K^{-1}$ | 1.4.3 |
| Sc | Schmidt number | | 2.5.2 |
| $S_{conf}$ | Configurational entropy | $J\ K^{-1}$ | 1.4.1 |
| $S_E$ | Seebeck coefficient | $V\ K^{-1}$ | 6.4.2 |
| $T$ | Temperature | °C or K | 2.2 |
| $t$ | Time | s | 1.4.1 |
| $t_B$ | Buckling time | s | 2.1.4 |
| $t_D$ | Drying time | s | 2.1.4 |
| $T_g$ | Glass-transition temperature | K | 1.4.1 |
| $t_{PS}$ | Phase-separation time | s | 2.1.4 |
| $T^*$ | Transmission | | 2.5.4 |
| $U_E$ | Electrical energy | J | 6.4.1 |
| $U_G$ | Free energy | J | 1.4.3 |
| $U_M$ | Mechanical energy | J | 1.4.2 |
| $V$ | Electric voltage (bias) | V | 2.1.1 |
| $\bar{v}$ | Specific molecular volume | $m^3$ | 1.4.1 |
| $V_0$ | Volume | $m^3$ | 1.4.1 |
| $v_C$ | Incompressible molecular volume | $m^3$ | 1.4.1 |
| $v_d$ | Velocity (drift) | $m\ s^{-1}$ | 2.1.1 |
| $V_{DS}$ | Source-drain voltage | V | 6.3 |
| $v_f$ | Free molecular volume | $m^3$ | 1.4.1 |
| $V_{GS}$ | Source-gate voltage | V | 6.3 |
| $v_j$ | Velocity (electrospinning jet) | $m\ s^{-1}$ | 2.1.2 |
| $V_{TH}$ | Threshold voltage in field-effect transistors | V | 6.3 |
| $v_x$, $v_y$, $v_z$ | Velocity (general coordinates) | $m\ s^{-1}$ | 1.4.2 |
| $V^*$ | Threshold voltage for conduction | V | 6.2 |

*(Continued)*

| *Symbol* | *Definition* | *Unit* | *First used* |
|---|---|---|---|
| $W$ | Channel width in field-effect transistors | m | 6.3 |
| $W_E$ | Electrodes width in field-effect transistors | m | 6.3 |
| $W_{EV}$ | Evaporation rate | m $s^{-1}$ | 2.1.4 |
| $\alpha$ | Absorption (or optical loss) coefficient | $m^{-1}$ | 2.5.4 |
| $\gamma$ | Strain | | 1.4.2 |
| $\dot{\gamma}$ | Shear rate | $s^{-1}$ | 1.4.2 |
| $\delta$ | Stress-strain phase angle | Rad | 1.4.2 |
| $\delta$ | Density | kg $m^{-3}$ | 2.1.2 |
| $\delta d$ | Perturbation shift | m | 2.5 |
| $\Delta E$ | Activation energy, energy barrier | J | 6.2 |
| $\Delta l$ | Indentation | m | 4.1 |
| $\Delta L$ | Deformation length (longitudinal) | m | 1.4.2 |
| $\Delta L_{\perp}$ | Deformation length (transverse) | m | 1.4.2 |
| $\varepsilon_0$ | Vacuum dielectric constant | $A^2s^4/(kg\ m^3)$ | 1.1 |
| $\varepsilon_r$ | Dielectric constant | | 1.1 |
| $\zeta_a, \zeta_j$ | Kinematic viscosity | $m^2\ s^{-1}$ | 2.5.2 |
| $\eta$ | Dynamic viscosity | Pa s | 1.4.1 |
| $\eta_E$ | Energy conversion efficiency | | 6.4.1 |
| $\eta_{EL}$ | Electroluminescence efficiency | | 5.3.1 |
| $\eta_P$ | Fractional guided power | | 5.4 |
| $\eta_{PL}$ | Photoluminescence efficiency | | 5.1.2 |
| $\Theta$ | Angle (general) | °, rad | 2.5 |
| $\theta$ | Heaviside, or unit, step function | | 1.4.1 |
| $\theta$ (or $2\theta$) | XRD diffraction angle | ° | 6.4.1 |
| $\theta_W, \theta_Y, \theta_{We}$ | Contact angle (wettability) | °, rad | 3.2 |
| $\kappa$ | Jet instability, dimensionless wavelength | | 2.5.3 |
| $\tilde{\kappa}^2$ | Dipole orientation factor | | 5.1.2 |
| $\kappa_T$ | Thermal conductivity | W/(K m) | 4.3.2 |
| $\Lambda$ | Inter-fiber distance | m | 4.2.2 |
| $\lambda, \lambda_c, \lambda_R$ | Wavelength | m | 1.2.1 |
| $\Lambda_B$ | Pattern period | m | 5.6 |
| $\lambda_E$ | Electrospinning longitudinal length | m | 2.5.2 |
| $\lambda_F$ | Mean free path | m | 4.3.1 |
| $\mu, \mu_0, \mu^*$ | Charge carrier mobility | $m^2\ V^{-1}\ s^{-1}$ | 6.1 |
| $\vec{\mu}$ | Electric dipole, transition dipole moment | C m | 2.1.5 |
| $\mu^{(linear)}$ | Linear mobility | $m^2\ V^{-1}\ s^{-1}$ | 6.3 |
| $\mu^{(satur.)}$ | Saturation mobility | $m^2\ V^{-1}\ s^{-1}$ | 6.3 |
| $\nu$ | Frequency | Hz | 2.3 |
| $\tilde{\nu}$ | Frequency wave number | $m^{-1}$ | 5.1.2 |
| $\nu_P$ | Poisson's ratio | | 1.4.2 |

(*Continued*)

| Symbol | Definition | Unit | First used |
|---|---|---|---|
| $\rho$ | Volumetric charge density | C $L^{-1}$, C $m^{-3}$ | 2.1.2 |
| $\rho_{P,Z}$, $\rho_{P,X}$ | Polarization ratio | | 5.1.2 |
| $\sigma$, $\boldsymbol{\sigma}$ | Stress | N $m^{-2}$ | 1.4.2 |
| $\sigma_C$ | Surface charge density | C $m^{-2}$ | 2.1.2 |
| $\sigma_F$ | Stress at failure | N $m^{-2}$ | 4.1 |
| $\sigma_s$ | Surface tension | N $m^{-1}$ | 2.1.2 |
| $\tau$, $\tau_D$ | Relaxation time | s | 1.4.1 |
| $\tau_{flow}$ | Flow timescale | s | 2.1.5 |
| $\Phi$ | Electrostatic potential | V | 2.1.2 |
| $\mathit{\Phi}$ | Orientation angle | | 2.1.5 |
| $\phi$ | Volume fraction | | 1.4.3 |
| $\chi$ | Electrical conductivity | S $m^{-1}$ | 2.1.1 |
| $\chi_0$ | Generalized static susceptibility | | 1.4.1 |
| $\chi_r$ | Electric susceptibility | | 2.2 |
| $\Psi$ | Relaxation function | | 1.4.1 |
| $\omega$ | Angular frequency | rad $s^{-1}$ | 1.4.2 |
| $\tilde{\omega}$ | Jet instability, growth-rate | $s^{-1}$ | 2.5.3 |
| $\omega_C$ | Cross-over frequency | rad $s^{-1}$ | 1.4.2 |
| $\varnothing$ | Pipette orifice or needle internal diameter | m | 2.2 |

# Subject Index